'This wonderful multinational and multidisciplinary collection is greater than the sum of its fascinating parts. Crystalline aliens, a mysterious Siberian explosion, silicon-based life forms, *Tintin*, *Thunderbirds*, *Star Trek* and Raëlians are just some of the many things which are examined in a brilliantly eclectic series of essays.'

—David Edgerton, Imperial College London

'With generous references to the scholarship and original sources, as well as its own intelligent and well-integrated contributions, this book establishes a comprehensive new field of research – "astroculture."'

—Michael G. Smith, Purdue University

'Europe too has a history of imagining outer space, distinct from yet inextricably linked with global cultures of perceiving and experiencing the universe. This splendid volume offers a fascinating panorama of visions of the future. Anyone interested in the complex relationship between technology, space and culture will garner much from this groundbreaking work.'

—Helmuth Trischler, Deutsches Museum

'Intriguing. [...] A book of essays filled with European perspectives on space and spaciness.'

—Alexis C. Madrigal, *The Atlantic*

'*Imagining Outer Space* is a brilliantly organized compendium of current scholarship at the intersection between space history and the popular cultures of science/fiction. It also sheds new light on the often underplayed European contributions to imagining outer space as a richly inhabited human realm. It successfully establishes "astroculture" as an energetic and growing area of scholarly production and debate.'

—De Witt Douglas Kilgore, *Science Fiction Studies*

'However peripheral Europe's contributions to the Space Age may have been, nothing was spared in the imagination. The matter was of exemplary global interest, after all. It is the details that count here, and the contributions in this volume offer plenty: crystalline aliens and Mars scenarios, spaceflight in comic strips and ghost rockets (a European equivalent to flying saucers), UFOs in postwar France and well-intentioned offers of interstellar communication.'

—Helmut Mayer, *Frankfurter Allgemeine Zeitung*

'*Imagining Outer Space* offers an interdisciplinary and transnational approach to the cultural and social history of the space age in Europe. It is its redrawing of the disciplinary boundaries of space history that should be most applauded. [...] Highly recommended not only to readers interested in the history of outer space and the Space Age.'

—Anke Ortlepp, *H-Soz-u-Kult*

'With its emphasis on multidisciplinarity, and its wide variety of contributions, topics, and themes, *Imagining Outer Space* demonstrates the rich potential that astrocultural studies holds for the field of the history of spaceflight, while at the same time, it truly contains something for everyone.'

—Janet Vertesi, *Quest: History of Spaceflight Quarterly*

'This is clearly an important contribution to the literature and a stimulus to ongoing and future debates and endeavours in the intertwining realms of culture, space and technology.'

—Derek Hall, *Space Policy*

'*Imagining Outer Space* offers rich potential in explaining the infatuation of spaceflight by Europeans of many different nationalities and cultures. [...] Without question, astrocultural investigation is one of the more interesting and original efforts to restructure spaceflight history in the early twenty-first century.'

—Roger D. Launius, *Technology and Culture*

'Together, the chapters survey an excellent variety of topics that fall under the "astroculture" umbrella. Further research into European astroculture would be a valuable contribution to other social and cultural histories of Europe and to wider understandings of human engagements with outer space. *Imagining Outer Space* is a giant leap in that direction.'

—Jason Beery, *European Review of History/Revue européenne d'histoire*

‘This is an eclectic, detailed and [...] revelatory set of essays that delve into how (mostly Western) Europeans portrayed outer space, spaceflight and space exploration. It certainly fills a gap.’

—Jon Agar, *British Journal for the History of Science*

‘This volume’s fifteen diverse essays, substantive introduction, and valuable epilogue all examine various aspects of “astroculture” by considering and configuring the cultural and social significance of the Space Age both to and within the Atomic Age. [...] Fascinating.’

—Pamela Gossin, *Isis*

Palgrave Studies in the History of Science and Technology

Designed to bridge the gap between the history of science and the history of technology, this series publishes the best new work by promising and accomplished authors in both areas. In particular, it offers historical perspectives on issues of current and ongoing concern, provides international and global perspectives on scientific issues, and encourages productive communication between historians and practicing scientists.

More information about this series at
http://www.palgrave.com/gp/series/14581

Also by Alexander C.T. Geppert

FLEETING CITIES
Imperial Expositions in *Fin-de-Siècle* Europe

Edited or Co-Edited

EUROPEAN EGO-HISTOIRES
Historiography and the Self, 1970–2000

ORTE DES OKKULTEN

ESPOSIZIONI IN EUROPA TRA OTTO E NOVECENTO
Spazi, organizzazione, rappresentazioni

ORTSGESPRÄCHE
Raum und Kommunikation im 19. und 20. Jahrhundert

NEW DANGEROUS LIAISONS
Discourses on Europe and Love in the Twentieth Century

WUNDER
Poetik und Politik des Staunens im 20. Jahrhundert

ASTROCULTURE AND TECHNOSCIENCE

OBSESSION DER GEGENWART
Zeit im 20. Jahrhundert

BERLINER WELTRÄUME IM FRÜHEN 20. JAHRHUNDERT

LIMITING OUTER SPACE
Astroculture After Apollo
(European Astroculture, vol. 2)

MILITARIZING OUTER SPACE
Astroculture, Dystopia and the Cold War
(European Astroculture, vol. 3) (forthcoming)

Alexander C.T. Geppert
Editor

Imagining Outer Space

European Astroculture in the Twentieth Century

Second Edition

European Astroculture
Volume 1

Editor
Alexander C.T. Geppert
New York University Shanghai
Shanghai, China

and

New York University, Center for
European and Mediterranean Studies
New York, NY, USA

Palgrave Studies in the History of Science and Technology
European Astroculture, Volume 1
ISBN 978-1-349-95338-7 ISBN 978-1-349-95339-4 (eBook)
https://doi.org/10.1057/978-1-349-95339-4

Library of Congress Control Number: 2017963534

Printed on acid-free paper

This Palgrave Macmillan imprint is published by Springer Nature
The registered company is Macmillan Publishers Ltd.
The registered company address is: The Campus, 4 Crinan Street, London, N1 9XW, United Kingdom

If one undertakes to discuss what man ought to do with the planets, one must first say what one thinks man ought to do with himself.

Olaf Stapledon, *Journal of the British Interplanetary Society* (1948)

Die Geschichtswissenschaft muß den Sprung in die planetarische Zukunft wagen.

Hermann Heimpel, *Frankfurter Allgemeine Zeitung* (25 March 1959)

Aujourd'hui, il s'agit de l'espace à l'échelle mondiale (et même au delà de la surface terrestre, de l'espace interplanétaire), ainsi que des espaces impliqués, à tous les échelons.

Henri Lefebvre, *La Production de l'espace* (1974)

Contents

Preface to the Paperback Edition

The re-issuing of *Imagining Outer Space: European Astroculture in the Twentieth Century* in paperback format six years after its original publication coincides with the advent of two companion volumes. *Limiting Outer Space: Astroculture After Apollo* (Palgrave Macmillan, 2018) and *Militarizing Outer Space: Astroculture, Dystopia and the Cold War* (Palgrave Macmillan, forthcoming) take up some of the problems raised and issues discussed in the present book. The second volume of this trilogy, *Limiting Outer Space*, focuses on a single decade in the history of imagining, thinking and practicing outer space – the 'long 1970s' – and foregrounds a single problem, that is the reconfiguration of sociotechnical imaginaries and human expansion scenarios during the decade after the moon landings, the so-called post-Apollo period. The third and final book, *Militarizing Outer Space*, explores the militant and violent dimensions of outer space in science fiction and science fact, thus exposing the 'dark' side of global astroculture.

All three volumes are the product of work conducted, choreographed or coordinated by the Emmy Noether research group 'The Future in the Stars: European Astroculture and Extraterrestrial Life in the Twentieth Century' at Freie Universität Berlin which I had the pleasure of directing from 2010 to 2016. As the publication of this *European Astroculture* trilogy was not planned from the outset but rather coalesced over the course of our collective expedition, there is no systematic rationale behind the thematic succession of these three volumes. What they have in common, however, is the endeavor to establish 'astroculture' as a new field of historical inquiry; the will to decenter space historiography by pushing its geographical focus beyond the borders of the two Cold War superpowers; and the quest to de-exoticize the history of outer space while allocating it the place it deserves within mainstream historiography of the twentieth century.

For the paperback edition a few factual errors were corrected and some minor improvements made. Web links have been checked, authors' biographies updated, and a limited number of references added to keep an already comprehensive bibliography as current as possible. Otherwise, all 17 contributions remain as they were originally published in 2012, unaltered in form and format.

Finally, I would like to express sincere gratitude to our funding body, the Deutsche Forschungsgemeinschaft (DFG), and, above all, to everyone who contributed to this enterprise's launch, lift-off and landing.

Shanghai
November 2017

Alexander C.T. Geppert

Acknowledgments

When contemplating outer spaces and other worlds, the 'pleasures of the imagination' are infinite – and the conceptualization, compilation and composition of the present volume has indeed proved such.[1] Early versions of all articles published here were originally presented at the first international conference on the cultural history of outer space in twentieth-century Europe, held on 6–9 February 2008 at the Zentrum für interdisziplinäre Forschung (ZiF) of Universität Bielefeld, Germany. Entitled *Imagining Outer Space, 1900–2000* and generously co-funded by the ZiF and Fritz Thyssen Stiftung, this conference congregated nearly 70 scholars from more than a dozen countries, with the common aim of historicizing outer space and analyzing its cultural significance in the European imagination, particularly since 1945.

Subsequent to that first gathering, the Deutsche Forschungsgemeinschaft has beneficently underwritten the establishment of an independent Emmy Noether research group, 'The Future in the Stars: European Astroculture and Extraterrestrial Life in the Twentieth Century,' located at the Friedrich-Meinecke-Institut of Freie Universität Berlin. Special thanks go to the core members of this group, Daniel Brandau and William R. Macauley. Together, we have already begun to take up and deepen many of the themes raised in this volume, with a view to integrating the cultural history of space into mainstream historiography of the twentieth century, so-called *Zeitgeschichte*. This collective venture will continue to propel such a 'leap into the planetary future' over the next five years.[2]

Organizing an event on this scale and preparing the ensuing book for publication entails the accumulation of unforeseen debts of gratitude. First of all, I would like to thank the ZiF and its former Managing Director, Ipke Wachsmuth, the late Johannes Roggenhofer, ZiF's Executive Secretary, as well as Barbara Jantzen, Scientific Assistant to its Board of Directors, for their trust in the intellectual potential of a certainly unusual, yet hardly exotic topic. Trixi Valentin, head of ZiF's conference office, proved to be the epitome of cordial professionalism and ensured that the event went off without a hitch. Without the Fritz Thyssen Stiftung's generosity, the enterprise could never have been launched. Claudia Schmölder's help on all matters concerned was absolutely central, her experience as always truly appreciated, and I am most grateful for her sage advice and long-standing sense of proportion. Finally, my superb research assistants Dorothee Dehnicke, Friederike Mehl, Tom Reichard, Katja Rippert, Magdalena Stotter and Ruth Haake proved as instrumental as possible. Jennifer Pierce and Severin Siebertz also helped substantially in preparing the manuscript for publication. I am truly obliged to all of them.

While this volume would not exist without the Bielefeld conference as its precursor, I should like to stress that the book by no means simply presents its 'proceedings.' Quite to the contrary: while I do regret that a strict, unsentimental selection of contributions was imperative, this book comprises a limited set of carefully chosen and thoroughly revised articles, painstakingly arranged in both thematic and largely chronological order, asking complementary questions and speaking directly to each other's concerns. I must also single out the help of several illustrious commentators whose much appreciated insights and criticism shaped the conference and, in turn, this volume. They include Peter Becker, Ralf Bülow, Paul Ceruzzi, Andreas W. Daum, Peter Davidson, Steven J. Dick, De Witt Douglas Kilgore, Kai-Uwe Schrogl, Angela Schwarz, Helmuth Trischler and Bernd Weisbrod. At Palgrave Macmillan, editors Michael Strang and Ruth Ireland gamely took on yet another lengthy manuscript and oversaw its publication with the same care that I had come to appreciate through previous collaboration. In a similar vein, cooperating with Penny Simmons proved again a true pleasure; I could not have wished for a more thoughtful and meticulous copy-editor. I also gratefully acknowledge the enthusiasm and encouragement of the two anonymous reviewers.

Imagining Outer Space endeavors to break new ground in the historicization of outer space by introducing the notion of 'astroculture,' inserting a distinctly (West) European element into the hitherto largely US- and USSR-centered historiography, elucidating the complex relationship between science and fiction, and emphasizing the significance of outer space as a site for the projection of competing versions of the future. The volume brings together original and innovative work by both junior scholars and some of the most distinguished experts in this small, but rapidly burgeoning field of historical research. Featuring 15 contributions – plus an introduction and an epilogue – from representatives of nine disciplines and eight countries, *Imagining Outer Space* is in itself an exercise in international transdisciplinarity. It is for this reason that the lion's share of my gratitude goes to the authors themselves, and it is with great respect that I acknowledge their unceasing willingness to travel thus far with me, both in time and space.

Berlin
August 2011

Alexander C.T. Geppert

Notes

1. This is, of course, John Brewer's term. See his *The Pleasures of the Imagination: English Culture in the Eighteenth Century*, London: HarperCollins, 1997.
2. Hermann Heimpel, 'Der Versuch mit der Vergangenheit zu leben: Über Geschichte und Geschichtswissenschaft,' *Frankfurter Allgemeine Zeitung* (25 March 1959), 11. See http://www.geschkult.fu-berlin.de/e/fmi/astrofuturismus/C_IOS/Hauptseite.html for the complete conference program; for further information on the Emmy Noether research group 'The Future in the Stars: European Astroculture and Extraterrestrial Life in the Twentieth Century,' consult http://www.geschkult.fu-berlin.de/astrofuturism.

List of Figures

Ever since American pilot Kenneth Arnold observed nine shiny 'saucer-like aircraft' flying in formation from Mount Rainier to Mount Adams in Washington State, USA, on 24 June 1947, such disc-shaped missiles have been known as 'flying saucers.' The cover image is based on German graphic designer Klaus Bürgle's dramatic 1971 interpretation of the UFO's founding myth. © Gösta Röver, Freie Universität Berlin.

Abbreviations

AI	Artificial Intelligence
ARD	Arbeitsgemeinschaft der öffentlich-rechtlichen Rundfunkanstalten der Bundesrepublik Deutschland
BBC	British Broadcasting Corporation
BIS	British Interplanetary Society
BNCSR	British National Committee on Space Research
BRD	Bundesrepublik Deutschland
CETI	Communication with Extraterrestrial Intelligence
CNES	Centre National d'Etudes Spatiales
COPERS	Commission Préparatoire Européenne de Recherche Spatiale
DDR	Deutsche Demokratische Republik
DEFA	Deutsche Film-Aktiengesellschaft
DFG	Deutsche Forschungsgemeinschaft
DVLR	Deutsche Versuchsanstalt für Luft- und Raumfahrt
ELDO	European Launcher Development Organization
ESA	European Space Agency
ESPI	European Space Policy Institute
ESRO	European Space Research Organisation
ET	Extraterrestrial
ETH	Extraterrestrial Hypothesis
ETI	Extraterrestrial Intelligence
EU	European Union
FAZ	*Frankfurter Allgemeine Zeitung*
GfW	Gesellschaft für Weltraumfahrt
HAEU	Historical Archives of the European Union
ID	Electromagnetic Identification
IGY	International Geophysical Year
IM	Inoffizieller Mitarbeiter
IONS	Institute of Noetic Sciences
ISS	International Space Station
JPL	Jet Propulsion Laboratory

MfS	Ministerium für Staatssicherheit (DDR)
NACA	National Advisory Committee on Aeronautics
NASA	National Aeronautics and Space Administration
NASM	National Air and Space Museum
NATO	North Atlantic Treaty Organisation
n.d.	No date
n.p.	No publisher/pagination
NRC	National Research Council
SDI	Strategic Defense Initiative
SED	Sozialistische Einheitspartei Deutschlands
SETI	Search for Extraterrestrial Intelligence
SF	Science Fiction
SRA	Self-Reproducing automaton
UFO	Unidentified Flying Object
UK	United Kingdom
UN	United Nations
USA	United States of America
USAF	United States Air Force
USSR	Union of Soviet Socialist Republics
VfR	Verein für Raumschiffahrt
ZDF	Zweites Deutsches Fernsehen

Notes on Contributors

Debbora Battaglia is Professor Emerita of Anthropology at Mount Holyoke College and Five College Fortieth Anniversary Professor. A sociocultural anthropologist, she specializes in alterity and world-making at intersections of science, technology and cosmology. Her books include the edited volume *E.T. Culture: Anthropology in Outerspaces* (2005), and she is the author of numerous articles on outer space in public and expert knowledge spheres. In progress are two book projects, *Seriously at Home in '0-Gravity'* and *Aeroponic Gardens and Their Magic*.

Thore Bjørnvig holds an MA in the History of Religions from the University of Copenhagen. A former associated member of the Emmy Noether research group 'The Future in the Stars: European Astroculture and Extraterrestrial Life in the Twentieth Century' at Freie Universität Berlin, his research focuses on intersections between science, technology and religion, with a particular emphasis on outer space. Recently published articles include 'The Holy Grail of Outer Space: Pluralism, Druidry, and the Religion of Cinema in *The Sky Ship*' (2012) and 'Outer Space Religion and the Ambiguous Nature of *Avatar*'s Pandora' (2013). Together with Roger D. Launius and Virgiliu Pop, Thore Bjørnvig has also co-edited a special issue of the journal *Astropolitics* on spaceflight and religion (2013). He blogs on astroculture for the Danish popular science news site videnskab.dk and the Nordic popular science news site sciencenordic.com.

Thomas Brandstetter is a historian of science and technology. He has published on a variety of subjects, including the history of astrobiology. Thomas Brandstetter's book publications include *Kräfte messen: Die Maschine von Marly und die Kultur der Technik* (2008).

Steven J. Dick was the 2014 Baruch S. Blumberg NASA/Library of Congress Chair in Astrobiology at the Library of Congress's John W. Kluge Center. From 2003 to 2009 he served as the NASA Chief Historian and Director of the NASA History Office; from 2011 to 2012 he held the Charles A. Lindbergh Chair in Aerospace History at the Smithsonian National Air and Space Museum in Washington, DC. Steven Dick is the author or editor of twenty books, including *Societal Impact of Spaceflight* (2007, co-ed.); *Discovery and Classification in Astronomy: Controversy and Consensus* (2013); and *The Impact of Discovering Life Beyond Earth* (2016, ed.). Minor planet 6544 Stevendick is named in his honor.

Rainer Eisfeld was Professor of Political Science at Osnabrück University from 1974 to 2006. Now emeritus, he continues to serve on the Board of Trustees of concentration camp Memorials Buchenwald and Mittelbau-Dora. His most recent publications are *Political Science in Central-East Europe: Diversity and Convergence* (2010, co-ed.); *Mondsüchtig: Wernher von Braun und die Geburt der Raumfahrt aus dem Geist der Barbarei* (1996, 2012); *Radical Approaches to Political Science: Roads Less Traveled* (2012); *Ausgebürgert und doch angebräunt: Deutsche Politikwissenschaft 1920–1945* (1991, 2013); and *Political Science: Reflecting on Concepts, Demystifying Legends* (2016).

Alexander C.T. Geppert holds a joint appointment as Associate Professor of History and European Studies and Global Network Associate Professor at New York University Shanghai as well as NYU's Center for European and Mediterranean Studies in Manhattan. From 2010 to 2016 he directed the Emmy Noether research group 'The Future in the Stars: European Astroculture and Extraterrestrial Life in the Twentieth Century' at Freie Universität Berlin. Recent book publications include *Fleeting Cities: Imperial Expositions in Fin-de-Siècle Europe* (2010, 2013); *Wunder: Poetik und Politik des Staunens im 20. Jahrhundert* (2011, co-ed.); *Obsession der Gegenwart: Zeit im 20. Jahrhundert* (2015, co-ed.); *Berliner Welträume im frühen 20. Jahrhundert* (2017, co-ed.); *Limiting Outer Space: Astroculture After Apollo* (2018, ed.); and *Militarizing Outer Space: Astroculture, Dystopia and the Cold War* (forthcoming, co-ed.). At present, Alexander Geppert is completing a cultural history of outer space in the European imagination, entitled *The Future in the Stars: Time and Transcendence in the European Space Age, 1942–1972.*

Henry Keazor is Professor for Early Modern and Contemporary Art History at Ruprecht-Karls-Universität Heidelberg. His research and publications focus on French and Italian painting of the seventeenth century, contemporary media and visual culture, especially on the French Baroque painter Nicolas Poussin, the reform in painting achieved by the Carracci towards the end of the sixteenth century in Italy, contemporary architecture and its relation to modern media, and the relationship between art and media, in particular the cartoon series *The Simpsons* and music video.

Pierre Lagrange teaches sociology of science at the Ecole Supérieure d'Art d'Avignon and is Associate Researcher at the Centre National de la Recherche Scientifique in Paris. He specializes in the study of 'belief' in the context of scientific and 'parascientific' controversies. Pierre Lagrange has published several books, mostly on UFO controversies (*La Rumeur de Roswell*, 1996), and on Orson Welles's 1938 invasion of Mars 'panic' broadcast (*La guerre des mondes a-t-elle eu lieu?*, 2005). He has also co-authored *L'Esoterisme contemporain et ses lecteurs: entre savoirs, croyances et fictions* (2006), a report for the public library of Beaubourg on readers of esoteric literature.

William R. Macauley is a postdoctoral research associate at the University of Manchester and a former member of the Emmy Noether research group 'The Future in the Stars: European Astroculture and Extraterrestrial Life in the Twentieth Century' at Freie Universität Berlin. His current research focuses on the portrayal of science and medicine in faith-based entertainment media products, notably evangelical Christian films from the postwar period to the present day. At present, William Macauley is working on two books, *Picturing Knowledge: NASA's Pioneer Plaque, Voyager Record and the History of Interstellar Communication, 1957–1977*, and *Science for the Soul: The Portrayal of Biosciences and Medicine in Faith-Based Entertainment Media*.

James I. Miller teaches at the Community College of Rhode Island. At present, he is working on a monograph that explores the politics of the reemergence of the traditional 'terroirs' in the construction of regional and national identity in France during the 1960s and 1970s, a period marked by increasing European integration, the end of Empire and the ongoing struggles of immigrants building new lives on this matrix.

Gonzalo Munévar is Professor Emeritus of Humanities and Social Sciences at Lawrence Technological University in Southfield, Michigan. His research interests include the epistemology of science, evolution, the philosophy of space exploration and neuroscience. His main publications include *Radical Knowledge: A Philosophical Inquiry into the Nature and Limits of Science* (1981); *Evolution and the Naked Truth* (1998); *The Master of Fate* (2000); and *Variaciones sobre Temas de Feyerabend* (2006). Gonzalo Munévar has also edited or co-edited several volumes, including *The Worst Enemy of Science? Essays in Memory of Paul Feyerabend* (2000); and *Sex, Reproduction and Darwinism* (2012, 2016). He is presently putting the final touches to *The Dimming of Starlight: The Philosophy of Space Exploration*.

Bernd Mütter is Program Director of the European culture channel ARTE and Head of Programming TV/New Media. He was a television author and editor at ZDF German television, Department for Modern History (Redaktion Zeitgeschichte) from 2003 to 2011, and, from 2008 to 2011, visiting lecturer at Justus Liebig-Universität in Gießen, Germany. Bernd Mütter's publications include a number of contributions to popular German history books.

Michael J. Neufeld is Senior Curator in the Space History Department of the Smithsonian National Air and Space Museum in Washington, DC, where he is responsible for the early rocket collection and for Mercury and Gemini spacecraft. Born and raised in Canada, he received his PhD from Johns Hopkins University in 1984. Michael Neufeld has written three books: *The Skilled Metalworkers of Nuremberg: Craft and Class in the Industrial Revolution* (1989); *The Rocket and the Reich: Peenemünde and the Coming of the Ballistic Missile Era* (1995), which won two book prizes; and *Von Braun: Dreamer of Space, Engineer of War* (2007), which has won three awards. He has also edited or co-edited Yves Béon's memoir *Planet Dora* (1997); *The Bombing of Auschwitz: Should the Allies Have Attempted It?* (2000); *Smithsonian National Air and Space Museum: An Autobiography* (2010); *Spacefarers: Images of Astronauts and Cosmonauts in the Heroic Era of Spaceflight* (2013); and *Milestones of Space: Eleven Iconic Objects from the Smithsonian National Air and Space Museum* (2014).

Philip Pocock is a Canadian artist living in Berlin. In the 1980s he exhibited photography at the Art Gallery of Ontario in Toronto and at the Cooper Union in New York where he was a faculty member of the International Center of Photography. He co-founded the desktop-published *Journal of Contemporary Art* in 1988 and relocated to Europe in 1991. In 1995 he co-produced a 'videoblog' *Arctic Circle*. In 1997 he was funded by Documenta X to create *A Description of the Equator and Some Øtherlands*, a Web 2.0 cinema. In 1999 he installed a *Humbot* at MOMA Paris and ZKM Karlsruhe, where he ran a lab with students for a decade, creating YouTube precursor *Unmovie* in 2002 and the *SpacePlace* 'app' in 2005, released before the iPhone appeared. Recently Philip Pocock installed *Aland* at the Nam June Paik Art Center in Korea.

Claudia Schmölders is a cultural scholar, author and translator. Her numerous book publications include *Die Kunst des Gesprächs: Texte zur Geschichte der europäischen Konversationstheorie* (1979, 1986); *Das Vorurteil im Leibe: Eine Einführung in die Physiognomik* (1995, 2007); *Gesichter der Weimarer Republik: Eine physiognomische Kulturgeschichte* (2000, with Sander Gilman); *Hitler's Face: Biography of an Image* (2005); *Das Vorurteil im Leibe: Eine*

Einführung in die Physiognomik (2007); *Balzac: Leben und Werk* (2007, co-ed.); and *Faust & Helena: Eine deutsch-griechische Faszinationsgeschichte* (2018).

Guillaume de Syon teaches modern European history and the history of technology at Albright College in Reading, Pennsylvania. He is also a research associate in the History Department at Franklin and Marshall College in Lancaster, Pennsylvania. He was previously a contributing editor to the Collected Papers of Albert Einstein and a visiting research fellow at the Graduate Institute of International Studies in Geneva. Guillaume de Syon's research interests include the history of visual and popular culture (especially postcards and comic books), and the history of aviation and space travel as expressed in culture and politics. He is the author of *Zeppelin! Germany and the Airship, 1900–1939* (2002, 2007); and *Science and Technology in Modern European Daily Life* (2008).

Tristan Weddigen is Professor of Art History at Universität Zürich. His publications focus on early modern art and art theory, and on methodological issues. Currently he is working on the iconology of the textile medium and on Heinrich Wölfflin. His publications include *Raffaels Papageienzimmer: Ritual, Raumfunktionen und Dekorationen im Vatikanpalast der Renaissance* (2006); *Metatextile: Identity and History of a Contemporary Art Medium* (2011, ed.); and *Paragone: Wettstreit der Künste* (2012, co-ed.).

Introduction

CHAPTER 1

European Astrofuturism, Cosmic Provincialism: Historicizing the Space Age

Alexander C.T. Geppert

'Outer Space' is an expanding subject.
D.J. Gibson, British Foreign Office (26 October 1959)[1]

Ubiquitous, limitless and ever-expanding as it may be, outer space has a history too. Over the course of the twentieth century, the dark, infinite and unfamiliar vastness that surrounds us has stimulated the human imagination to an extent hitherto unknown. Numerous ventures to 'explore,' 'conquer' and 'colonize' the depths of the universe in both fact and fiction must be read as attempts to counter the prevailing *horror vacui*, the fear of empty spaces and voids of infinity felt and explicitly formulated since the sixteenth century. They all aim at overcoming what Sigmund Freud (1856–1939) termed in 1917 humankind's 'cosmological mortification,' the humiliating decentering of the earth effected by Nicolaus Copernicus's (1473–1543) heliocentric cosmology. Three decades and two world wars after Freud's observation, influential British futurist and science-fiction writer Arthur C. Clarke (1917–2008) identified a related 'desire to know, whatever the consequences may be, whether or not man is alone in an empty universe' as the one key motive underlying all human efforts to overcome gravity and reach out beyond humankind's natural habitat on planet Earth.[2]

Imagining and re-imagining space and furnishing it time and again with one artifact after another, be they mental or material, has had a doubly paradoxical

Alexander C.T. Geppert (✉)
New York University Shanghai, Shanghai, China
New York University, New York, NY, USA
e-mail: alexander.geppert@nyu.edu

Alexander C.T. Geppert (ed.), *Imagining Outer Space*
European Astroculture, vol. 1
https://doi.org/10.1057/978-1-349-95339-4_1

effect. As outer space became increasingly cluttered, it simultaneously became more and more concrete, and, concomitantly with such imaginary colonization, regarded in ever more spatial terms. An entire geography of outer space developed that presented itself as a continuation, if not a logical extension of earlier geographies of imperial expansion and colonial domination (Figure 1.1).[3] At the same time, outer space developed into one of the major sites of twentieth-century utopian thinking, where relations vis-à-vis science, technology and the future were positioned, played out and negotiated as nowhere else. In the process, outer space was transformed into a place in its own right. In 1974 cosmic jazz musician Sun Ra (1914–93) was timely when famously proclaiming with His Astro Intergalactic Infinity Arkestra that 'Space is the Place.' For much of the twentieth century, it was indeed.

Even an ever-expanding space, however, is subject to limitations. As numerous other observers – no less insightful than Sigmund Freud, Arthur C. Clarke and Sun Ra – have noted time and again, in defiance of all grand rhetoric and despite all arduous, piecemeal steps into the often glorified and frequently kitschified 'unknown,' to date the so-called Copernican revolution has still not been fully consummated. While versions and visions of outer space, extraterrestrial life and alien worlds – 'where no man has gone before' – have become increasingly elaborate, multifarious and competing, they have not succeeded in completely transcending life as we have long known and lived it, notwithstanding considerable cultural repercussions and societal impact. The more far-fetched these outlooks have become, the more geocentric they remain.[4] When the Allensbach-based Institut für Demoskopie, the oldest German polling institution, found during the Space Race's heyday that the proportion of West German citizens believing in the existence of extraterrestrial intelligence had declined from 42 to 28 percent between June 1954 and May 1967, it aptly termed such a seemingly counterintuitive diagnosis 'cosmic provincialism.'[5] Space enthusiasm and terrestrial geocentrism are two faces of the same coin. Aiming to observe and to comprehend rather than to believe, to preach or even to predict, it is particularly imperative that space historians find the right measure of benevolent, yet critical, distance from historical actants and propagandists of spaceflight and extraterrestrial expansion, the powerful promises they made, and the time-tested rhetorics they employed.

A truism for some, politically undesirable for others, the historicity of outer space and its human-made character is patently good news for the historian, permanently on the prowl for past forms of human self-expression. Historical visions of a future in outer space, imagined encounters with extraterrestrial civilizations and changing conceptions of alien life forms seem deeply characterized by their insurmountable anthropomorphism, insofar as they, quite unsurprisingly, always reveal more about their author's societies than about 'them' or any 'other.' If so, then the comprehensive historicization of outer space and extraterrestrial life must not only be intensified and advanced at once, but also instantly acquitted from all potential charges of exoticism, arcaneness and, hence, political irrelevance. Quite to the contrary,

THE ILLUSTRATED LONDON NEWS,

The World Copyright of all the Editorial Matter, both Illustrations and Letterpress, is Strictly Reserved in Great Britain, the British Dominions and Colonies, Europe, and the United States of America.

SATURDAY, SEPTEMBER 15, 1951.

THE FIRST STEP TOWARDS THE CONQUEST OF SPACE: AN UN-MANNED SATELLITE STATION CIRCLING THE EARTH.

Figure 1.1 In the fall of 1951, the cover page of the reputable *Illustrated London News* featured a 'generally recognized' concept for the 'first step towards the conquest of space.' The upper image shows an unmanned satellite station circling the earth in its orbit as a communication device. A solar mirror, pointing towards the sun and focused on a central heating coil, is integrated to produce electricity, while an earthward-oriented arm carries a radio transmitting system and receiving instruments. The lower image details the placing of three such space stations in earth orbit and their radio interconnections. Largely based on Arthur C. Clarke's far-reaching concept of 'extra-terrestrial relays,' published in the October 1945 issue of *Wireless World*, the aim was to establish the kind of global communication system considered indispensable in a world society yet to come.
Source: G.H. Davis with Eric Burgess and Arthur C. Clarke, *Illustrated London News* (15 September 1951), 393.

far from being outlandish or restricted to obscure elite discourses, ideas and images of outer space have been inextricable from the self-ascribed technoscientific modernity of the twentieth century as exemplified by that outdated yet still alluring notion, the Space Age.[6]

When such a Space Age occurred, how long it endured, and when it ceased to exist – or whether we still live in its midst – are valid questions still open to debate. Irrespective of such periodization problems, it is entirely indubitable that outer space was, for several decades in the postwar era, intimately bound with notions of modernity and utopian visions of human progress. 'Our present-day world and our present-day human existence is most profoundly influenced and shaped by the fact of spaceflight,' philosopher Günther Anders (1902–92) noted in 1970.[7] As the 15 contributions to this volume demonstrate time and again, for a limited, surprisingly short-lived time, outer space became the epitome of modernity – comparable only to that other major technoscientific project of the twentieth century, nuclear power. The Space Age and the Atomic Age went hand in hand, yet the former's radiance remains largely unacknowledged compared to its modern iconic 'evil twin,' bomb culture. It is necessary, but not nearly sufficient, to explain fears of alien invasion by evoking a Cold War context and employing the notion of Cold War *Angst*. Space enthusiasm, fantasies of spatial expansion and visions of interplanetary colonization are older and more all-encompassing, and should not be reduced to a collective, psychosis-like defensive complex.[8]

In his introduction to the standard work, ...*The Heavens and the Earth*, the award-winning political history of this period published more than a quarter-century ago, historian Walter McDougall identified three structural forces necessary to launch the American space program: an economy prosperous enough to finance the endeavor; the availability of appropriate technological means; and, more hazily, yet suggestively, 'imagination.' Within this triad, the present book focuses on the third vector, what McDougall described with sociologist Daniel Bell as 'culture, the realm of symbolism that explores the existential questions facing all human beings all the time – death, love, loyalty, tragedy.'[9] Unlike the bulk of existing historiography, contributions in this book do not set out to examine political, diplomatic and technological aspects of space history. Rather, they explore the socio-cultural rationales behind these efforts and their relationship to the imaginary, from both individual and collective perspectives. Three core questions drive this book: First, how did the idea of outer space, spaceflight and space exploration develop over the course of the twentieth century into a central element of the project of Western and, in particular, European modernity? Second, how was outer space represented and communicated, imaged, popularized and perceived in media as varied as print and film, as well as a diverse array of narrative conventions including historical fiction and institutional reporting, all in their own ways contributing to the imaginary bestowal of the universe? And, third, in what way have these conceptions of the cosmos and extraterrestrial life been affected by the continual exploration of outer space, and vice versa?

I Defining astroculture

On 11 July 1969, towards the end of the period under scrutiny in this book, British pop musician and actor David Bowie (1947–2016) released 'Space Oddity,' a song produced to coincide with the Apollo 11 lunar mission (Figure 1.2). Used in conjunction with the BBC's coverage of the first moon landing nine days later, 'Space Oddity' combined futuristic electro sounds with ethereal strings and more familiar rock timbres. Reaching number five in the British charts, it became Bowie's first commercial hit. Firmly grounded in established motifs, 'Space Oddity' was inspired by Stanley Kubrick (1928–99) and Arthur C. Clarke's 1968 landmark science-fiction film *2001: A Space Odyssey*, as the pun in its title overtly signaled. Yet it also added cultural references to the repertoire that would recur in future attempts at making sense of outer space, notably a new fictive hero, the soon-to-be legendary astronaut Major Tom, whose remains are ostensibly floating indefinitely through the universe. Tom has indeed traveled far – if not to the physical limits of the galaxy, at least into the depths of international pop culture. Bowie's own productions frequently drew on this space trope, such as in 'Ashes to Ashes' (*Scary Monsters*, 1980) and 'Hallo Spaceboy' (*Outside*, 1995). Bowie's Major Tom has also been prominently evoked by Def Leppard ('Rocket,' 1987), Peter Schilling ('Völlig losgelöst,' 1983)

Figure 1.2 UK cover of David Bowie's 1969 hit record 'Space Oddity,' his portrait superimposed on a work by the French-Hungarian Op-Art artist Victor Vasarely (1906–97), consisting of blue and violet spots on a green background.
Source: Courtesy of Vernon Dewhurst.

– a key protagonist of the so-called *Neue Deutsche Welle* in early 1980s pop music – and numerous others.[10]

Although the song's lyrics comprise only 35 lines, on closer inspection one finds a surprising number of astral sub-themes addressed, many of which are featured in contributions to this volume. They include the science/fiction complex ('Take your protein pills and put your helmet on/ [...] Commencing countdown, engines on'); the intricate commerce/media/public triangle ('And the papers want to know whose shirts you wear'); the so-called overview effect, that is, the view back onto the earth usually associated with the epoch-making 1968 spaceflight of Apollo 8 and the standard argument that humankind's thrust into outer space would, ultimately, constitute a return to itself ('For here am I sitting in a tin can/ Far above the world/ Planet Earth is blue/ And there's nothing I can do'); as well as religious-spiritual implications and references to a spatial-transcendental beyond that only the blessed and chosen astronaut is capable of approximating by ascending into heaven ('May God's love be with you').

Bowie's 'Space Oddity' is exemplary for another reason. In the twentieth century outer space, futurism and alien images permeated contemporaneous culture and society to an unprecedented extent. His hit epitomizes a specific complex of space-related cultural products that have gained considerable momentum since the Second World War, furthered by actants in politics, mass media and popular culture. Analytically, their complicated alliances and interconnections are hard to disentangle, not least because of the sheer lack of a widely recognized standard terminology. As a remedy, this book examines the cultural significance and societal repercussions of outer space and space exploration under the new label of 'astroculture.' How have human beings used their creative powers to render the infinite vastness of outer space conceivable? Far from intending to establish yet another academic subdiscipline, astroculture constitutes an umbrella concept to ease McDougall's terminological difficulties in referring to an underspecified and barely studied field of historical research. To remain within and augment his vocabulary: astroculture comprises a heterogeneous array of images and artifacts, media and practices that all aim to ascribe meaning to outer space while stirring both the individual and the collective imagination.[11]

At the same time, this superordinate concept is designed as an explicitly culture-related counterpart to such better known and firmly established notions as 'astrophysics,' 'astropolitics' – evidenced by the founding of an academic journal by this title in 2003 – or 'astrosociology.'[12] When historicizing outer space, for reasons of practicality, inclusiveness and connectivity, astroculture is to be preferred over other umbrella notions. The obvious and conceivably encompassing, yet far too imprecise choice, 'space culture(s),' is unsuitable due to the equivocality of the term 'space' itself, thus inviting conceptual misunderstandings from other fields like urban studies or entire disciplines such as geography. Further alternative suggestions include, for instance, Margaret Mead's and Donald N. Michael's largely inconsequential mid-1950s 'Man-Into-Space'

(MIS) program for the social sciences, launched well *before* the first artificial satellite, or the more recent, narrower 'extraterrestrial (ET) culture' as developed by American anthropologist Debbora Battaglia.[13] Astroculture as a novel concept does share some of the defining features of ET culture, including an emphasis on lived experience, the objective of de-exoticizing the alien, and its self-understanding as an exploratory project. Yet, there are also distinct differences. Not all astroculture revolves around alien life or extraterrestrial technology, anthropocentric and terrestrial as those may be, but comprises a wider range of images, artifacts and activities conducted by a broader range of expert and amateur actants. Different as the so-called 'space,' 'science fiction,' 'ET,' 'UFO' and other related communities are – the first, *mutatis mutandis*, focusing on applied science, the second on fantasy, the third on humans and the fourth on alien technology – their agendas, concerns and practitioners overlap and compete to such an extent that any separating, non-integrative approach seems unduly self-limiting from the outset and would require particular justification. Taking seriously the umbrella concept of astroculture leads to analyzing similarities and commonalities *before* possibly re-establishing differences and boundaries between the various subcultures. Hence, the entire range of supposedly obscure and frequently exoticized phenomena, including UFOs, the 'technological wing of the ET imaginary'; early contact claims, alien abduction experiences and 'starship memories'; or Erich von Däniken's so-called pre-Astronautics fall as well under the purview of astroculture, as do space mirrors, space elevators, space stations and space colonies.[14]

Strenuously exempting these phenomena from historicization as a consequence of their 'pseudoscientific' character or rejecting them as 'frivolous speculation' would be a rash and grave intellectual error. The Space Age cannot be thoroughly historicized without taking debates about the epistemic-ontological status of claims regarding space exploration and extraterrestrials into account. Research on the history of astroculture does not aim at providing definitive answers regarding the reality or fiction of space-related phenomena. Instead, it critically focuses on the intentions, actions, categories and explanations provided by actants themselves, because they are part and parcel of the ways in which human beings attempt to come to terms with and make sense of the infinite universe that surrounds us. And vice versa: viewed from the opposite perspective, that of historiography, it is hoped that the formulation of this new umbrella concept of astroculture will lead to the controlled import of elsewhere long-established analytical key categories such as 'language,' 'consumption,' 'representation,' 'appropriation,' 'memory,' 'materiality,' and, above all, 'meaning,' in addition to numerous others into space history, where they have played no more than a minor, dramatically undervalued role.

II Introducing Europe

In addition to proposing the concept of astroculture and demonstrating exemplary ways in which its concerns can be historicized, the present volume pursues a second, hardly less ambitious objective. Introducing and foregrounding a specifically (West) European perspective, it aims to find an analytical 'third way' or middle course between West and East, and address, if not solve, the European paradox of comprehensive space enthusiasm despite decades-long abstinence from manned spaceflight.

Since 1945, Western Europe's contribution to the physical exploration of outer space has been peripheral and, for many years, a secondary priority at best. As a concomitant of the rapidly emerging US-USSR polarization during the Cold War, much of Europe's cultural hegemony was lost. Making a virtue of necessity and in order to profit from the rising prestige of technoscience, the concept of Europe as the 'third space power' – under French leadership – was invented as a political convenience, proving to be of particular political attraction to President Charles de Gaulle (1890–1970) who announced plans to establish a French orbital space program as early as 1959. On a supranational level, the institutional prehistory and inner-European unification process of what would become in 1975 the European Space Agency, seated in Paris, proved tremendously complex, taking almost two decades for the organization to be formed by merging ELDO (European Launcher Development Organization) and ESRO (European Space Research Organization), both set up in 1964 (Figure 1.3).[15]

The reason for which autonomy – understood as independent human launch capability – has always been the central theme of the European space program was quite simply its absence.[16] European spaceflight had begun with unmanned satellites, as it had with the USSR and the USA. Ariel 1, the first international earth satellite, was launched on 26 April 1962, as a joint project of the British and American space agencies; the Italian-American San Marco 1 satellite followed two years later. On 26 November 1965, France became the third nation to orbit a satellite, Astérix, with its own Diamant rocket, launched from Hammaguir, a remote site in central Algeria still under French control. ESRO only managed to launch its first satellite in 1968. Yet, manned spaceflight proved a different matter. The first non-Soviet European human to fly in space was the Czech Vladimír Remek (1948–) in March 1978, with the French spationaut Jean-Loup Chrétien (1938–) to follow four years later. These flights came 17 and 21 years, respectively, after those of Yury Gagarin (12 April 1961) and John Glenn (20 February 1962), the first human and the first American, respectively, to orbit planet Earth.[17]

What was different in Europe, then, was the long time-lag of roughly two decades between unmanned (1962) and manned spaceflight (1978/1983), the latter still today attracting media coverage and public attention of an incomparable magnitude and hence generally treated as the only truly worthwhile form of spaceflight. Together with its civilian use, the absence of manned space activities in Western Europe may also help to explain why an organized anti-space movement has never evolved, not even an intermittent, anti-space discourse among the intellectual elites. Such an absence is all the more conspicuous when compared to the widespread opposition to atomic power and the large-scale anti-nuclear weapons movements of the late 1950s

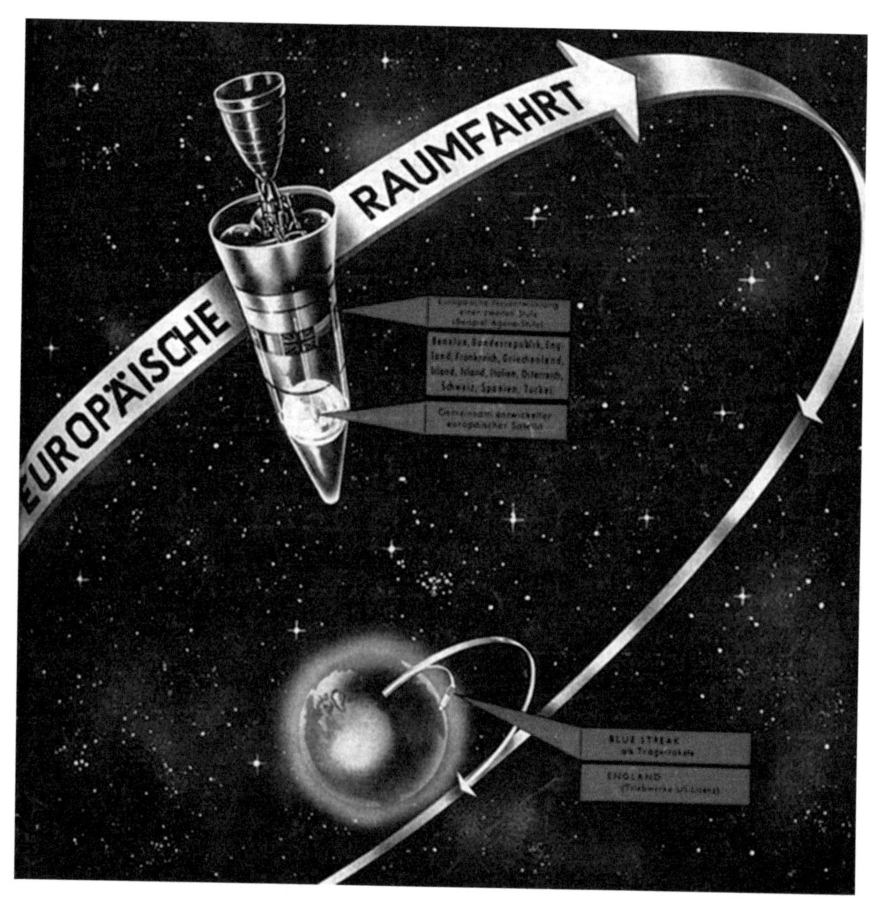

Figure 1.3 Illustration to explain and publicize the West European space effort, 1961. In this picture, a satellite commonly developed by the six original member states of the European Economic Community (Belgium, France, West Germany, Italy, Luxembourg, Netherlands) in addition to England, Greece, Ireland, Iceland, Austria, Switzerland, Spain and Turkey (as indicated by the small flags on its body and the accompanying text box) is seen circling the globe, having been placed there by its carrier vehicle, the British Blue Streak ballistic missile. Ironically, the European Space Agency (ESA) was founded only in 1975 after repeated attempts to develop such an independent European launcher system had ended in failure.
Source: *Weltraumfahrt* 12.3 (May/June 1961), cover image.

and early 1960s, particularly in Great Britain and West Germany, triggered by the threat of nuclear war and its lethal radioactive after-effects.[18]

As the chapters in this volume testify, popular interest in outer space and its presence in everyday life was nonetheless tremendous during the 1950s, 1960s and 1970s, and it remains so through today. It will require considerable effort to adequately explain this European paradox of overwhelming space enthusiasm simultaneous with such an extended period of abstinence from independent manned spaceflight activities. Without doubt, a broader, Europeanized historical perspective can only be achieved by forging a transdisciplinary and transnational approach that takes all necessary transatlantic references and transcontinental interdependencies into account. While in principle as worthwhile as any such internationalizing and hence widening move, current calls for writing a 'global history' of space exploration by shifting attention to the relationship between 'spaceflight and national identity' risk the danger of stating the obvious. What's more, such pleas cannot convince, at least until this intermediate, hitherto missing perspective, namely the West European, has been conceptually and empirically explicated as a necessary counterweight to the overbearing focus on US and USSR histories. Provincializing Europe is always a neat feat, yet hardly feasible as long as space historians do not quite know when and what 'Europe in space' was.[19]

Historiographically, such a discrepancy between American and Soviet/Russian space history on the one hand, and its underdeveloped European counterpart on the other, is a direct consequence of their respective institutional settings. Especially in the United States, the concerted activities, resources and unparalleled research programs of NASA's History Program Office, founded in 1958, and the Smithsonian National Air and Space Museum, reopened in the new building along the National Mall in Washington, DC, in July 1976, have effectively made space history a respectable academic topic.[20] Together with an interplay of persons, ideas and funds, these institutions have defined and structured a new field of historical research. Heavily invested in NASA's history-making powers, their establishment has proved a self-fulfilling prophecy. Indirectly at least, and somewhat ironically, scholarship on Eastern Europe and the 'Russian Space Age' – in particular Sputnik and the space *persona* of Yury Gagarin (1934–68) – has also benefited from such an unprecedented institutional shaping.[21] Today, both American and Soviet/Russian space history present themselves as open and expansive, yet comparatively well-established and structured fields, in spite of contrary claims and the inevitable degree of research-gap rhetorics.

In Western Europe, space history is by comparison a much smaller, more fragmented and underdeveloped affair, frequently exoticized and occasionally ridiculed by mainstream historians. Unfortunately, an institutional equivalent to NASA's History Program Office does not exist, and neither does the corresponding position of a Chief Historian. ESA's outreach activities into academic territory remain woefully limited, particularly as far as the humanities and social sciences are concerned. Having commissioned a small group of top-class historians under the direction of John Krige and Arturo Russo to author its institutional history in 1990, ESA subsequently extended this first self-historicizing initiative by commissioning 40 additional 'History Study Reports' with individual authors treating a total of 16 countries in overview-oriented booklets of 30–100 pages, in addition to general aspects of space study, such as satellite programs, the history of sounding rockets or international cooperation.[22] Yet, a mere accumulation of one national space history after the other – from Austria and Belgium to Switzerland and the United Kingdom – cannot compensate for a genuinely European history that treats the continent as a geographical setting *and* makes the question of Europeanness its central heuristic concern. By focusing exclusively on institutional, political, diplomatic and technical aspects of the European space effort, ESA has underestimated and neglected significant larger questions. Societal impact and cultural repercussions have not played a significant role in its historical self-assessment. What's worse, since the completion of the so-called History Project and despite the successful establishment of the European Space Policy Institute (ESPI) in Vienna in 2003 – a largely policy-oriented think tank – active promotion of non-science, non-applied research has come to a standstill. ESA's interest in its own past and position within European society, while inherently forward-looking, remains parochial, displaying almost its own version of 'cosmic provincialism.'

That said, the cultural history of Europe in space and space in Europe is a problem that this volume can pose with great verve, yet by no means solve. The current state of research on these topics is too divergent, uneven and disconnected to yield conclusive results. As a consequence – and, at this early stage, possibly an inevitability – there can be no doubt that the present volume possesses a certain British/French/West German bias, with the Scandinavian countries, for instance, or the wider Mediterranean world, in particular Italy and Francoist Spain, not receiving the kind of attention that they deserve. Competition and cooperation, comparisons and connections between individual countries within capitalist Western and communist Eastern Europe are themes that several contributions pursue, but which the volume as a whole does not squarely confront.[23] It will require a good deal of additional research before a fully fledged, empirically grounded and theoretically informed answer can be advanced as to the existence of a specifically West European perspective on outer space between 1945 and the early 1970s. While contributions to this volume are confident in staking out a new field, they do not claim to offer more than the highlighting of a number of viable paths along which to address the European space history paradox.

III Fictionalizing science, scientizing fiction

To a large extent, the collective imagination of outer space relies on the power of images, both still and filmic. In the last book published before her death, literary theorist and public intellectual Susan Sontag (1933–2004) observed that civilians' understanding and envisioning of violent conflict is a direct product of photographic images of war. A parallel argument can be seamlessly applied here. It is virtually impossible to experience outer space in a direct, unmediated manner. So far only 12 men have walked on another celestial body, and while space tourism is becoming increasingly popular, it still remains limited to a handful of affluent aficionados willing to spend a fortune for a few days in low-earth orbit on board the International Space Station (ISS). As a consequence, popular understanding of outer space is chiefly a product of images and representations, and their composition into narratives such as the ones analyzed in this book.[24]

Making the complex relationship between 'realities' and 'visions,' between 'science' and 'fiction' the third focus of this volume is, then, perhaps not entirely original, yet seems an almost unavoidable choice (Figure 1.4). The theme as such is long familiar to all space historians, first employed in 1944 by the German-born science popularizer and space expert Willy Ley (1906–69), taken up by Clarke in a comprehensive paper read to the British Interplanetary Society (BIS) in April 1950, and subsequently expanded and elaborated by numerous other advocates and activists dabbling in amateur historiography.[25] The reasons for pursuing such vested interests on the part of protagonists and propagandists alike were as simple as they were straightforward: For a long time the 'spaceships of the mind' were the only ones existing. Members of the early spaceflight movement found themselves in dire need of a *longue durée* perspective in order to counter contemporaneous appeals against their allegedly dubious expertise, as well as rebuffing public scrutiny and hostile criticism toward the respectability of this contested, then newly developing field.[26] The launch of Sputnik 1 in October 1957 – so the widely accepted standard periodization and oft-repeated master narrative asserts – marked the beginning of the eagerly awaited Space Age, and over the course of the ensuing 'Space Race' the 'visionary' or 'pioneering' era of spaceflight was finally superseded by 'real' spaceflight, with the 'exploration' and ensuing 'conquest' of space being gradually, yet continually advanced. According to proponents of this view, it was during this historic and revolutionary process that 'science fiction' became increasingly substituted by 'science fact,' sooner or later ceding much of its historical significance to the 'right stuff.'[27]

While evidently not questioning the power of fiction, be it scientific or not, the present volume does not partake in these debates about primacy and substitution. Science fiction has never been a 'blueprint' for anything, and

Figure 1.4 A fictitious, satirical 'Map of Mars containing all information so far obtained by Astronomers, Astronauts & other Observers,' published by the British weekly magazine *Punch* in April 1956, claimed to chart all existing knowledge of the Red Planet's geography onto the most comprehensive map yet created. Without exception, the five listed authors – Edgar Rice Burroughs, Arthur C. Clarke, Charles Chilton, Ray Bradbury and Herbert George Wells – were British and American science-fiction authors, neither scientists nor engineers. An accompanying article published in the same *Punch* issue explained that Mars, first visited in 1866, had since become the most widely-explored of all planets, and one on which the manufacture of flying-machines was especially well developed. ▶

Source: William Hewison, *Punch* 230 (11 April 1956), 411.

A Map of
MARS
SOUTH POLE
VALLEY DOR
OTZ MOUNTAINS
M. CIMMERIUM
M. TYRRHENUM
M. ERYTHRAEUM
SOLIS LACUS
SYRTIS MAJOR
MARE ACIDALIUM
Thark
Zodanga
Helium
Edgar Rice Burroughs
Arthur C. Clarke
Charles Chilton
Ray Bradbury
Herbert George Wells
fecerunt

neither has 'science' evolved out of purely fictitious systems of thought. To be sure, there are differences between science in fiction, fiction in science and science fiction. Science fiction and science fact do overlap and continually influence each other, yet neither one has ever fully subsumed or eclipsed the other. Assuming a linear development 'from imagination to reality' – as the venerable British Interplanetary Society's motto still reads – leads too easily to a naïve endorsement of the type of teleological master narratives that professional historiographical scholarship must avoid by all means.[28]

Arguing that 'science fiction' and 'science fact' are not contradictory but complementary, this book questions whether it is simply their different epistemologies and alternative modes of representation that configure the pivotal difference. If we interpret the science versus fiction problematic not as one narrative successfully replacing the other, but as a simultaneous coexistence with intersecting waves and continuous, mutual repercussions between the two, the core question is no longer one of primacy but about contact points, interrelations and their 'in-betweens.' Such an approach allows for encompassing historicization: Which scientific fictions became, at what point in time, predominant and were then realized and/or transformed into actual science? Which others 'failed' by remaining 'merely' fictitious, though by no means insignificant or ineffectual? And vice versa: what effects did science have on the conceptualization and design of fiction? Many science-fiction authors in the 1930s, for instance, felt it was their duty to write 'realistic' science fiction so that it could serve as an inspiration to contemporaneous scientists. Analyzing the conditions and contexts, consequences and crossovers of science and fiction is as significant as examining the multifarious socio-cultural effects these 'scientific fictions' had in different historical settings. Contributions in this volume strive to balance both perspectives. Taken together, they constitute a prime example of how cultural history can help to question and effectively overcome long-established standard periodizations that, upon revision, suddenly forfeit much of their conventional logic.

IV Transcending the future

In addition to defining astroculture, introducing a West European perspective and exploring the science/fiction complex, this book pursues two additional objectives.

First, *Imagining Outer Space* argues that changing conceptions of outer space and extraterrestrial civilizations must be read as historical expressions of earthly ideas of the spatialized beyond and past expectations of planetary futures. For approximately three decades, from the aftermath of the Second World War through the mid-1970s, it was widely assumed that the future was destined to play out in outer space. In a few years, experts agreed, gigantic space mirrors, nuclear wonder weapons, manned space bases and numerous other imagined technologies would be positioned in the near-earth orbit, while the permanent colonization of the moon, followed by

Mars, and later the cosmic unknown beyond our solar system was believed to be only a matter of time. This is the same discursive complex for which American literary scholar De Witt Douglas Kilgore has coined the notion of 'astrofuturism,' here understood as a specific subcategory forming part of astroculture. The present volume explores the concept's usefulness by applying it empirically and historically within a defined geographical setting, that is, Western Europe.[29] How is the tight connection between outer space imaginaries and future visions to be explained, particularly prominent during the 1950s and 1960s? And does the observation hold that, by the mid-1970s, space was no longer 'the place,' that the promises of the Space Age began to lose their popular appeal at precisely the same time when faith in technology as a trustworthy engine of social change was on the wane as well? It is a standard historical argument that, with the global oil crisis of 1973, general expectations about the future underwent correspondingly radical shifts, with the Sex Pistols's 'No Future' (1977) becoming the slogan of the day.[30]

In addition to such a futuristic, later often explicitly utopian strand, there is, second, a strong transcendental element to be found at work within astroculture at large, directly connecting it to much older debates on the epistemologies of the supernatural and the theological beyond (Figure 1.5). This latter strand is often used to explain man's continuing and inescapable fascination with outer space, when confronted with the infinite and inconceivable breadth of the abyss. Freud skeptically discussed this phenomenon under the term of *ozeanisches Gefühl* (oceanic feeling), considered by some of his (and our) contemporaries as nothing less than the basis of religion. Likewise, in a *Playboy* interview undertaken four decades later, director Stanley Kubrick went so far as to associate and explain 'the grandeur of space' with 'the myriad mysteries of cosmic intelligence' to be found therein.[31]

Thus, exploring imaginaries of outer space and conceptions of other worlds eventually leads to analyzing their strong, yet all too often obscured, affiliations with transcendental beliefs and the spiritual beyond. How did changing images of outer space and the entire cosmos impinge on religion?[32] Such a diagnosis goes well beyond obvious episodes like Pope Pius XII declaring, at the Seventh International Astronautical Congress in Rome in 1956, that humankind's efforts to explore the 'whole of creation,' that is, the entire universe, were 'legitimate before God'; astronaut Frank Borman (1928–) reading the Bible aboard Apollo 8 on Christmas Day 1968; Pope Paul VI's praising the moon landing as an 'advance for all mankind'; or Pope Benedict XVI's conversing live with 12 astronauts on board the ISS on 21 May 2011, lauding them as 'our representatives spearheading humanity's exploration of new spaces and possibilities for our future, going beyond the limitations of our everyday existence.'[33]

In the end, *Imagining Outer Space* argues that the twentieth century's most radical version of alterity, namely its evolving conceptions of alien life forms, an 'other' unlike any before, cannot be analyzed without taking the

Figure 1.5 Future encounter with extraterrestrial intelligences on a distant planet as depicted by German graphic designer Klaus Bürgle (1926–2015) in 1965 for a supplement to the popular science journal *Das Neue Universum*. A settlement can be detected in the background, while a reception committee of insect-like creatures is already approaching the space travelers. 'This much is certain, the vital spark of creation has not only developed on our Earth,' the original caption read: 'The globe is not more than just a living space in the vast world building (*Weltgebäude*).'
Source: Courtesy of Klaus Bürgle, 'In fremden Sphären,' *Das Neue Universum* 82 (1965), unpaginated foldout.

transcendental component of such encounters into account. Historicizing the Space Age, then, promises to shed new light on the modernity of an allegedly secularized century that, for several decades, held fast to the possibility of redemption by translocating its earthly obsessions into the infinite vastness of the universe, with the hope of thereby retrieving cosmic transcendence in the imagined, secularized spatial beyond of the twentieth century.

V Structuring this volume

Tackling a century that shaped and was shaped by outer space to an unprecedented degree, this book analyzes European imaginaries as they formed world narratives and laid out interplanetary futures. Its 15 chapters – in addition to this introduction and a comprehensive epilogue – trace the current thriving interest in spatiality and space to earlier attempts at exploring worlds other than our own. Contributions do not analyze the actual scientific findings or technological feats, but focus on the cultural significance and imaginative repercussions of outer space and extraterrestrial life. Despite their different disciplinary provenances, they all share a cultural-historical perspective, take an interpretative approach and aim at overcoming space history's self-chosen 'splendid isolation,' with a view to integrating it more closely into mainstream social and cultural historiography.

All authors were asked to address the following three questions, or to seize a combination thereof in their contributions:

1. *Western Europe.* Was there a specifically European perspective on outer space, in particular between 1945 and the mid-1970s? How do we address – and, eventually, explain – the 'European paradox' of comprehensive space enthusiasm concomitant with a decades-long abstinence from manned spaceflight?
2. *Science/fiction.* How has the complex relationship between 'science' and 'fiction' evolved over time, in particular within the European imagination? Does the argument hold that science and fiction must be understood as complementary and relational, not antithetic, even if they are obviously both subject to their own rules, conventions and paces?
3. *The future.* How is the close connection between outer space and visions of the future to be explained, by many long believed to be inevitable and imminent? To what extent is Kilgore's notion of 'astrofuturism' analytically helpful? And is the argument historically correct that by the mid-1970s the idea of a utopia in outer space had lost much of its former compellingness and widespread appeal?

Arranged in a simultaneously thematic and largely chronological order – reaching from the *fin-de-siècle* through the present day, some even daring to speculate further ahead – the contributions give particular emphasis to the three decades between 1945 and the mid-1970s. Bracketing the entire *hausse* of Western cosmic enthusiasm, this period encompasses the so-called 'golden age of space travel' *before* the stationing of Sputnik 1 through the last Apollo landing on the moon in December 1972 and the establishment of the European Space Agency in 1975. Divided into five distinct parts – 'Narrating Outer Space,' 'Projecting Outer Space,' 'Visualizing Outer Space,' 'Encountering Outer Space,' and 'Inscribing Outer Space' – consisting of three chapters each, contributions historicize outer space from an interdisciplinary and transnational perspective. They focus on a wide range of prominent activists, momentous cases, specific sites, pertinent type of media and historical problems of particular significance.

Part I – 'Narrating Outer Space' – comprises a broad overview in Chapter 2 by former NASA chief historian Steven J. Dick on the role of the imagination in the making of outer space; a detailed reading in Chapter 3 by literary scholar Claudia Schmölders of the so-called Tunguska event, the ominous meteor strike in Siberia in June 1908, and its literary, scientific, metaphysical and pictorial impact; and in Chapter 4 an exploration by philosopher of science Thomas Brandstetter into images of, and debates about, crystalline aliens, that is, inorganic life forms on other planets, in twentieth-century science and fiction.

Part II – 'Projecting Outer Space' – encompasses Chapter 5 by political scientist Rainer Eisfeld on the changing human projections on planet Mars since the mid-nineteenth century, distinguishing between an 'Arcadian,' an 'Advanced,' a 'Frontier' and a 'Cold War' Mars; Chapter 6, an analysis of a largely unsuccessful 1960s East German print and film campaign against the American rocket engineer of German origin, Wernher von Braun (1912–77) and his controversial Nazi past by historian Michael J. Neufeld; and Chapter 7 on another prototypical space *persona*, the aforementioned British science-fiction author Arthur C. Clarke, and the powerful, yet carefully subdued transcendental strand in his all-embracing space thought by historian of religion Thore Bjørnvig.

Chapters in Part III – 'Visualizing Outer Space' – focus on West European conceptions of outer space in different media contexts. Chapter 8 by historian Bernd Mütter compares the space coverage in West German newspapers and science television shows between 1957 and 1987; in Chapter 9 historian Guillaume de Syon studies popular Franco-Belgian comic strips such as Hergé's well-known *Tintin* albums *Objectif lune* and *On a marché sur la lune* of 1953/54, but also *Buck Danny* and *Dan Cooper*, two comic series with a similar space theme; and in Chapter 10 art historian Henry Keazor submits the popular British television series *Space: 1999*, launched in the mid-1970s after *Star Trek* (1966–69) but before *Star Wars* (1977), to a close reading.

Part IV – 'Encountering Outer Space' – focuses on terrestrial contacts with extraterrestrial civilizations. Anthropologist Debbora Battaglia in Chapter 11 juxtaposes an analysis of a US National Research Council project on alien life forms and its hidden investment in century-old colonial projects with a reading of Werner Herzog's 2005 docu-fantasy film *The Wild Blue Yonder* and the neo-creationist origin myth of Raëlism, a contemporary UFO religion; Chapter 12 by sociologist Pierre Lagrange revisits the way in which sociologists have (mis)represented and (mis)attributed the appearance of so-called flying saucers in the global skies after 1947 to a Cold War context; and in Chapter 13 historian James Miller analyzes postwar UFO sightings in Quarouble, a small village in northern France, following the subsequent activities and media career of young metalworker Marius Dewilde, prime observer and alleged extraterrestrial contact.

Finally, Part V – 'Inscribing Outer Space' – features in Chapter 14 an article by philosopher Gonzalo Munévar on the impossibility of exploring the depths of the universe by infinitely self-reproducing probes, and the consequences that such a technology would have for the search for extraterrestrial life; an analysis in Chapter 15 of the famous NASA Pioneer plaque and its iconic interstellar message by historian of science and technology William

R. Macauley; and Chapter 16 by art historian Tristan Weddigen on the calibration target that noted British artist Damien Hirst created for ESA's Mars lander Beagle 2 in 2002. Finally, Philip Pocock draws this volume to a finale with his wide-ranging epilogue, part commentary, part analysis, by historicizing space art from the perspective of a practicing artist.

Imagining Outer Space looks at Europe in light of its preoccupation with the outer limits of the spatial; analyzes contact points between science and fiction; and critically examines sites and situations where images and technologies contributed to the omnipresence of fantasmatic thought and translocated futures in the popular imagination of the twentieth century. Taken together, the contributions that follow aim to expand contemporary understandings of 'outer space' such that astroculture becomes a new field of modern European historiography.

Notes

1. The National Archives of the UK (TNA), FO 371/140426, IA 19/4, 1. For comments and criticism I would like to thank Debbora Battaglia, Steven J. Dick, Till Kössler, William R. Macauley, Bruce Mazlish, Michael J. Neufeld, the two anonymous reviewers and, above all, Anna Kathryn Kendrick.
2. Sigmund Freud, 'Eine Schwierigkeit der Psychoanalyse' [1917], *Gesammelte Werke*, vol. 12, Frankfurt am Main: Fischer, 1999, 3–26, here 7: 'kosmologische Kränkung'; Arthur C. Clarke, 'The Conquest of Space,' *The Fortnightly* 999 (March 1950), 161–7, here 167. The three standard works on the so-called Copernican Revolution remain Alexandre Koyré, *From the Closed World to the Infinite Universe*, Baltimore: Johns Hopkins University Press, 1957; Thomas S. Kuhn, *The Copernican Revolution: Planetary Astronomy in the Development of Western Thought*, Cambridge, MA: Harvard University Press, 1957; and Hans Blumenberg, *Die Genesis der kopernikanischen Welt*, Frankfurt am Main: Suhrkamp, 1981.
3. This countervailing historical development is also the reason for which I insist on using the somewhat old-fashioned term 'outer space' for the infinite, vacuous void beyond the earth's atmosphere, while the notion of 'space' remains reserved for 'spatiality,' when used in a more abstract, geographical sense.
4. The two *loci classici* are Günther Anders, *Der Blick vom Mond: Reflexionen über Weltraumflüge* [1970], 2nd edn, Munich: C.H. Beck, 1994; and Archibald MacLeish, 'A Reflection: Riders on Earth Together, Brothers in Eternal Cold,' *New York Times* (25 December 1968), 1. Space analyst Dwayne A. Day has suggested that the famous *Star Trek* phrase 'to boldly go where no man has gone before' might indeed have been borrowed from an official White House booklet, *Introduction to Outer Space* (Washington, DC: Government Printing Office), issued on 26 March 1958. On its first page, the booklet referred to 'the compelling urge of man to explore and to discover, the thrust of curiosity that leads men *to try to go where no one has gone before*' (my emphasis); see Dwayne A. Day, 'Boldly Going: *Star Trek* and Spaceflight,' *The Space Review* (28 November 2005), http://www.thespacereview.com/article/506/1 (accessed 1 October 2017).
5. 'Kosmischer Provinzialismus: Immer mehr Menschen halten sich für die einzig denkenden Lebewesen im Weltall,' *Allensbacher Berichte* (1967), 1–4, here 3.

Such a decline was only temporary: By 1976, the number of believers in the existence of extraterrestrial intelligence rose to 38 percent, and by 1985 had climbed back to 40 percent. See 'Hallo Nachbarn! Im Weltall nicht allein?,' *Allensbacher Berichte* 24 (1976), 1–7, here 5; 'Der Kosmos gehört uns nicht allein,' ibid. 26 (1985), 1–8, here 4; and 'Andere Sterne,' *Jahrbuch der öffentlichen Meinung* 5 (1968–73), 155.

6. The term 'Space Age' is older than the Space Age itself, if conventionally defined, and is not of American, but of British origin. Its first usage can be found on the January 1946 cover of the popular journal *Everybody's Weekly*, promoting an article by journalist Harry Harper (1880–1930) that explained how the man of the future would 'penetrate the stratosphere and conquer outer space.' The term featured also in the title of a book-length study, *The Dawn of the Space Age*, that Harper published later that year. See Harry Harper, 'The Space Age,' *Everybody's Weekly* (19 January 1946), cover and 8–9; and *Dawn of the Space Age*, London: Sampson Low & Co., 1946.
7. Anders, *Der Blick vom Mond*, 11: 'Unsere heutige Welt und unser heutiges menschliches Dasein [wird] durch die Tatsache der Raumflüge aufs tiefste mitbeeinflußt und mitgeprägt.' This introduction is not an adequate place to present an overview and discuss all existing scholarship on the history of outer space, spaceflight and extraterrestrial life, but see the comprehensive bibliography at the end of this volume for an attempt at identifying the most relevant titles within this nascent but growing transdisciplinary field of research. The ten most significant core studies would have to include, in chronological order: William Sims Bainbridge, *The Spaceflight Revolution: A Sociological Study*, New York: John Wiley, 1976; Karl S. Guthke, *Der Mythos der Neuzeit: Das Thema der Mehrheit der Welten in der Literatur- und Geistesgeschichte von der kopernikanischen Wende bis zur Science Fiction*, Bern: Francke, 1983; Walter A. McDougall, *...The Heavens and the Earth: A Political History of the Space Age*, New York: Basic Books, 1985; Michael J. Neufeld, *The Rocket and the Reich: Peenemünde and the Coming of the Ballistic Missile Era*, New York: Free Press, 1995; Hans Blumenberg, *Die Vollzähligkeit der Sterne*, Frankfurt am Main: Suhrkamp, 1997; Steven J. Dick, *The Biological Universe: The Twentieth Century Extraterrestrial Life Debate and the Limits of Science*, Cambridge: Cambridge University Press, 1996; Howard E. McCurdy, *Space and the American Imagination*, Washington, DC: Smithsonian Institution Press, 1997; Jodi Dean, *Aliens in America: Conspiracy Cultures from Outerspace to Cyberspace*, Ithaca: Cornell University Press, 1998; De Witt Douglas Kilgore, *Astrofuturism: Science, Race, and Visions of Utopia in Space*, Philadelphia: University of Pennsylvania Press, 2003; and Steven J. Dick and Roger D. Launius, eds, *Societal Impact of Spaceflight*, Washington, DC: NASA, 2007, in particular the afterword by Martin Collins, 'Production and Culture Together: Or, Space History and the Problem of Periodization in the Postwar Era,' in ibid., 615–29. For two thorough and helpful reviews of the existing historiography, see Roger D. Launius, 'The Historical Dimension of Space Exploration: Reflections and Possibilities,' *Space Policy* 16.1 (February 2000), 23–38; and Asif A. Siddiqi, 'American Space History: Legacies, Questions, and Opportunities for Future Research,' in Steven J. Dick and Roger D. Launius, eds, *Critical Issues in the History of Spaceflight*, Washington, DC: NASA, 2006, 433–80. Unfortunately, neither discusses much non-American literature, nor works in languages other than English.

8. On Cold War culture, for instance, Paul S. Boyer, *By the Bomb's Early Light: American Thought and Culture at the Dawn of the Atomic Age*, New York: Pantheon, 1985; Stephen J. Whitfield, *The Culture of the Cold War*, Baltimore: Johns Hopkins University Press, 1991; and Patrick Major and Rana Mitter, 'Culture,' in Saki R. Dockrill and Geraint Hughes, eds, *Palgrave Advances in Cold War History*, Basingstoke: Palgrave Macmillan, 2006, 240–62. For two comprehensive reviews of much of the recent literature in Cold War history, see Melvyn P. Leffler, 'The Cold War: What Do "We Know Now"?,' *American Historical Review* 104.2 (April 1999), 501–24; and Thomas W. Zeiler, 'The Diplomatic History Bandwagon: A State of the Field,' *Journal of American History* 95.4 (March 2009), 1053–73.
9. McDougall, ...*The Heavens and the Earth*, 12; Daniel Bell, 'Technology, Nature and Society: The Vicissitudes of Three World Views and the Confusion of Realms,' in idem, *The Winding Passage: Essays and Sociological Journeys, 1960–1980*, Cambridge, MA: Abt Books, 1980, 3–33. McCurdy, *Space and the American Imagination*, 29, makes the same reference.
10. Michael Wale, 'David Bowie: Rock and Theatre,' *The Times* (24 January 1973), 15. There are repeated references to outer space and extraterrestrial beings in Bowie's comprehensive *oeuvre*, culminating in his portrayal of the space traveller Thomas Jerome Newton in Nicolas Roeg's 1976 film *The Man Who Fell to Earth*; see for instance 'Life on Mars,' *Hunky Dory* (1971); 'Starman,' *The Rise and Fall of Ziggy Stardust and the Spiders from Mars* (1972); 'Loving the Alien,' *Tonight* (1985); and 'Looking for Satellites,' *Earthling* (1997). Bowie's second extraterrestrial *persona* and *alter ego* was the rock superstar Ziggy Stardust, first introduced in 1972. The history of space as a prominent leitmotiv of pop music and the defining element of various subgenres – including 'space rock' (ca. early 1970s, with a brief revival in the early 1990s); Sun Ra's 'afrofuturism' (ca. early to mid-1970s), later taken up by funk musician George Clinton; and 'space disco' (ca. 1977–80) – remains to be written. Bowie's 'Space Oddity' is an early example of the former. For a first, largely inventorial discussion of space, alien- and technofuturistic themes in popular music, see Ken McLeod, 'Space Oddities: Aliens, Futurism and Meaning in Popular Music,' *Popular Music* 22.3 (October 2003), 337–55.
11. See also Steven Dick's discussion in Chapter 2 of this volume. It is, admittedly, unfortunate that 'culture' is in itself such a broad, catch-all term, but there is no better.
12. See *Astropolitics: The International Journal of Space Politics and Policy*, Philadelphia: Taylor & Francis, 2003–. Unfortunately, the journal's editors chose to define their title term very broadly when outlining the scope of *Astropolitics* as 'the role of space in politics, economics, commerce, culture and security.' A few years later, one member of the journal's editorial board, Jim Pass, declared that he had 'set out to develop astrosociology as a new sociological subdiscipline,' yet seems not to have generated much academic resonance, possibly because Pass proclaimed the necessity of such a disciplinary addition prior to undertaking any empirical research to demonstrate its practical fruitfulness. See Everett C. Dolman and John B. Sheldon, 'Editorial,' *Astropolitics* 1.1 (2003), 1–3, here 1; and Jim Pass, 'Astrosociology as the Missing Perspective,' *Astropolitics* 4.1 (2006), 85–99.

13. Donald N. Michael, 'Man-Into-Space: A Tool and Program for Research in the Social Sciences,' *American Psychologist* 12.6 (June 1957), 324–8; John Lear, 'Dr. Mead and the Red Moons,' *New Scientist* 2.52 (14 November 1957), 20; Debbora Battaglia, ed., *E.T. Culture: Anthropology in Outerspaces*, Durham: Duke University Press, 2005. See also Joseph M. Goldsen, *Research on Social Consequences of Space Activities*, Santa Monica: Rand Corporation, 1965; Charles P. Boyle, *Space Among Us: Some Effects of Space Research on Society*, Washington, DC: Aerospace Industries Association of America, 1974; William I. McLaughlin, ed., *The Impact of Space on Culture*, London: British Interplanetary Society, 1993 (=*Journal of the British Interplanetary Society* 46.11); and Alvin Rudoff, *Societies in Space*, New York: Peter Lang, 1996. As early as 1965, MIT historian Bruce Mazlish came to the foresighted conclusion that the space program's philosophical impact, albeit at 'the farthest remove from an intended primary aim,' might ultimately be one of its most significant effects, and that it could be 'treated under the heading of "imagination"'; see Bruce Mazlish, 'Historical Analogy: The Railroad and the Space Program and Their Impact on Society,' in idem, ed., *The Railroad and the Space Program: An Exploration in Historical Analogy*, Cambridge, MA: MIT Press, 1965, 1–52, here 41.
14. Debbora Battaglia, 'Insiders' Voices in Outerspaces,' in *E.T. Culture*, 1–37, here 1–2, 6, 19; see also her contribution, Chapter 11 in this volume. The notion of 'pre-Astronautics' refers to supposed extraterrestrial impact on early human civilization, taken up and popularized by the Swiss best-selling author Erich von Däniken (1935–) as the so-called 'ancient astronaut hypothesis.' Among von Däniken's countless publications, with sales exceeding 60 million, see in particular his *Erinnerungen an die Zukunft: Ungelöste Rätsel der Vergangenheit*, Düsseldorf: Econ, 1968; *Chariots of the Gods? Unsolved Mysteries of the Past*, London: Souvenir, 1969.
15. For the foundational document that declared space 'a field of research so enormous and important that it far surpasses anything that can be imagined today,' see the memorandum 'Introduction to the Discussion on Space Research in Europe,' 30 April 1959, by physicist and scientific statesman Edoardo Amaldi (1908–89), Historical Archives of the European Union (HAEU), Florence, Italy, COPERS 0001. John Krige, Arturo Russo and Lorenza Sebesta, *A History of the European Space Agency*, 2 vols, Noordwijk: ESA, 2000, here vol. I, 19–20, 91. See also John Krige, 'Building a Third Space Power: Western European Reactions to Sputnik at the Dawn of the Space Age,' in Roger D. Launius, John M. Logsdon and Robert W. Smith, eds, *Reconsidering Sputnik: Forty Years Since the Soviet Satellite*, Amsterdam: Harwood, 2000, 289–307, here 301–2, 395; and Walter A. McDougall, 'Space-Age Europe: Gaullism, Euro-Gaullism, and the American Dilemma,' *Technology and Culture* 26.1 (January 1985), 179–203.
16. For a contemporaneous debate between a German astronomer and space critic, NASA's Deputy Director for international affairs, a French geophysicist and a German senior civil servant see Rudolf Kühn, Arnold W. Frutkin, Jean Coulomb and Max Mayer, 'Herausforderung "Weltraum" – Europas Antwort,' *Dokumente: Zeitschrift für übernationale Zusammenarbeit* 20.3 (1964), 201–22; and Orio Giarini, *L'Europe et l'espace*, Lausanne: Centre de Recherches Européennes, 1968. In the late 1980s, a joint policy report by five renowned European research institutions proclaimed such space autonomy – defined as the 'capability to reach, to

operate in and to return from space, and to do so, not on sufferance of friend or foe, but according to its own perception of what is to the common good' – 'Europe's stated goal.' The report also went so far as to declare outer space a 'major area in which Europe can consolidate a common identity and develop its unity.' See Forschungsinstitut der Deutschen Gesellschaft für Auswärtige Politik (Bonn), Institut Français des Relations Internationales (Paris), Istituto Affari Internazionali (Rome), Nederlands Instituut voor Internationale Betrekkingen 'Clingendael' (The Hague) and Royal Institute of International Affairs (London), *Europe's Future in Space: A Joint Policy Report*, London: Routledge & Kegan Paul, 1988, 181, 187, 3.

17. The first American in space, Alan Shepard (1923–98), did not orbit the earth during his 15-minute flight on 5 May 1961; see Brian Harvey, *Europe's Space Programme: To Ariane and Beyond*, London: Springer Praxis, 2003, 249–50.
18. See, for instance, Holger Nehring, 'National Internationalists: British and West German Protests Against Nuclear Weapons, the Politics of Transnational Communications and the Social History of the Cold War, 1957–1964,' *Contemporary European History* 14.4 (2005), 559–82.
19. Asif A. Siddiqi, 'Competing Technologies, National(ist) Narratives, and Universal Claims: Toward a Global History of Space Exploration,' *Technology and Culture* 51.2 (April 2010), 425–43, esp. 426, 442.
20. See W.D. Kay, 'NASA and Space History,' *Technology and Culture* 40.1 (January 1999), 120–7, here 120; and, as a valuable research aid, Steven J. Dick, Stephen J. Garber and Jane H. Odom, eds, *Research in NASA History: A Guide to the NASA History Program*, 3rd edn, Washington, DC: NASA, 2009. As an institution, the museum predates its current spectacular Space Age building. Established as 'The National Air Museum' in 1946, the supplement 'and Space' was added in 1966; see Michael J. Neufeld and Alex M. Spencer, eds, *Smithsonian National Air and Space Museum: An Autobiography*, Washington, DC: National Geographic, 2010.
21. Again, it is impossible to discuss the much more extensive literature on Soviet and East-European history in all desirable detail here. Interested readers should consult such works as Paul R. Josephson, 'Rockets, Reactors, and Soviet Culture,' in Loren R. Graham, ed., *Science and the Soviet Social Order*, Cambridge, MA: Harvard University Press, 1990, 168–91; Svetlana Boym, 'Kosmos: Remembrances of the Future,' in Adam Bartos and Svetlana Boym, *Kosmos: A Portrait of the Russian Space Age*, New York: Princeton Architectural Press, 2001, 82–99; James T. Andrews, *Science for the Masses: The Bolshevik State, Public Science, and the Popular Imagination in Soviet Russia, 1917–1934*, College Station: A&M University Press, 2003; Matthias Schwartz, *Die Erfindung des Kosmos: Zur sowjetischen Science Fiction und populär-wissenschaftlichen Publizistik vom Sputnikflug bis zum Ende der Tauwetterzeit*, Frankfurt am Main: Peter Lang, 2003; Igor J. Polianski and Matthias Schwartz, eds, *Die Spur des Sputnik: Kulturhistorische Expeditionen ins kosmische Zeitalter*, Frankfurt am Main: Campus, 2009; and the publications by Asif A. Siddiqi, esp. *The Red Rockets' Glare: Spaceflight and the Soviet Imagination, 1857–1957*, Cambridge: Cambridge University Press, 2010.
22. Krige, Russo and Sebesta, *A History of the European Space Agency*. These 40 'ESA History Study Reports' are available at http://www.esa.int/esapub/pi/hsrPI.htm (accessed 1 October 2017). For a summary of the activities undertaken within

this project, see Karl-Egon Reuter and Johann Oberlechner, 'The ESA History Project,' *ESA Bulletin* 119 (August 2004), 48–54; also available at www.esa.int/esapub/bulletin/bulletin119/bul119_chap6.pdf (accessed 1 October 2017). Popular volumes, such as *The Impact of Space Activities Upon Society*, ed. International Academy of Astronautics and European Space Agency, Noordwijk: ESA, 2005, do not constitute an exception to this rule. For further reflections on the long overdue Europeanization of space history, see Alexander C.T. Geppert, 'Flights of Fancy: Outer Space and the European Imagination, 1923–1969,' in Dick and Launius, *Societal Impact of Spaceflight*, 585–99.

23. See, for example, the contributions by Claudia Schmölders (Chapter 3), Michael J. Neufeld (Chapter 6) and Pierre Lagrange (Chapter 12) in this volume.
24. Susan Sontag, *Regarding the Pain of Others*, New York: Farrar, Straus & Giroux, 2003, 21: 'Creating a perch for a particular conflict in the consciousness of viewers exposed to dramas from everywhere requires the daily diffusion and rediffusion of snippets of footage about the conflict. The understanding of war among people who have not experienced war is now chiefly a product of the impact of these images.' The vast, largely US-oriented literature on the complex interplay between science fact and science fiction within feature film includes Vivian Sobchack, *Screening Space: The American Science Fiction Film*, 2nd edn, New Brunswick: Rutgers University Press, 1997; Errol Vieth, *Screening Science: Contexts, Texts, and Science in Fifties Science Fiction Film*, Lanham: Scarecrow Press, 2001; J.P. Telotte, *Science Fiction Film*, Cambridge: Cambridge University Press, 2001; and, most recently, David A. Kirby, *Lab Coats in Hollywood: Science, Scientists, and Cinema*, Cambridge, MA: MIT Press, 2011.
25. See, for instance, in chronological order Willy Ley, *Rockets, Missiles, and Space Travel* [1944], 3rd edn, New York: Viking, 1951; Arthur C. Clarke, 'Space-Travel in Fact and Fiction,' *Journal of the British Interplanetary Society* 9.5 (September 1950), 213–30, reprinted in Arthur C. Clarke, *Greetings, Carbon-Based Bipeds! Collected Essays, 1934–1998*, New York: St. Martin's Press, 1999, 84–98; Eugene M. Emme, ed., *Science Fiction and Space Futures: Past and Present*, San Diego: American Astronautical Society, 1982; and Frederick I. Ordway and Randy Liebermann, eds, *Blueprint for Space: Science Fiction to Science Fact*, Washington, DC: Smithsonian Institution Press, 1992. Early critical literary studies include, in chronological order, Christof Junker, *Das Weltraumbild in der deutschen Lyrik von Opitz bis Klopstock*, Berlin: Matthiesen, 1932; Edwin M.J. Kretzmann, 'German Technological Utopias of the Pre-War Period,' *Annals of Science* 3.4 (October 1938), 417–30; Marjorie Hope Nicolson, *A World in the Moon: A Study of the Changing Attitude Toward the Moon in the Seventeenth and Eighteenth Centuries*, Northampton: Smith College, 1935–6; James Osler Bailey, *Pilgrims Through Space and Time: Trends and Patterns in Scientific and Utopian Fiction*, New York: Argus, 1947; Martin Schwonke, *Vom Staatsroman zur Science Fiction: Eine Untersuchung über Geschichte und Funktion der naturwissenschaftlich-technischen Utopie*, Stuttgart: Ferdinand Enke, 1957; and Roger Lancelyn Green, *Into Other Worlds: Space-Flight in Fiction, from Lucian to Lewis*, London: Abelard-Schuman, 1958. Two important contemporary works on science-fiction literature and criticism are Adam Roberts, *The History of Science Fiction*, Basingstoke: Palgrave Macmillan, 2006; and Fredric Jameson, *Archaeologies of the Future: The Desire Called Utopia and Other Science Fictions*, London: Verso, 2005.

26. 'Spaceships of the Mind' was the title of a 1978 BBC TV series produced by Dick Gilling and presented by Nigel Calder; see Nigel Calder, *Spaceships of the Mind*, New York: Viking Press, 1978.
27. Tom Wolfe, *The Right Stuff*, New York: Farrar, Straus & Giroux, 1979. For a more detailed analysis, see Alexander C.T. Geppert, 'Space *Personae*: Cosmopolitan Networks of Peripheral Knowledge, 1927–1957,' *Journal of Modern European History* 6.2 (2008), 262–86. That the pendulum could be said to have swung back towards 'fiction' in recent years, caused by 'factual' disillusionments such as the Space Shuttle *Columbia* disaster in February 2003; the limited public appeal of the most expensive civilian project ever undertaken, the International Space Station; or the cancellation of America's Constellation and Space Shuttle programs in February 2010 and July 2011, respectively, might be a noteworthy observation beyond the scope of this essay.
28. As early as 1972, science-fiction author Isaac Asimov (1920–92) raised similar doubts inspired by the Apollo moon landings: '[…] so ist es amüsant festzustellen, daß viele meinen, nachdem die Astronauten auf dem Mond gelandet sind, habe die Wissenschaft die Science-fiction eingeholt. Denn nicht die Science-fiction-Bagatelle der Mondlandung selber ist bedeutungsvoll, sondern die gesellschaftliche Wirkung der Raumfahrt' ('[…] it is amusing to note that many believe, now that the astronauts have landed on the moon, that science should have caught up with science fiction. For it is not the science fiction-bagatelle of the moon landing itself that is momentous, but the societal impact of spaceflight'); see his 'Plädoyer für Science-fiction,' *Der Spiegel* 11 (6 March 1972), 138–9, here 139.
29. Kilgore defines astrofuturism as 'an escape from terrestrial history. Its roots lie in the nineteenth-century Euro-American preoccupation with imperial expansion and utopian speculation, which it recasts in the elsewhere and else*when* of outer space. […] [I]t is also the space of utopian desire. Astrofuturist speculation on space-based exploration, exploitation, and colonization is capacious enough to contain imperialist, capitalist ambitions and utopian, socialist hopes. […] While [astrofuturism is] an American phenomenon anchored by the nation's mid-century commitment to the space race, its roots and membership are international'; see *Astrofuturism*, 1, 3. Kilgore does not elaborate on this international perspective. For a helpful review essay, see Joan Gordon, 'Ad Astra Per Aspera,' *Science Fiction Studies* 32.3 (November 2005), 495–502.
30. Standard works on the history of the future include Joseph J. Corn, ed., *Imagining Tomorrow: History, Technology, and the American Future*, Cambridge, MA: MIT Press, 1986; Georges Minois, *Histoire de l'avenir: des Prophètes à la prospective*, Paris: Fayard, 1996; and Lucian Hölscher, *Die Entdeckung der Zukunft*, Frankfurt am Main: Fischer, 1999. In the present context see in particular Brian Horrigan, 'Popular Culture and Visions of the Future in Space, 1901–2000,' in Bruce Sinclair, ed., *New Perspectives on Technology and American Culture*, Philadelphia: American Philosophical Society, 1986, 49–67; and Roger D. Launius, 'Perfect Worlds, Perfect Societies: The Persistent Goal of Utopia in Human Spaceflight,' *Journal of the British Interplanetary Society* 56.5 (September/October 2003), 338–49. There is no European equivalent to the American 1970s pro-space movement and its inherent spaceflight utopianism that Launius analyzes but see the contributions to Alexander C.T. Geppert, ed., *Limiting Outer Space: Astroculture After Apollo*, London: Palgrave Macmillan, 2018 (= *European Astroculture*, vol. 2).

31. Sigmund Freud, 'Das Unbehagen in der Kultur' [1930], *Gesammelte Werke*, vol. 14, Frankfurt am Main: Fischer, 1999, 419–506, here 422 ('ein Gefühl wie von etwas Unbegrenztem, Schrankenlosem, gleichsam "Ozeanischem"') and 430; Eric Nordern, 'Interview with Stanley Kubrick,' *Playboy* 15.9 (September 1968), 85–96, 158, 180–95, here 94. See also William B. Parsons, 'The Oceanic Feeling Revisited,' *Journal of Religion* 78.4 (October 1998), 501–23; and Thore Bjørnvig's contribution, Chapter 7 of this volume.
32. For initial ventures into the conceptual history of the 'beyond,' see the contributions to Lucian Hölscher, ed., *Das Jenseits: Facetten eines religiösen Begriffs in der Neuzeit*, Göttingen: Wallstein, 2007; and Colleen McDannell and Bernhard Lang, *Heaven: A History*, New Haven: Yale University Press, 1988.
33. John Hillaby, 'Astronauts Get Blessing of Pope,' *New York Times* (21 September 1956), L; 'Astronautics in Britain,' *Spaceflight* 3 (1967), 234; 'European Interest in Apollo Dwindles,' *New York Times* (10 February 1971), 24; 'Pope Benedict XVI Greets Shuttle, Station Crew,' *NASA Television* (21 May 2011), http://youtu.be/81jAmb_e1pg (accessed 1 October 2017); *Frankfurter Allgemeine Zeitung* (23 May 2011), 9.

PART I

Narrating Outer Space

CHAPTER 2

Space, Time and Aliens: The Role of Imagination in Outer Space

Steven J. Dick

I The cultural history of outer space

The role of personal and collective imagination in the Space Age – both in making spaceflight possible and in its reverse effect on individuals and culture – is a complex subject fraught with difficulty. Even when I contemplate my own career, it is not easy to separate the effect of events in the real world from youthful imagination and the cultures in which both are embedded. It is perhaps useful to begin by relating my personal experience as an entrée to the large issues of the subject.

I was 7 years old when Sputnik was launched, an event that undoubtedly had some impact in launching my own imagination. When I was 11, I spent the summer of 1961 in Karlsruhe, then in the western part of divided Germany. One of the indelible memories of that summer was a science-fiction movie that I still remember as *Venus Won't Answer*. That movie further whet my appetite for space, but my interest was undoubtedly generated in the first place by real events then taking place in space – only a few months after the first human spaceflights by Yury Gagarin, Alan Shepard and Gus Grissom. In this way imagination and reality feed on each other symbiotically, an eternal entanglement difficult to deconstruct, precisely because in many ways they may be understood as complementary and intertwined.

Further contemplation of this formative personal experience also reveals the difficulties of determining influences and how they may operate unconsciously. The movie was shot in East Germany, directed by Kurt Maetzig (1911–2012), co-produced with a Polish company, titled *Der schweigende Stern* (*The Silent*

Steven J. Dick (✉)
Ashburn, VA, USA
e-mail: stevedick1@comcast.net

Alexander C.T. Geppert (ed.), *Imagining Outer Space*
European Astroculture, vol. 1
https://doi.org/10.1057/978-1-349-95339-4_2

Star) and released in East Germany in February 1960. The movie was also released in West German theaters in September 1960 under the title *Raumschiff Venus antwortet nicht* (Spaceship Venus Won't Answer), which accounts for me seeing it the following summer. An Americanized version was released in the United States in 1962 under the title *First Spaceship on Venus* (Figure 2.1).[1] It turns out that Maetzig's movie was based on the first novel of none other than the great Polish science-fiction writer Stanisław Lem (1921–2006), titled *The Astronauts* (1951), thus the Polish co-production.[2] The novel was translated into many languages, but never into English. Three decades later Stanisław Lem had a great influence on me through his novels *Solaris* and *His Master's Voice*, but I failed to realize until recently that he had unknowingly influenced me already in 1961, at the age of 11, through *Raumschiff Venus antwortet nicht*. This experience emphasizes the many different levels to the theme of 'space and the imagination.' One of them is how any particular individual is influenced, which is not always easy to determine even by the individual. Another level is how the European imagination can affect the American imagination, and by extension how one culture can affect another.

Yet another lesson emerges from the actual content of this movie – the influence of cultural context on the imagination as represented in the film. Born in 1911, Kurt Maetzig was an East German film director, and during the Second World War a member of the anti-Nazi German Communist Party. At the time he made the film, he lived in Soviet-occupied East Germany. It is not surprising Maetzig took up the popular novel of Lem, who was living under Soviet occupation in Poland, and had to portray earth as a social utopia in order to get his novel published. Lem's novel tells of a Venusian artifact found buried near Tunguska, the Siberian site of the famous extraterrestrial impact in 1908, with data indicating the Venusians will irradiate the earth and take over. Earth officials send the spaceship Kosmokrator to Venus, where scientists find the remains of a warlike civilization that perished in a nuclear war. In the movie version (reflecting the new technology of radio telescopes), a radio signal with greetings is sent to Venus, but there is no reply (thus 'Venus antwortet nicht'). Kosmokrator (equipped with a vacuum tube computer) travels to Venus with its international crew of scientists, and finds only advanced machines, programmed to carry out the goals of the original Venusians. The film is full of communist ideology and anti-American sentiment, removed from the 82-minute Americanized version. In politics, nuclear war and technology, the novel and the different versions of the movie reflect the cultural context of their time. That context, including the incipient Space Age, fed the imagination of these two European artists, who, even while under totalitarian rule, absorbed the space aspirations of the Soviet Union.[3]

I begin with this personal experience because it illustrates in a concrete way just how complex the subject of this volume – a cultural history of outer space and space exploration – can be. It is central first of all to know what we mean by 'culture,' or, more accurately, how difficult it is to know what we mean by that word. 'Culture' and its derivatives are infused with multiple meanings by

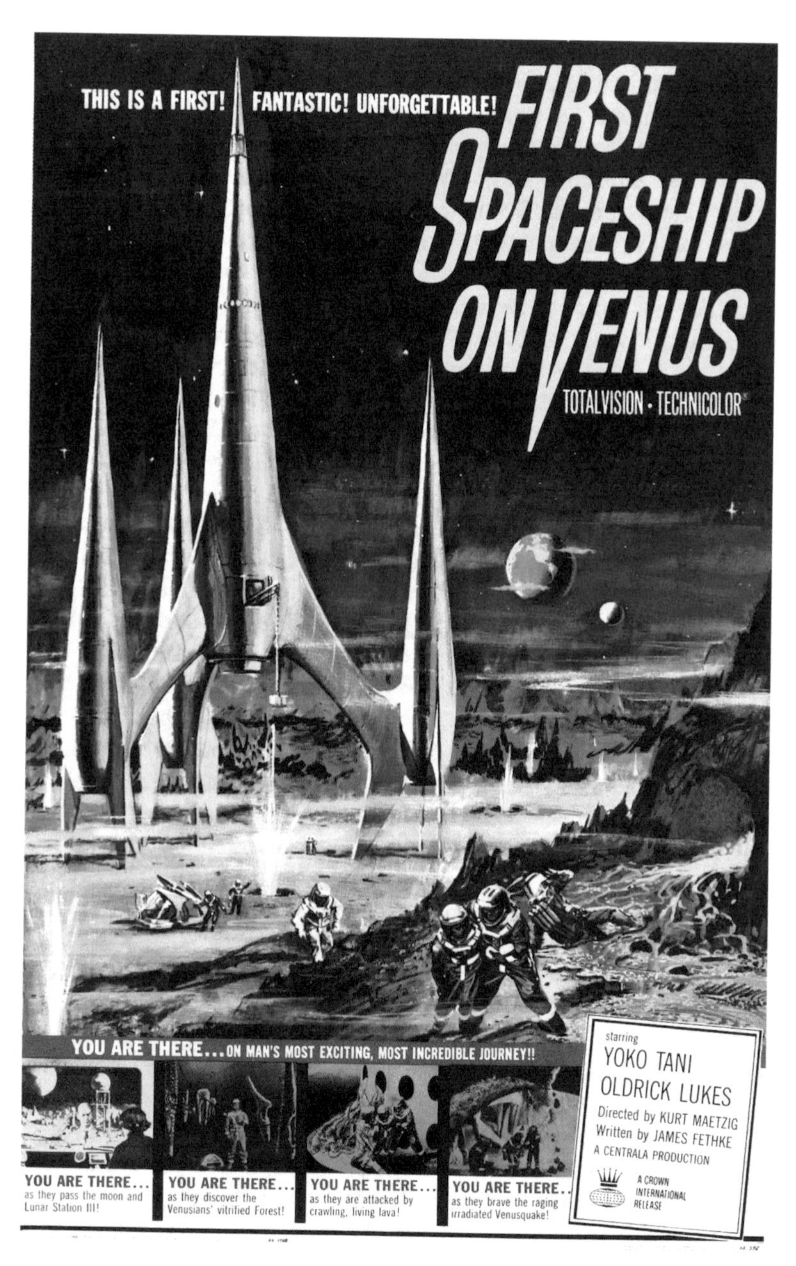

Figure 2.1 Poster for *First Spaceship on Venus*, the American version of *Der schweigende Stern*, directed by Kurt Maetzig, DDR/PL 1960 (VEB DEFA Studio für Spielfilme/DEFA Gruppe Roter Kreis/Film Polski/Iluzjon). The movie was released in the United States in 1962. The poster shows the spaceship Kosmokrator on the surface of Venus, which the Mariner 2 spacecraft revealed at the end of 1962 was extremely hot due to the greenhouse effect – already conjectured in this poster due to the relative proximity of Venus to the sun.
Source: Courtesy of DEFA-Stiftung.

different individuals and different professions, and not only among countries but also within countries. Whereas American scholars are more apt to refer to 'cultural evolution,' for example, many Europeans prefer 'social evolution,' perhaps because 'cultural anthropology' has roots as a discipline in the United States, while 'social anthropology' was born in Europe. Though the two disciplines can have quite different meanings – the former referring to more concrete cultural variation among human societies over time and the latter to social behaviors – the two have grown closer together over the decades.[4]

Such difficulties have not kept scholars from trying to define the term. More than 50 years ago two anthropologists collapsed 164 distinct definitions of culture into one: 'Culture is a product; is historical; includes ideas, pattern, and values; is selective; is learned; is based upon symbols; and is an abstraction from behavior and the products of behavior.'[5] Perhaps a brave attempt at a scholarly definition, but hardly one that yields a concrete intuitive grasp of what culture really is. Two decades later anthropologist Clifford Geertz (1926–2006), a giant in the field, defined culture more understandably as 'an historically transmitted pattern of meanings embedded in symbolic forms by means of which men [people] communicate, perpetuate and develop their knowledge about and attitudes toward life.'[6] According to Harvard biologist E.O. Wilson – famed for his work on sociobiology – each society creates culture and is created by it.[7] In short, the idea of 'culture' is a moving target, evolving with time and in space (and perhaps literally in outer space); not only do the understandings of the concept differ in Chinese and Western cultures, both were more different 50 years ago than they are now. So spaceflight is a manifestation of culture, a product of culture, but it is also embedded in culture. The influences travel both ways, and it is well to recognize this at the outset. And in the cosmic context, our terrestrial ideas of 'culture' may be expanded if we discover cosmic civilizations, in which case the natural history of cultural evolution and its theoretical underpinnings will be taken to a new level.[8]

If, as Wilson says, society creates culture, then there is the question of what is 'society'? This too is problematic – it is a law of nature that any time academics focus on a word or concept it becomes problematic – but the question of the difference between society and culture is an important and venerable topic of discussion among anthropologists. A recent book on the key concepts in social and cultural anthropology put it this way: 'Throughout the modernist period, a concept of society has underpinned the construction of all social theory, whatever its hue or denomination. If the concept of culture has played the role of queen to all analytic categories of the human sciences, the notion of society has been king. It is the master trope of high modern social thought.'[9] It is, the authors said, a treacherous friend, a necessary term, but a term to be used at one's risk.

Similarly, 'imagination' has been the subject of both theoretical and descriptive study, and comes in many forms: the personal imagination of creative writers, artists and scientists; the collective imagination of a given culture, as in the 'American imagination,' the 'European imagination,' or the 'Russian imagination,' each formed by the distinctive history and experiences of a specific culture; or the perhaps distinct (because so self-consciously explicit) imagination of science fiction, often characterized and even flouted as a literature of the imagination – so much so that even the best science fiction is still not accepted as sophisticated literature in some circles. Each of these forms of imagination is at work in any general study of outer space and the imagination. Their complex nature and interaction remain largely uncharted waters in the field of space history. But the richness awaiting researchers is evident in Howard McCurdy's book *Space and the American Imagination*, where McCurdy shows how the American space program took advantage of elements deeply ingrained in the American imagination, notably the exploration imperative, the search for extraterrestrial life, and the idea of 'the last frontier.' Similarly, McCurdy and Roger Launius have shown how the imagery of space, from *Buck Rogers* and *Flash Gordon* to the art of Chesley Bonestell and real images beamed from outer space, have inspired the imagination and had a real effect on public and scientific interest in space. In a broader sense Harvard historian of science Gerald Holton has shown how the imagination of the scientist, rather than objective criteria, is often important in the early stages of a scientific idea.[10] In short, imagination is not to be trifled with, but constitutes a real force with real-life consequences.

Although it is counterproductive to spend too much time on definitions, it is important to realize that the cultural history of outer space and the role of imagination are not simple problems precisely because of the vagueness of the terms. Other historical subfields suffer from the same conceptual problem, but few fields are as expansive as the physical extent of outer space, or as complex as the mental terrain of the imagination. Perhaps it is best to say what the cultural history of outer space is not: it is less about the political, diplomatic and technological aspects of spaceflight, than about the socio-cultural rationale for spaceflight, a term that nicely circumvents the differences between the social and the cultural by combining the two terms.[11] In addition to rationale, it is also about socio-cultural impact, belief and visions of the future. Given such expansive mental and spatial terrain, it is hardly surprising that approaches to the subject may (and should) be expansive as well.

Given the difficulties with the concepts of culture, society and imagination, and the difficulties of determining the exact role of imagination on any one individual, much less on society and culture, we nevertheless boldly proceed into the unknown. We can divide the question of outer space and the imagination into three parts: First, how has space exploration affected the imagination and society? Second, and conversely, how has imagination historically affected space exploration? And third, what is the effect of spaceflight on our world view, our *Weltanschauung*, to use the great theory-laden German word? This is a large subject, and can be only faintly illuminated here.

II Space and the imagination: how has space affected our imagination?

The question of how changing ideas of space and time over the twentieth century have affected our imagination is related to, but distinct from, the question of space and our *Weltanschauung*, to which I shall return to at the end of the chapter. The first observation that must be made is that it is no less than astonishing how much our ideas of space have changed over the last century. In terms of spatial extent, at the beginning of the twentieth century Alfred Russel Wallace (1823–1913), the co-founder with Darwin of the theory of natural selection, offered a model of the universe only 3,600 light years across. In supporting it at length in his well-known volume *Man's Place in the Universe* (1903), Wallace claimed that he was simply espousing the view of the most eminent astronomers of his day, a reasonable claim. When Wallace wrote, all stars, and indeed all observable phenomena in the universe, were widely believed to be part of a single system perhaps several thousand light years in diameter (compared to the 100,000 light years now estimated), with the moon in a nearly central position. The 'island universe' theory, which postulated many such systems, had been in gradual decline since the 1860s and had completely fallen from favor by the late 1880s. It is therefore not surprising that Wallace viewed the universe as a single system of stars with our solar system at the approximate center.[12]

Though *Man's Place in the Universe* went through seven editions by 1908 and another in 1914, and was translated into German in 1903 and French in 1907, it had little influence beyond the second decade of the twentieth century. The reason is not far to seek. Within 15 years of Wallace's death in 1913, most of his central assumptions had been rendered obsolete by an emerging new cosmology. In 1918 the American astronomer Harlow Shapley (1885–1972) reported, based on his study of the distribution of globular clusters of stars, that our solar system was located in a very eccentric position in the galaxy, at its periphery rather than its center. This proved to be one of the great shifts in our cosmological world view, from the geocentric to the heliocentric to the galactocentric, as Shapley himself called his revolutionary new view.[13] By 1924 Edwin P. Hubble (1889–1953) had demonstrated to the satisfaction of most astronomers that many other galaxies exist outside our own, galaxies that he showed a few years later are fleeing from one another in what could be interpreted as an 'expanding universe.'[14] We now know from the Hubble Space Telescope and other observations that we live in a universe billions of light years in extent, characterized by an interrelation among parts and the whole that astronomers characterize by the term 'cosmic evolution.' Though Wallace recognized the evolution of the stars based on the contemporary work of astronomers, neither he nor they could have known the extent of full-blown cosmic evolution, ranging from the Big Bang to the present and covering some 13.7 billion years of time.[15] As Olaf Stapledon and many other science-fiction writers have commented, this greatly

Figure 2.2 About 1,500 galaxies are visible in this deep view of the universe, allowing the Hubble Space Telescope to stare at the same tiny patch of sky for ten consecutive days in 1995. The image covers an area of sky only about the width of a dime viewed from six meters away.
Source: Courtesy of Robert William, the Hubble Deep Field Team (STScI) and NASA.

enlarged universe gives vast scope for imagination and for action, whether by humans or extraterrestrials, conjuring the warp speeds of *Star Trek* in order to traverse its domain (Figure 2.2).

In addition to an expanded concept of space, the idea of cosmic evolution represents another dimension – the dimension of time – which has become very important to our world view, and will be even more so in the future. The idea of cosmic evolution only gradually came to be realized during the twentieth century, as the Big Bang cosmology gained greater acceptance and as the idea of the explosive beginning of the universe gave greater force to a coherent story of the universe. The evolution of stars was known from the work of the astronomer George Ellery Hale (1868–1938) among others, and a broader idea of cosmic evolution was occasionally discussed by the followers of Herbert Spencer's (1820–1903) evolutionary world view and by a few scientists such as Lawrence J. Henderson (1878–1942). But it was only in the late 1950s, with the writings of Harvard astronomer Harlow Shapley, that the modern idea of cosmic evolution was fully enunciated and sustained.[16] It became a major driving force taken up at NASA, first in its search for extraterrestrial intelligence (SETI) program, then in its exobiology program, and finally in its origins and astrobiology programs. With the discovery of cosmic background radiation in the 1960s, and its detailed analysis by the COBE and

WMAP spacecraft, we now know that the universe is 13.7 billion years old, with an uncertainty of only 1 percent, or about 100,000,000 years. The fact of cosmic evolution is inherent in most of the work done in space science by the national space agencies, which may be seen as filling in the gaps in the history of cosmic evolution, the ultimate master narrative of the universe. But only a few scholars, pre-eminently astronomer Eric Chaisson, have analyzed the idea of cosmic evolution across its full astronomical, biological and cultural breadth.[17]

Along with these expanded notions of the *extent* of space and time, we have had an equally revolutionary change in our perception of the very *nature* of space and time. The work of Albert Einstein (1879–1955) yielded the concept of a space-time continuum, demonstrating that Newtonian ideas were incomplete at relativistic speeds and cosmological distances. Einstein's personal imagination was essential for his signal scientific advances; we need only recall Einstein's thought experiments, in which he imagined himself riding on moving trains, or on a lightbeam or in an enclosed chamber in free-fall, in order to arrive at his radical ideas of simultaneity and relativity. At the same time Einstein was greatly affected by a signal development in the late nineteenth and early twentieth centuries – attempts to synchronize timepieces across increasingly larger areas of the earth, and with increasing accuracy.[18] In the wake of Einstein, the old concepts of absolute space and absolute time were no longer viable. The result was a new concept of space and time that in turn altered our world view and fed the imagination of science-fiction writers. Consonant with the nature of science, the Einsteinian world view may itself someday be subsumed under a more general theory, and with it the parameters of imagination will change once again.

In addition to changing conceptions of space and time over the century, the possibility of life beyond earth has been a continuous and often spectacularly popular theme. The so-called Drake Equation, originated by the American astronomer Frank Drake (1930–) in 1961 in the wake of the first search for artificial extraterrestrial radio signals, is the iconic image for extraterrestrial intelligence.[19] The Drake Equation tries to assess the number (N) of technological civilizations in the galaxy, and in doing so it represents various parameters of astronomical, biological and cultural evolution. Depending on the values inputted, N might be millions, as Carl Sagan (1934–96) and Frank Drake opined, or only one – us. Despite decades of research and speculation since the first radio search for intelligent signals beyond earth, we do not yet know if there is any life beyond earth, primitive or intelligent, and this lack of an answer leaves open for the imagination the question of whether the universe is for aliens, for humans or for both. This has indeed proven a fertile source of imagination, a playing ground for countless profound and less-profound thinkers, especially in science-fiction literature.[20]

How has the new view of space, time and aliens affected culture, even with all the ambiguities of that term? That is an enormous question, so let us narrow our inquiry into how a few selected science-fiction writers, representing specific cultures, were affected by the new views induced by Shapley, Hubble and Einstein, among others. Because aliens have been a favorite theme of science-fiction literature at least since H.G. Wells (1866–1946) and Kurd Lasswitz (1848–1910) at the end of the nineteenth century, it is obvious that the new view of the cosmos was not *required* for the depiction of aliens. But *The War of the Worlds* was only a local battle within our own parochial solar system, with Martians invading earth. And Kurd Lasswitz's more peaceful Martians in *Auf zwei Planeten* were also localized in their interaction with earth. Although the effect of world views on cultures requires a comprehensive approach, here we may look briefly at four of the most influential science-fiction writers of the twentieth century – two in Britain, one in the United States, and one in Poland – to illustrate how much the scope of alien literature was expanded by the new view of space, time and aliens. Olaf Stapledon, Arthur C. Clarke, Isaac Asimov and Stanisław Lem each represent a different aspect of the question, and each shows the effect of the new world view on the imagination in different ways.

In 1930, at the age of 44, the British philosopher Olaf Stapledon (1886–1950), a graduate of Oxford University in history and Liverpool University in philosophy, took up the writing of fiction, in which aliens immediately played an essential role. In his novels *Last and First Men* (1930) and *Star Maker* (1937), political, religious and philosophical ideas dominate rather than adventure. Stapledon lived in an era when the immensity of the cosmos was well known, and his novels appropriately cover billions of years. He knew of Edwin Hubble's work, and for his conception of the size of the cosmos he cited the astronomer Willem J. Luyten's (1899–1994) *The Pageant of the Stars*. Still, 'immensity is not itself a good thing,' Stapledon wrote: 'A living man is worth more than a lifeless galaxy. But immensity has indirect importance through its facilitation of mental richness and diversity [...] though spatial and temporal immensity of a cosmos have no intrinsic merit, they are the ground for psychical luxuriance, which we value. Physical immensity opens up the possibility of vast physical complexity, and this offers scope for complex minded organisms.'[21] This is a direct statement of how space affected the imagination of Olaf Stapledon, for his novels were played out on this immense tapestry of infinite space and billions of years of time (though not yet Einsteinian space-time). His characters were a variety of amazing and evolutionarily connected life forms, a fertile source of imagination for future science-fiction authors. Through his novels Stapledon taught us to think long-term about space, time and aliens, and the richness of this thought over these time spans is still in many ways unsurpassed.

Arthur C. Clarke (1917–2008) possessed a more technical background than Stapledon, and in fact served as chairman of the British Interplanetary Society. The composition of his early stories overlap in time with Stapledon, who was one of his main influences. Virtually all of his novels are filled with aliens, and their themes are human interaction with aliens, as in *Childhood's End*, *Rendezvous with Rama* or *2001: A Space Odyssey* and its sequels. 'The idea that we are the only intelligent creatures in a cosmos of a hundred million galaxies is so preposterous that there are very few astronomers today who would take it seriously,' he wrote in 1972, 'It is safest to assume, therefore, that They are out there and to consider the manner in which this fact may impinge upon human society.'[22] Clarke believed that extraterrestrials gave a true perspective on humanity, 'true' meaning in the broadest context of the possibilities inherent in the new universe, dwarfing even the globalists of the day. It was this perspective that was the main theme of almost all of Clarke's novels.

The prolific American science-fiction writer Isaac Asimov (1920–92) took a very different approach, the opposite side of the coin of humanity's role in the newly expanded space and time. With one exception, a novel titled *The Gods Themselves*, aliens are not at all prominent in his science fiction, which is nevertheless considered some of the best of the twentieth century. The famous original *Foundation* trilogy, its subsequent prequels and sequels, and Asimov's robot novels as well, have no aliens at all, but show how the new ideas of space and time have greatly expanded the scope for human and robotic action.[23]

Meanwhile, as we have seen, in continental Europe the Polish physician and writer Stanisław Lem (1921–2006) had taken up science fiction at mid-twentieth century with his novel *Astronauci*, absorbing the Soviet fascination with space despite running into trouble with Soviet Lysenkoism. Despite the movie treatment of that first novel, it was Lem's novel *Solaris*, published in 1961, that spread his fame. By this time, Lem had read Clarke and Asimov, as well as Ray Bradbury. Although affected by those authors, Lem's treatment of the alien was very different, allowing him to play out themes in an alien setting unlike anything produced in the West. Solaris is a planet with an ocean that is alive, 'a monstrous entity endowed with reason, a protoplasmic ocean-brain enveloping the entire planet and idling its time away in extravagant theoretical cognitation [*sic*] about the nature of the universe.' The monologue of this living being, however, was beyond the understanding of humans. While the ultimate purpose of Lem's novel is to use the cosmos to learn about humans, it may also be read at a different level as an argument against attempting contact before humans understand themselves: 'Man has gone out to explore other worlds and other civilizations without having explored his own labyrinth of dark passages and secret chambers, and without finding what lies behind the doorways that he himself has sealed.'[24] Lem seems to be saying that, bold as the new universe may be, humans may after all remain its central mystery (or at least a central mystery), imparting the message that our fate may lie not in the stars, but in ourselves. But the mere possibility of Lem's non-humanoid aliens expanded the scope of human imagination and illuminated age-old human questions. *Solaris* was first filmed in 1971, and subsequently received other movie treatments both in the Soviet Union and the United States. Multiplied hundreds of times in sophisticated or shallower treatments, ideas of the alien have spread rapidly throughout popular culture, affecting individuals and cultures in countless, though not always quantifiable, ways.

The comparison of these thinkers in relation to the new view of the universe highlights an important point. Although they were both affected by the same new view of the universe, Clarke and Asimov provide two views of human destiny – one in which humans interact with aliens beings, and one in which human destiny is to expand throughout the galaxy for its own purposes, without having to deal with pesky aliens. Lem believes we may have to deal with aliens, even though alien minds may be incommensurable with ours – and we had better learn to understand ours better. In addition to these very different views of human destiny, Stapledon constantly reminds us of the necessity of thinking over billions of years. Applying this kind of Stapledonian thinking to cultural evolution in the cosmos, and taking cultural evolution as a serious and dominating integral of cosmic evolution, the long time spans over which extraterrestrial intelligence may have existed implies that they are nothing like humans. They may in fact have evolved beyond flesh and blood biologicals, giving rise to postbiologicals, perhaps in the form of artificial intelligence (AI).[25] Although that idea is not new, the idea of cultural evolution over eons of time as an integral part of cosmic evolution gives it new force. Thus, we may live in a postbiological universe full of machines, and this may have implications for the SETI scientists, who should be looking for machines rather than biologicals like us.

While this may seem far-out speculation, it is our knowledge of space and time, together with the real possibility of aliens, that leads us to such a vision. Based as it is on current ideas of terrestrial cultural evolution, the likelihood is that it is too conservative rather than too speculative.

III The imagination and space: how has imagination affected space exploration?

In examining whether imagination has affected space exploration we turn, first, to one of space exploration's most perceptive historians, Walter McDougall. McDougall has argued that imagination is one of three structural forces necessary for spaceflight, along with funding and technology. There is no doubt that spaceflight pioneers were imaginative thinkers. As McDougall himself put it in his Pulitzer prize-winning book …*The Heavens and the Earth:* 'The great pioneers of modern rocketry – Tsiolkovsky, Goddard, Oberth and their successors Korolyov, von Braun, and others – were not inspired primarily by academic or professional interest, financial ambitions, or even patriotic duty, but by the dream of spaceflight. To a man they read the fantasies of Jules Verne, H.G. Wells and their imitators, and the rocket for them was only a means to an end.'[26]

This much is well known, but how much can it be generalized? From personal experience I can say that not only was my entry into astronomy affected by imagination in the form of science fiction, but also that a good percentage of my colleagues at NASA and other space agencies around the world were (and still are) influenced by science fiction, and that therefore imagination played a role in their entry into careers in astronomy and spaceflight. But this is certainly not true of all pioneers and practitioners of spaceflight. Let us take the case of the three pioneers of Explorer 1, the first US satellite, which recently passed its fiftieth anniversary. These pioneers, Wernher von Braun (1912–77), William Pickering (1910–2004) and James Van Allen (1914–2006), are familiar from the iconic photo of the three at the early morning press conference following the successful launch of Explorer 1 on 31 January 1958 (Figure 2.3). All three are the subject of recent exhaustive biographies.[27]

Figure 2.3 The three men responsible for the success of Explorer 1, America's first earth satellite which was launched 31 January 1958. At the left is Dr. William H. Pickering, former director of the Jet Propulsion Laboratory, which built and operated the satellite. Dr. James A. Van Allen, center, of the State University of Iowa, designed and built the instrument on Explorer 1 that discovered the radiation belts which circle the earth. At the right is Dr. Wernher von Braun, leader of the Army's Redstone Arsenal team which built the first stage Redstone rocket that launched Explorer 1.
Source: Courtesy of NASA.

There is no doubt that one of von Braun's influences was the so-called 'father of German science fiction,' Kurd Lasswitz. Lasswitz was a philosopher and historian, a Kantian who was steeped in the school of German idealism and wrote a biography of Gustav Fechner. In the first English translation of Lasswitz's science-fiction novel *Auf zwei Planeten* (On Two Planets), published in 1971 during the Apollo program, von Braun wrote 'I shall never forget how I devoured this novel with curiosity and excitement as a young man.'[28] One can safely assume it was one of the complex of factors that propelled von Braun forward to the stars.

But it was quite different for Pickering and Van Allen, as is evident from their recent biographies. Neither one of these physicists was influenced by science fiction, but rather more by technology. Douglas J. Mudgway, the author of the new Pickering biography, when asked whether Pickering was at all influenced by science fiction from his native New Zealand, wrote that Pickering 'definitely was not. He was greatly attracted to the things around him in his country town, radio crystal sets, the town electric generator that ran only four hours per day and the telephone switchboard and telephone system in his town. Later at secondary school he became fascinated with amateur radio.'[29] Similarly for Van Allen, who gives no evidence of science -fiction influence.[30] It is therefore important to realize that sources of inspiration exist other than science fiction, in this case a fascination with technology, quite different from 'imagination,' or at least a different kind of imagination. Even from such a small sample we can conclude that imagination in the science-fiction sense is neither necessary nor sufficient for space exploration. It is not necessary because it drove only some of the spaceflight pioneers, and it is not sufficient because imagination cannot propel any nation to the moon in the absence of McDougall's other two factors, funding and technology.

We may also look at a second area, not spaceflight itself, but exobiology and the search for extraterrestrial intelligence (SETI), a subject taken up by NASA already in the 1960s, though at a relatively low level of funding. More than a decade ago in *The Biological Universe*, my history of the twentieth-century extraterrestrial life debate, I included an entire Chapter on 'The Role of Imagination.' There I concluded that an understanding of the alien in science fiction was essential to understanding why it held such a grasp on popular culture, and even why it was taken up by some scientists. During the twentieth century, I found:

> science and science fiction increasingly complemented each other: speculative science fiction provided the perfect outlet for scientists who wished to go beyond science. Not only did scientists exercise their imaginations in science fiction, science fiction also inspired them to tackle questions in the real world. Many of the pioneers in exobiology and SETI grew up on science fiction and were led to their careers by its imaginative lure. Having nurtured science fiction, science now received in return some of the rewards of imagination.[31]

In the SETI arena David Swift's book of interviews, *SETI Pioneers*, is revealing. Swift found that Philip Morrison, famous for his 1959 paper on interstellar communication, was influenced by H.G. Wells; similarly, Freeman Dyson 'read a good deal of science fiction' and was especially influenced by H.G. Wells. Carl Sagan, around 7 or 8 years old, read the Edgar Rice Burroughs novels, and of course later himself wrote his own science-fiction novel *Contact*. One of Barney Oliver's 'more profound influences' as a youth was Hugo Gernsback's *Amazing Stories*; Oliver went on to write the famous book on Project Cyclops and helped direct NASA's SETI program. Ron Bracewell, author of *The Galactic Club*, first thought about extraterrestrials when, like Carl Sagan, he read Edgar Rice Burroughs, as well as Gernsback's *Amazing Stories*. Jill Tarter 'loved science fiction and read enormous amounts of it,' especially Robert Heinlein. In the Soviet Union Nikolay Kardashev, well-known for his typology of civilizations, also read science fiction.[32]

There were, of course, some SETI scientists not influenced by science fiction; Frank Drake and Iosif Shklovsky are among them. Nevertheless, certain fields related to outer space have a more imaginative component, and we can conjecture that the relationship is directly proportional: the more imaginative the field, the more its practitioners have been influenced by science fiction and other imaginative drivers.

There is room here for more interesting research on how particular fields differ in the role of the imagination, from general categories like scientists versus engineers, to specific categories like SETI scientists. In any case, it is clear that imagination has historically affected spaceflight, but one needs to be nuanced in just how general that claim can be made, in what areas, and in particular, to what effect.

IV Space and our *Weltanschauung*: how has space exploration affected our world view?

The evolution of our ideas of space, time and aliens in the twentieth century has affected more than just our imaginations. It has also affected our individual and collective world views, our *Weltanschauung*, our society and culture however one wishes to define them. And, I would argue, our new knowledge of the universe should affect our world views even more. Our new knowledge of cosmic evolution demonstrates for the first time in an empirical way our true place in the universe, both in space and time, with the question of aliens still very much open, perhaps the greatest question remaining in the history of science. The new view of cosmic evolution is already affecting us in numerous ways, though the diffusion rate of cosmic ideas into popular culture is in many ways agonizingly and remarkably slow. In science – arguably one of the primary drivers of culture – there is no doubt that cosmic evolution is now the master narrative, the subject of scholarly books, public broadcasting television treatment, and most importantly of all, research programs. It is clear that the photos from the Hubble Space Telescope and the other great observatories have fired the popular imagination, but they can also be seen as pieces in the story of cosmic evolution – the story that leads (in a non-teleological way) to humans and to the question of life beyond earth. That question continues to fascinate the public, and to draw in an increasingly diverse audience of scholars into the fields of astrobiology and SETI. Over the last few decades the astrobiology and SETI communities have convened special groups to discuss the societal impact of the discovery of life in the universe. Not surprisingly, they have concluded that there will be a multitude of reactions to the discovery of extraterrestrial life, depending on the scenario and the society.[33]

Cosmic evolution has also made small inroads into a number of academic disciplines. In history, it has specifically spawned the movement known as 'Big History.' Big History, pioneered by David Christian and Fred Spier, views history in the context of 13.7 billion years of cosmic evolution, rather than in the traditional mode of thousands of years of wars and politics.[34] A continuation of the 'cosmic calendar' used to great affect by Carl Sagan and others, Big History has the potential to revolutionize the teaching of history even beyond the current and more advanced trends toward global history. Just as global history expands the individual's *Weltanschauung*, so cosmic history views global history as just one example among many possible worlds, and explicitly questions parochial terrestrial assumptions in history, philosophy and all areas of human thought. While this method of teaching history is not yet widespread, nor the cosmic mode of thought embodied in the cosmic calendar internalized in most people's lives, it very likely will be in the future.

Similarly, small inroads have been made in anthropology. For the last several years, there have been SETI sessions at the annual meetings of the American Anthropological Association, drawing anthropologists into the subject from a variety of viewpoints. A cover story in the British publication *Anthropology Today* recently emphasized how anthropology can be applied to cultures beyond the earth (either human or alien); it has been used by anthropologists to guide thinking on interstellar migration, and anthropologists have even written alien science fiction.[35]

Aside from science, history and anthropology, small inroads have also been made in religion, theology and philosophy. The Space Age spawned considerable discussion about theological implications, especially if life were to be actually found. The papers from a 1998 Templeton Foundation meeting exemplify the theological implications of the new universe. An article by the British biochemist and Anglican priest Sir Arthur Peacocke calls cosmic evolution 'Genesis for the third millennium' and argues that 'any theology – any attempt to relate God to all-that-is – will be moribund and doomed if it does not incorporate this perspective [of cosmic evolution] into its very bloodstream.' Another article in the same volume argues that a 'cosmotheology' that takes into account what we know about the universe could greatly expand current terrestrial theologies. A German volume of essays on the same subject indicates that the possibility of a cosmotheology is not an idea confined to one culture.[36] Similarly, a few scholars have begun to discuss cosmophilosophy, asking what part of our knowledge is necessary, what is contingent, and how the traditional problems of terrestrial philosophy might be broadened by the expanded outlook afforded by space exploration. This, in the end, is the great benefit of the Space Age, providing a much broader perspective, making us realize that all our earthly knowledge may be only a single instance of a much more generalized knowledge.[37] The diffusion of the cosmic perspective into academic disciplines has been excruciatingly slow. Yet, slowly but surely, it is making its mark, and it will likely gather momentum over the next decades as our cosmic consciousness increases. It is already increasingly seeping into consciousness through curricula actually based on cosmic evolution.[38]

Still, the chief impact today has been not on these academic disciplines, but mainly in popular culture through science fiction, the debate over unidentified flying objects (UFOs), and popular interest in *Star Trek*, *Star Wars*, and the visual media that stimulate the imagination and from which much of the public takes their ideas of science. Taken together, science fiction, the UFO debate and their depiction in media and the arts may be seen as one way that popular culture absorbs this new world view of a biological universe, expanded in space and time and perhaps replete with aliens.

The immediate impact of the Space Age, however, is far more diverse than the ultimate discovery of life in space. Even if no aliens are found, space has already impacted, and will continue to impact our civilization in surprising and not always evident ways. In her recent book *Rocket Dreams: How the Space Age Shaped Our Vision of a World Beyond*, Marina Benjamin argues that space exploration has shaped our world views in diverse ways. She argues that 'the impact of seeing the earth from space focused our energies on the home planet in unprecedented ways, dramatically affecting our relationship to the natural world and our appreciation of the greater community of mankind, and prompting a revolution in our understanding of the earth as a living system.' Benjamin thinks it is no coincidence that the first Earth Day on 20 April 1970 occurred in the midst of the Apollo program; or that one of the astronauts developed a new school of spiritualism; or that people 'should be drawn to an innovative model for the domestic economy sprung free from the American space program by NASA administrator James Webb.'[39] Nor is Benjamin the first, or strongest, proponent of this argument, which has been made since the first Apollo days by poets like Archibald MacLeish and authors such as Frank White.[40] Space exploration shapes world views and changes cultures in unexpected ways. So does lack of exploration.

As Isaac Asimov foretold in his *Foundation* series, eventually humans will spread into the cosmos at large. Some see space in utopian terms, as the new frontier, or a place to start over for a new and better world. The *Star Trek* mission 'to boldly go where no man has gone before' is the clarion call of those who see space exploration as a necessary part of human evolution, not a luxury. Historians and social scientists have analyzed this kind of argument, and not all agree that the utopian ideal of spreading humanity to outer space is a valid reason for going, or that utopia is what we will build when we get there. Others have demonstrated the complex relation of such space goals to social, racial and political themes. One such study is De Witt Douglas Kilgore's book *Astrofuturism: Science, Race, and Visions of Utopia in Space.* In this book Kilgore examines the work of Wernher von Braun, Willy Ley, Robert Heinlein, Arthur C. Clarke, Gentry Lee, Gerard O'Neill and Ben Bova, among others in what he calls the tradition of American astrofuturism.[41]

In the end we must also realize that the impact of space exploration on our world view will also vary according to individuals and cultures – coming back now full circle to that problematic term 'culture' and the relation of the individual imagination to it. Howard McCurdy's *Space and the American Imagination* critically analyzes ideas such as the new frontier, progress, the exploration imperative and the search for extraterrestrial life as part of American culture.[42] Although he does not discuss it, his book implicitly raises the question 'What is the role of space in the European imagination, or the Chinese or Russian imagination?' How do different cultures affect the imagination, and how does the imagination affect cultures differently? And just how important has space exploration been as one among many sources of imagination in the twentieth century like atomic power and other wonders of science? In undertaking these studies, we need to remember that we both produce culture and are a product of culture. We need to remember that in doing history our remembrance of things past is inevitably colored and clouded by the geographical separations in space, by the passage of time, and by our own minds that at times seem alien to each other. Comparative studies should certainly be undertaken on these subjects, and this volume is an opening contribution toward that goal in the European context.

Notes

1. *Der schweigende Stern*, directed by Kurt Maetzig, was a co-production of the DDR and Poland, undertaken by DEFA-Studios. It was released in East Germany on 26 February 1960 and in West Germany on 9 September 1960 with the new title.
2. The Polish title is *Astronauci*. Though Lem in later life did not think highly of his first novel, its success encouraged him to write more fiction in this vein. On the novel in the context of Lem's life, see Peter Swiarski, *A Stanislaw Lem Reader*, Evanston: Northwestern University Press, 1997, esp. 3–4.
3. See James T. Andrews, 'In Search of a Red Cosmos: Space Exploration, Public Culture, and Soviet Society,' in Steven J. Dick and Roger D. Launius, eds, *Societal Impact of Spaceflight*, Washington, DC: NASA, 2007, 41–52. See also Claudia Schmölders's contribution, Chapter 3 in this volume.
4. See Nigel Rapport and Joanna Overing, *Social and Cultural Anthropology: The Key Concepts*, London: Routledge, 2000, vii.
5. Alfred L. Kroeber and Clyde K.M. Kluckhohn, *Culture: A Critical Review of Concepts and Definitions*, Cambridge, MA: Peabody Museum, 1952, 656.

6. Clifford Geertz, *The Interpretation of Cultures*, New York: Basic Books, 1973, 289. For more on the debate over the nature and meaning of culture, see Adam Kuper, *Culture: The Anthropologists' Account*, Cambridge, MA: Harvard University Press, 1999. For debated differences between the concepts of culture and society, a good starting point is Rapport and Overing, *Social and Cultural Anthropology*, entries on 'culture' and 'society,' 92–102, 333–43.
7. E.O. Wilson, *Consilience: The Unity of Knowledge*, New York: Alfred A. Knopf, 1998.
8. Steven J. Dick and Mark L. Lupisella, eds, *Cosmos and Culture: Cultural Evolution in a Cosmic Context*, Washington, DC: NASA, 2010; Steven J. Dick, 'Anthropology and the Search for Extraterrestrial Life: An Historical View,' *Anthropology Today* 22.2 (April 2006), 3–7.
9. Rapport and Overing, *Social and Cultural Anthropology*, 333.
10. Howard E. McCurdy, *Space and the American Imagination*, Washington, DC: Smithsonian Institution Press, 1997; Roger D. Launius and Howard E. McCurdy, *Imagining Space: Achievements, Predictions, Possibilities, 1950–2050*, San Francisco: Chronicle Books, 2001; Gerald Holton, *The Scientific Imagination: Case Studies*, Cambridge: Cambridge University Press, 1978; and for Holton's typology of imagination in science, 'Imagination in Science,' in idem, *Einstein, History and Other Passions*, New York: Addison-Wesley, 1996, 78–102.
11. Alexander C.T. Geppert, 'Imagining Outer Space, 1900–2000: An International Conference,' 6–9 February 2008, conference description and the introduction to this volume.
12. Robert Smith, *The Expanding Universe: Astronomy's Great Debate, 1900–1931*, Cambridge: Cambridge University Press, 1982; Richard Berenzden, Richard Hart and Daniel Seeley, *Man Discovers the Galaxies*, New York: Science History Publications, 1976. On Wallace, see Steven J. Dick, *The Biological Universe: The Twentieth Century Extraterrestrial Life Debate and the Limits of Science*, Cambridge: Cambridge University Press, 1996, 36–58; and idem, 'The Universe and Alfred Russel Wallace,' in Charles H. Smith and George Beccaloni, eds, *Natural Selection & Beyond: The Intellectual Legacy of Alfred Russel Wallace*, Oxford: Oxford University Press, 2008, 320–40.
13. On this change in world view, see, in addition to the literature above, Bart J. Bok, 'Harlow Shapley and the Discovery of the Center of Our Galaxy,' in Jerzy Neyman, ed., *The Heritage of Copernicus: Theories 'More Pleasing to the Mind,'* Cambridge, MA: MIT Press, 1974, 26–62.
14. Smith, *Expanding Universe*, 97–146.
15. Eric Chaisson, *Cosmic Evolution: The Rise of Complexity in Nature*, Cambridge, MA: Harvard University Press, 2001; and *Cosmic Dawn: The Origins of Matter and Life*, Boston: Little, Brown, 1981.
16. Dick, *Biological Universe*; idem, 'Cosmic Evolution: History, Culture, and Human Destiny,' in idem and Lupisella, *Cosmos and Culture*, 25–59.
17. Notably Chaisson, *Cosmic Evolution*; and idem, *Cosmic Dawn*.
18. On Einstein's thought experiments, see, for example, Walter Isaacson, *Einstein: His Life and Universe*, New York: Simon & Schuster, 2007, 122–7. On the effect of time synchronization technology on Einstein, see Peter Galison, *Einstein's Clocks, Poincaré's Maps: Empires of Time*, New York: Norton, 2003.
19. I have given this history in *The Biological Universe*, 399–472.

20. On the history of the extraterrestrial life debate, see Dick, *Biological Universe*; Michael J. Crowe, *The Extraterrestrial Life Debate, 1750–1900: The Idea of a Plurality of Worlds from Kant to Lowell*, Cambridge: Cambridge University Press, 1986; and Karl S. Guthke, *The Last Frontier: Imagining Other Worlds from the Copernican Revolution to Modern Science Fiction*, Ithaca: Cornell University Press, 1990.
21. Olaf Stapledon, 'A Note on Magnitude,' in *Last and First Men/Star Maker*, New York: Dover Publications, 1968, 435.
22. 'When the Aliens Come,' in Arthur C. Clarke, *Report on Planet Three and Other Speculations*, New York: New American Library, 1972, 89–102; see also Dick, *Biological Universe*, 254–6.
23. The original *Foundation* trilogy is Isaac Asimov, *Foundation* (1951), *Foundation and Empire* (1952) and *Second Foundation* (1953), subsequently published in numerous editions. The robot novels are *Caves of Steel* (1954), *The Naked Sun* (1957), *The Robots of Dawn* (1983) and *Robots and Empire* (1985).
24. Stanisław Lem, *Solaris*, New York: Berkley Publishing Company, 1970, 28, 165.
25. Steven J. Dick, 'Cultural Evolution, the Postbiological Universe, and SETI,' *International Journal of Astrobiology* 2.1 (2003), 65–74; reprinted under the title 'Bringing Culture to Cosmos: The Postbiological Universe,' in idem and Lupisella, *Cosmos and Culture*, 463–87.
26. Walter McDougall, ...*The Heavens and the Earth: A Political History of the Space Age*, New York: Basic Books, 1985, 20.
27. Michael J. Neufeld, *Von Braun: Dreamer of Space, Engineer of War*, New York: Alfred A. Knopf, 2007; Doug Mudgway, *William H. Pickering: America's Deep Space Pioneer*, Washington, DC: NASA, 2008; Abigail Foerstner, *James Van Allen: The First Eight Billion Miles*, Iowa City: University of Iowa Press, 2007.
28. Wernher von Braun, Introduction to Kurd Lasswitz, *Two Planets* (*Auf zwei Planeten*), trans. Hans H. Rudnick, Carbondale: Southern Illinois University Press, 1971, epigraph. On Lasswitz, see Dick, *Biological Universe*, 227–30, and on the influence of Lasswitz on the Space Age, see Willy Ley, *Rockets, Missiles, and Men in Space*, New York: Viking, 1969, 65–9.
29. Doug Mudgway, personal communication; and idem, *William H. Pickering*.
30. Foerstner, *James Van Allen*.
31. Dick, *Biological Universe*, 266.
32. David W. Swift, *SETI Pioneers: Scientists Talk About Their Search for Extraterrestrial Intelligence*, Tucson: University of Arizona Press, 1990; for Morrison, see 22; Dyson, 323; Sagan, 211; Oliver, 88–9; Bracewell, 140–1; Tarter, 350–1; and Kardashev, 180.
33. John Billingham et al., eds, *Social Implications of the Detection of an Extraterrestrial Civilization: A Report of the Workshops on the Cultural Aspects of SETI, held in October 1991, May 1992 and September 1992 at Santa Cruz, California*, Mountain View: SETI Press, 1999; Allen Tough, ed., *When SETI Succeeds: The Impact of High-Information Contact*, Bellevue: Foundation for the Future, 2000; and Connie Bertka, Nancy Roth and Matthew Shindell, *Workshop Report: Philosophical, Ethical, and Theological Implications of Astrobiology*, Washington, DC: American Association for the Advancement of Science, 2007; full proceedings published in Constance M. Bertka, ed., *Exploring the Origin, Extent, and Future of Life: Philosophical Ethical, and Theological Perspectives*, Cambridge: Cambridge University Press, 2009. See also Albert Harrison, *After Contact: The Human Response to Extraterrestrial Life*, New York: Plenum, 1997.

34. On Big History, see David Christian, 'The Case for "Big History,"' *Journal of World History* 2.2 (Fall 1991), 223–38; David Christian, *'Maps of Time': An Introduction to 'Big History,'* Berkeley: University of California Press, 2004; David Christian, 'History and Science after the Chronometric Revolution,' in Dick and Lupisella, *Cosmos and Culture*, 441–62; as well as Fred Spier, *The Structure of Big History: From the Big Bang Until Today*, Amsterdam: Amsterdam University Press, 1996; and Marnie Hughes-Warrington, 'Big History,' *Historically Speaking* 4.2 (November 2002) 16–17, 20.
35. Dick, 'Anthropology and the Search for Extraterrestrial Life.' Among alien science fiction by anthropologists, see Mary Doria Russell, *The Sparrow*, New York: Fawcett, 1996; and idem, *Children of God*, New York: Villard, 1998.
36. Arthur Peacocke, 'The Challenge and Stimulus of the Epic of Evolution to Theology'; and idem, 'Cosmotheology: Theological Implications of the New Universe,' both in idem, ed., *Many Worlds: The New Universe, Extraterrestrial Life and the Theological Implications*, Philadelphia: Templeton Foundation Press, 2000, 89–118, 191–210. A similar and more international collection is Tobias Daniel Wabbel, ed., *Leben im All: Positionen aus Naturwissenschaft, Philosophie und Theologie*, Düsseldorf: Patmos, 2005.
37. On the relation between alien and terrestrial knowledge, see the section on extraterrestrial epistemology in Edward Regis Jr., ed., *Extraterrestrials: Science and Alien Intelligence*, Cambridge: Cambridge University Press, 1985, 83–128. Also, Steven J. Dick, 'Extraterrestrials and Objective Knowledge,' in Tough, *When SETI Succeeds*, 47–8.
38. Dick, 'Cosmic Evolution,' in idem and Lupisella, *Cosmos and Culture*, 25–59.
39. Marina Benjamin, *Rocket Dreams: How the Space Age Shaped Our Vision of a World Beyond*, New York: Free Press, 2003, 4.
40. On the history of this argument, see Robert Poole, *Earthrise: How Man First Saw the Earth*, New Haven: Yale University Press, 2008.
41. De Witt Douglas Kilgore, *Astrofuturism: Science, Race, and Visions of Utopia in Space*, Philadelphia: University of Pennsylvania Press, 2003.
42. McCurdy, *Space and the American Imagination*.

CHAPTER 3

Heaven on Earth: Tunguska, 30 June 1908

Claudia Schmölders

> In the early morning hours of 30 June 1908 tens of thousands in central Siberia were able to observe an extraordinary natural phenomenon. A gleaming white ball rose in the heavens which moved quickly from South East to North West. It traversed the airspace over most of the Yennissey Department – a distance of over 500 kilometers – and shook the earth underneath its flight path, rattled window panes; the plaster fell off the walls. [...] People thought the world was coming to an end. Shortly after the disappearance of the glowing ball, a giant pillar of fire rose over the horizon. Within a radius of 750 kilometers there were detonations. At all the meteorological stations in Europe and America, seismographers registered the tremors of the earth's crust.[1]

I The Tunguska Event

This is how the Polish author Stanisław Lem (1921–2006) described the enormous explosion in the stony Tunguska region in June of 1908 in his novel *Astronauci*, some 40 years after the fact. Most of his knowledge was drawn from reports by Leonid Kulik (1883–1942), a mineralogist at the Soviet Academy of Science, who undertook the first explorations into the region only 15 years later in 1922. Consequently, most of the descriptions of what was likely the largest 'impact event' in recent human history resemble one another. Modern estimates put the force of the explosion, which took place between 5 and 10 kilometers above the Podkamennaja Tunguska River in central Siberia, at more than a thousand times that of the atom bomb dropped on Hiroshima. The blast also felled nearly 80 million trees in an area measuring more than 2,000 square kilometers (Figures 3.1 and 3.2).

Claudia Schmölders (✉)
Humboldt-Universität zu Berlin, Berlin, Germany
e-mail: c.schmoelders@online.de

Alexander C.T. Geppert (ed.), *Imagining Outer Space*
European Astroculture, vol. 1
https://doi.org/10.1057/978-1-349-95339-4_3

Figure 3.1 Map showing the approximate location of the 1908 Tunguska Event.
Source: https://en.wikipedia.org/wiki/Tunguska_event (accessed 1 October 2017).

Figure 3.2 Trees felled by the Tunguska explosion. Photograph taken by the Leonid Kulik Expedition.
Source: Courtesy of NASA.

Witnesses of the catastrophe, from a variety of vantage points, described it alternately as being like cannon shots, storms followed by columns of fire and lightning with thunderclaps. Tents, storage huts and cattle were hurled aloft and/or incinerated. The apocalyptic event took some ten minutes (though variously reported as lasting from between two minutes to an hour); one man even went to a bathhouse to wash himself so as to be clean when death came. Even from a distance of 65 kilometers people could still feel the heat on their faces, evidence of a very high gas flame rather than the impact of a comet or meteorite. Other indices of a gas flame were the bright nights observed in Europe and Western Asia, starting late on 29 June and lasting until midnight on 2 July 1908.[2] In fact, during that same year people were also able to see with the naked eye the Morehouse Comet, which did not hit the earth but came so near that it could be photographed all over the world.

The most brilliant image was taken by the Heidelberg astrophysicist Maximilian Wolf (1863–1932) on 16 November 1908.[3] But, despite the reams of scientific analysis and reports, especially those by Lem's source Leonid Kulik, written between 1922 and 1930, the 'Tunguska Event' has never been fully explained.[4] Though Kulik hypothesized that it was a meteor impact, no meteoric debris or significant amounts of radiation have ever been found at the site.

The mystery provoked numerous pseudo-scientific explanations and made Tunguska into something of a tourist attraction. In the 1980s a geologically consistent explanation was put forward, according to which tectonic shifts had released large quantities of natural gas, which then exploded. If this theory is correct, there was no meteor or comet at all.[5] Further research of more recent origin has yielded contradictory results, but little attempt has been made to harmonize the conclusions. Each solution to the riddle would also, of course, mean a loss of tourist dollars. In 2006 a Tunguska museum opened in the nearby town of Kraznoyarsk, but the following year it supposedly lost a three-ton rock that may have proven to be key evidence. In 2008, on the centennial of the Tunguska event, Italian scientists under the supervision of Giuseppe Longo, a professor of astrophysics at the University of Naples, announced another search for the lost meteor, this time in the nearby crater-shaped Lake Cheko, but the search never began (Figures 3.3 and 3.4).[6]

There are many ways to read an event such as this. Scientists would usually begin by researching the geological facts and physical causes for such a massive explosion. Such exploration falls to those practicing hard sciences such as geology and astrophysics. Thus, the description of these efforts would be the domain of the history of science. A second perspective would locate this particular event within the global history of major catastrophes and the apocalyptic reactions that they produce. This perspective belongs to the history of ideas or intellectual history.[7] A third approach would analyze the supposed meteoric impact as a striking *symbol* for the physical 'meeting' of heaven and earth, that is, the intrusion of physical outer space into the inner biosphere, in terms of 'astronoëtic' reasoning and also therefore as part of the history of ideas. German philosopher Hans Blumenberg (1920–96) coined the term 'astronoëtics' to capture this approach, which encompasses most of the philosophical, cultural and religious thought on the subject since antiquity.[8] One of the most detailed intellectual histories of secular astronoëtic thought is in French historian Alexandre Koyré's book *From the Closed World to the Infinite Universe*, published just prior to the launch of Sputnik 1 in October 1957.[9] Many other authors have also contributed to this, often speculative, perspective, with a notable debate occurring recently between French philosopher Edgar Morin and the astrophysicist Michel Cassé.[10] A fourth perspective on Tunguska would be an analysis of the event's religious semantics, an endeavor also located within the history of ideas. The vision of heaven and earth communicating in continuous exchange is deeply rooted in many religious traditions. All over the world, human beings have imagined mortals, gods and goddesses exchanging their places in space. Here, we have not only a key motif of shamanism, but also of the human impulse to fly, starting with Icarus

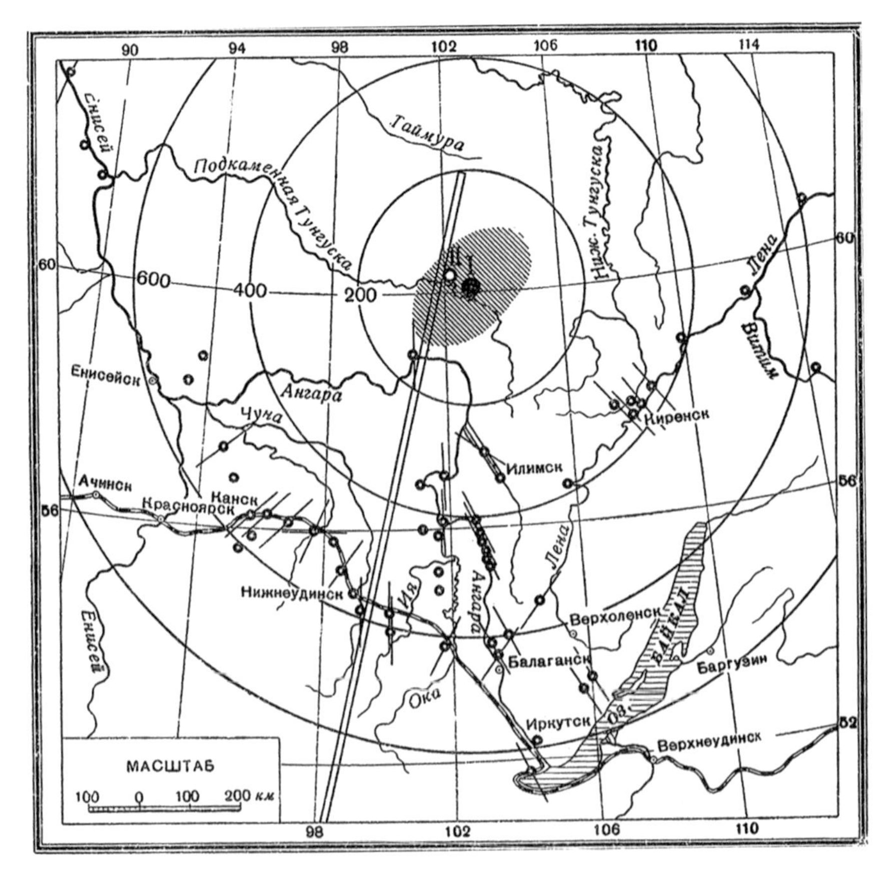

Figure 3.3 A map of the Tunguska Event with the geographical distribution of correspondents who sent their reports to the Irkutsk Observatory in July 1908. The map was first published by A.V. Voznesensky in 1925 and then reproduced by Evgeny Leonidovtich Krinov in 1949.

Source: A.V. Voznesensky, 'Padenie meteorita 30 iyunya 1908 g. v verkhoviyah reki Hatangi,' *Mirovedenie: Izvestija Russkogo Obščestva Ljubitelej Mirovedenija* 14.1 (1925), 25–38; reproduced in Evgeny Leonidovitch Krinov, *Tungussky meteroit*, Moscow: Izdatelstvo Akademii Nauk SSSR, 1949, 20–1.

in Greek mythology and continuing, unbroken, to the present day. Taken together, all four perspectives mold the Tunguska Event into a very particular *lieu de mémoire* of the year 1908.

II The literary perspective or, the narrative

Not surprisingly, the Tunguska Event has inspired a copious body of literature and fictional accounts in print, film and other mass media. In his 2007 essay on the Tunguska lore, William Hartwell rightly places it in a long series of spectacular cosmic impact or airburst events recorded over the course of

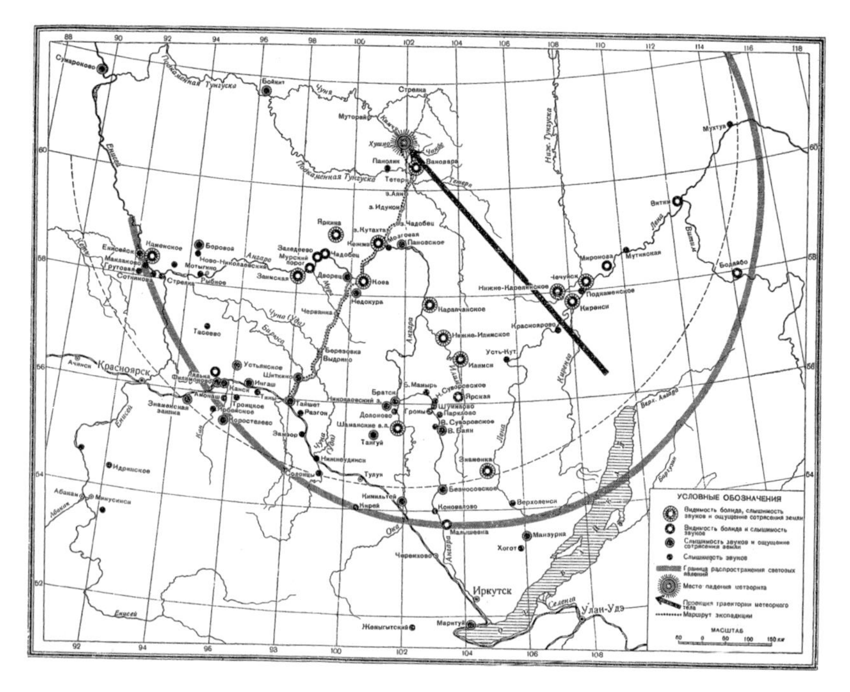

Figure 3.4 In the years following, further statements on the meteorite's flight path suggested a more northwesterly route. Subsequent maps were altered correspondingly. *Source*: Evgeny Leonidovitch Krinov, Tungussky meteroit, Moscow: Izdatelstvo Akademii Nauk SSSR, 1949, 28–9.

centuries and the many narratives they generated.[11] Without doubt, these stories are largely symbolic and require individual analysis. Most of them thematize bad luck, universal suffering and the end of the world in apocalyptic scenarios. For instance, in his 1893 novel *La Fin du monde* (The End of the World) the famous French astronomer and science-fiction author Camille Flammarion (1842–1925) described horrible catastrophes as the result of a comet impacting the earth's surface.[12] But Hartwell shows as well that the nightmare scenarios dramatically increased in the 1950s. Possibly the bleakest of these apocalyptic stories, Larry Niven's *Lucifer's Hammer*, appeared in 1977.[13]

Yet, interestingly enough, despite all these horror stories, Tunguska has also led to the creation of the opposite kind of lore, built around the observation that, despite the destructive force of the event, it did relatively little harm. There were no deaths or complete destruction of homes, no communication dangerously interrupted, and no culture erased. As a catastrophe without any catastrophic effects, the Tunguska explosion seemed like an ideal experiment secretly arranged by nature itself, an interpretation that might well explain a number of divergent plots in other narratives during the second half of the twentieth century, beginning with Aleksandr

Kazantsev's 1946 science-fiction story *Vzryv: Rasskazgipoteza* (The Explosion).[14] The Soviet writer (1906–2002) had visited Hiroshima after the dropping of the atomic bomb on 6 August 1945 and interpreted Tunguska later as a nuclear explosion. In his story, a nuclear-powered spaceship from Mars explodes during an attempted emergency landing. Not surprisingly, a Siberian shaman plays a central role, for flight is an important motif of shamanism. Kazantsev did not receive much attention when his book was published, but his idea has been connected with those of Soviet scientist and Slavic studies expert Boris Ljapunov (1921–72), who also concluded that only an atomic spaceship, and thus extraterrestrial intelligence, could have caused the explosion. Ljapunov's essay was widely read and taken at face value.[15]

Kazantsev as well as Ljapunov may have inspired Lem's *Astronauci*, in which an alien spacecraft explodes at the Tunguska site. In Lem's plot, set in 2003, invaders from Venus are planning to destroy planet Earth but are killed in an accident. They leave a message that can be deciphered by experts thanks to a newly invented giant computer. Humans then travel to Venus to negotiate a peace agreement, but upon their arrival, they find the planet entirely destroyed by fire and conclude the Venusians exterminated themselves in a nuclear holocaust. The story becomes – and remains – an allegorical warning about nuclear weapons. Neither Kazantsev nor Lem focused on a catastrophe on earth, but instead used the explosion as a motive for their characters to leave the planet and search for intelligent extraterrestrial life. Both stories are simultaneously utopian and cautionary tales. The Tunguska Event was likewise interpreted in the 1960 film adaptation of Lem's story, the first science-fiction movie made in communist East Germany: Kurt Maetzig's *Der schweigende Stern* (*The Silent Star*), which was released shortly before the Cuban Missile Crisis when the threat of nuclear conflict loomed especially ominously.[16] *Der schweigende Stern* was not the only film inspired by the Tunguska explosion. Since the 1950s, dozens of movies, computer games, cartoons, stories, TV shows, and even songs have interpreted the historical material. The best-known film is George Pal's *When Worlds Collide* (1951), which Steven Spielberg is rumored to be considering remaking.

Real events as well as fictive scenarios are susceptible to multiple interpretations. While political readers of science fiction could interpret the use of nuclear powers in Lem's novel as a parable for the Cold War, others saw it as a harbinger of the Atomic Age or as a confirmation of the utopian vision of extraterrestrial intelligence. The dangerous alien visitors often represent a highly developed culture of the kind whose allure was felt especially strongly during the Soviet period and which continues to fuel our evolutionary imagination. The Soviet Union, as well as the United States and even Germany, were obsessed with molding humans to fit a cosmic future in outer space. This was also the rationale in the 1950s, when Gotthard Günther (1900–84), the highly respected German philosopher,[17] located the 1949 novel *The Incredible Planet*, written by science-fiction author John W. Campbell (1910–71), within the history of ideas.[18] Günther found in Campbell's novel a Hegelian spirit motivating human beings to build a new universe all of their own, with their own planets and stars. The message was that the true human avant-garde was trying to leave terrestrial war and suffering behind. Günther's plot was aimed at Americans, who were at the time working with physicists like Wernher von Braun trying to realize actual space missions. Günther's reasoning was particularly appealing to Germans after the Second World War since his vision drew heavily on the ideas not only of Hegel, but also of the German philosopher Oswald Spengler.

A mixture of literary and technological enthusiasm concerning the exploration of outer space reigned, especially in Eastern Europe. Literary historian Matthias Schwartz has analyzed the cosmological and cosmonautic excitement in Soviet science fiction between 1957 and 1965. According to Schwartz, political outer space propaganda during the Cold War was clearly aimed at mobilizing both the powers of imagination as well as of strategic planning. After the death of cosmonaut Yury Gagarin in 1968, however, a sense of depression set in. Authors such as the Russian brothers Strugatsky later brilliantly transformed the waning of Russia's cosmological fascination into widely read suspense literature. But probably the best fictional use of the Tunguska explosion to come out of this region was published only a few years ago. Vladimir Sorokin's *Trilogiya*, published between 2002 and 2006, describes an immense cosmo-psychic arc from the year 1908 to the millennium.[19] Its composition is too complex to be recounted here in detail, but the first volume contains a clear presentation of the history of German-Russian obsession with outer space. The plot focuses on a sect of ethereal beings who fell from grace and were condemned to an earthly existence. With the help of icy matter from the Tunguska meteor, they search for one another among millions of ordinary human beings in order to return to space as a group. One does not need to be a historian to recognize that Sorokin's sect of blond and blue-eyed creatures in this novel satirizes German 'Aryanism' and Hanns Hörbinger's so-called *Welteislehre* (world ice theory) of the 1920s, cabbalistic speculations, and, of course, the cosmic fantasies of Russian astronoëtics as envisioned around 1900 by Konstantin Tsiolkovsky (1857–1935). But Sorokin finally transports this congeries of themes into the capitalistic present. At the end of the first novel, creative entrepreneurs divest the high pathos of ideology and sell Tunguska ice as a health product. The novel concludes with an eloquent caricature: a child playing with one of the Tunguskan ice-cubes. But it is not the child that is cold. Instead, the ice cube pleads for warmth. The child takes the cube to bed – it's time for a 'thaw,' for a bit of humanity.

III The scientific perspective or, the space

Such a domestic turn at the very end of Sorokin's novel connects it to the most widely read literary tale of astronautic lore during the twentieth century. Among the most antithetical to any Tunguskan catastrophe, *Le Petit Prince* (The Little Prince) was written by the French pilot and poet Antoine de Saint-Exupéry (1900–44) and published in 1943.[20] In this story, planets are but distant homes of strange creatures. The accent is on 'homes': the little prince talks of domestic spaces, of an atmosphere just as cozy as in stories by Jules Verne, rather than of an alien void. The plot derived from the experience of a pilot who symbiotically lived with and within his machine, traversing vast expanses of space.

The idea of outer space as cozily domestic seemed even more plausible as a consequence of the actual technological progress being made at the time. In 1945, two years after the *Little Prince*'s appearance, the British physicist and author Arthur C. Clarke was designing the geo-stationary communication satellites that today beam the images of earth as a fragile blue tiny planet into our living rooms.[21] Conversely, the picture of a beautiful living room shows up in the last scene of Kubrick's film *2001: A Space Odyssey*. Outer space seems to have lost its hostile, threatening character, a perspective echoed in the philosophical spatial approach to the Tunguska Event of 1908. Its scenario of earth coming into contact with a dangerous outer space may be read as a contribution to the contradictory theories of space evolving at the time. In fact, old-fashioned ideas of private and personal space were confronted with scientific explorations of physical space in terms of geological as well as electrical matter. The latter, in particular, led to new modes of terrestrial communication such as television, the Internet, cyberspace and so forth – modes of intellectual existence, that today in turn threaten the idea of privacy itself. Thus, by 1908, space was a central category of reasoning. Coincidentally or not, it was introduced into the arts with so-called Cubism in that year, and four of the most important spatial sciences also emerging at that time, two of which are located in the natural sciences: biology and geochemistry, and two in the humanities: sociology and radio history. The biological work of Johann Jakob von Uexküll (1864–1944) belongs more to the latter than to the former. In 1909 he wrote his first – and bestselling – book, *Umwelt und Innenwelt der Tiere* (Environment and the Inner World of Animals).[22] Here, he demonstrated that animals have their own densely textured environment, a private sphere, as it were, combining a so-called *Merkwelt* with a *Wirkwelt*. The term '*Merkwelt*' anticipates the term 'semiosphere' coined by the later Russian literary scholar Yuri Lotman (1922–93). Already in 1925, Uexküll founded an institute for 'environmental studies' at the University of Hamburg. A second seminal work from 1908 also made a case for a private space, this time a human one. Georg Simmel's *Soziologie: Untersuchungen über die Formen der Vergesellschaftung*

(Sociology: Investigations on the Form of Sociation) does not define space biologically, but rather in terms of civil law and psychology.[23] For Simmel (1858–1918), the decisive human mark in terrestrial space is the limit or border. Spatial borders, which social subjects draw for themselves and others, allow for and symbolize phenomena such as protection, discretion, property and neighborhood and are thus part of the very core of everyday life.

Though neither the biological nor the social conception of private space lends itself immediately to translation into planetary dimensions, two other interpretations of space around 1908 sought to do just this. These two interpretations, the geochemical and the electronical, were formulated in the United States and Russia, the two countries that would eventually set out to conquer outer space. The Russian theory was authored by geologist and mineralogist Vladimir Vernadsky (1863–1945), who served from 1915 as the head of the same commission on Russian resources for which Leonid Kulik undertook his expeditions into the Tunguska region. In 1908 Vernadsky proposed a new science, geochemistry, which was no longer exclusively a terrestrial, but rather a planetary discipline analyzing samples from meteorites.[24] Vernadsky is also known for having coined the word 'biosphere.' During his lectures at the Sorbonne in 1922, he could well have met the French Jesuit and paleoanthropologist Pierre Teilhard de Chardin (1881–1955) for an exchange of ideas. Teilhard too was a fervent geologist, and it seems reasonable to assume that the two men discussed the first reports by Leonid Kulik from his first Tunguska expedition in 1922.

In any case, both expanded the idea of the biosphere into that of the 'noösphere,' perhaps the most brilliant expression of astronoëtic thought. Vernadsky wrote about this term in his last essay, *Some Words About the Noösphere*, published in 1943, as had Teilhard already in the late 1930s.[25] Both scientists considered matter (*Materie*) a kind of spiritual substratum – insofar as it shows human work molding earth – and geology a *Leitwissenschaft* – a leading science. This is perhaps not surprising for a Russian materialist scientist, but it amounts to something of a scandal coming from a Catholic priest, for whom the 'noösphere' constituted an intense Christian telos. Linking the whole process of evolution not to science, but rather to the person and message of Jesus Christ made him one of the foremost propagators of existential faith in outer space experience.

The fourth doctrine of space from around 1908 defined outer physical space as materialistically as Vernadsky's, but centered on space as an electromagnetic field. Hugo Gernsback (1884–1967), radio pioneer, writer, and sometimes hailed as the 'father of science fiction,' imagined a society based on wireless communication and set up the first technical magazine on amateur radio called *Modern Electrics*. In 1911, this time as a science-fiction writer, Gernsback also conceived of a space station able to receive radio waves.[26] With his vision of the earth's atmosphere as permeable to radio waves and thus also to outer space, Gernsback provided a first literary model of cosmic communication. Most of his contemporaries, on the other hand, proved incapable of turning this physical insight into literature. Their concepts of personal inner and physical outer space could not be brought into one single consistent narrative. They remained imaginatively constrained through the First World War and its bloody conflicts. There were, however, a few exceptions with characteristically cultural connotations.

IV The perspective of light or, metaphysical reasoning

Certainly at that time the scientific community in Europe was also discussing electrical waves – most prominently the light waves of which Einstein spoke in his famous 1905 essay.[27] And this leads to a third aspect of the Tunguska Event, even more philosophical and visual than the previous two. This third perspective focuses on the image as such and, above all, its prerequisite: light. All the eyewitnesses to the 1908 explosion spoke of seeing a light that lasted for days and was visible as far away as London. In terms of contemporaneous culture, such reports echo the cults of sun fanatics and the overwhelming enthusiasm for light that gripped Europe and Russia in the wake of electrification.

Light was, then, to become a dominant theme not only in cultural contexts but also in astronoëtic theory.[28] In a recent study of the prehistory of Einsteins's theory, art historian Karl Clausberg has analyzed the Breslau law professor Felix Eberty's 1846 study on *Die Gestirne und die Weltgeschichte* (The Stars and World History).[29] Even though there is no direct connection with the Tunguska Event, this small booklet provided an important vantage point from which to view this radiant explosion. Eberty took actual visual technology into account. A mere decade after the invention of photography, he already imagined a cosmic 'photographic archive,' born from the observation that light from space stores a picture of every moment in the earth's history, because objects on earth reflect that light and send it back. Were we to have a large central eye, a panopticon, somewhere, Eberty argued, it would be able to receive the images after millions of years of travel and to read their story. Thoughts like these oscillate between astrophysics, religion and poetry, and Eberty was fluent in all three of these intellectual idioms.

Around 1900, several German authors followed this path, among them Paul Scheerbart (1863–1915), the first 'planetary poet' of the century, who propagated the end of all military activity on terrestrial ground and envisioned a futuristic world of light and glass, of transparency and luminescence. Also important was Ludwig Klages (1872–1956) who wrote his platonic book *Der kosmogonische Eros* (The Cosmogonian Eros) in 1922.[30]

All of them set aside the idea of light *invading* our world. Educated along the lines of Greek and German Goethean *Bildung*, they preferred the idea of light *touching* the earth, the notion of a tender touch of cosmos. Light may intrude, but destroys nothing, at least insofar as it is filtered by the atmosphere. Light gives life to the biosphere: it is meant for the eye – and not just the human eye – and the eye is meant and made for seeing it. A legacy of the Platonic philosopheme, according to which the eye is made up of solar elements, this metaphysics of light was transmitted via Goethe into the twentieth century. The idea had nothing to do with actual conquest, but rather stylized the relationship of human being and heaven into a mutual gaze, a cosmic eye-to-eye.

No Western author favored this kind of reciprocity more than the aforementioned Jesuit priest Pierre Teilhard de Chardin. Deeply impressed by the French philosopher Henri Bergson, Teilhard made Bergson's 1905 book *L'Evolution créatrice* (Creative Evolution) his lifelong spiritual guide and tried to synthesize Christian and biological thought.[31] 'Through incarnation God descended down to earth in order to overspiritualize it and get it back up to his own heavenly place,' was one of his central formulas.[32] He did not conceive of life as light, but instead of life as illumination. Teilhard became the most fervent defender of man's anthropological, that is real bodily, destination in heaven, that is, outer space. He applauded all evolutionary speculation and physical research. For Teilhard, even the military exploration of atomic power meant nothing less than a step forward in human spiritual growth. In the same year that Alexandr Kazantsev, inspired by Hiroshima, published his story on the Tunguska Event in the Soviet Union, Teilhard commented in China on the first nuclear test in the Bikini Islands of 30 June 1946. His report read strikingly similar to Lem's first sentences cited above:

> A little over a year ago, in the early morning, a blinding flash of exceptional brilliance in the desert of Arizona illuminated the farthest mountain tops and eclipsed the first rays of the sun. There followed a terrifying tremor. [...] It has happened. For the first time, a nuclear fire, lit by the science of man, had consumed space for a second. But once the deed was done, once the dream of making lightning had been realized, man, numbed by his success, soon returned to himself – and in the flash of that lightning that had come out of his hand, he tried to realize what his work had made out of him. His body was whole. But what had happened to his soul?[33]

In the following pages Teilhard again develops his idea that man, with his newfound power, is on the way to his real destination in heaven/space as designed and intended by God himself. The human catastrophe of Hiroshima is not mentioned.

Teilhard de Chardin's idea of reciprocity is religious in a very deep sense, but also inconsistent with scientific ideas about space. Trying to explain this inconsistency, German cultural philosopher Peter Sloterdijk has emphasized that all astronoëtic narratives oscillate between two entirely opposing conceptions of our cosmic position.[34] There is, on the one hand, the Ptolemaic narrative of a circular order of planets – centric, static and entirely removed from human meddling. This is the forerunner of all scientific concepts of space: even if one does not consider the earth the center of planetary organization, the paths of the stars' movements remain unchanged. On the other hand, however, there is also a vertical hierarchy of the religious cosmos with its dramatic ups and downs as prefigured in light exchange itself. This *dialectic* of ascent and descent has always been part of our religious cosmo-psychological 'construction.' We cannot see it when the heavens are viewed only as a two-dimensional pictorial surface, as for instance in astrology, but rather when we speak of heaven and hell. The rising and falling of souls is a common theme in all religions, a vertical movement up to superior worlds or down to hell, and conversely, the entrance of gods into our terrestrial domain and their departure again for heaven. Judeo-Christian iconography has a famous allegory for this: Jacob's Ladder, which shows only angels, no humans, ascending and descending. This is Teilhard's vision – with one significant difference. In order to establish a connection with modern science, he tried to explain away the image's symbolic character in reading religious thought as a biological program: 'What could serve as a better background and basis for the epiphanies of a Christogenesis than an ascending anthropogenesis?'[35] Certainly, the difference between the symbol and its symbolized entity is a great theme in the history of ideas. Its loss also marks the development of literature insofar as science fiction tends to fuse with scientific experimental thought. Many of the astronauts and astrophysicists of the present grew up reading science-fiction literature, just as their ancestors did in the 1920s when they read Jules Verne, H.G. Wells, Hugo Gernsback and others. Flying to the moon or mastering the powers of the universe as in atomic explosion thus found a cultural playground, not only in literature but also in art history and the history of ideas, which brings us to our fourth and final perspective on Tunguska.

V The pictorial perspective or, the contribution of the arts

The most interesting reflection on the prospect of earth getting into touch with outer space and vice versa emerged in 1908 from the world of art, more precisely from the German cultural philosopher Oswald Spengler (1872–1956). His bestselling work *Der Untergang des Abendlandes* (The Decline of the West), already outlined in 1912, reached hundreds of thousands of readers after its publication in 1918, that is, after the First World War.[36] Spengler's approach was deeply connected with the space theories of his time, especially with Eberty's theory of light as it had been adapted by popular books on astronomy. In contrast to these ideas, Spengler translated the cosmic experience into some inner spiritual creation, proclaiming the idea of a measureless inner gaze:

> The 'visual world' is the totality of light-resistances, since vision depends on the presence of radiated or reflected light. The Greeks took their stand on this and stayed there. It is the Western world-feeling that has produced the idea of a limitless universe of space – a space of infinite star-systems and distances that far transcend all optical possibilities – and this was the creation of the inner vision, incapable of all actualization through the eye, and, even as an idea, alien to and unachievable by the men of a differently disposed culture.[37]

Today, we know Spengler mostly as a trailblazer of the Nazi dictatorship with its dilettante biologist view of cultural history. He compared his way of defining cultural growth and cultural decay with the change of perspective that Copernicus had brought for astronoëtics. As Copernicus devaluated earth's position in the cosmos, Spengler devaluated Western culture in the context of cultural, world history: there was no epicenter of progress. Moreover, Spengler can also be seen as part of the history of art, especially of the Renaissance. His focus on the idea of spatial limitlessness, that is outer space, as the visionary capacity of inner human gaze echoes another prominent invention of the Renaissance, the invention of perspective around 1430 by learned artists such as Leon Battista Alberti (1404–72). In essence, this invention meant the intrusion of physical space into the semiosphere or symbolic space of painting, an event widely discussed ever since.[38]

Ernst Cassirer (1874–1945) and Erwin Panofsky (1892–1968), both renowned art historians, have dealt in depth with this intrusion. Though not directly related to the Tunguska Event, the idea of an explosive energy intruding on earth from an invisible depth of an unheard-of outer space compares with a similar revolutionary change of visual consciousness in the world of painting. Never before had outer space touched earth as catastrophically, and never before in the relative absence of God, gods and religious emotions. To the contrary, the Tunguska Event met a secular condition of space consciousness and thus a scientific and even technical one. None of its contemporaries was more efficient than Spengler, who infused his idea of the 'depth of space' with the enthusiasm of the early rocket engineers. Taking his points of departure as the image and the gaze rather than ideas of gravity, real space or rocket technology, Spengler counters the powerful invasion of physical space on earth through light with equally powerful fantasies about real departure into outer space, thus molding the religious dialectics of rising and descending in a literal sense. His motivation, Spengler's biographers tell us, grew out of Germanic fantasies of conquest. Consequently, he tried to relate the discovery of central perspective not to the Italians, but to the so-called German Gothic, because the Gothic entailed for him Goethe's *Faust* or the so-called Faustian bargain:

> This is the outward- and upward-straining life-feeling – true descendent, therefore, of the Gothic – as expressed in Goethe's Faust monologue when the steam-engine was yet young. The intoxicated soul wills to fly above Space and Time. An ineffable longing tempts him to indefinable horizons. Man would free himself from the Earth, rise to the Infinite, leave the bonds of the body, and circle in the universe of space amongst the stars.[39]

At the time, sentences like these were not just floating in space, so to speak. In fact, Spengler's view of the Gothic drew on a brief passage from Wilhelm Worringer's famous 1908 book *Abstraktion und Einfühlung* (Abstraction and Empathy).[40] It was Worringer (1881–1965) who first spoke of an alien German drive toward the infinite, even of a 'barbarian extravagance,' and ascribed a vertiginous tendency to verticality to the German Gothic in particular. Primitive art, he argued, usually draws on the very opposite, namely a pronounced claustrophobia (*Raumscheu*). This main thesis of Worringer's book could well have inspired Spengler to develop his counter-thesis of the Faustian 'lust for space' while he was still working as a schoolteacher in Hamburg in 1908.

Spengler completed his work in 1922; in 1923, a completely revised version appeared as one volume. The same year saw a new edition of Felix Eberty's work, now with a preface by Albert Einstein. Last but not least, the most important book concerning the opening up of space, Hermann Oberth's legendary *Die Rakete zu den Planetenräumen* (The Rocket into Interplanetary Space), was also published in 1923. It was the first authoritative work on the technical conditions for actual and not merely fictional space travel.[41] It was Oberth's book that inspired a young Wernher von Braun to start his research on space travel via rockets. Outer reality had taken over inner vision.

In a certain sense then, Teilhard de Chardin followed Spengler's dynamic vision. After the Second World War and the experiences with Hiroshima and Nagasaki, however, the militant optimism vanished. Teilhard did not experience the actual triumph of the first man in the moon or even the launch of Sputnik – he died in 1955. Two years later, the German poet Arno Schmidt (1914–79) published his deeply pessimistic 1957 novel *Die Gelehrtenrepublik* (The Egghead Republic) – a story set in a devastated earth in the year 2008 – while the French philosopher and historian of science Gaston Bachelard (1884–1962) turned his interest back toward the earth. Bachelard's 1957 *Poétique de l'espace* (Poetry of Space) dealt not only with poetry, but with all forms of private and closed forms of space, such as houses and nests. It was as if it had been inspired by Uexküll's 1909 work on the *Umwelt und Innenwelt der Tiere* or even by Saint-Exupéry's *Little Prince*.[42]

In the same year, 1957, Bachelard's colleague Alexandre Koyré published his seminal historical book on astronoëtic thought from Nicholas of Cusa through Leibniz. Under the title *From the Closed World to the Infinite Universe*, he put aside any notions of (religious) privacy or mental anxiety but replaced the idea of God with a so-called *intelligentia supra-mundana*. About ten years later, and in spite of the astonishing progress of factual space travel, German philosopher Hans Blumenberg returned to the pessimistic view. In his great 1965 essay on the Copernican revolution he tried to install the literary approach at the center of his astronoëtic approach. The idea that the earth revolves around the sun rather than vice versa, Blumenberg wrote, affects society not simply as a theoretical process, 'but as a metaphor: the reconstruction of the world-edifice became the sign for a transformation in human self-understanding, for a new mode of locating the human being in the totality of nature or the loss of its localizability and for the meaninglessness of a particular place in the world.'[43]

Metaphors exist only in and through language, and human self-understanding in Blumenberg's sense can only come about in the linguistic mode, as Blumenberg himself discussed in his famous 'Metaphorology.' Metaphoric thinking may contribute to our philosophical reasoning but can also be understood as a kind of playful practice aimed at coping with real threats. The category of the 'as if,' a central code of nineteenth-century philosophy, provides the foundation of the notion of 'thinking as *Probehandeln*' (behavior in rehearsal), to use Sigmund Freud's term. But Freud, too, understood the Copernican revolution as a monstrous affront to humanity's self-estimation and self-esteem. The question is whether this reading has ever held true and, if so, whether it still does. Perhaps we should realize that on the contrary, our experimental confrontation with Copernicus's new planetary constellation actually conditioned us to *conquer* space, spurring us on to action rather than engendering humility.

Blumenberg wrote and published his book after the first human beings had already traversed space astronautically and even 'spacewalked,' like birds in the air. Since then space travel has continuously advanced, and there seems to be little to humble our cosmic emotions. Severe setbacks due to technological failure and even disasters do not slow down the pace of exploration. On the contrary, we have domesticated the cosmos with various ingenious experiments that have long surpassed the realm of metaphor. Scholars, engineers and designers plan hotels and subdivisions for the moon and Mars, invent space elevators, sell space trips, and so forth. Furthermore, the Copernican shock has dissipated in a major way because of the Internet. Thanks to outer space and satellite-based communication, any private computer owner can use the technology of Google Earth to look up terrestrial addresses, as though they were looking in at earth from outer space. Conversely, Google Sky is opening up a reverse perspective into the depths of space, based in large part on Hubble Space Telescope images.

The two phenomena are thus interrelated. Not only do we actually go up and into space, we have also turned the on-screen simulation of actions, that is to say the metaphoric process, into our visual mode of life. Thanks to technology, we have downloaded space into our offices and living rooms, rather than allowing it to scare, let alone humiliate us. During the 1960s, the same period in which Blumenberg wrote about the Copernican revolution, Adorno explicitly attacked television as a tool for domesticating and trivializing all content. How much more vigorously would he have condemned the Internet, with its continuous trivialization and domestication of the feeling of ultimate homelessness connected with thoughts of outer space?[44]

VI Traditional hermeneutics as a framework

Describing the physical catastrophe of the year 1908, that is the Tunguska Event, from four different perspectives – the historical, the literary, the scientific and, finally, the visual – means writing history in a very old-fashioned way, alluding to the so-called theory of the fourfold sense of the text. In fact, this theory stems from the oldest technique of interpretation, namely Biblical hermeneutics, with its categories of historical, allegorical/metaphorical, moral and anagogical or even mystical sense. Interestingly, this heuristic instrument has been strongly developed in the twentieth century by several authors, chiefly by the literary historian and former priest Northrop Frye and then by the historians Hayden White and Jörn Rüsen, the former director of the Kulturwissenschaftliches Institut in Essen.[45] All these authors agree that in constructing historical or literal meaning we must use *all* perspectives, or at least keep in mind the other three if we are working with the fourth, since no author keeps only to one mode of narration.

Reflecting upon the scope of cultural studies the idea gains a new credence. In this field, we have fostered the theory of so-called turns. For a few years now, there has been a 'spatial turn,' which followed the former 'pictorial' and even older 'linguistic turn,' only to prepare itself for another as-yet-unnamed turn. Supposedly, endlessly changing 'turns' are constitutive for cultural studies. Yet 'turns' in this sense mean nothing other than perspectives and even a superficial glance at the topic at hand reveals that it cannot be reduced to one single perspective only. Precisely for the study of the space imaginary, we need all of the existing, and perhaps many more, perspectives.[46]

Naturally, this leads to the origin of secularized hermeneutics, to the eighteenth-century historian Johann Martin Chladenius (1710–59). For the optical paradigm of looking at an object from different points of view, Chladenius coined the term 'Sehepunkte' – the perspectives of different eyewitnesses.[47]

Chladenius put forward the idea that historical events should be viewed from several perspectives in order to arrive at the truth, thus hearkening to the hundreds of eyewitness statements collected by Soviet scientists in the 1950s and early 1960s describing the Tunguska Event. People told of their impressions of light, of thunder, of heat, of pain and of horror. All this data could be used for different scientific approaches and has, in fact, been used by NASA.[48] Collecting statements of eyewitnesses developed during the 1930s into a scientific approach of its own, namely oral history. Oral history is akin to the 'linguistic turn,' the turn to literature in the sense of 'oral tradition' or even 'oral poetry.' Where, then, does a simple, though uncanny event of short duration such as that of Tunguska fit in? Looking at Tunguska as a piece of oral tradition may mean classifying it as a folktale. Tunguska would then be the name of a saga since many so-called etiological sagas deal with the origin of some territorial special feature of a precisely bounded region. This is the reason behind the spread of Tunguska stories about felled trees, explosions of light, the sound of thunder, the memory of comparable events, fears about the world coming to an end, and so on. Last but not least, an event like Tunguska could be located within the narrative genre of riddles since the phenomenon has still never been fully explained.[49]

Compared to the massive dimensions of the real event, the folktale may seem an inappropriately trivial genre. Within the context of so-called higher literature, one would probably locate the catastrophe within the form of the novella, characterized by Goethe as an *unerhörte Begebenheit* (unprecedented occurrence). Indeed, one of the most renowned German novellas has a geocatastrophic motif, Heinrich von Kleist's *Das Erdbeben von Chile* (The Earthquake in Chile). Here we have not only the earth shaking but also the shaking of human destinies in a sense of ordeal. Exactly because it lacks this background of heavenly fate the explosion of Tunguska would not be narratable as a novella in the classical sense.

In fact, the apocalyptic genre comes from biblical narration or pagan mythology. Both seem to have been absorbed by the genre of science fiction, which excels in both dystopian and utopian plots. Precisely because science-fiction literature deals with technical progress, its vision of the future is not really apocalyptic but, above all, human-centered. All that mankind is able to do with its creative powers of imagination and intellect could lead to either catastrophe or happiness. It was the American science-fiction author Philip Gordon Wylie's 1933 novel *When Worlds Collide* that demonstrated that humans might survive a terrestrial catastrophe on earth by leaving it for outer space.[50] In the end, then, it makes sense to return to the multi-perspective horizon of old Chladenius whose nephew, Ernst Florens Chladni (1756–1827), published in 1794 the first description of meteorites as cosmic detritus falling piecemeal from the solar system.[51] Here, we see science realizing that cosmic material had been intruding upon the earth since the beginning of creation, eliciting the first invocation of an alien heaven on earth.

Notes

1. Stanisław Lem, *Astronauci*, Warsaw: Czytelnik, 1951; *Der Planet des Todes*, Berlin (Ost): Volk und Welt, 1957, 7. Lem has never allowed an English translation to be published, possibly because of the rather similar subject in George Pal's movie *When Worlds Collide*, directed by Rudolph Maré and George Pal, USA 1951 (Paramount Pictures). In 1950 Immanuel Velikovsky's *Worlds in Collision* (New York: Macmillan) appeared as well, a bestselling book with general theories about outer space material impacting the earth. I would like to thank Adrian Daub and Jeffrey Chase for editorial help and linguistic assistance, but most thanks go to Alexander Geppert who encouraged me to approach a famous astrophysical conundrum in a philological way.
2. Peter T. Bobrowsky and Hans Rickman, eds, *Comet/Asteroid Impacts and Human Society: An Interdisciplinary Approach*, Heidelberg: Springer, 2007. This volume contains the most informative analysis of the Tunguska explosion.
3. Bajac Quentin, Aurélien Barrau, Denis Canguilhem, Agnès de Gouvion Saint-Cyr, Peter Hingley and Françoise Launay, eds, *Dans le champ des étoiles: Les photographes et le ciel, 1850–2000*, Paris: Réunion des musées nationaux, 2000, 83 (exhibition catalogue, Musée d'Orsay, Paris, 16 June–24 September 2000; and Staatsgalerie, Stuttgart, 23 December 2000–1 April 2001).
4. John Baxter and Thomas Atkins, *Wie eine zweite Sonne: Das Rätsel des sibirischen Meteors*, Düsseldorf: Econ, 1977; *The Fire Came By*, New York: Doubleday, 1976.
5. Wolfgang Kundt, 'Tunguska (1908) and Its Relevance for Comet/Asteroid Impact Statistics,' in Bobrowsky and Rickman, *Comet/Asteroid Impacts and Human Society*, 331–9.
6. Giuseppe Longo, 'The Tunguska Event,' in ibid., 303–30.
7. Olaf Briese, *Die Macht der Metaphern: Blitz, Erdbeben und Kometen im Gefüge der Aufklärung*, Stuttgart: J.B. Metzler, 1998.
8. Blumenberg was interested in outer space philosophy from the start. While his most renowned book is *Die kopernikanische Wende* (Frankfurt am Main: Suhrkamp, 1965; *The Genesis of the Copernican World*, Cambridge, MA: MIT Press, 1989), he also set new standards for terminology. 'Astronoëtics' comes from Greek 'aster' = star and 'nous' = intellect or mind. The term appears first in Blumenberg's *Die Vollzähligkeit der Sterne*, a posthumously published collection of essays (Frankfurt am Main: Suhrkamp, 1997). The term was meant ironically as an intellectual answer to the so-called Russian 'astronautics'; that is, the Russian technical exploration of the moon, but it evolved into a term to describe the whole tradition of thinking or reasoning about outer space mainly without instruments and technology. From Thales to Copernicus, from Bruno to Kepler to Kant, the progress of astronomy had been made mainly by reasoning alone; that is, astronoëtics, and this is why the early science of outer space is on the same level as early mythology, religion and literature about this subject. Only with the development of technology were the two divided into different spheres of practice: science fiction on the one hand, astronautics and space research on the other.
9. Alexandre Koyré, *From the Closed World to the Infinite Universe*, Baltimore: Johns Hopkins University Press, 1957.

10. Michel Cassé and Edgar Morin, 'Himmelskunde: Leere, Licht, Materie – über Phänomene und Gesetze des Kosmos,' *Lettre 67* (Winter 2004), 94–105.
11. William T. Hartwell, 'The Sky on the Ground: Celestial Objects and Events in Archaeology and Popular Culture,' in Bobrowsky and Rickman, *Comet/Asteroid Impacts and Human Society*, 71–87.
12. Camille Flammarion, *La Fin du monde*, Paris: Flammarion, 1894.
13. Larry Niven and Jerry Pournelle, *Lucifer's Hammer*, Robinsdale: Fawcett, 1933.
14. Aleksandr Kazantsev, 'Vzryv: Rasskaz-gipoteza' (Explosion: A Hypothetical Story), *Vokrug Sveta* (Around the World) 1 (1946), 39–46.
15. Boris Ljapunov, 'Iz glubiny Vselennoj' (From the Depths of Outer Space), *Znaniesila* 10 (1950), 4–7. See also Matthias Schwartz, *Die Erfindung des Kosmos: Zur sowjetischen Science Fiction und populärwissenschaftlichen Publizistik vom Sputnikflug bis zum Ende der Tauwetterzeit*, Frankfurt am Main: Peter Lang, 2004.
16. See also Steven Dick's contribution, Chapter 2 in this volume.
17. Gotthard Günther was located in the tradition of Hegel, Spengler and Heidegger. In 1937 he left Germany, and in 1960 joined the University of Illinois where he worked in the electrical engineering department together with Heinz von Foerster, Humberto Maturana and others. Günther became famous with *Idee und Grundriß einer nicht-aristotelischen Logik*, Hamburg: Meiner, 1959.
18. John W. Campbell, *Der unglaubliche Planet*, Düsseldorf: Karl Rauch, 1952; *The Incredible Planet*, 1949. Campbell was a founding figure of science-fiction literature, especially as editor of the *Astounding Science-Fiction* series since 1937.
19. Vladimir Sorokin, *Trilogija* [*Put*, 2002; *Bro*, 2005; *Loed*, 2006], Moscow: Zacharov, 2006; Schwartz, *Erfindung des Kosmos.*
20. Antoine de Saint-Exupéry, *Le Petit Prince* (The Little Prince), San Diego: Harcourt Brace Jovanovich, 1943.
21. Arthur C. Clarke, 'Extra-Terrestrial Relays: Can Rocket Stations Give World-Wide Radio Coverage?,' *Wireless World* 51.10 (October 1945), 305–8.
22. Jakob von Uexküll, *Umwelt und Innenwelt der Tiere*, Berlin: Springer, 1909. See also idem, 'Environment and Inner World of Animals,' in Gordon M. Burghardt, ed., *The Foundations of Comparative Ethnology*, New York: Van Nostrand, 1985, 222–45.
23. Georg Simmel, *Soziologie: Untersuchungen über die Formen der Vergesellschaftung*, Munich: Duncker & Humblot, 1908.
24. See George S. Levit, *Biogeochemistry – Biosphere – Noosphere: The Growth of the Theoretical System of Vladimir Ivanovich Vernadsky*, Berlin: VWB Verlag, 2001.
25. Vladimir I. Vernadsky, 'Some Words About the Noösphere' [1945], *21st Century* 5 (Spring 2005), 16–21; Pierre Teilhard de Chardin, *Le Phénomène humain*, Paris: Editions du Seuil, 1955, 201.
26. See Everett Bleiler, ed., *Science-Fiction: The Gernsback Years*, Kent: Kent State University Press, 1998; Jörg Dünne and Stephan Günzel, eds, *Raumtheorie: Grundlagentexte aus Philosophie und Kulturwissenschaften*, Frankfurt am Main: Suhrkamp, 2006.
27. Albert Einstein, 'Über einen die Erzeugung und Verwandlung des Lichtes betreffenden heuristischen Gesichtspunkt,' *Annalen der Physik* 17.132 (1905), 132–48.

28. Wolfgang Schivelbusch, *Lichtblicke: Zur Geschichte der künstlichen Helligkeit im 19. Jahrhundert*, Munich: Hanser, 1982; see also Arthur Zajonc, *Catching the Light: The Entwined History of Light and Mind*, Oxford: Oxford University Press, 1995.
29. Karl Clausberg, ed., *Zwischen den Sternen: Lichtbildarchive*, Berlin: Akademie, 2006.
30. Ludwig Klages, *Der kosmogonische Eros*, Munich: Georg Müller, 1922.
31. Henri Bergson, *Evolution créatrice*, Paris: Alcan, 190; *Creative Evolution*, New York: Henry Holt, 1911.
32. Pierre Teilhard de Chardin, *L'Energie humaine*, Paris: Editions du Seuil, 1962, 220: 'Par l'incarnation, Dieu est descendu dans la Nature pour la sur-animer et la ramener à Lui: voilà le dogme chrétienne dans sa substance.'
33. Pierre Teilhard de Chardin, *L'Avenir de l'homme*, Paris: Editions du Seuil, 1970, 179: 'Il'y a un peu plus d'un an, au petit jour, dans les "mauvaises terres" de l'Arizona, une lueur éblouissante, d'un éclat insolite, illumine les cimes les plus lointaines, éteignant les premiers rayons du soleil levant. Puis, un ébranlement formidable [...]. C'est fait. Pour la première fois sur terre un feu atomique venait de brûler l'espace d'une seconde, industrieusement allumé par la science de l'homme. Or, le geste une fois accompli, son rêve une fois réalisé de créer une foudre nouvelle, l'homme. Etourdi par son succès, s'est bientôt retourné sur soi; et, à la lumière de l'éclair qu'il venait de faire jaillir de sa main, il a cherché à comprendre ce que son œuvre avait fait de lui-même. Son corps était sauf. Mais, à son âme, que venait-il d'arriver?'
34. Peter Sloterdijk, *Sphären I–III*, vol. 2: *Globen*, Frankfurt am Main: Suhrkamp, 1999, 467–9.
35. Chardin, *L'Energie humaine*, 221: 'Quoi de mieux qu'une ascendante anthropogénèse pour servir d'arrière-plan et de base aux illuminations descendantes d'une Christogénèse?'
36. Oswald Spengler, *Der Untergang des Abendlandes: Umrisse einer Morphologie der Weltgeschichte*, Munich: C.H. Beck, 1919/1922; *The Decline of the West*, New York: Alfred A. Knopf, 1939.
37. Idem, *Untergang*, 222: 'Die "gesehene Welt" ist die Gesamtheit von *Lichtwiderständen*, weil das Sehen an das Vorhandensein von strahlendem oder zurückgestrahltem Licht gebunden ist. Die Griechen blieben auch dabei stehen. Nur das abendländische Weltgefühl schuf die *Idee* eines grenzenlosen Weltraums mit unendlichen Fixsternsystemen und Entfernungen, die weit über alle optischen Möglichkeiten hinausgeht – eine Schöpfung des *inneren* Blickes, die sich jeder Verwirklichung durch das Auge entzieht und Menschen anders fühlender Kulturen selbst als Gedanke fremd und unvollziehbar bleibt' (emphases in original).
38. Most recently by Hans Belting, *Florenz und Bagdad: Eine westöstliche Geschichte des Blicks*, Munich: C.H. Beck, 2008.
39. Spengler, *Decline of the West*, 503; cf. idem, *Untergang*, 1189: 'Es ist das hinaus- und hinaufdrängende und eben deshalb der Gotik tief verwandte Lebensgefühl, wie es in der Kindheit der Dampfmaschine durch die Monologe des Goetheschen Faust zum Ausdruck gelangte. Die trunkene Seele will Raum und Zeit überfliegen. Eine unnennbare Sehnsucht lockt in grenzenlose Fernen. Man möchte sich von der Erde lösen, im Unendlichen aufgehen, die Bande des Körpers verlassen und im Weltraum unter Sternen kreisen.'

40. Wilhelm Worringer, *Abstraktion und Einfühlung*, Munich: Piper, 1908.
41. Hermann Oberth, *Die Rakete zu den Planetenräumen*, Munich: Oldenbourg, 1923. It is unlikely, however, that Oberth ever read Spengler, even though he studied in Munich where Spengler's book was published.
42. Arno Schmidt, *Die Gelehrtenrepublik*, Karlsruhe: Stahlberg, 1957; Gaston Bachelard, *Poétique de l'espace*, Paris: Presses universitaires françaises, 1957.
43. Blumenberg, *Kopernikanische Wende*, 100: 'Der kopernikanische Umsturz ist nicht als theoretischer Vorgang Geschichte geworden, sondern als Metapher: die Umkonstruktion des Weltgebäudes wurde zum Zeichen für den Wandel des menschlichen Selbstverständnisses, für eine neue Selbstlokalisation des Menschen im Ganzen der gegebenen Natur oder für den Verlust dieser Lokalisierbarkeit und für die Bedeutungslosigkeit einer Weltstelle.'
44. Theodor W. Adorno, 'How to look at Television,' *The Quarterly of Film, Radio and Television* 8 (Spring 1954), 214.
45. Northrop Frye, *The Anatomy of Criticism*, Princeton: Princeton University Press, 1957; Hayden White, *Metahistory: The Historical Imagination in Nineteenth-Century Europe*, Baltimore: Johns Hopkins University Press, 1973; Jörn Rüsen, 'Die vier Typen historischen Erzählens,' in Reinhart Koselleck, Heinrich Lutz and Jörn Rüsen, eds, *Formen der Geschichtsschreibung*, Munich: Deutscher Taschenbuch Verlag, 1982, 514–606.
46. On the history of these consecutive turns in cultural studies, see Doris Bachmann-Medick, *Cultural Turns: Neuorientierungen in den Kulturwissenschaften*, Reinbek: Rowohlt, 2006.
47. Claudia Schmölders, '"Sinnreiche Gedancken": Zur Hermeneutik des Chladenius,' *Archiv für Geschichte der Philosophie* 58.3 (January 1976), 240–64.
48. See Giuseppe Longo, 'The Tunguska Event,' in Brobowsky and Rickman, *Comet/Asteroid Impacts and Human Society*, 314–15. The total number of testimonies amounts to more than 1,000.
49. See André Jolles, *Einfache Formen*, Halle/Saale: Niemeyer, 1930. This book has become a classic for European literary studies. It presented the first theoretical framework for oral poetry in terms of genres: *Mythe* (myth), *Sage* (saga), *Legende* (legend), *Rätsel* (riddle), *Sprichwort* (proverb), *Kasus* (case), *Memorabile* (memoir), *Märchen* (fairy-tale) and *Witz* (joke) are the nine forms that constitute oral tradition. Certainly it is impossible to use all of those genres explicitly in everyday conversation, but some of them, namely riddle, joke, proverb or fairy tale, are prevalant.
50. Philip Gordon Wylie, *When Worlds Collide*, New York: Frederick A. Stokes, 1933.
51. Ernst Florens Chladni, *Über den kosmischen Ursprung der Meteorite und Feuerkugeln* [1794], 2nd edn, Leipzig: Geest und Portig, 1982.

CHAPTER 4

Imagining Inorganic Life: Crystalline Aliens in Science and Fiction

Thomas Brandstetter

The alien as we know it first appeared at the end of the nineteenth century in the literary works of H.G. Wells (1866–1946) and Kurd Lasswitz (1848–1910). Before then, inhabitants of other planets, like the Saturnians of Voltaire's *Micromégas*, were part of a satirical tradition that used skewed reflections of humans as a means to comment on social, moral or political issues.[1] With Lasswitz and Wells, however, aliens were not so much a reflection of a certain status quo as projections of possible future developments of life. The rarefied humanoids of Lasswitz's *Auf zwei Planeten* (On Two Planets) show the potential of ethical progress, while the tentacled creatures of Wells's *War of the Worlds* exhibit all characteristics of a degenerated species dependent on technology.[2] Both authors were eager to convey a plausible image of their extraterrestrials by fleshing out the physiological details: the big eyes, delicately chiseled features and lean limbs of Lasswitz's Martians are incorporations of a universal ideal of beauty and virtue, while the reduced anatomy of Wells's invaders, containing only the brain and nerves leading to the tentacles, are expressions of a cruel, instrumental efficiency in the struggle for survival. These aliens were certainly inspired by evolutionary biology. Wells, especially, a former student of Thomas Henry Huxley (1825–95), applied principles and motives like 'the struggle for existence' to the design of his extraterrestrial creatures as well as to his plotline.[3]

However, these aliens can also be seen as literary instances of the concept of the plasticity of living matter. This idea was most prominently voiced by the biologist Jacques Loeb (1859–1924), whose search for purely mechanical

Thomas Brandstetter (✉)
Vienna, Austria
e-mail: tbrandstetter@monochrom.at

Alexander C.T. Geppert (ed.), *Imagining Outer Space*
European Astroculture, vol. 1
https://doi.org/10.1057/978-1-349-95339-4_4

causes for animal development led him to probe the limits of manipulability of living tissue. In a letter to Ernst Mach he voiced his conviction that 'man himself can act as a creator, even in living nature, forming it eventually according to his will.'[4] Experiments in hybridization led Loeb to believe that 'the number of species existing today is only an infinitely small fraction of those which can originate and possibly occasionally do originate, but which escape our notice because they cannot live and reproduce.'[5] The notion of a fundamental malleability of living matter was taken up by Wells in his 1895 essay *The Limits of Individual Plasticity*, where he argued that transplant surgery and other medical technologies show that living matter could be molded and modified at will.[6] Dr. Moreau, the protagonist of another of his novels, was an embodiment of the power to stretch and reshape the matter of life – a power that, one could argue, was even more radically wielded by the author himself when he designed his alien invaders of *The War of the Worlds*. The Martians were a species that, like the experimentally grafted heteromorphic *Antennularia* of Loeb, could be created by man, albeit not in the laboratory but in the text of Wells and the imagination of the reader.[7]

The aliens of Lasswitz and Wells were literary experiments in morphology: instances of an imagination that twisted the flesh and altered the form so as to produce species hitherto unseen. While Lasswitz's Martians show features quite similar to human ones, Wells's Martians were 'the most unearthly creatures it is possible to conceive.'[8] However, they still showed a physiology that could be recognized by human researchers: a brain, nerves, eyes, ears and tentacles, as well as lungs, a mouth and a heart. Even though the author altered their form, he left the basic structure as well as the basic stuff of life intact. The Martians were but beings made of flesh; or, to be more precise, beings made of protoplasm, which was held to be the basic substance of life.[9]

There was, however, another, more radical possibility. This chapter traces the idea of life forms made up of minerals as articulated in science and fiction from around 1900 to the end of the twentieth century. Crystalline aliens are a rare species in literature; however, they nevertheless address fundamental questions pertaining to our conception of extraterrestrial life.

Necessarily, speculations about extraterrestrial life as yet are fictitious. Therefore, there is no clear border between fiction and science. Imagination becomes the main tool of scientists and literary writers alike. By constructing alien life forms, literary texts can pose scientific questions and scientific texts can indulge in flights of fancy. Of course, imaginary alien creatures are always dependent on the historical context. Therefore, they offer us an opportunity to investigate the presuppositions as well as the limits of the biological thought of an epoch. As Stefan Helmreich has shown, the project of astrobiology is traversed by the search for a definition of life itself, as researchers in the field are themselves aware that our search for alien life presupposes categories based on our own domestic forms of life.[10] However, I want to argue that crystalline aliens offer even more than just an insight into the historicity of conceptions and definitions of life: they also allow us to see in which way the question of life itself was posed at different moments in history. 'What is life?' is not a universal question. There is always a reason why it emerges at a certain moment in history, and the way it is articulated depends on the context and the specific aims. I want to argue that throughout the history of biology and science fiction, imaginations of inorganic crystalline life offered a place for self-reflection. By discussing the possibilities of such life forms, writers reflected on the framework of contemporary definitions of life and pondered the ways in which the question of life itself could be posed and answered. Crystalline aliens therefore not only allow us to better grasp the perspective-dependency of definitions of life, they also show us how this perspective-dependency itself was theorized at certain historical moments.

My investigation will concentrate on crystalline (or silicon-based) life forms as they appear in scientific as well as fictitious texts. I do not want to suggest that there is no difference between science and fiction; however, in the case of extraterrestrial life, both necessarily share a highly speculative approach. I will, however, show how in science, speculation increasingly becomes linked to experimental work. This in turn affects not only the literary renderings of crystalline aliens, but also the questions they raise. How could we investigate beings that are fundamentally different from ourselves? What would it mean for our definitions of life? Which methodologies and practices would be apt to recognize and analyze alien life, and in which way would these shape the way we pose the question 'What is life?'

I will start with two short stories by the French writer J.H. Rosny. In the first story especially, the encounter with crystalline aliens leads the protagonist to formulate something like a proto-exobiological project: the conviction that even the strangest creatures could be grasped by experiment and reason. The second section will deal mainly with early scientific theories about silicon-based life. As these never left the speculative level, they offered scientists and literary writers alike an opportunity to dwell on imaginative renderings of such creatures. In the third section, I want to show how, at the beginning of the twentieth century, the experimental approach of synthetic biology probed the lines of demarcation between the living and the inanimate and led to a fundamental unsettling of definitions of life. This perspective was taken up by science-fiction writers like Stanley Weinbaum, who used crystalline aliens as an opportunity for a self-reflective stance that drew attention to the limits of his contemporaries' knowledge. As the fourth section will elaborate, after the Second World War cybernetics claimed to offer an exit out of this impasse and to formulate a new, universal definition of life. On the basis of these theories, crystalline aliens in science fiction from the 1960s up to the 1980s lost their otherness and became partners for communication. However, as I will demonstrate in the sixth and final section, this did not dispose of the provocative potential of silicon-based life forms. Exobiology proper and the planning for unmanned probes to explore Mars led to a new perspective on the probability of life: technical constraints necessitated a pragmatic stance, which acknowledged the blind spots of every seemingly universal definition. As the example of Jacques Monod shows, this led to a self-reflexive turn in which crystals were no more alien than humans themselves.

I Crystalline aliens enter the scene

In 1888, the prolific French author Joseph Henri Böex (1856–1940) published the short story *Les Xipéhuz* (The Xipéhuz) under his pen-name J.H. Rosny. The tale was set in prehistoric Babylon at the dawn of mankind, in which a tribe of nomads encounters an assembly of strange forms: cones, cylinders and bark-like layers.[11] These immediately attack the humans, causing many casualties by what seems to be some kind of telepathic force. From the beginning on, Rosny leaves no doubt that the strange geometrical forms are intelligent creatures: they are described as showing a change of colors when they first perceive the tribe, and their attack is depicted as ordered and planned. When later on the main protagonist, Bakhoûn, conducts an ethnological field study to find a way to destroy the Xipéhuz, he recognizes communication among them, different traits of character, expression of feelings and other unmistakable signs of intelligence. However, it is also clear from the beginning that these beings constitute an absolute enemy, an adversary that has to be fought to the death. Rosny stages the encounter between man and alien as a turning point in the history of the planet: would it belong to the forms, or to mankind? In his story, there is no possibility of means of understanding or cohabitation between such completely different life forms, and he makes the future of mankind dependent on whether the humans manage to extinguish all the Xipéhuz. In his later story *La Mort de la terre* (The Death of the Earth), published in 1910, Rosny once more returned to this topic. This time, we witness the demise of mankind in the far future, when the last human enclaves run out of water and are destroyed by violent earthquakes. But already, a new form of life is spreading, the 'ferromagnétaux' – mineral beings whose complex crystal-like structures stretch out over the deserted plains: 'For human eyes, the Earth was horribly dead. Yet the other life already prospered there, for this was its time of genesis.'[12] Like the Xipéhuz, the ferromagnetics are natural enemies of humans, killing them by deprivation of red blood cells. Again, Rosny shows us a turning point of the history of earth; but this time, it is mankind that is doomed to perish.

Before the last man sacrifices himself to the ferromagnetics at the end of the novel, he has a dream in which he watches the whole process of evolution: from the moment of life's beginning in the oceans to the conquest of land by reptiles and insects and finally to the development of mammals, a grand narrative unfolds before his inner eye and which culminates in the rise of mankind as the dominating species. In this short paragraph, Rosny depicts evolution as a continuous chain of beings, a process whose coherence is guaranteed by heredity transmission. And this process ends with the last man, while a completely different process starts: 'la Vie Nouvelle,' the new life of beings based on mineral and metals instead of carbon composites. This final scene of the story works for Rosny as his key argument: life is not to be reduced to the historical process of evolution. Rather, evolution as we know it is only one possible manifestation of life, while the Xipéhuz and the ferromagnetics together represent altogether different manifestations.

This amounts to a radical diminution of the place of man in the universe. While many nineteenth-century recipients of Darwin still considered man to be the apex of evolution, Rosny offers the possibility of an evolution without an end, where man is no more than an episode in the ongoing drama of life. Mineral life forms present him with the opportunity to imagine life beyond the constraints of a linear process of development, be it refinement or degeneration, as described by Lasswitz, Wells and others. Life is not restricted by the limits of plasticity in organic, carbon-based matter; it is something mystical, a force pervading the universe, capable of animating even minerals. However, life also means strife, and the rise of one form means the fall of another. Rosny uses the Xipéhuz and ferromagnetics to show that life as we know it may come to an end, but that life itself will always persist. This mystical vision points to a concept of life that is defined by its transcendental nature: life is a unifying power pervading the universe and tangible only in the sometimes fantastic and bizarre forms it creates.[13]

The creatures described in the two stories by Rosny represent the first appearances of crystalline aliens. Their coming into being is clearly situated in the context of the scientific romance, a form of imagination explicitly controlled by contemporary scientific theories. Rosny draws on Darwinism for the framework that drives the narrative: the struggle for survival between two different species.[14] Furthermore, he uses the hero of *Les Xipéhuz* to convey a particular perspective on nature. Bakhoûn embodies the values of the age of positivism: by relying on observations and field studies instead of traditional lore and superstition, he is able to develop a strategy against the mineral entities. Even the greatest mysteries of life can be analyzed by scientific reason and experiment. By inventing the crystalline alien, Rosny at the same time invents the concept of exobiology as the science of alien life forms. His story contains not only a description of such creatures but also a methodology for research on them. Other authors of the time did not bother to include a depiction of the process of gathering knowledge on aliens: while Lasswitz's Martians are similar enough to explain themselves, the spiritual entities of Camille Flammarion are beyond the limited reasoning of human beings, and the vastly superior invaders of H.G. Wells cause nothing but shock and awe, leaving the protagonists of the novel no opportunity for research. By contrast, Rosny explicitly integrates a self-reflexive level, which enables him to reconcile science and imagination: nothing the imagination can conjure is beyond the grasp of scientific reason, not even mineral-based life. This, in turn, opens a discursive field where scientific speculations about 'weird life,' and especially crystalline life, can take place.[15]

II Speculating about another basis for life

In his stories, Rosny showed no interest in speculations about the actual chemical structures that would make mineral life possible. Scientists, however, had been discussing that question since the end of the nineteenth century. Perhaps the first one to ponder the possibility of life based on mineral compounds was the physiologist William T. Preyer (1841–97). The immediate context of his contribution was the debate about spontaneous generation: Can living beings originate from matter, or is there an unbridgeable gap between life and matter? At that time, the discussion was shaped by the protoplasmic theory of life. This theory argued that the content of cells, a substance called protoplasm, forms 'the physical basis of life,' as Thomas Huxley had stated.[16] It was this concept that, as I want to show, presents us with the origin and driving force of the scientific discussions about mineral life.

In his 1877 text on spontaneous generation, Preyer questioned established thinking about the origin of life by stating that life could only be generated by life itself: 'omne vivum e vivo.' With this phrase, he alluded to the famous statement by Rudolf Virchow, 'omnis cellula e cellula' (every cell from a cell). For Virchow, the cell constituted the smallest living entity; life was no further reducible than to this level.[17] This made life a phenomenon that was intrinsically tied to a specific substance: protoplasm.

Since the 1860s protoplasm had become a testing ground for the mechanist interpretation of life.[18] In the mid-nineteenth century, it had been identified as the substance common to the cells of both plants and animals, and usually it was held to be the substratum of all vital activities. In a famous talk in 1868, Thomas Huxley emphasized this view, but gave it a radical twist as he maintained that protoplasm was composed of purely chemical elements and that the so-called vital forces were nothing more than 'molecular forces.'[19] This was meant as a rebuttal of the vitalistic position, which claimed that the phenomenon of life was brought about by some 'vital force' not reducible to chemical or physical forces and independent from the matter through which it acted. For Huxley, life was nothing immaterial, but the effect of a certain chemical compound consisting, as could be shown by analysis, of the elements carbon, hydrogen, oxygen and nitrogen. The presence of protoplasm, a material substance, had become the defining criterion of life.

Preyer elaborated on this view and expanded it. According to him, experiments on spontaneous generation had always presupposed that life as we know it – and therefore protoplasm as we know it – was the only possible form living entities could take. But what if, in the history of earth, there had been precursors to that substance, other substances that were chemically different but which also exhibited vital activities? One candidate for such a substance was silicon: 'Who knows if after substituting a part of the carbon in the protoplasm, for example by silicon, and a part of the hydrogen by metals, one might not obtain another protoplasm, another which had existed and lived?'[20]

By assuming that other compounds might also be able to sustain life, Preyer detached the phenomenon of life from the narrow material basis Huxley had given it while, at the same time, trying to avoid the notions of a downright vitalism. No one singular compound should be taken to be the material basis of life, but life, understood as a 'complex of certain phenomena of motion which are highly dependent on temperature,' could appear as a result of different configurations of matter.[21] In this context, the word 'protoplasm' had acquired a new meaning: it no longer referred to a substance that could be identified by its chemical composition, but to any matter that sustained vital processes. Life was capable of materializing in different kinds of protoplasm – like one made up of silicon and metals. Preyer concluded that everything lives or has lived; inorganic substances are only the dead remains of once living processes. Life does not at all times adhere only to animals and plants, there might have existed other material carriers before.[22]

For some researchers, crystals were promising candidates for life. A German doctor residing in Naples, Otto von Schroen (1837–1907) made such a claim at the turn of the century after carefully examining their growth, arguing that crystals are organized beings subject to a continuous development like plants and animals, and that they have their own distinct biology and pathology. Like Preyer, Schroen also departed from Huxley's understanding of protoplasm, inventing the name 'bioplasm' to refer to any 'matter that constitutes living beings.'[23] Schroen distinguished between different kinds of bioplasms, each of which sustains a different form of life: phitoplasm for plants, zooplasm for animals, anthropoplasm for humans and petroplasm for minerals. Minerals are attributed with different properties of living beings, like movement, reproduction or struggle for survival, even reaction to external stimuli. Schroen, in contrast to Preyer, invoked the actions of a 'vital force' to account for the vivification of matter, thereby showing that speculations about mineral life could be adopted to support any philosophical position: vitalism, hylozoism and mechanism.

In 1893, the chemist James Emerson Reynolds (1844–1920) addressed the British Association for the Advancement of Science with a lecture on silicon. By comparing the activities of this element with carbon, he recognized 'remarkably close analogies.'[24] Both elements easily form compounds, and some of the compounds show structural similarities. Of course, silicon can perform these tasks only at very high temperatures; however, Reynolds argued, it may be possible that in the early stages of earth's history, silicon was the dominant element, constituting 'Nature's earliest efforts in building compounds similar to those suited for the purpose of organic development.'[25] It was H.G. Wells who pursued this scientific idea by means of fiction. In 1894, he published an essay entitled 'Another Basis for Life,' where, after explaining the findings of Reynolds, he let his imagination run freely: 'One is startled towards phantastic imaginings by such a suggestion: visions of silicon-aluminum organisms – why not silicon-aluminum men at once? – wandering through an atmosphere of gaseous sulphur, let us say, by the shores of a sea of liquid iron some thousand degrees or so above the temperature of a blast furnace.'[26] Even though he mentioned an implication of Reynolds hypothesis, namely the possibility of extraterrestrial silicon life, Wells never elaborated on this idea in his later writings, and it is interesting to note that even his silicon organisms bear resemblance to men.

As another example shows, biological speculation could employ similar, if not more daring, imaginative strategies. In a book that summarized evidence for an analogy between the process of crystallization and the formation of basic organisms, the Austrian neurologist Moriz Benedikt (1835–1920) advanced some thoughts on the possibilities of life on other celestial bodies.[27] On the sun, he reasoned, organisms could exist in a searing-hot and liquid state. Their brain cells could consist of quartz crystals able to receive thoughts by Hertzian waves. This idea was, of course, inspired by contemporary radio receivers that used crystals like iron disulfide as detectors to pick up radio transmission. However, the idea of crystalline aliens is also connected to a far older analogy between crystals and living beings. This analogy dates back at least to the seventeenth century, but, as I will show in the next section, it was at the end of the nineteenth century that it acquired a new meaning in biology. In this context, the question for mineral life ceased to be solely a field for speculation and began to be approached experimentally. Crystalline life was no longer limited to theoretical deliberation and literary fiction: it was actually brought to life in the laboratory.

III Living crystals

In his book, Moriz Benedikt refers to the work of scientists who tried to actually create life-like artifacts from inorganic matter. Such a constructionist approach to understanding life came into being during the first years of the twentieth century under the name of synthetic biology.[28] The aim of researchers in this field was to reproduce basic functions of life by artificial means, thereby showing how chemical or physical processes such as osmosis or crystal growth could explain life. Especially in developmental biology, the self-organizing growth of crystals and their highly structured nature made them attractive models for explaining the coming into being of organisms without having to invoke special vital properties or forces.[29] Synthetic biology was a heterogeneous field of research, and the interpretations of the experimental results were highly contested. While only a few researchers claimed, like Schroen, that their creations were actual living beings, most argued, nonetheless, that they were more than mere models. For protagonists such as Hans Przibram (1874–1944), Stéphane Leduc (1853–1939) or Otto Lehmann (1855–1922), the objects they created had a special ontological status as they occupied a position in-between dead matter and organisms.[30] For example, the plant-like structures produced by Leduc, the liquid crystals discovered by Lehmann (Figure 4.1) or the regenerating crystals of Przibram were held to exhibit several properties of living beings. To lend weight to their arguments, experimenters usually used a peculiar literary strategy: phenomena observed in, say, crystals of mecon acid were described in terms of biological phenomena like ‘poisoning,’ even if the growth of the crystals was only disturbed by another substance, such as aniline dye.[31]

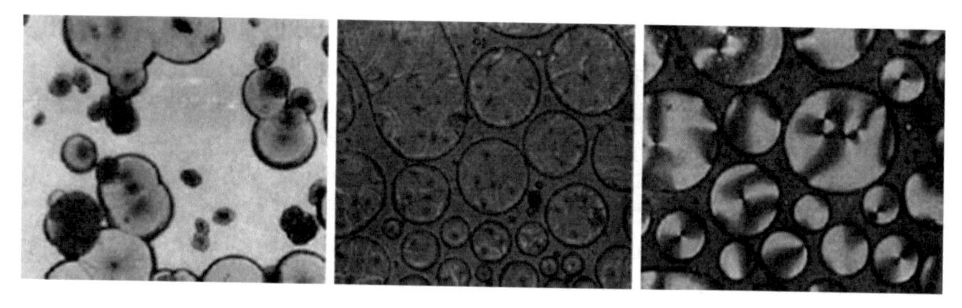

Figure 4.1 Otto Lehmann (1855–1922) argued that liquid crystals like these exhibit certain properties of life.
Source: Ernst Haeckel, *Kristallseelen: Studien über das anorganische Leben*, Leipzig: Kröner, 1917, 32–3.

This wording had a twofold effect: Firstly, by applying the notions customary for the description of vital activities to the processes observed in inorganic substances, it rendered inorganic structures similar to organisms and blurred the borders between them. Secondly, it redefined the notions themselves, depleting them of their established meaning.[32] As exponents of synthetic biology were eager to show that no absolute dividing line between the realms of the living and the unenlivened (inanimate) could be drawn, the zone of indeterminateness opened up by such descriptions was thoroughly intended. Synthetic biology undermined traditional definitions of life. Adversaries and skeptics who were not convinced that crystals and similar structures could justly be called living were forced to bring forward new definitions, thereby exhibiting their own perspective-dependency and putting presuppositions up for discussion.[33] The question 'What is life?' was no longer an abstract one. Instead, it had to be answered in face of the new entities produced in the laboratory.

Science fiction was ready to take up the challenge and incorporate the self-reflective stance of synthetic biology into its repertoire.[34] One example is a novel by the German author Annie Francé-Harrar (1886–1971). *Der gläserne Regen* (The Glassy Rain) begins with the solitary experiments of Frank Neal, who breeds plant-like silicon structures in his laboratory in the Australian desert. Judging from the description the author gives of these *Silicinen*, they bear a remarkable similarity to the osmotic growths achieved by Stéphane Leduc in the 1910s, and Francé-Harrar's book is full of allusions to liquid crystals, colloids and other favorite subjects of researchers in this area.[35] However, in contrast to Leduc's exemplars, the ones described in the novel begin to show signs of life when exposed to the light of the moon: they move, glow and emit sound. However, these creatures, which are classified by another protagonist as 'in-between organic and inorganic life,'[36] soon lose their mystery as Neal is contacted by entities living on the moon. It becomes clear that a higher form of crystal life exists there and that the *Silicinen* are only a preliminary stage.

During the first 250 pages or so of the rather lengthy book, the discourse of Francé-Harrar follows the scientific discussion about an alternative basis of life. Her hero Frank Neal, who is generally depicted as a rational-minded researcher, elaborates on analogies between silicon and carbon, like their abilities to create diverse compounds. However, matters take a new course when, around the world, silicon starts to grow abnormally, and Frank Neal establishes contact with lunar crystalline aliens. These inform him that earth is to be terraformed (or lunaformed) as their own planet is no longer capable of sustaining their lives. In the end, Neal is able to save the earth from such a fate by providing the crystals with an alternative source of energy.

In the course of the novel, Francé-Harrar leaves the scientific discourse behind, and her story becomes increasingly hackneyed and soaked with awkward ideological undertones.[37] The confrontation between Frank Neal and the inhabitants of the moon is depicted as a battle between the cold, rational and emotionless way of the crystals and the feeling nature of human beings: 'the Lunarians [...] are nothing other than the perfect, no, the highest in its self-perfected matter. Matter without spirit and without soul [...].'[38] The crystal aliens are reduced to symbols for perfect, yet limited forms of life, organized in a society grounded solely on functional principles.

While Francé-Harrar misses the chance to employ literary means to probe the limits of discussions about a definition of life, another author was more successful in this direction. In 1934 the American science-fiction writer Stanley Weinbaum (1902–35) published the short story *A Martian Odyssey*, which is still heralded as a landmark for its imaginative rendering of a strange world and its even stranger inhabitants.[39] Presented as the tale of a crash-landed astronaut making his way back to the base, the narrative not only features a friendly, yet inapprehensible bird-like creature called 'Tweel,' but also an even stranger entity that pushed the topic of mineral life onto a new stage. While crossing the Martian desert, the protagonist and his companion Tweel encounter a long row of pyramid-shaped structures made up of silica bricks and continuously increasing in size. By the amount of weathering of the bricks, it is evident that the smallest of these structures are about half a million years old. Understandably, the explorers are quite surprised when they watch 'a nondescript creature' crawling out of the last pyramid: 'body like a big gray cask, arm and a sort of mouth-hole at one end; stiff, pointed tail at the other – and that's all. No other limbs, no eyes, ears, nose – nothing!'[40] The lack of any (recognizable) sensory organ already indicates that they are confronted with a very primitive organism. But what is it? The answer comes fast when the protagonists watch as the creature takes bricks from its mouth-hole, putting them onto the ground: 'The beast was made of silica! There must have been pure silicon in the sand, and it lived on that.'[41] Weinbaum's heroes encounter a mineral entity, a being composed of silicon instead of carbon. But in contrast to Rosny, Weinbaum was not just describing this entity as an altogether different form of life. Instead, he followed the approach of synthetic biology and addressed the conditions for posing the question of life itself: 'there the thing was, alive and yet not alive.'[42] This is especially clear in his short story *The Red Peri*, where the protagonists encounter 'crystal crawlers' on Pluto, entities that are assigned a status on the 'borderline' between the living and the non-living.[43] Weinbaum's depiction of silicon aliens was not so much concerned with ascribing life to such strange beings and thereby just expanding the realm of life to include any kind of matter. Rather, he entered an epistemological discussion about what life is and how we can approach this question at all.

Weinbaum had a background in science as he had graduated in chemical engineering at the University of Wisconsin, and the arguments his protagonists bring forward resemble the arguments that were raised by scientists working in the field of synthetic biology. When discussing the status of the 'crystal crawlers,' the old professor in *The Red Peri* states: 'Alive? I don't know. Crystals are as close as inorganic matter comes to life. They feed; they grow.'[44] And he goes on to lecture on the criteria of life, enumerating the different arguments and proving that each of them is ultimately inconsistent. Movement, growth, reproduction are disqualified, as they are also properties of fire and crystals.[45] He acknowledges that irritability and adaptation may be unique to life, thereby identifying the crystal crawlers as living, as they show both of these properties. This, however, does not settle the issue. Later in the story the space pirate known as the Red Peri, who knows these creatures having lived on the planet since childhood, suggests that they are '[n]ot exactly alive. They're – well – on the borderline. They're chemical-crystalline growths, and their movement is purely mechanical.'[46] But this is nothing more than a phenomenological description and does not provide a basis on which one could start to classify these entities.

Weinbaum used his crystal aliens to mix up the established order of nature. They provide him with an opportunity to enter into a self-reflexive discussion about the possibility of an unequivocal definition of life. However, his conclusion is discouraging. There is no way to arrive at such a definition, as one can always find one example that contradicts it. The crystal aliens serve a negative function: they are neither machine nor animal, but they also do not constitute anything above this classification, some unifying principle that would allow subsuming both kinds of beings. They are, as it were, epistemological hurdles, warning us lest we should have final, all-too ready answers about the nature of life.

IV The cybernetic view on silicon life

One way out of this impasse concerning the definition of life was offered by the self-proclaimed universal science of cybernetics.[47] Norbert Wiener (1894–1964), one of the main proponents of a cybernetic world view, propagated a view of life focusing on the notion of information. He wanted to do away with the quarrels between vitalism and mechanism: the whole controversy should be relegated to the 'limbo of badly posed questions.'[48] As the title of his 1948 manifesto, *Cybernetics or Control and Communication in the Animal and the Machine*, already indicates, concepts like 'feedback mechanism' served to undermine the differences between organisms and machines.

In this context, crystals acquired a special status because they had the property of self-reproduction, a phenomenon that became central to the cybernetic approach to biology. Since the 1960s, artificial self-reproducing automata undermined the distinction between the organic and the inorganic as well as between the living and the non-living in a new way: by desisting from ontological quarrels and concentrating solely on functional issues.[49] In this context, crystal analogies were widely employed. The pioneer of computing, John von Neumann, developed his cellular model of self-reproducing automata after the mathematician Stanisław Ulam (1909–84) drew his attention to the structural features of crystal lattices.[50] Lionel Penrose (1898–1972), an English geneticist who experimented with mechanical self-reproducing devices, referred to them as 'crystals,' and in the 1964 novel by Stanisław Lem (1921–2006), *Niezwyciężony* (The Invincible), the insect-like self-reproducing machines threatening the crew of a spaceship are described as having a crystalline structure.[51]

It is interesting to note that the idea of crystalline aliens based on cybernetics seemed to be more fascinating for science-fiction writers from communist countries. Not only the Pole Lem, but also the Russian Aleksandr Meyerov featured them in a novel.[52] Cybernetics had won ground in Soviet science since the late 1950s, and during the 1960s, the classics of the field became available in Russian translations.[53] Soviet scientists seem to have been particularly fascinated by the idea of artificial organisms and proposed the building of automata that would display no essential difference to living creatures. As the mathematician Andrey Kolmogorov (1903–87) said in a public lecture in April 1961: 'If such qualities of a material system as "being alive" or "capable of thinking" are defined in a purely functional way [...], then one would have to admit that in principle living and thinking beings can be *created artificially*.'[54] Researchers in this field were confident that life could be analyzed in its entirety with methods employing information theory and the theory of feedback mechanisms. Life, therefore, was no longer a property of a special kind of substance or chemical compound. Rather, whether something could be considered to be alive or not depended on the organization and function of its elements.

In 1968 the Soviet chemist and writer Aleksandr Meyerov developed a consistent cybernetic view on crystalline aliens in his novel *Sirenevyj kristall* (The Lilac Crystal).[55] The book starts with an archaeological mystery: at the South Seas archipelago Pautoo, an expedition finds the remains of a civilization that seems to have controlled a force enabling it to turn different materials into stone. This, they learn, was due to the priests' ability to control silicon plasma extracted from the remains of a crashed meteorite. The scientists, with help from the Soviet Academy of Sciences and despite problems caused by the rival efforts of an unscrupulous capitalist, succeed in activating the plasma, which pours out as an amorphous mass of mineral substance turning everything it touches into stone. But this substance is only the first step in the evolution of mineral life they are about to witness. The second step consists of mineral seeds which, when activated, develop into gigantic tortoises (called 'Rodbarids' after their discoverer). After a while, these activate 'Jusgorids,' flying cucumber-shaped creatures that display several traits of higher intelligence. Both of these strange mineral creatures are significantly different from the ones we have encountered so far: neither are they so alien as to be the ultimate enemy of humankind, as Rosny's Xipéhuz, nor do they inhabit a zone in-between the living and the inanimate, as the crystal crawlers of Weinbaum.

Instead, in *The Lilac Crystal*, silicon life forms have acquired a new function. The scientists do not ask ontological questions. Their prime issue is not what these organisms actually are; rather, they want to know what they do and what their purpose is. After the establishment of a sort of natural preserve and an observation post, the scientists watch the Rodbarids rooting up the ground to mine minerals. They use their booty to nourish one special individual that finally brings forth the Jusgorids. When discussing these activities, the researchers identify the tortoise-like creatures as workers 'blindly obeying a higher order, or perhaps some stored program or an old instinct.' Later on they specify this judgment by referring to them as 'alien cybernetic entities.'[56] The mineral aliens have become materializations of a cybernetic view of life. They constitute a class of beings exhibiting the properties common to men, animals and machines: the functional logics of communication and control. After some initial perplexity, their way of thinking becomes intelligible: Under the guidance of the lilac crystal, which turns out to be a program storage unit, they first build a dish antenna to communicate with their home planet and then a spaceship in which they leave earth, inviting the humans to follow in another rocket.

Meyerov's novel evades the self-reflexive stance of Weinbaum. His depiction of the exobiological research on the silicon life forms is firmly grounded in a confidence in the cybernetic world view. His protagonists are not interested in what it means to ascribe life to these entities. Refraining from any ontological question, they are only concerned if the aliens exhibit purposeful behavior, which is enough to establish a relation with them. Crystalline aliens are no longer just objects that confront us with fundamental questions about the nature of life. Now they have become subjects and possible partners for interaction. For Meyerov, the silicon life forms serve to convey a message: even if they appear to be very different from what we encounter on earth, the aliens have acquired a peaceful, well-organized society capable of carrying out technological feats still out of reach for humans. But the foundations these achievements rest upon are comprehensible for any intelligent being, as they consist of the universal rules of logic worked out by cybernetics; therefore, humans can take them as models to improve their own society.[57]

Using silicon life as a means to deconstruct human prejudices and exhibit the universality of logical thought was not the sole realm of Soviet science fiction. The 1967 episode of the TV series *Star Trek*, *The Devil in the Dark*, features a silicon life form calling itself the 'Horta' threatening workers at a distant mining outpost. However, its behavior is shown to be fully rational

Figure 4.2 Mr. Spock communicating with the silicon-based life form Horta.
Source: *Star Trek* episode *The Devil in the Dark*, USA 1967.

after Mr. Spock establishes communication and finds out that the creature is only trying to protect its eggs (Figure 4.2). Consequentially, the humans understand its motives and promise to respect them. Silicon life is no longer something special: the Horta may look strange, but it embraces the universal rules of thinking based on logic – or as Spock states: 'The Horta has a very logical mind. And after close association with humans, I found that curiously refreshing.' Mineral aliens finally have become our siblings.

One final example from science-fiction literature may further stress this point. In 1984 Alan Dean Foster (1946–) published the novel *Sentenced to Prism*. The storyline revolves around the technician Evan Orgell, who is ordered to inspect a research outpost on the mysterious planet Prism. He discovers a world full of silicon beings, making friends with an intelligent species living in a functionally differentiated and highly specialized society. Foster takes some time describing the environment of Prism, making every effort to conjure up the vision of a very strange world. For a short moment, his protagonist even poses the, by now familiar, question about the status of its inhabitants: 'You couldn't even say that such an automaton was alive, in the normal sense of the term.'[58] However, as soon as one of the creatures establishes communication via a sort of telepathic link, Azure, as the caterpillar-like being is called, turns out to be quite a chatty fellow. The author shows the disintegration of prejudices in the face of the alien, who is no longer an absolutely foreign being but shares fundamental characteristics with his human companion; foremost, in good cybernetic tradition, the ability to communicate. Foster's novel shows, despite its rhetoric of otherness, how silicon beings have become an unproblematic inventory for conventional science-fiction storytelling. Everything gets explained: the ecology of Prism as well as the abilities of its intelligent inhabitants and their society.[59] The latter topic even allows Foster to insert a political allegory: when on his search for a lost member of the research outpost, the hero encounters a rare organic being that shows intelligence, but immediately attacks Orgell and engulfs him in its structure. By telepathy it informs the startled hero that it is the 'Integrator' whose aim is to incorporate as many living beings as possible. The debate that follows is clearly political. Orgell, being reminded of 'a common cry to many would-be tyrants and dictators stretching far back into the depths of human history,' accuses the Integrator of being unorganized and muzzling all individual freedom.[60] The answer Foster puts in the creature's mouth reminds one of totalitarian governments: 'There can be no individuality within a true Associative.'[61] In contrast to Francé-Harrar, who depicted the crystal society as a rational, emotionless and static order, Foster represents the very antithesis. Under the paradigm of cybernetics, the functionally structured society of the silicates is the equivalent of an efficient but free society, while the organic 'Integrator' is an allegory of totalitarianism.

The examples from Meyerov and Foster, as well as from TV shows like *Star Trek*, show that in the age of cybernetics, crystalline aliens were no longer epistemological hurdles. Associated with topics like self-reproduction, universal communication and the efficient organization of society, they were no longer the absolute other but potential actors in a social relationship that included all kinds of strange creatures.

V Recognizing life

This does not mean that they had altogether lost their affiliation to fundamental biological questions. But this time, it was biology itself, which used scenarios from science fiction to give these questions a new form and a hitherto unknown radicalism. Up to the last quarter of the twentieth century even such weird aliens as mineral entities were defined in relation to man, and their strangeness was a function of their distance from our own nature. With the planning of actual missions to search for life on other planets and the birth of exobiology as a distinct science with its own research program, it became increasingly unclear who, in fact, the alien was.[62] This change of perspective entered mainstream biology in a book by the French molecular biologist and Nobel Prize laureate Jacques Monod (1910–76). Monod challenged his readers to imagine a probe launched by a civilization on Mars to look for life on earth. How can such a device be programmed to discern artificial objects from natural ones and living beings from inanimate structures?[63] The criteria that first come to mind, like regularity and repetition, are not sufficient to identify life. If the probe also takes into account the way the examined objects were produced, it will separate artificially produced structures as well as structures made by chance, like rocks, from living beings, as the first are brought forward by the action of external forces, while organisms develop by inner morphogenetic forces. Apart from animals and plants there is, however, another class of objects the probe would also classify as living: crystals.

In Monod's scenario, two distinct discourses come together to form a new perspective on the question of life. One is molecular biology, where crystals have played an important part since the 1940s, when Erwin Schrödinger (1887–1961) associated the material basis of heredity with 'aperiodic crystals': aggregates that are not based on repetition, like normal crystals, but in which every atom plays an individual role.[64] Monod explicitly refers to that term to show that the secret of life rests in the molecular structure of DNA and proteins, and that this structure is sufficient to explain the function of organisms. From this perspective, there is no qualitative difference between crystals and organisms – they only differ in the quantity of information transmitted in the process of reproduction.[65]

While molecular biology provides the content of the scenario and thereby determines the special meaning of crystals, another discourse provides the form of the question: exobiology. Monod's rendering of an automated probe visiting a foreign planet and looking for life was directly derived from the discussions surrounding the missions to other planets planned by NASA since 1960, when the Jet Propulsion Laboratory in Pasadena was commissioned to conduct a study on a spacecraft designed to land on Mars.[66] In the context of space exploration, the question of life had taken on a pragmatic and technical meaning. The decisive issue was to get a machine to identify what was called 'signs of life.' From the 1950s on, several devices were developed to achieve this feat.[67] Their design was accompanied by discussions about reliable criteria for the detection of life. Although NASA did promote theoretical biology, many scientists were aware of the fact that no formal and universally valid definition of life was available. As the microbiologist Joshua Lederberg (1925–2008), who coined the term 'exobiology,' stated already in 1960: 'Our only consensus so far is that such a definition must be arbitrary.'[68] In view of the planned launches of probes to Mars, scientists started to favor pragmatic definitions over abstract ones. That is to say, they looked for criteria that could be implemented in an automated probe. The question 'What is life?' would now have to be solved experimentally. However, an experiment can only answer the question it is designed to answer, and even then its result is seldom unequivocal. Furthermore, in the case of spaceflight, technical and economic considerations have to be taken into account: a probe can only carry a very limited array of instruments and it is operated from afar. Therefore, researchers had to acknowledge that not every theoretically conceivable possibility could be tested, as 'it would seem pointless and very uneconomic to send a space probe to detect a speculative life-form.'[69]

By getting involved in the technicalities of planning for space missions, biologists were confronted with practical constraints that made them conscious of their blind spots. The development of methods for recognizing alien life forced them to lay bare their own presumptions. Therefore, the concepts developed for defining life carried indices, which denoted the specific context of use and the perspective of the generalization in question.

It required just one more step in imaginative reasoning to mirror such a stance back on terrestrial life, like Monod did in his thought experiment about the Martian probe which was unable to distinguish organisms from crystals. Philosopher Carol Cleland recently employed a similar reversal of perspective. In a provocative essay, she argues that there might be alien life forms on earth we were unable to recognize as such up to now, because they do not fit into the paradigms that determine our concept of life. Drawing attention to phenomena like desert varnish, a mysterious mineral coating found on rocks (Figure 4.3), she shows how the discourse on possible forms of life is framed by our own preconceptions: 'The upshot is that intelligent, silicon-based life from a very unearth-like environment might well draw an analogous conclusion about the possibilities for carbon-based life, namely that carbon isn't capable of forming macromolecules with the requisite degree of

Figure 4.3 Desert varnish seen through an electron microscope.
Source: Courtesy of M. Spilde and P. Boston, UNM Scanning Electron Microscope Images.

complexity, stability, and versatility.'[70] This prosopopoeia of silicon life confronts us not only with a mirror image that destabilizes the apparently firm ground of our own existence, but also exposes us to the contingencies that lie at the heart of every definition of life.

VI Conclusion

This voyage along the strange paths of crystalline aliens through science and fiction has shown how imaginations about inorganic life forms led to fundamental questions concerning criteria for defining and recognizing life, as well as to an insight into the irreducible perspective-dependency of such questions. Furthermore, two more general conclusions can be drawn. The first concerns the role of the imagination. Imagination has always played an important part in science, and during the nineteenth and twentieth century numerous scientists and philosophers have argued for the importance of speculative reasoning and thought experiments. The invention of scientific romance and science fiction as a literary genre has provided a cultural place for appropriating and probing the limits of scientific theories, a place that was often also used by scientists.[71] Cases such as the concept of extraterrestrial life show that science and fiction sometimes blended, exchanging knowledge and techniques to create hypothetical forms of alien life. In the course of the twentieth century, such imaginings of the alien have had a profound impact on the imagination of the public and thereby shaped the stance toward the endeavor of space exploration.[72] A historical and epistemological investigation of crystalline aliens therefore contributes to an archaeology of the relationship between science and fiction.

A second general point to be made is the influence of the material culture of science upon the processes of imagination. The experimental stance of synthetic biology at the beginning as well as the technicalities of space travel at the second half of the century have determined the way speculations about life were approached. The experiments of Lehmann, Leduc and others forced scientists to acknowledge the impossibility of a clear demarcating line between living and inanimate matter. The designs of devices to be carried on automated probes have led to a modest and pragmatic approach towards the scope of definitions of life. Scientists had to choose between many different possible tests for the detection of life, leaving out those that were impracticable or too expensive and thereby admit to their own perspective-dependency. Although the question 'What is life?' seems to be an ahistorical one, and answers advanced often claim to be universally valid, the history of imaginary alien life shows that they, in fact, are determined by technological possibilities as well as by constraints imposed by the cultural context. The case of crystalline aliens shows how the confrontation of imaginations with experiments and technologies can lead to a self-reflexive stance, making visible blind spots of our own presuppositions and taking into account the irreducible perspective-dependency of our definitions of life.

Notes

1. Steven J. Dick, *The Biological Universe: The Twentieth-Century Extraterrestrial Life Debate and the Limits of Science*, Cambridge: Cambridge University Press, 1996, 223–38; Voltaire, *Le Micromégas: Avec une histoire des croisades et un nouveau plan de l'histoire de l'esprit humain*, Berlin 1753.
2. Kurd Lasswitz, *Auf zwei Planeten*, Weimar: Treber, 1897; H.G. Wells, *The War of the Worlds*, Leipzig: Tauchnitz, 1898.
3. Dick, *Biological Universe*, 233.
4. Quoted in Philip J. Pauly, *Controlling Life: Jacques Loeb and the Engineering Ideal in Biology*, New York: Oxford University Press, 1987, 4.
5. Jacques Loeb, *The Mechanistic Conception of Life*, Cambridge, MA: Harvard University Press, 1964, 27.
6. See Hannah Landecker, *Culturing Life: How Cells Became Technologies*, Cambridge, MA: Harvard University Press, 2007, 9.
7. Loeb experimented with the substitution of one organ for another (heteromorphosis) in the hydroid species Antennularia; see Loeb, *Mechanistic Conception*, 80.
8. Wells, *War of the Worlds*, 199.
9. See Gerald L. Geison, 'The Protoplasmic Theory of Life and the Vitalist-Mechanist Debate,' *Isis* 60.3 (Fall 1969), 273–92.
10. Stefan Helmreich, 'The Signature of Life: Designing the Astrobiological Imagination,' *Grey Room* 23 (Spring 2006), 66–95.
11. J.H. Rosny Aîné, 'Les Xipéhuz,' in *La Mort de la terre*, Paris: Denoel, 1958, 11–60, here 16.
12. Ibid., 164. Unless otherwise stated, all translations are mine.
13. See Jean-Pierre Vernier, 'The SF of J.-H. Rosny the Elder,' *Science Fiction Studies* 2.2 (July 1975), 156–63.
14. See Jean-Marc Gouanvic, *La Science-fiction française au XXe siècle (1900–1968): Essai de socio-poétique d'un genre en émergence*, Amsterdam: Rodopi, 1994, 49–51.
15. For the concept of 'weird life,' see Debbora Battaglia's contribution, Chapter 11 in this volume.
16. Thomas Henry Huxley, 'On the Physical Basis of Life,' *Fortnightly Review* 11 (February 1869), 129–45.
17. William T. Preyer, 'Kritisches über die Urzeugung,' *Kosmos* 1 (1877), 377–87, here 382. For Preyer and the debate on spontaneous generation, see John Farley, *The Spontaneous Generation Controversy from Descartes to Oparin*, Baltimore: Johns Hopkins University Press, 1977, 152; for Virchow, see William Coleman, *Biology in the Nineteenth Century: Problems of Form, Function, and Transformation*, Cambridge: Cambridge University Press, 1977, 32–3.
18. Geison, 'Protoplasmic Theory.'
19. Huxley, 'On the Physical Basis of Life,' 140.
20. Preyer, 'Kritisches,' 386.
21. Ibid., 382.
22. Ibid., 387.
23. I am drawing on an article published in 1904 by Brazza and Pirenne; however, Schroen had been working on this theory for quite some time. F. di Brazza and P. Pirenne, 'La vie dans les cristaux,' *Revue scientifique* 1 (23 April 1904), 518–23, here 518.

24. James Emerson Reynolds, 'Opening Address,' *Nature* 48 (1893), 477–81, here 479.
25. Ibid., 481.
26. H.G. Wells, 'Another Basis for Life,' in *Early Writings in Science and Science Fiction*, Berkeley: University of California Press, 1975, 144–7, here 146.
27. Moriz Benedikt, *Krystallisation und Morphogenesis: Biomechanische Studie*, Vienna: Perles, 1904, 65.
28. For the claims and discussions surrounding synthetic biology, see Evelyn Fox Keller, *Making Sense of Life: Explaining Biological Development with Models, Metaphors and Machines*, Cambridge, MA: Harvard University Press, 2003, 15–49. This approach was continued during the 1950s and 1960s in origin of life studies; see Steven J. Dick and James E. Strick, *The Living Universe: NASA and the Development of Astrobiology*, New Brunswick: Rutgers University Press, 2005, 71–2.
29. See J. Lorch, 'The Charisma of Crystals in Biology,' in Yehuda Elkana, ed., *The Interaction Between Science and Philosophy*, Atlantic Highlands: Humanities Press, 1974, 445–61; Donna Haraway, *Crystals, Fabrics, and Fields: Metaphors of Organicism in Twentieth-Century Developmental Biology*, New Haven: Yale University Press, 1976.
30. Hans Przibram, 'Kristall-Analogien zur Entwicklungsmechanik der Organismen,' *Archiv für Entwicklungsmechanik der Organismen* 22.1 (1906), 207–87; Stéphane Leduc, *La Biologie synthétique, étude de biophysique*, Paris: A. Poinat, 1912; Otto Lehmann, *Flüssige Kristalle und die Theorien des Lebens*, Leipzig: Barth, 1906.
31. For this example, ibid., 12–13.
32. An especially avid user of this strategy was Ernst Haeckel; see his book *Kristallseelen: Studien über das anorganische Leben*, Leipzig: Kröner, 1917.
33. See, for example, Wilhelm Roux, 'Die angebliche künstliche Erzeugung lebender Wesen,' *Die Umschau* 10 (1906), 141–5.
34. At least once, synthetic biology also turns up in 'serious' literature. The 1947 novel *Doctor Faustus* by Thomas Mann features a short discussion about laboratory-made structures that are clearly drawn from the works of Ludwig Rhumbler and Stéphane Leduc. See Thomas Mann, *Doktor Faustus: Das Leben des deutschen Tonsetzers Adrian Leverkühn, erzählt von einem Freunde*, Frankfurt am Main: Fischer, 2003, 27–9.
35. Annie Francé-Harrar, *Der gläserne Regen*, Hamburg: Toth Verlag, 1948, 27 (Silicinen), 28 (liquid crystal), 176 (colloids).
36. Ibid., 204.
37. For example, the lunar crystal society is organized hierarchically by some kind of racial difference, the white crystals being the leaders and the black ones forming the lowest kind. Ibid., 378.
38. Ibid., 450.
39. Stanley G. Weinbaum, 'A Martian Odyssey,' in idem, ed., *Interplanetary Odysseys*, London: Leonaur, 2006, 7–31.
40. Weinbaum, 'Martian Odyssey,' 20.
41. Ibid.
42. Ibid., 21.
43. Stanley G. Weinbaum, 'The Red Peri,' in idem, *Interplanetary Odysseys*, 164–213, here 190.

44. Ibid., 170.
45. Fire had been used as an analogy to life up to the nineteenth century.
46. Weinbaum, 'Red Peri,' 190.
47. For an overview on cybernetics and its claims, see Claus Pias, 'Zeit der Kybernetik – Eine Einstimmung,' in idem, ed., *Cybernetics: The Macy-Conferences 1946–1953*, vol. 2: *Essays und Dokumente*, Zurich: Diaphanes, 2004, 9–41.
48. Norbert Wiener, *Cybernetics or Control and Communication in the Animal and the Machine*, Cambridge, MA: MIT Press, 1994, 44 (first English edition 1948, first Russian edition 1958).
49. See Andrew Pickering, 'A Gallery of Monsters: Cybernetics and Self-organisation, 1940–1970,' in Stefano Franchi and Güven Güzeldere, eds, *Mechanical Bodies, Computational Minds: Artificial Intelligence from Automata to Cyborgs*, Cambridge, MA: MIT Press, 2005, 229–45.
50. Stanisław Ulam, *Adventures of a Mathematician*, Berkeley: University of California Press, 1991, 241.
51. Lily E. Kay, *Who Wrote the Book of Life? A History of the Genetic Code*, Stanford: Stanford University Press, 2000, 112; Stanisław Lem, *The Invincible*, Harmondsworth: Penguin, 1976.
52. One reason for the literary interest in crystal life may have been a genuine literary tradition. For example, the novella 'The Ethereal Trail' by Andrey Platonov elaborates on the theory that electrons are actual living beings, and that minerals can be grown in the earth. While it was written already in 1927, it was published only in 1968. See Andrey Platonov, 'Efirnyi trakt,' in *Fantastika 1927: Vypusk i-yi*, Moscow: Molodaya gvardyia, 1968, 247–302.
53. Slava Gerovitch, *From Newspeak to Cyberspeak: A History of Soviet Cybernetics*, Cambridge, MA: MIT Press, 2002, 199.
54. Ibid., 215.
55. As I do not read Russian and as there is no English translation, I will refer to the German edition: Alexander Mejerow, 'Der fliederfarbene Kristall,' in *Vetorecht/ Der fliederfarbene Kristall: Zwei phantastische Romane*, Berlin: Volk und Welt, 1979, 5–250.
56. Ibid., 211 and 215.
57. The improvement of society had been an aim of Soviet cybernetics since the early 1960s, see Gerovitch, *From Newspeak to Cyberspeak*, 253–92.
58. Alan Dean Foster, *Sentenced to Prism*, London: Hodder & Stoughton, 1988, 98.
59. The episode *Home Soil* (1988) of *Star Trek: The Next Generation* also uses crystalline aliens to discuss prejudices and environmental issues.
60. Foster, *Sentenced to Prism*, 212.
61. Ibid., 213.
62. For the history of exobiology, see Dick and Strick, *Living Universe*, 30–5, 57.
63. Jacques Monod, *Le Hasard et la nécessité: Essai sur la philosophie naturelle de la biologie moderne*, Paris: Editions du Seuil, 1970, 19–25.
64. Erwin Schrödinger, *What is Life? The Physical Aspect of the Living Cell*, Cambridge: Cambridge University Press, 1948, 61.
65. Monod, *Le Hasard et la nécessité*, 25.
66. Edward Clinton Ezell and Linda Neuman Ezell, *On Mars: Exploration of the Red Planet 1958–1978*, Washington, DC: NASA, 1984, 51–80; also available at http://history.nasa.gov/SP-4212/on-mars.html (accessed 1 October 2017).
67. Dick and Strick, *Living Universe*, 30–5, 57.

68. Joshua Lederberg, 'Exobiology: Approaches to Life Beyond the Earth,' *Science* 132 (12 August 1960), 393–400, here 394.
69. James E. Lovelock, 'A Physical Basis for Life Detection Experiments,' *Science* 207 (7 August 1965), 568–70, here 568.
70. Carol E. Cleland, 'Epistemological Issues in the Study of Microbial Life: Alternative Terran Biospheres?,' *Studies in the History and Philosophy of Biological and Biomedical Sciences* 38.4 (December 2007), 847–61, here 851.
71. See Thomas Macho and Annette Wunschel, eds, *Science & Fiction: Über Gedankenexperimente in Wissenschaft, Philosophie und Literatur*, Frankfurt am Main: Fischer, 2004. Scientists who wrote science fiction include the geneticist J.B.S. Haldane (1892–1964), the mathematician Norbert Wiener (1894–1964) and the astronomer Fred Hoyle (1915–2001); furthermore, many professional science-fiction writers have a background in science, like the above-mentioned Stanley Weinbaum (1902–35).
72. See Debbora Battaglia, ed., *E.T. Culture: Anthropology in Outerspaces*, Durham: Duke University Press, 2005; and Dick, *Biological Universe.*

PART II

Projecting Outer Space

CHAPTER 5

Projecting Landscapes of the Human Mind onto Another World: Changing Faces of an Imaginary Mars

Rainer Eisfeld

Reporter: Is there life on Mars?
Returning Astronaut: Well, you know, it's pretty dead most of the week, but it really swings on Saturday night.

Popular NASA joke

I Deceptive world

For centuries, the planet Mars continued to deceive terrestrial observers like no other celestial body in our solar system. Believing to discern ever more distinct features on Mars through earthbound telescopes, astronomers designated these as continents, oceans, even canals, to which they gave names. With exceedingly rare exceptions, however, these markings did not correspond to geomorphological, or rather areomorphological, structures. Actually, they originated from the different reflectivity of bright and dark surface regions changed in its turn by wind activity which has continued to transport and deposit fine dust across the planet. Space probes, rather than telescopes, were needed to explain these processes and to shed light on the Red Planet's true characteristics.[1]

Until robotic explorers arrived, no other planet seemed to offer such clues for educated guessing – first to the conjectural astronomy of the nineteenth

Rainer Eisfeld (✉)
Universität Osnabrück, Osnabrück, Germany
e-mail: rainer.eisfeld@uni-osnabrueck.de

Alexander C.T. Geppert (ed.), *Imagining Outer Space*
European Astroculture, vol. 1
https://doi.org/10.1057/978-1-349-95339-4_5

century, subsequently to the science fiction of the latter part of that period and the twentieth century. Conjectural astronomy was the term used, in the wake of Bernard de Fontenelle's 1686 *Conversations on the Plurality of Worlds* and Christiaan Huyghens's 1698 *The Celestial Worlds Discover'd*, or Conjectures Concerning the Habitants, Plants and Productions of the Worlds in the Planets, to denote that branch of the discipline which engaged in hypothesizing on 'the living conditions and natural environments of other celestial bodies.' While expected to be not directly contradicting astronomical observations, such suppositions were, to a high degree, matters of interpretation, often based on 'few definitely established and unambiguous data.'[2]

In contrast to the discipline's mathematical branch, conjectural astronomy was intended to bridge the widening rift of mutual incomprehension between the humanities and the sciences. From the seventeenth to the nineteenth century, the encyclopedic outlook on learning, so central to the Enlightenment, included both the spiritual and the material world. Inexorably, however, the progress of scientific research fostered specialization. Conjectural astronomy, in contrast, increasingly resorted to manifest speculation, relegating stellar and planetary astronomy to the role of ancillary sciences in the service of a preconceived, stoutly held idea, based on philosophical considerations: that intelligent life existed throughout the universe, including the solar system's planets.

German astronomers Wilhelm Beer (1797–1850) and Johann Heinrich Mädler's (1794–1874) mid-nineteenth-century assumption that it would 'not be too audacious to consider Mars, also in its physical aspects, as a world very akin to our earth,'[3] went unchallenged in its time. By 1906, however, when American astronomer Percival Lowell (1855–1916) published his spectacular – and highly speculative – interpretation *Mars and its Canals*, scientists were debating issues such as the composition of the Martian atmosphere or the planet's climate controversially and much more skeptically. Within a year, a devastating rebuttal by British biologist Alfred Russell Wallace (1823–1913) appeared under the title *Is Mars Habitable?* Wallace answered the question in the negative: Realistic temperature estimates precluded animal life; low atmospheric pressure would make liquid water – let alone Lowell's supposed irrigation works – impossible. Science and fiction were irrevocably parting ways.

A mere decade after Beer and Mädler had published their treatise on the solar planets' physical properties, the term 'Science Fiction' was introduced in 1851 by British essayist William Wilson in his work *A Little Earnest Book upon a Great Old Subject*. When coining the expression, Wilson referred to a 'pleasant story,' 'interwoven with [...] the revealed truths of Science,' itself 'poetical and true.' By the 1890s, the emerging genre included not merely pleasant, but definitely unedifying tales putting mankind at the mercy of technically superior beings from other celestial bodies. The planetary novel was coming into its own: no longer were planets conceived as self-contained distant places. Rather, their inhabitants might seek out other worlds with either benevolent or inimical intent.[4]

Mars, supposedly older than the earth (according to what was then believed about the formation of the solar system), particularly fired the imagination. Intersecting around the turn of the century, conjectural astronomy and science fiction served as vehicles for succeeding generations to 'project [their] earthly hopes and fears' onto Mars.[5] These pipe dreams and nightmares came to vary, not least according to the economic, social and political upheavals that would figure uppermost in men's minds during successive periods. Two examples:

In the wake of the October Revolution, Soviet author Aleksey Tolstoy (1883–1945) and movie director Yakov Protazanov (1881–1945) imagined during the early 1920s that it would take the arrival by spaceship of a terrestrial revolutionary, Gusev, to whip the exploited workers of Mars into a proletarian uprising against their despotic ruler Tuskub: 'Follow me, Martian Comrades, and organize a society of workers. The Union of the Soviet Socialist Republics of Mars' (*Aelita*, 1924). The 'world' to be revolutionized did not need to be identical with earth.

By the mid-1950s, with female emancipation considered a dire threat in many quarters, a British film portrayed Nyah, *Devil Girl from Mars*, landing her flying saucer by a country tavern, telling the male customers that the birth rate on her home planet had fallen alarmingly after the introduction of matriarchy. For breeding purposes, her planet needed men, she explained. Rather than, in post-Victorian resignation, 'closing their eyes and thinking of Mars,' however, the British males put up embittered resistance.[6]

Looking at the 'mainstream' of the ways in which successive generations of astronomers and science fiction – their treatises, novels, short stories, movie scripts – depicted an imaginary Mars, we may discern a sequence of faces attributed to the planet on which this chapter will subsequently focus: An *Arcadian Mars* (1865 ff.) exhibiting 'all the various kinds of scenery which make our earth so beautiful'; a highly civilized *Advanced Mars* (1895 ff.) crisscrossed by immense canals; a forbidding *Frontier Mars* (1912 ff.) where the rugged adventurer might again come into his own; a *Cold War Mars* (1950 ff.), source of an assault on the earth, or haven for refugees after our planet would have perished from nuclear war; finally, a *Terraformed Mars* (1973 ff.), again with strong frontier undertones, lending itself to human colonization and exploitation. While these 'types' would often overlap – with the frontier metaphor, in particular, persisting into the present – each type set the tone for a generation.

II Arcadian Mars

'Life, youth, love shine on every world [...] This divine fire glows on Mars, it glows on Venus.'[7] With unmatched fervor and elegance of style, Camille Flammarion (1842–1925) argued the case for intelligent extraterrestrial life during the second half of the nineteenth century, bolstered by the authority of the renowned astronomer who, in 1887, founded the Société Astronomique de France. Flammarion's description of the Martian environment, in his very first work *La Pluralité des mondes habités*, is quoted *in extenso* here, because it would inform the astronomical and popular discourse on the Red Planet for nearly a generation:

> The atmospheres of Earth and Mars, the snowfields seasonally expanding and shrinking on both planets, the clouds intermittently floating over their surfaces, the similar apportionment of continents and oceans, the conformities in seasonal variations: all this makes us believe that both worlds are inhabited by beings who physically resemble each other. [...] In our mind's eye, we behold, here and there, intelligent beings, united into nations, vigorously striving for enlightenment and moral betterment.[8]

In 1840, Beer and Mädler had drawn the first chart of Mars. Capital letters denoted observed 'regions' – darker spots on bright ground.[9] The letters used by Beer and Mädler remained in use for two and a half decades, until Richard Proctor replaced them by the names of Mars observers on the map he composed in 1867. Proctor also 'improved' on the way his compatriot John Phillips (1800–74) had, three years earlier, designated darker parts as 'seas' and brighter, reddish tracts as 'lands.' Proctor's chart showed continents and islands, oceans and seas, inlets and straits. These features had a suggestive effect. They seemed to portray a second – albeit smaller – earth, with just a different division into zones of land and water.

The suggestion was deliberate. Proctor depicted Mars as a 'miniature of our earth,' waxing hardly less rhapsodically than Flammarion about the prettiness of the place:

> The mere existence of continents and oceans on Mars proves the action of [...] volcanic eruptions and earthquakes, modeling and remodeling the crust of Mars. Thus there must be mountains and hills, valleys and ravines, watersheds and water-courses. [...] And from the mountain recesses burst forth the refreshing springs which are to feed the Martia[n] brooklets [...].[10]

And in a brilliant phrase, which Percival Lowell would later reclaim for entitling his final book, Proctor called Mars 'the abode of life' – for if no life existed, 'all these things would be wasted.'

Proctor (1837–88) was an Honorary Secretary of the Royal Astronomical Society. Like Flammarion's work, his study of our solar system's planets, subtitled 'under the light of recent scientific researches,' continued to be reprinted until the advent of the twentieth century. Public fascination was spurred further when the Mars opposition of 1877 led to the discovery of two small moons by Asaph Hall (1829–1907) – and to the observation, by Giovanni Schiaparelli, of markings that the Italian astronomer took for *canali*, channels furrowing the planet's surface, some of which he compared to 'the Strait of Malacca, the very oblong lakes of Tanganyika and Nyassa, and the Gulf of California.'[11] After Schiaparelli reported that some of the lines he had sighted between 1877 and 1882 ran for 4,800 kilometers, attaining a width of 120 kilometers, it came as no surprise that Flammarion was among the first to comment: 'One may resist the idea, but the longer one gazes at [Schiaparelli's] drawing, the more the interpretation suggests itself [...] [that] we are dealing with a technological achievement of the planet's inhabitants.'[12]

In the minds of some of the period's foremost astronomers, the image of a lush and youthful Arcadian Mars would soon begin to give way to that of a much more ancient world – possessing no natural water-courses, but rather artificial waterways surpassing anything so far constructed on earth.

III Advanced Mars

As judged by a present-day astronomer, after Giovanni Virginio Schiaparelli (1835–1910) taught a whole generation of observers how to see Mars, it became eventually 'impossible to see it any other way. Expectation created illusion.'[13] If channels discernibly divided Mars to the extent of making its topography 'resemble that of a chessboard,' if several such *canali* even 'form[ed] a complete girdle around the globe of Mars' – could they any longer be interpreted as natural attributes, 'like the rilles of the moon'?[14] Might they not more convincingly be explained as non-natural features, as *canals* serving a purpose which had to be derived from the planet's characteristics?

The landscape of Advanced Mars, which from 1895 was construed by Percival Lowell in response to Schiaparelli's revelations, differed dramatically from that of Arcadian Mars. No more stately oceans, impetuous rivers, swift water courses or refreshing springs. A much grimmer environment predominated on earth's neighbor-world: 'The rose-ochre enchantment is but a mind mirage. [...] Beautiful as the opaline tints of the planet look, [...] they represent a terrible reality [...] [a] vast expense of arid ground [...], girdling the planet completely in circumference, and stretching in places almost from pole to pole.'[15]

Erudite descendant of a wealthy Boston family, excelling in mathematics and literature, composing Latin hexameters at 11 and using his first telescope at 15, Percival Lowell became enthusiastic about Flammarion's impressive compilation *La Planète Mars* (1892) and his views on the habitability of the planet. In 1894 he founded his own observatory near Flagstaff, Arizona Territory, with the express purpose of studying the conditions of life on other worlds, particularly on Mars. From his first 12 months of observations, Lowell drew conclusions which he immediately published in a book that 'influence[d] and shape[d] the imagination of writers' such as Wells and Lasswitz.[16] The darker regions of Mars he took to be 'not water, but seasonal areas of vegetation,' with the planet depending, for its water supply, 'on the melting of its polar snows.' Then came the clincher:

> If, therefore, the planet possesses inhabitants, […] irrigation, upon as vast a scale as possible, […] must be the chief material concern of their lives […] paramount to all the local labor, women's suffrage, and [Balkan] questions put together.[17]

After the ironic aside, Lowell turned his attention to the canals which, he held, were dug precisely for such 'irrigation purposes':

> What we see is not the canal proper, but the line of land it irrigates, dispos[ing] incidentally of the difficulty of conceiving a canal several miles wide. […] What we see hints at the existence [… of] a highly intelligent mind […] of beings who are in advance of, not behind us, in the journey of life.[18]

Much later, Carl Sagan (1934–96) would famously quip that, most certainly, intelligence was responsible for the straightness of the lines observed by Lowell. The problem was just 'which side of the telescope the intelligence is on.'[19]

While the nineteenth was turning into the twentieth century, canals – like automobiles, dirigibles and airplanes – had come to symbolize progress, the triumph of technology over nature. In 1869, the Suez Canal had reduced the sea route to India by 10,000 kilometers, permitting Phileas Fogg and Passepartout to accomplish their imaginary journey around the world in 80 days. Work on the Panama Canal had begun in 1880, and even if the first French effort had foundered, a second American construction attempt was under way. Canals, whether on earth or (supposedly) on another world, continued to make for headlines: On 27 August 1911, the *New York Times* captioned a one-page article. 'Martians Build Two Immense Canals in Two Years,' its headline read: 'Vast Engineering Works Accomplished in an Incredibly Short Time by Our Planetary Neighbors' (Figure 5.1).

And Lowell's arid, aging Mars offered a further fascinating perspective: A community that had forsworn armed conflict, unified by a common endeavor, valiantly fighting its imminent doom, demonstrated to war-torn earth what a civilization might achieve once it had overcome strife and hate.[20] Sunlight might be converted into electricity on Mars's high plateaus, stones into bread by extracting protein and carbohydrates from rocks, soil, air and water. And material advancement might release additional energies needed for moral improvement: Such was the vision offered by Kurd Lasswitz (1848–1910) in his 1897 novel *Auf zwei Planeten*, intended to confront the imperialist powers of Europe with the notion of a world governed by reason. While an abridged English translation would only appear by 1971, the book was immediately translated into a number of other European languages and a popular German edition, which continued to be reprinted, was published in 1913. Until the Nazis branded the book as 'un-German,' the novel sold 70,000 copies in Germany.[21]

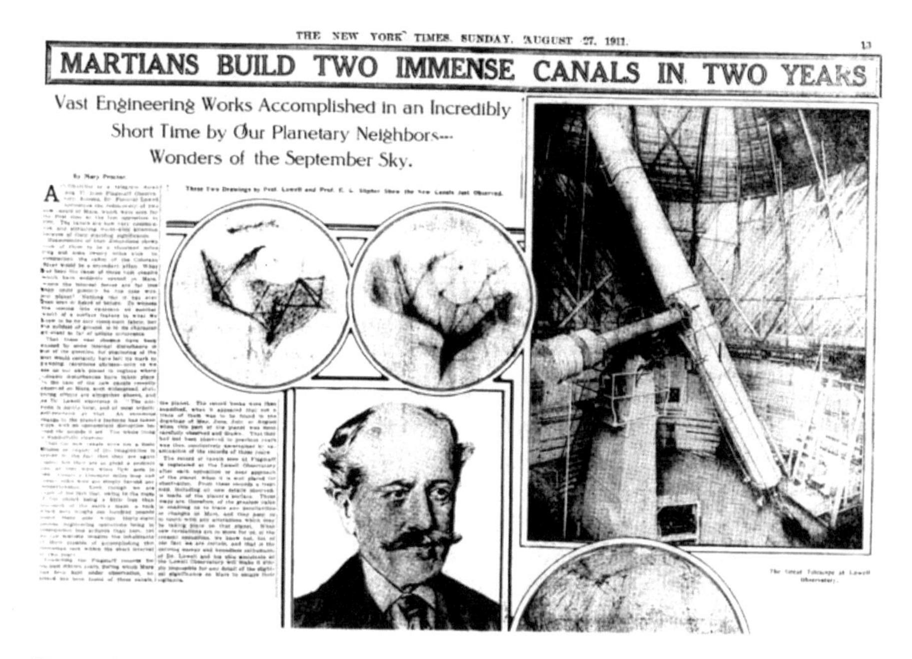

THE NEW YORK TIMES. SUNDAY, AUGUST 27, 1911. 13

MARTIANS BUILD TWO IMMENSE CANALS IN TWO YEARS

Vast Engineering Works Accomplished in an Incredibly Short Time by Our Planetary Neighbors---Wonders of the September Sky.

Figure 5.1 Astronomer Percival Lowell (1855–1916) could hardly have hoped for more favorable publicity. On 27 August 1911, the *New York Times* waxed eloquently on two Martian 'canals' recently identified by Lowell Observatory.
Source: Courtesy of Lowell Observatory Archives, Flagstaff, Arizona.

Lasswitz portrayed a Mars on which the 'colossal effort' required by irrigation had united the original 154 states into a single league. Thanks to the canal system – and here Lasswitz sounded like pure Lowell – 'the desert region was traversed by fertile strips of vegetation nearly 100 kilometers wide which included an unbroken string of thriving Martian settlements.'[22] A one-year mandatory labor service for both sexes helped maintain the network of canals. The discovery of anti-gravity had made Martians 'the masters of the solar system,' permitting them to construct a wheel-shaped space station 6,356 kilometers above earth's North Pole. Due to terrestrial arrogance, the first contact between men and 'Nume' ended in the occupation of Europe by the league of Martian states and the establishment of a protectorate aimed at 're-educating' mankind.

Wielding power over the earth, however, worked to morally corrupt the Martian conquerors. When they threatened to extend their protectorate to the United States, American engineers secretly succeeded in copying Martian arms and taking over their space station. Faced with a choice of violating their highest values by resorting to a war of extinction, or leaving the earth, the Martians chose to depart. Terrestrial nations formed not only an alliance, but went on to adopt new constitutions in a Kantian 'spirit of peace, liberty and human dignity.'[23] A peace treaty with Mars ensured coexistence on the basis of equality.

However, an alternative scenario might be imagined, derived from the hypothesis that Martians had failed 'in attempting to safeguard the habitability of their planet.' In that case, might not beings with minds 'vast and cool and unsympathetic' feel tempted to resort to aggression, pitilessly exterminating mankind in search of 'living space'? Rather than Lasswitz's pacifist vision, the result would be the social-Darwinist *War of the Worlds* that Herbert George Wells (1866–1946) envisioned in the same year. Skillfully, Wells gave the debate about the significance of the surface features on Mars a new twist. 'Men like Schiaparelli,' he wrote, 'failed to interpret the fluctuating appearances of the markings they mapped so well. All that time the Martians must have been getting ready.'[24]

Contrary to what a cursory reading of his tale might suggest, Wells did not depict the inhabitants of Mars – a Mars, it should be repeated, much older than the earth, according to prevailing opinion – as alien monstrosities. Rather, regarding their appearance, he projected on them those 'characters of the Man of the remote future' which he had predicted as the final stage of human evolution in an earlier essay:[25] an expanding brain and head, diminishing bodies and legs, unemotional intelligence, nourishment by absorption of nutritive fluids – blood in the case of the Martian invaders – atrophy of ears, nose and mouth, the latter 'a small, perfectly round aperture, toothless, gumless, jawless.' Wells's Martians were not so much invaders from space as invaders from time, 'ourselves, mutated beyond sympathy, though not beyond recognition.'[26]

Lasswitz, in a Kantian vein, had intended to confront Europe's imperialist powers with the alternative notion of a world governed by reason and peace. Wells's *War of the Worlds* remorselessly held the mirror up to contemporary colonialism (Figure 5.2). During 1897–98, Imperial Germany occupied the Chinese port of Jiaozhou; China had to cede a further part of Hongkong to Great Britain; France consolidated its position in West Africa; the United States annexed the Hawaiian Islands. In Asia, in Africa, in the Pacific, native populations were being subjugated or pushed back. 'Are we such apostles of mercy,' Wells asked rhetorically, 'as to complain if the Martians warred in the same spirit?'[27]

Finally, Wells could count not only on an audience turned receptive by a spate of recent novels – such as George Chesney's *The Battle of Dorking* (1871), William Butler's *The Invasion of England* (1882), William Le Queux's *The Great War in England* (1894) – to the notion of French and (more frequently) German raids on England. Moreover, these authors had already begun to explore a theme on which Wells focused his attention in *The War of the Worlds*: the disappearance of any distinction between battle fronts and zones where civilians might feel reasonably safe, the expansion of mechanized 'total' warfare to engulf entire populations.[28]

Such total war was raging in China 40 years later, after Japanese armies had invaded the country in 1937. For a brief moment, it had been avoided in Europe after Czechoslovakia had yielded, under British and French pressure, to the Munich Agreement. The war scare was still fresh in many Americans'

Figure 5.2 The action of *War of the Worlds* was moved to California for the 1953 film version by producer George Pal (1908–80) and director Byron Haskin (1899–1984). Promotional posters depicted the alien's arm as symbolizing the Martian attempt to seize and colonize the earth.
Source: Author's archive.

minds when CBS, on 30 October 1938, aired *The War of the Worlds* as a 60-minute radio play, directed by Orson Welles (1915–85), with the action transferred to New Jersey. Presented as a series of increasingly ominous news bulletins, the first half of the broadcast produced mass hysteria: All over the United States, people 'were praying, crying, fleeing frantically [...] Some ran to rescue loved ones. Others [...] sought information from newspapers or radio stations, summoned ambulances and police cars.'[29] An estimated 250,000 people believed the United States to be under attack by either Germany, Japan – or indeed from Mars. In a bewildering world troubled by prolonged economic depression, wars and political crises, many Americans thought anything might happen.

By that time, H.G. Wells had turned social reformer, slowly despairing of men's folly. His last ideas about an invasion from earth's 'wizened elder brother' Mars, published under the title *Star Begotten* shortly before Orson Welles's broadcast, differed considerably from his first – though not without a self-deprecating glance back:

> Some of you may have read a book called *The War of the Worlds* – I forget who wrote it – Jules Verne, Conan Doyle, one of those fellows. But it told how the Martians invaded the world, wanted to colonize it and

> exterminate mankind. Hopeless attempt! They couldn't stand the different atmospheric pressure, they couldn't stand the difference in gravitation. [...] To imagine that the Martians would be fools enough to try anything of the sort. But –[30]

But if they resorted to cosmic rays instead? Modifying the genetic structure of unborn children, creating new beings that were, in fact, *their* spiritual children? That was the obsessive idea with which the tale's protagonist wrestled, until he discovered that his wife, their son – that he himself was star begotten, a changeling. The change, however, was benevolent, meant to salvage mankind – 'a lunatic asylum crowded with patients prevented from knowledge and afraid to go sane'[31] – from stupidity and immaturity by making humans more flexible, more open-minded, more innovative. Mature Martian civilization emerged as a *deus ex machina* for solving, by imperceptible intervention from outside, those pressing problems which mankind found itself unable to surmount.

IV Frontier Mars

Implying, as it did, that the Red Planet's inhabitants would beat humans to accomplishing spaceflight, the idea of Advanced Mars ran counter to deeply engrained expansionist impulses of the imperialist age. Small wonder the tabloid journalist Garrett Putnam Serviss (1851–1929) immediately responded to Wells's tale with a serial in the sensationalist *New York Evening Journal*, published in 1898. Entitled *Edison's Conquest of Mars*, it depicted the 'wizard of Menlo Park,' aided by Lord Kelvin and Wilhelm Conrad Röntgen, as devising both a disintegrator ray and an electric spaceship (admittedly based on the operating principles of the Martian machines). Financed by the great powers, 100 spaceships – armed with 3,000 disintegrators – were built and flew to Mars, where they wreaked havoc by forcibly opening the 'floodgates of Syrtis Major,' thereby deluging the planet's equatorial regions. Lasswitz had already attributed the defeat of his Martian conquerors to American engineering talent and 'daring.' By presenting an entire arsenal of innovative weapons, Serviss left no doubt about America's claim to global leadership: technologically superior, the 'new world' had outrivaled the 'old' as the torchbearer of progress.

As regards the image of the Red Planet, Serviss's tale did not set a new trend. Lowell's arid Mars was taking hold in the public mind, and Serviss's vast oceans and floodgates were just too wildly off that mark. Concerning stylistic treatment of his subject and the moral of the story, however, Serviss's lurid account marked a significant change away from 'highbrow' European-style literature. The new perspective was fully brought to bear by American cowboy, gold miner, salesman and – in the end – novelist Edgar Rice Burroughs (1875–1950) who wrote *A Princess of Mars* in 1912 (Figure 5.3). He rechristened Mars, gave it the name Barsoom – and

Figure 5.3 Edgar Rice Burroughs's 'Princess of Mars,' as imagined on the cover of the 1925 German translation.
Source: Edgar Rice Burroughs, *Eine Mars-Prinzessin: Dreiundvierzig Millionen Meilen von der Erde*, Stuttgart: Dieck, 1925, cover image.

henceforth Mars exploration would be 'as much a re-creation of the past as a vision of the future.'[32]

With Burroughs, American science-fiction authors started to build on a 'forceful [...] cultural tradition' that would eventually inspire the US space program no less than it initially spurred 'romantic vision[s]' of exploring, even colonizing Mars: the myth of America's western frontier.[33] 'Lowbrow' pulp fiction would henceforth provide the medium for two generations of writers including, subsequent to Burroughs, most prominently Leigh Brackett (1915–78). Due to the existence of a virtual 'entertainment industry'[34] with international outlets, American pulp writers took the lead from European authors in projecting their fantasies onto the Red Planet. Burroughs and his heirs retained Lowell's deserts and canals, but discarded the idea of a sophisticated Martian civilization. Instead, they fantasized towns, ancient beyond imagination lying in the southern hemisphere of Mars, their outskirts touching the shores of the dried-up Low Canals that once discharged their waters into the now dust-blown bed of a long-vanished ocean. The towns, once ruled by pirate kings, bore names such as Jekkara, or Valkis, or Barrakesh. Their women – partly resembling Indians, partly Mexicans – wore tiny golden bells chiming temptingly. Barbarian tribes came to these places from distant deserts, such as Kesh and Shun. A Terran spaceport did exist at Kahora, not far from Olympus Mons. But only hard-boiled adventurers dared approach the Low Canals, after having galloped across the Drylands on half-wild saurians. They had 'the rawhide look of the planetary frontiers about them' and wore their ray-guns low in their holsters. Because Barrakesh, Jekkara, Valkis were towns outside the law.[35]

The scenario was Leigh Brackett's, dreamt up during the 1940s. Like its American counterpart, the 'planetary frontier' signified no demarcation line, as the term was understood by Europeans, but rather the advancing rim of settlement, site of the violent clash between savagery and civilization. After the US government had announced, in 1890, the 'closing' of the frontier in its statistical meaning of less than six inhabitants per square mile, the frontier – 'by transcending the limitations of a specific temporality' – came to be projected from the past into the present and even the future. Creating a specific 'moral landscape,' depicting the course of American history as progress through violence (or, as Burroughs would have it in *A Princess of Mars*), the myth of the frontier has continued to provide patterns of identification and legitimization for individual and collective attitudes and behavior to the present day.[36] The rugged individualist, the onward-thrusting pioneer, the hardy adventurer, all armed *and morally justified* to shoot or to slash in a stereotyped black-and-white situation of good versus evil: These are the vivid images evoked by the frontier metaphor. They re-emerged in 'the "space opera" (as opposed to "horse opera")' with the 'typical structures and plots of westerns,' but the 'settings and trappings of science fiction.'[37]

Burroughs virtually defined the sub-genre, creating the quintessential space opera character: Captain John Carter, a 'gentleman of the highest type' and former plantation owner from Virginia, who had proved his prowess in the Civil War, and who was magically teleported from Arizona – where he had been battling Apaches – to Mars.[38] Burroughs made no effort to conceal that John Carter was modeled on Captain John Smith, a seventeenth-century Virginian colonist who figured prominently in another American legend – the narrative of Pocahontas, Indian 'princess' of the Powhatan tribe. Supposedly, Pocahontas (at the tender age of 12 or 13) had become enamored of Smith and had rescued him from torture by her tribe. After arriving on a Mars peopled by warlike black, red, green and yellow races, Burroughs's John Carter met and married the 'incomparable' princess Dejah Thoris, daughter of the Jed (ruler) of Helium, chief of a red-skinned people that exhibited 'a startling resemblance [...] to [...] the red Indians of [...] earth.'[39] In *The Princess of Mars* and Burroughs's subsequent Mars novels, it was Carter's task to repeatedly save Dejah Thoris, with the extraordinary physical powers lent to him by Mars's lesser gravity, from a fate 'worse than death.' To leave not the slightest doubt about the tradition he was embracing, Burroughs chose this context to revive another stereotype of frontier melodrama: With 'a cold sweat,' his main protagonist reflected that if he should fail, it would be 'far better' for Dejah to 'save friendly bullets [...] at the last moment, as did those brave frontier women of my lost land, who took their own lives rather than fall into the hands of the Indian braves.'[40]

Frontier Mars became a place where only-too-familiar characters lounged in the doorways of earth's latest colony – Northwest Smith for one, created in 1933 by writer Catherine L. Moore (1911–87), 'tall and leather-brown, hand on his heat-gun'; where everybody understood the 'old gesture' when that gun was drawn with a swift motion, sweeping 'in a practiced half-circle'; where John Carter, Northwest Smith and their likes fought human or half-human tribes; where conflicts were invariably 'resolved' by resorting to weapons. An 'extension of our original America,' with 'Martians await[ing] us' whom 'we [could] assimilate to our old myths of the Indian,' Frontier Mars was destined to remain a very parochial planet, familiar rather than alien.[41]

V Cold War Mars

'Watch the skies!' moviegoers were counseled in 1951 at the end of the science-fiction film *The Thing from Another World*. The Cold War had turned hot in Korea. Who knew what the communists, 'masters of deceit' (J. Edgar Hoover), aggressively pushing from outside, subversively boring from within, threatening 'the continuance of every home and fireside,'[42] might have up their sleeves?

Two years later, the 10-year-old stargazing protagonist of *Invaders from Mars* did watch the skies at night, only to observe a flying saucer landing and burrowing in the sandy ground across from his home. Everyone who investigated next morning – the child's father, his mother, a neighbor girl, two policemen, finally the local chief of police – was 'transformed' in succession, displaying an implant in the neck and behaving robot-like. *Invaders from Mars* recounted not just an invasion, but a 'conspiracy,' an emerging 'fifth column' of concealed infiltrators. Neither parents nor friends could be trusted anymore – a patent allusion (including the unfeeling attitudes displayed by affected adults) to rampant paranoia about the supposed subversion of American life by Communists. The invaders themselves were depicted as puppets, telepathically controlled by a 'supreme intelligence.' As might be expected, the military – alerted by the boy's school psychologist and her friend, an astrophysicist – arrived in time to save the day and blow up the Martian saucer.

By the beginning of the decade, movies were America's most popular entertainment; only from the mid-1950s would they be outranked by television. *Destination Moon* (1950) made the idea of space travel not only plausible but fascinating. *The Thing from Another World* (1951) brought the idea of creatures from other planets coming here to vivid life. From 1950, too, the screen added Martian landscapes to those portrayed in the printed media. Both *Rocketship XM*, Kurt Neumann's bleak movie of humans arriving on a Mars destroyed by nuclear war, and *The Martian Chronicles*, Ray Bradbury's seminal novel of the Red Planet's colonization against the backdrop of atomic war eventually engulfing earth, came out during that year. In *Rocketship X(pedition) M(oon)*, the first manned spaceflight to earth's satellite was thrown off course by a swarm of meteors and forced to land on Mars. The crew found themselves in a post-nuclear wasteland, deducing 'from artifacts and ruins so radioactive they can't approach them that there had once been a high civilization on Mars, but that atomic warfare reduced the Martians to savagery.' Mutated Martians attacked the expedition, killing two and wounding a third crew-member. The rest of the crew escaped, but the rocket ran out of fuel on its return flight and crashed. A year before, the Soviet Union had detonated its first nuclear device. President Truman had ordered development of the hydrogen bomb in early 1950. 'The idea that we now had the potential to wipe out civilization entirely was beginning to permeate mass culture' – and was projected onto Mars by 'the first film to expound such a grim warning about our possible future.'[43]

The Soviet explosion and Truman's announcement drew an immediate response from a 30-year-old writer, Ray Bradbury (1920–2012), who felt that man might 'still destroy himself before reaching for the stars. I see man's self-destructive half, the blind spider fiddling in the venomous dark, dreaming mushroom-cloud dreams. Death solves it all, it whispers, shaking a handful of atoms like a necklace of dark beads.'[44] On 6 May 1950, *Collier's* magazine published one of Bradbury's most powerful stories, 'There Will Come Soft Rains.' It had no human protagonists. Rather, it focused on the final 'death' of an electronically programmed house, left standing empty among glowing radioactive ruins, after its occupants had perished, their images – as had happened in Hiroshima – 'burnt on the wood in one titanic instant.'[45] The story was included by Bradbury as a chapter in his loosely-knit classic of the same year, *The*

Martian Chronicles, intended by the author to 'provide a mirror for humanity, its faults, foibles, and failures [...] an allegory transplanted to another world.'[46]

Before being killed off by chicken pox, which American colonists had introduced to Mars, the planet's golden-eyed 'natives' had inhabited crystal houses at the edge of the canals that – attuned to nature – 'turned and followed the sun, flower-like.'[47] The settlers not only brought chicken pox. They also brought gas stations, luggage stores and hot-dog stands. With their hammers, they 'beat the strange world into a shape that was familiar, they bludgeon[ed] away all the strangeness. [...] In all, some ninety thousand people came to Mars.'[48] But the majority left again when flashing light-radio messages from earth reported that there was war and that everybody should come home. To those who had remained on Mars, the night sky soon offered a horrible sight: 'Earth changed [...] It caught fire. Part of it seemed to come apart in a million pieces. [...] It burned with an unholy dripping glare for a minute, three times normal size, then dwindled.' Humans had turned two worlds, Mars and earth, into 'tomb planet[s].'[49]

However, Bradbury – influenced by both Burroughs and Brackett – had also decided 'that there would be certain elements of similarity between the invasion of Mars and the invasion of the Wild West.'[50] The frontier myth held that America and its democracy would be reborn at every new frontier between the Atlantic and Pacific – and beyond. One family, more fortunate than the folks annihilated in 'There Will Come Soft Rains,' had escaped the inferno on earth (with rumors maintaining that a second one had also made it to Mars). The father had promised the children that they would set out for a picnic and would see Martians. Now they were gazing at their reflections in a canal – and the Martians stared back at them. The implication was evident. Bradbury's 'intensely critical examination' of the frontier myth – of 'the shallow and mercurial properties of America's predominant cultural construct' – notwithstanding, Mars emerged as another 'virgin land' (Henry Nash Smith) where America might both survive and regenerate.[51] The Frontier Mars image, in other words, had proved its adaptability to the hydrogen bomb age, reducing Cold War Mars to a mere variant of an already familiar theme. And, as would soon become evident, the frontier metaphor had not exhausted its usefulness.

VI Terraformed Mars

For American engineer Robert Zubrin (1952–), the writing presently 'is on the wall': 'Without a frontier from which to breathe life,' Zubrin holds, 'the spirit that gave rise to the progressive humanistic culture that America for the past several centuries has offered to the world is fading.' The engineer is convinced that the creation of a new frontier presents itself 'as America's and humanity's greatest need.' And he believes 'that humanity's new frontier can only be Mars.' Zubrin, and the Mars Society, formed in 1998 on his initiative, consider privately funded Mars flights and the establishment of a permanent Mars base as just initial steps. To fulfill the planet's mission of reinvigorating terrestrial civilization, its atmospheric and surface conditions need to be dramatically changed by a long-term project. Mars must be 'terraformed.'[52]

According to the *Shorter Oxford English Dictionary*, terraforming implies a process of planetary engineering, aimed at creating an extraterrestrial environment that would be habitable for humans. First use of the term has been credited to science-fiction writer John Stewart ('Jack') Williamson in a 1942 novella. The concept started to gain a certain scientific acceptability after Carl Sagan had published an article in 1961 on introducing algae into the atmosphere of Venus to slowly change that planet's extremely hostile conditions. In 1973 Sagan followed with a piece 'Planetary Engineering on Mars,' kicking off the debate with regard to the Red Planet. To terraform Mars, both atmospheric pressure and surface temperature would have to be raised. The 'global warming' process – basically comparable to that which earth is presently experiencing – would require an increase in 'greenhouse gasses,' such as carbon dioxide or more powerful fluorocarbons, for which several ways have been proposed, and the subsequent build-up of a hydrosphere providing the water necessary to sustain life. The idea was, of course, picked up by science fiction – most elaborately by Kim Stanley Robinson in his trilogy *Red Mars/ Green Mars/Blue Mars* (1992–96). The work focused on the century-long conflict between 'Greens,' whose sense of mission prompted them to contaminate the Red Planet with robust mosses and lichens at every opportunity, and the 'Red' environmentalists who were finally driven underground.[53]

As before, such imaginary landscapes have revealed more about the desires, the hopes, the anxieties of those who designed them, than about any future 'green' or 'blue' Mars. While Zubrin took care to link the emergence of a terraformed Martian frontier to the promotion of values such as individualism, creativity and belief in the idea of progress, his basic approach was far more hard-nosed:[54]

> If the idea is accepted that the world's resources are fixed, then each person is ultimately the enemy of every other person, and each race or nation is the enemy of every other race or nation. The inevitable result is tyranny, war and genocide. Only in a universe of unlimited resources can all men be brothers.

Put differently: Either a new frontier will be opened up – or containment, rather than self-containment, will become the 'natural' order of things .... Objections against such reasoning were the exception. In terms reminiscent of Bradbury, but more starkly, historian Patricia Limerick in her contribution to the 1992 volume *Space Policy Alternatives* emphasized the social-Darwinist consequences of 'rugged individualism' that had shaped the 'conquest' of the American West, including greed and corruption, violence against 'aliens' (Indians and Mexicans), environmental destruction. She rejected the simplified picture of westward expansion painted in 1893 by historian Frederick

Jackson Turner (1861–1932) – the famous Turner thesis again extolled by Zubrin – because it had 'denied consequences and evaded failure.'[55]

Because of 'America's pioneer heritage, technological pre-eminence, and economic strength, it is fitting that we should lead the people of this planet into space,' the Paine Commission had stated in 1986. Chaired by an earlier NASA Administrator, it included UN Ambassador Jeane Kirkpatrick, former test pilot Charles Yeager and retired Air Force General Bernard Schriever (who had directed IRBM Thor and ICBM Atlas development). In their report, tellingly entitled *Pioneering the Space Frontier*, the members had proposed to 'stimulate individual initiative and free enterprise in space,' and had resolved that 'from the highlands of the moon to the plains of Mars,' America should 'make accessible vast new resources and support human settlements beyond earth orbit.'[56] This was no space opera. This was a National Commission on Space, appointed by the President of the United States, issuing a declaration that was 'vintage 1890s [...] with a fervent optimism and cheeriness that might well have made Frederick Jackson Turner himself a bit ill-at-ease. [...] a picture of harmony and progress where historical reality shows us something closer to a muddle.'[57]

Abstracting and reducing from reality, the frontier myth has created a historical cliché. Clichés, as Richard Slotkin – among others – has reminded us, may serve to interpret new experiences as mere recurrences of familiar happenings, reflecting a refusal to learn. Identifying Mars as merely another 'frontier,' projecting a moral purpose on the adoption of that so-called planetary frontier to human settlers' needs, tops a tradition of invoking a cultural stereotype that must be classed as highly problematic.

Notes

1. Victor R. Baker, *The Channels of Mars*, Austin: University of Texas Press, 1982, 3–4. This chapter's argument is partly based on Rainer Eisfeld and Wolfgang Jeschke, *Marsfieber*, Munich: Droemer, 2003. Much like Robert Markley's subsequent *Dying Planet: Mars in Science and the Imagination*, Durham: Duke University Press 2005, the book discusses both the imagined Red Planet and the actual Mars progressively unveiled by robotic missions.
2. Wilhelm Beer and Johann Heinrich Mädler, *Beiträge zur physischen Kenntniss der himmlischen Körper im Sonnensysteme*, Weimar: Bernhard Friedrich Voigt, 1841, vii.
3. Ibid., 124–5.
4. Brian Aldiss and David Wingrove, *Trillion Year Spree: The History of Science Fiction*, London: Paladin, 1988, 603 n. 47; Martin Schwonke, *Vom Staatsroman zur Science Fiction: Eine Untersuchung über Geschichte und Funktion der naturwissenschaftlich-technischen Utopie*, Stuttgart: Ferdinand Enke, 1957, 43.
5. Carl Sagan, *Cosmos*, New York: Random House, 1980, 106.
6. This brief reference to the British film goes back to Eisfeld and Jeschke, *Marsfieber*, 163. *Devil Girl on Mars* was subsequently discussed by Robert Markley, *Dying Planet*, 227–9.
7. Camille Flammarion, *Les Terres du ciel*, Paris: Marpon & Flammarion, 1884, 208.

8. Idem, *Die Mehrheit bewohnter Welten*, Leipzig: J.J. Weber, 1865, 51–2, 71. Flammarion published the book as a 20-year old.
9. Before the Mariner 9 space probe permitted production of the first 'reliable map,' more than 130 years would elapse. See Oliver Morton, *Mapping Mars: Science, Imagination, and the Birth of a World*, New York: Fourth Estate, 2002, 37–8.
10. Richard A. Proctor, *Other Worlds than Ours* [1870], London: Longmans, Green, 3rd edn 1872, 85, 109–10.
11. Giovanni Schiaparelli, *Astronomical and Physical Observations of the Axis of Rotation and the Topography of the Planet Mars: First Memoir*, 1877–78, trans. William Sheehan, MS, Flagstaff: Lowell Observatory Flagstaff (Archives), 1994, 124.
12. Camille Flammarion, 'Découvertes nouvelles sur la planète Mars,' *Révue d'Astronomie populaire* 1.7 (July 1882), 218; Camille Flammarion, 'La Planète Mars,' ibid. 1.7 (July 1882), 216.
13. William Sheehan, *The Planet Mars*, Tucson: University of Arizona Press, 1996, 85.
14. Schiaparelli, *Astronomical and Physical Observations*, 123–4.
15. Percival Lowell, *Mars as the Abode of Life*, New York: Macmillan, 1908, 134.
16. Mark R. Hillegas, 'Martians and Mythmakers, 1877–1938,' in Ray B. Browne, Larry N. Landrum and William K. Bottorf, eds, *Challenges in American Culture*, Bowling Green: Bowling Green University Popular Press, 1970, 150–77, here 156.
17. Percival Lowell, *Mars*, Boston: Houghton Mifflin, 1895, 122, 128–9.
18. Ibid., 165, 208–9.
19. Carl Sagan, 'Hypotheses,' in Ray Bradbury, Arthur C. Clarke, Bruce Murray, Carl Sagan and Walter Sullivan, eds, *Mars and the Mind of Man*, New York: Harper & Row, 1973, 13.
20. Percival Lowell, *Mars and its Canals*, New York: Macmillan, 1906, 377.
21. Franz Rottensteiner, 'Kurd Lasswitz: A German Pioneer of Science Fiction,' in Thomas D. Clareson, ed., *SF: The Other Side of Realism*, Bowling Green: Bowling Green University Popular Press, 1971, 289.
22. Kurd Lasswitz, *Auf zwei Planeten* [1897], Frankfurt am Main: Zweitausendeins, 1979, 98.
23. Ibid., 875.
24. H.G. Wells, *The War of the Worlds*, New York: Pocket Books, 1953, 2.
25. Idem, 'The Man of the Year Million' [1893], in David Y. Hughes and Harry M. Geduld, eds, *A Critical Edition of the War of the Worlds*, Bloomington: Indiana University Press, 1993, Appendix III, 291–3.
26. Frank McConnell, *The Science Fiction of H.G. Wells*, Oxford: Oxford University Press, 1981, 128, 130.
27. Ibid.
28. Ibid., 132–3.
29. Hadley Cantril, *The Invasion from Mars: A Study in the Psychology of Panic*, Princeton: Princeton University Press, 1940, 47.
30. H.G. Wells, *Star Begotten*, London: Chatto & Windus, 1937, 50–1.
31. Ibid., 167–8.
32. Howard E. McCurdy, *Space and the American Imagination*, Washington, DC: Smithsonian Institution Press, 1997, 2.
33. Ibid., 233–4.

34. Benjamin S. Lawson, 'The Time and Place of Edgar Rice Burroughs's Early Martian Trilogy,' *Extrapolation* 27.3 (March 1986), 203–20, here 209.
35. Leigh Brackett, *The Secret of Sinharat*, New York: Ace Books, 1964, 8.
36. Richard Slotkin, *Gunfighter Nation: The Myth of the Frontier in Twentieth-Century America*, New York: Atheneum, 1992, 4–5, 6–7, 14, 24.
37. Lawson, 'Time and Place,' 213.
38. Edgar Rice Burroughs, *A Princess of Mars* [1912], New York: Random House, 2003, xxiii–iv, 14–15.
39. Ibid., 152.
40. Ibid., 75.
41. Catherine L. Moore, 'Shambleau' [1933], in *Northwest Smith*, New York: Ace Books, 1981, 2, 3; Leslie A. Fiedler, *The Return of the Vanishing American*, London: Paladin, 1972, 25; Lawson, 'Time and Place,' 208.
42. J. Edgar Hoover, *Masters of Deceit: The Story of Communism in America and How to Fight It*, New York: Pocket Books, 1958, vi.
43. Bill Warren, *Keep Watching the Skies: American Science Fiction Movies of the Fifties*, vol. 1: *1950–1957*, Jefferson: McFarland, 1982, ix, 2, 11.
44. Ray Bradbury, quoted in William F. Nolan, 'Bradbury: Prose Poet in the Age of Space,' *Magazine of Fantasy & Science Fiction* 24.5 (May 1963), 8.
45. Ray Bradbury, *The Martian Chronicles*, New York: Bantam Books, 1951, 185.
46. Sam Weller, *The Bradbury Chronicles*, New York: William Morrow, 2005, 156, 159.
47. Bradbury, *Martian Chronicles*, 2.
48. Ibid., 86.
49. Ibid., 158, 172.
50. Bradbury, as quoted in Weller, *Bradbury Chronicles*, 155.
51. Gregory M. Pfitzer, 'The Only Good Alien is a Dead Alien: Science Fiction and the Metaphysics of Indian-Hating on the High Frontier,' *Journal of American Culture* 18.1 (Spring 1995), 51–67, here 58.
52. Robert Zubrin, *The Significance of the Martian Frontier* [1994], www.nss.org/settlement/mars/zubrin-frontier.html (accessed 1 October 2017).
53. Carl Sagan, 'Planetary Engineering on Mars,' *Icarus* 20.4 (December 1973), 513–14; Christopher P. McKay, Owen B. Toon and James F. Kasting, 'Making Mars Habitable,' *Nature* 352 (8 August 1991), 489–96; Christopher P. McKay, 'Restoring Mars to Habitable Conditions: Can We? Should We? Will We?,' *Journal of the Irish Colleges of Physicians and Surgeons* 22.1 (January 1993), 17–19.
54. Zubrin, *Significance*.
55. Patricia Nelson Limerick, 'Imagined Frontiers: Westward Expansion and the Future of the Space Program,' in Radford Byerly Jr., ed., *Space Policy Alternatives*, Boulder: Westview Press, 1992, 249–62.
56. Paine Commission, http://history.nasa.gov/painerep/parta.html (accessed 1 October 2017).
57. Limerick, 'Imagined Frontiers,' 253–4, 256–7.

CHAPTER 6

'Smash the Myth of the Fascist Rocket Baron': East German Attacks on Wernher von Braun in the 1960s

Michael J. Neufeld

Late in 1962, a West Berlin correspondent of Wernher von Braun (1912–77), the world-famous Director of NASA Marshall Space Flight Center in Huntsville, Alabama, sent him a series of hostile articles that had just appeared in East Germany. They turned out to be excerpts from the forthcoming book *Geheimnis von Huntsville: Die wahre Karriere des Raketenbarons Wernher von Braun* (Secret of Huntsville: The True Career of Rocket Baron Wernher von Braun).[1] Written by a popular East German author of non-fiction spy books, Julius Mader (1928–2000), *Geheimnis* heralded a Soviet-bloc attempt to destroy von Braun's reputation by unmasking the depths of his involvement with the Nazi regime, the SS and its concentration camps. Von Braun had certainly been a tempting target for communist press attacks – the United States' leading rocket specialist was an ex-Nazi who led the development of one of Hitler's terror weapons, the V-2 ballistic missile, before changing sides literally overnight at the end of the war. But little effort had been expended in the Warsaw Pact in uncovering and propagating the details of his service to the National Socialist regime. Mader threatened to change all that.[2]

It is well to remember how famous and popular Wernher von Braun was at the time, especially in the United States and West Germany. Already well known in the mid-1950s because of his efforts to sell spaceflight through books, magazine articles and appearances on Walt Disney's TV program, he became the vindicated prophet of astronautics after the Soviets orbited Sputnik on 4 October

Michael J. Neufeld (✉)
Smithsonian National Air and Space Museum, Washington, DC, USA
e-mail: NeufeldM@si.edu

Alexander C.T. Geppert (ed.), *Imagining Outer Space*
European Astroculture, vol. 1
https://doi.org/10.1057/978-1-349-95339-4_6

1957. Less than four months later he ascended to the status of national hero in those two countries after his US Army team launched the first American satellite. Several Hollywood studios immediately wanted to make a 'biopic' about him; the winner, Columbia Pictures, took over an already existing West German project, leading to the co-produced *I Aim at the Stars*. It premiered in Munich in August 1960, sparking protest over his connections to the Nazis and to nuclear weapons. Ultimately the movie bombed at the box office, at least in the United States, because it was so mediocre. But judging by the ongoing hero-worship in the America and West German press in the 1960s, not to mention that of less-friendly countries like France, *I Aim* had little lasting impact on his role as an icon of the Cold War Space Race against the Soviet Union (Figure 6.1).[3]

During the 1950s, von Braun, his associates and the US government had largely neutralized his Nazi problem through a selective use of history and through a conspiracy of silence about his SS officer status and the V-2 program's extensive abuse of concentration camp labor.[4] Those were certainly two things that Mader tried to expose; his book became a big seller throughout the Warsaw Pact. *Geheimnis von Huntsville* went on to spawn a major East German motion picture, which opened in spring 1967 with the otherwise unnamed 'Rocket Baron' (von Braun was in fact a Prussian baron) as one of the central villains. Several months later, a West German court opened a trial of three SS men from the Mittelbau-Dora concentration camp, which supplied the labor to the underground V-2 plant. Although this proceeding was not about von Braun, the East German co-counsel ultimately succeeded in having him called as a witness, presumably with the primary intent of embarrassing him. That marked the third East German attempt to undermine the rocket engineer.

Yet these attacks ultimately failed to make much of an impression on von Braun's reputation in the West. Major contributing factors appear to be: first, the relatively small resources East Germany invested in this effort compared to campaigns against West German politicians; second, the primary focus on shaping East-bloc opinion rather than Western attitudes; third, the bitter German division, which built a high wall in the West against East German propaganda; fourth, fear of lawsuits, which hindered export of the movie to the West, and perhaps also the book; fifth, the US government's successful classification of damaging documents and their unavailability to the East German secret police; and, sixth, the great value Western governments and the media placed upon von Braun because of his Space Race role. As a result, the Mittelbau-Dora camp and his SS membership remained largely unknown, especially in America, until the US declassification of damaging information about the German rocketeers in 1984. The East German assault on von Braun thus ironically only reinforced the incompatible discourses about him and the Nazi rocket program on either side of the Cold War divide. Because of his Third Reich past, von Braun remained a contested figure even in the West, but on quite different terms than in the East, where he was vilified as a Nazi war criminal.[5]

Figure 6.1 The West German version of *I Aim at the Stars*, released in fall 1960, probably sparked East German author and Stasi collaborator Julius Mader's (1928–2000) research into Wernher von Braun's Nazi past.
Source: Courtesy of Columbia Pictures.

I Julius Mader's attack on von Braun

By the time Mader published his article series in 1962 in the organ of the East German communist youth group, he had become a covert officer in the Ministry of State Security (the Ministerium für Staatssicherheit or MfS), the secret-police and foreign-intelligence agency colloquially known as the 'Stasi.' Thus a key question about Mader's book is whether it was the opening round in an East German state campaign against von Braun. In the sense that everything Mader did was an official action, as he needed approval from his Stasi superiors and all his publishers, who were owned by the state, the ruling party, or its organs, certainly yes. But it may also be possible that Mader as a successful author possessed significant autonomy to choose the subject of his works and that he campaigned without extensive support. Unfortunately, the Stasi files about Mader, at least those discovered so far, are just too thin to be much help in deciding the degree to which the anti-von Braun initiative was suggested by his superiors and pushed by them beyond supporting the research for his book.[6]

Born in 1928 into a lower middle-class German family in Czechoslovakia, Mader had been a junior Hitler Youth leader. Departing the Sudetenland immediately after the end of the war, together with so many other expellees, he had helped organize an 'antifascist Front' in a Saxon town and had joined the new liberal party in the Soviet occupation zone at age 18. That party was soon forced into a National Front dominated by the Socialist Unity Party of Germany (Sozialistische Einheitspartei Deutschlands or SED), as the Communist Party renamed itself in 1946 after it absorbed the Social Democrats in the East. But Mader left the liberals after a year and did not join the SED until 1961. His career track leaves the impression of an opportunist, but he steadily rose as economic administrator, then as a journalist and editor for economic organizations, as well as a trade union leader in the Deutsche Demokratische Republik (DDR) formed in October 1949. By 1960, his activist record and propagandistic publications leave the impression that he had become a true believer.[7]

Mader's connection to the Stasi arose from his research for his 1959 anti-CIA book *Allens Gangster in Aktion* – Allen being Director of Central Intelligence Allen Dulles (1893–1969). The British intelligence historian Paul Maddrell rates it as part of the Stasi propaganda campaign connected to Soviet leader Nikita Khrushchev's attempt to pressure the Western Allies out of West Berlin in 1958–59. Mader's cooperation with the Stasi began in 1958 when the ministry heard he was working on the book. The Stasi took him on board as a paid covert collaborator (*inoffizieller Mitarbeiter* or IM) of the Agitation Section as of 1 January 1960, allowing him to quit his editor's job and pose as a freelance journalist. He followed that book with *Die graue Hand* (The Grey Hand) in 1960, an attack on the West German intelligence service as a nest of ex-Nazis (which in fact many of its leaders were). In a late 1961 evaluation of Mader leading to his appointment as an officer, two section chiefs praised him highly for his contributions to 'MfS-organized agitation campaigns – unmasking enemy plans and attempts at provocation, unmasking Bonn Nazis [Bonn was the West German capital], [and] unmasking the activity of enemy secret services.' On 21 April 1962, Mader was sworn in as a captain, with the designation 'officer in special service' (*Offizier im besonderen Einsatz* or OibE) for the Agitation Section. He maintained his independent office and front as a journalist.[8]

The origin of Mader's decision (or that of his superiors) to go after von Braun is unknown, but it seems likely to have been stimulated by *I Aim at the Stars*, which was released in West Germany in a dubbed version, *Wernher von Braun: Ich greife nach den Sternen*, in fall 1960. The first trace of Mader's researches is a letter he wrote to von Braun in Huntsville in July 1961, under the pretext of researching an article on Otto Skorzeny (1908–75), an infamous SS commando, including his alleged connection to the manned version of the V-1 cruise missile. Von Braun could assert that he had no responsibility for that project and directed him to the Western biographical and V-2 literature then available in English and German. By the fall, it was clear that Mader was researching him and his family, as von Braun exchanged letters with his father over the Mader missives both had received. Wernher von Braun advised no further correspondence because 'one never knows with these chaps what's behind it all.'[9]

Von Braun finally found out when Mader's articles arrived in Huntsville in late 1962. He described them to his West Berlin correspondent as 'lies, fabrications and grotesque distortions [...] in part skillfully linked to facts.' He had a point. The book (like the excerpts that preceded it) had two openings. First, a preface describing Huntsville, based on West German newspaper articles, which emphasized von Braun's secluded personal life and the protective attitude that the city took toward him. It included false details like rockets launching from the Army and NASA area – an impossibility given the local geography – because Mader had no chance to visit the place, and simply made things up. More egregious was the prologue, a completely invented scene of a summer 1931 meeting with key members of German Army Ordnance at von Braun parents' country house in the Silesian hills of eastern Germany (now Poland). Mader used it to underline the engineer's roots in the reactionary Prussian aristocracy and army officer corps. Out of this meeting allegedly came the then-university student's work for the army on rockets as weapons; in fact it would be another year before von Braun had any serious contact with Ordnance.[10]

Several less fictional, but equally heavy-handed sections of the book betray its often crude, propagandistic character. With access to the Humboldt University archive in East Berlin, Mader was able to use von Braun's doctoral records to make it look like the army forced the university to give the rocket engineer a physics degree without proper review or qualifications. In another case, Mader depicted the exclusion of Hermann Oberth (1894–1989) and Rudolf Nebel (1894–1978), two rocket pioneers in the Weimar Republic, from active participation in the army program as persecution of the real pioneers of the technology. The fact that both were ill suited to a serious rocket-engineering program and extreme right-wingers sympathetic to the Nazis were facts that Mader probably knew, but left out in the interests of propaganda. A final, particularly egregious example: late in the book, Mader depicts engineers Klaus Riedel (1907–44) and Helmut Gröttrup (1916–81) (the latter the eventual leader of the German rocket group that went to the Soviet Union) as persecuted too, notably in the case of their brief arrest by the Gestapo in March 1944. Yet nowhere does he mention that von Braun was arrested with them.[11]

Still, there was enough dynamite in Mader's work to make it potentially devastating to von Braun's career. From the Stasi's Nazi files used to investigate, blackmail or embarrass individuals in East and West, Mader received a couple of documents about Wernher von Braun's SS memberships. He had the enrollment sheet (pictured in the book) that showed that the engineer joined the SS-*Reitersturm* (riding unit) in Berlin on 1 November 1933, while a student at the university (and secretly working for the army on rockets). An SS officer list from the Second World War showed von Braun's promotion to *Sturmbannführer* (major) in June 1943. Lacking the SS officer record that had fallen into American hands and been classified, plus the secret declarations that von Braun had made to the US Army in 1946–47, which showed that he quit the *Reitersturm* in mid-1934 and only rejoined the SS in spring 1940 under indirect pressure from *Reichsführer*-SS Heinrich Himmler (1900–45), Mader was able to depict von Braun as a committed SS man throughout almost the whole history of the Third Reich. Even if one did not accept that reading of the evidence, the mere fact that he had been an SS officer was a revelation, as it was essentially unknown to anyone except his German colleagues and the US officials who had been his superiors or who had worked on his case. Thanks to a shared desire to protect von Braun, no one talked about it, and he certainly was not foolish enough to mention it.[12]

Equally potentially damaging was Mader's discussion of the scandal of concentration camp labor in the V-2 program, most notably at the underground Mittelwerk rocket factory near Nordhausen and its associated Mittelbau-Dora camp. Although the survivors of that place had certainly not forgotten the beatings, the hangings and the hellish conditions that led to the deaths of thousands of prisoners, in less than two decades after the war the camp had been virtually written out of history, especially in the West. All the memoirs of the key participants in the program like von Braun and his military chief, General Walter Dornberger (1895–1980), either barely mentioned the place or completely ignored it. Western journalists followed their lead, basically because they knew little about the V-2 story, but they also were motivated to protect the rocketeers who were now Cold War assets. So Mader's account was potentially a revelation, even if it was freighted with communist rhetoric. As specific evidence against von Braun, Mader produced the affidavits of two prominent East-European survivors, who swore that they saw him in the tunnels, sometimes in a brown uniform, that he had to have walked by the piles of dead bodies at one tunnel entrance, and that he witnessed prisoner hangings in the tunnel using the overhead crane that lifted the rockets into position for testing. Most of these assertions are almost certainly untrue, but at a minimum, it was hard to deny that von Braun had been in the tunnels several times, had been in a responsible position, and was aware of the murderous conditions.[13]

Thus *Geheimnis von Huntsville*, which finally appeared in June 1963, threatened to become von Braun's worst nightmare. On the dust jacket was an accurate drawing of him in a black SS-*Sturmbannführer*'s uniform, skull and crossbones on his cap and the Knight's Cross around his neck that Hitler had awarded him in late 1944 (Figure 6.2). Von Braun decided to studiously ignore Mader – a strategy he had already announced to his German correspondents in winter 1962–63. However, late in 1963, he was shaken enough by a letter from a German-American correspondent finally to bring it to the attention of his Washington bosses. Shortly before Christmas, NASA chief James Webb (1906–92) wrote von Braun: '[Deputy Administrator Hugh] Dryden, [Associate Administrator Robert] Seamans, and I have discussed the subject matter of our conference on December 17th, namely, the series or articles and book published in East Germany by a communist writer which represent your activities in Germany in the past and now in the US as militaristic and bloodthirsty.' Von Braun had particularly sought

Figure 6.2 This illustration from a pre-publication leaflet for Julius Mader's anti-von Braun book *Geheimnis von Huntsville* has the probable jacket cover of the 1963 first edition.
Source: Courtesy of the American Institute of Physics, Niels Bohr Library.

his counsel regarding 'the letter that you had received from Peter L. Krohn of Easton, Pennsylvania, referring to a radio broadcast repeating some of the East German allegations.' The implication of Krohn's letter – that the story could break out in America – obviously disconcerted the rocket engineer. The NASA Administrator counseled him not to respond unless publicly questioned, in which case he was to answer that 'everything related to my past activity in Germany [...] is well known to the US Government.' He had become a citizen and therefore was 'not disposed to enter into discussion of events of many years ago.' Webb thereby confirmed von Braun's no-comment policy.[14]

Von Braun and Webb must have breathed a sigh of relief in 1964, when the story went nowhere outside the Soviet bloc. A rare exception was a very short, unrevealing and inaccurate *Washington Post* report in August that the official Moscow newspaper *Izvestia* had attacked him for the treatment of prisoners in an 'underground Nazi rocket base in Poland.' This assault was likely connected to the translation of Mader's work into Russian. For the United States and its Allies, and in much of the rest of the world, von Braun's heroic, quasi-official biography remained intact; indeed, several new books in the mid-1960s only reinforced it. The incompatible discourses about von Braun and the V-2 in East and West remained firmly in place, anchored by the competing ideologies and little effective two-way penetration of press and publications across the 'Iron Curtain.'[15]

Whatever the impact of Mader's book in the West, in the Warsaw Pact it was, like his earlier works, a major success. A Hungarian excerpt appeared in 1963, a full Czech edition in 1964, and the Soviets apparently printed 365,000 copies in Russian, 15,000 in Latvian and 10,000 in Ukrainian in 1965. His DDR publisher issued two more editions in 1965 and 1967. The last's jacket states that 438,000 copies had been issued by then (including the Soviet translations, apparently) and that excerpts had appeared in ten languages. We do not know how much Mader's propaganda shaped the opinions of ordinary people in the Soviet bloc – a very difficult thing to know in any case – but he had successfully created an alternate, communist history of von Braun and the V-2, one that survived to the end of the Soviet empire.[16]

Clearly the Stasi and other party and state authorities supported Mader's research and the dissemination of his work, but does it deserve to be called an official campaign against von Braun? It is somewhat a matter of definition. The tactic of campaigning against individuals in the West by digging up damaging information about their Nazi pasts – with the exception of this case, almost exclusively prominent West Germans – had taken on a new form and dimension in the late 1950s, as the DDR not only worked in concert with Soviet objectives but also sought desperately to defend its legitimacy, notably to its own population. Until the backdoor to the West was slammed shut on 13 August 1961 with the construction of the Berlin Wall, hundreds of thousands of East Germans left to escape the SED's dysfunctional, bureaucratized economy, not to mention its secret-police repression and relentless ideological control. The primary means to DDR self-legitimization was to depict the West German Federal Republic as Nazism and militarism reborn, a threat to world peace, a creature of giant capitalist monopolies and launching pad for American imperialism, whereas the DDR was the true heir to the anti-fascist resistance, of course always led by communists. As part of that propaganda campaign certain high West German politicians, administrators, judges and generals were targeted as ex-Nazis, with some success. In a few cases, these campaigns escalated into carefully staged news conferences, document releases (occasionally including papers faked by the Stasi) and even trials in absentia. Albert Norden (1904–82), the SED Politbüro member responsible for 'West work' took a leading role in steering these campaigns.[17]

Mader's attacks on von Braun were clearly built on this model, but by comparison seem small-scale. Publishing, plus occasional opportunities to speak on East-bloc radio and television, were his only outlets. There is no indication that Norden or the party leadership took any interest in von Braun.[18] Nor is there any indication of a serious attempt to penetrate Western, especially American, opinion about him. No English translation was ever published, unlike Mader's *Who's Who in the CIA* (1968), which gained some currency in the third world. Historian Paul Maddrell concludes that Mader's 'writings were principally aimed at public opinion in the DDR and the rest of the Eastern Bloc.'[19] *Geheimnis* was probably hard to obtain in the Federal Republic, as the barrier between East and West was then so complete. Fear of a von Braun lawsuit may have also blocked publication of the book in the West, since such considerations played out in the motion picture that was to come out of it. In short, Mader's attack was a significant success – but only in the Soviet bloc.

II *Frozen Lightning*

In the DDR, *Geheimnis von Huntsville* certainly stimulated new interest in the history of the V-2 and Mittelbau-Dora. By the end of 1963, Professor Walter Bartel (1904–92), chair for modern history at the Humboldt University and a communist leader in the underground resistance group at the Buchenwald concentration camp, formed a special student cooperative to begin studying the topic. The industrial dimensions of the rocket program and Dora seemed to allow ample opportunity to prove the Marxist-Leninist 'state monopoly capitalism' thesis that giant 'monopolies' (corporations) were the controlling interests in the Third Reich. Several pioneering, if jargon-laden, theses and dissertations came out of that group in the 1960s, at a time when no scholarly work appeared on the topic in West Germany. Mader's book probably also contributed to the creation of a small concentration camp memorial at the old camp site outside Nordhausen. It opened in 1964 and focused predictably on the heroes of the camp resistance led by the martyred communist leader Albert Kuntz (1896–1945), while ignoring all politically incorrect dimensions of the story, such as the culpability of the local population or how much the Soviet rocket program had profited from occupying the underground V-2 plant next to the camp.[20]

But by far the most expensive by-product of Mader's book came out of the East German official film studio DEFA: a major motion picture about the V-2. The initiative came from the director, János Veiczi (1924–87), and the screenwriter, Harry Thürk (1927–2005), of a recent smash hit in the DDR, *For Eyes Only* (original title in English). It was a film in the spirit of Mader's earlier and later work: it dramatized the 'worrying facts about the war preparations of the West German and American secret services' against the East. Looking for their next project, late in 1963 they found Mader's book and immediately saw its film potential. As a model they also had in mind the huge Second World War epics then appearing in the West, like *The Longest Day* (1962). The name they hit on was one of Mader's chapter titles: *Die gefrorenen Blitze* (*Frozen Lightning*) – the name locals gave to the mysterious zigzag contrails that appeared in the sky after the first V-2 launches from the Baltic-coast rocket center of Peenemünde (Figure 6.3).[21]

A fundamental question from the outset for the two, and for DEFA's *Dramaturge* for this project, Dieter Wolf, was how central would be the objective of 'smash[ing] the myth of the fascist rocket baron.'[22] A handwritten concept document from winter 1963–64, probably in the handwriting of Veiczi (a Hungarian who had been a forced laborer in Nazi Germany), calls von Braun a 'key figure in special weapons production for Hitler's Germany, who today as a US citizen continues his devilish and self-willed task with undiminished energy'; that is, the development of rockets as weapons of mass destruction. The proposed movie would be 'the counter-film' to *Wernher von Braun: Ich greife nach den Sternen*. But a purely negative movie did not make a credible concept, nor did it serve the ideological purposes of the SED. Wolf's February 1964 position paper stated rather that '[o]ur theme is the international resistance fight against a weapon of mass annihilation.' Through a series of dramatic links, the film would implant 'the basic concept of international solidarity in the antifascist liberation struggle.' Picking episodes from Mader, that meant scenes of resistance activities in Germany, France and Poland, British intelligence' discovery of the V-weapons, and ultimately, sabotage in the underground plant by prisoners led by Kuntz, ending in their martyrdom.[23]

Still, von Braun was central to the first draft script completed in June 1964. Their picture of him in many ways inverted reality – although I suspect that Veiczi, Thürk and Wolf, good communists all, actually believed that picture to be true from their reading of Mader, Ruth Kraft's successful DDR novel about Peenemünde, *Insel ohne Leuchtfeuer* (Island Without a Lighthouse), and whatever else came to hand. Central to the plot was the relationship between von Braun and a fictional Dr. Grunwald, who dreams of spaceflight and is apolitical at the outset. Von Braun (who in real life was the space-travel obsessive) attracts him to Peenemünde early in the war to work on engines, but tells him, 'somewhat bemused' by his interest in spaceflight: 'In order to prevent misunderstandings: here in Peenemünde our objective is the development of a weapon. And we need it *before* the war is over.' If they

Figure 6.3 Julius Mader's 1963 book provided the basis for a major East German film about the V-2, *Die gefrorenen Blitze* (*Frozen Lightning*), which was released in spring 1967.
Source: Courtesy of Bundesarchiv Berlin and DEFA-Stiftung.

got their rocket, 'there won't be any more enemies' – then they could go into space. Several pages later von Braun waxes enthusiastic about the potential annihilation of London; on screen, the cratered moon morphs into the cratered moonscape of the British capital – a scene soon discarded.[24]

As the war progresses, Grunwald gradually becomes more and more disillusioned, eventually secretly cooperating with the prisoner resistance in the Mittelwerk. Von Braun, on the other hand, is comfortable with concentration camp labor and is chummy with the SS leadership. The arrest of Gröttrup, Riedel and von Braun in real life becomes the arrest of Grunwald in the film because he will not watch the hanging of prisoners from the overhead crane, and von Braun, who does watch, is the one who gets him out because he is useful, telling the SS that he is a 'dreamer.' In fact Gröttrup – because he led the German group in the USSR – was the film-makers' heroic model for Grunwald. In contrast, the filmic von Braun is shown at Los Alamos in the closing scene watching a movie of the first A-bomb test, and is practically rubbing his hands with glee at the prospect of putting nuclear warheads on his missiles.[25]

As the film evolved, the fundamental characterizations of von Braun and Grunwald did not change much, but the film-makers dropped von Braun's name from the script, as well as that of his diplomat brother Sigismund, who played a subsidiary part in the (fictional) secret negotiations to bring Wernher over to the Americans. Wernher von Braun instead became 'the Doctor,' and in the final version 'the Rocket Baron,' although his true identity was transparent. The reason was the desire to export the film to offset its great expense, combined with a fear of a libel lawsuit by the von Braun family. On 10 June 1965, Veiczi and Wolf met with the famous East German lawyer, Friedrich Karl Kaul (1906–81), who acted as co-counsel and representative of East-bloc plaintiffs in West German war crimes trials. He advised the film-makers that the chances of a suit in the East were zero, but that in the West there had been precedents created by other DEFA films. They had two choices: give up on exporting to the West, or take out the names of real people and put in a disclaimer about all characters being fictional. So they did the latter.[26]

By 1965–66, the film-makers' ambition to make a Second World War anti-fascist epic issued into rapidly escalating costs and delays, causing controversy inside the DEFA. There were technically demanding sets and special effects needed to show rocket development, V-2 launches at Peenemünde and missile production in the underground plant, plus scenes to be filmed in English, French, Polish and Russian, featuring actors from those countries in intelligence and resistance roles. There were over 100 speaking parts. Extensive support was needed from East German and Polish institutions and the Soviet Army in the DDR.[27]

There was also pressure from several quarters to increase the antifascist dimension of the film. In late 1964, Bartel's students studying the Dora camp met Veiczi and Wolf and criticized the first script as not being appropriate to the 'honor and the reputation of the resistance fighters.' Among other things they reacted prudishly to sex scenes, including one set in a bordello. Veiczi was furious. In 1965 a prominent Polish documentary-maker more substantively argued that the script was still dominated by the anti-von Braun theme, and that such a negative purpose did not make for a very good movie. The pressure continued to the very end, when the Ministry of Culture's review committee asked the film-makers, shortly before release of *Die gefrorenen Blitze* in spring 1967, to add even more documentary footage featuring the Soviet Army in the Second World War in the interludes between scenes.[28]

The result was a film that cost 5.1 million DDR marks – one of the most expensive films ever made in that country. It ran for three hours, including an intermission, was loosely structured and sometimes hard to follow. Several scenes were brilliantly realized, however, and it remains the best feature film ever made on the V-2 program, although given competition like *I Aim at the Stars*, that is not saying much. When it premiered in East Berlin on 13 April 1967, the East German press greeted it with significant coverage and restrained enthusiasm. By mid-August, it had sold over 632,000 tickets in the DDR (a country of only 17 million people) and several Soviet bloc allies bought film rights. Still, it appears likely that it lost a lot of money.[29]

One reason was the failure to export much to the West, largely because the anti-von Braun theme again came back to haunt the film-makers. Late in editing *Die gefrorenen Blitze*, as they struggled to finish it and rationalize its form to DEFA review bodies, Veiczi, Thürk and Wolf again played up the movie's value in undermining the 'myth' (*Legende*) surrounding Wernher von Braun – as did some press articles after its appearance. Dramaturge Wolf indignantly noted an opinion survey among West German youth, who rated von Braun as a hero and role model on a par with Albert Schweitzer, the sainted Swiss missionary doctor in Africa, no doubt because West Germany's press continued to run fawning profiles of the rocket engineer. But when the East German Ministry of Culture sent it for legal review regarding Western export, the lawyer again noted the dangers of a lawsuit by the von Brauns, particularly in West Germany; eliminating their names from the movie had apparently not been enough. He could only justify it if 'professional historians confirm the predominantly historically correct presentation in the film of events around Wernher von Braun.' Not willing to take the risk of losing precious hard currency fighting a legal action, the ministry banned export of the film to the Federal Republic, which cast a pall over attempts to sell it elsewhere in the West.[30]

There is a curious postscript to this story. Two years later, the DEFA did manage to convince French state television to buy the rights, but that led to an incident when the network, to protect itself, showed it to the new West German ambassador to France – who happened to be Sigismund von Braun. He naturally objected to several scenes, including the entirely fictional ones involving him, and sent notes about the movie to his brother. The network cancelled the broadcast scheduled for March 1969. After a press controversy, the Parafrance film company did run the re-titled film in its 40 cinemas following a prestigious Paris première, but only when DEFA made a bowdlerized version. Dieter Wolf, in a July 1969 critique, noted that the shortened version was technically accurate, tighter and more suspenseful, but politically gutted: 'Episodically only an anonymous chief of the rocket project turns up, one who is, however, not politically engaged, but apparently only fulfills orders. He no longer takes part in the liquidation of prisoners in concentration camp Dora, and his presence at Führer headquarters and his decoration [by Hitler] are suppressed. All disagreements between him and the humanistic scientist Grunwald have been cut,' as was the closing scene of von Braun in Los Alamos. Moreover, the communist resistance role was ironically diminished through other cuts. Effectively *Die gefrorenen Blitze* became 'two completely different movies' in Eastern and Western Europe, the Western one stripped of its anti-von Braun content. That further guaranteed that, just like the Mader book, it would have very little influence on his reputation in the West.[31]

III The Dora trial in Essen

Just as DEFA was preparing the film's première in spring 1967, DDR 'star attorney' Kaul geared up for a new trial – the first one in nearly two decades on the Mittelbau-Dora camp. After multi-year preliminary investigations typical of West German trials, leading to sometimes justified DDR complaints about the slowness and inadequacy of the Western judicial response to Nazi crimes, prosecutors in the Ruhr industrial city of Essen filed charges against three individuals: Helmut Bischoff (1908–93), an SS and Gestapo officer and chief of security for the V-2 program; Ernst Sander (1916–90), an SS sergeant in the Gestapo under Bischoff; and Erwin Busta (1905–82), a camp guard infamous at Dora as '*Stollenschreck*' (the 'terror of the tunnels'). Kaul failed to get appointed to the trial as prosecuting co-counsel (*Nebenkläger*) for the family of the martyred Albert Kuntz, as he could not produce specific evidence that any of the accused were involved, but soon reached that position on behalf of East-bloc survivors who became plaintiffs. As always, Kaul coordinated his activity through the SED leadership, with the primary objective of fighting another battle for the recognition of the DDR, which the West formally did not acknowledge. There were ideological objectives as well: 'effectively supporting the struggle against neo-Nazism in the Federal Republic' and demonstrating that the same giant corporations that allegedly controlled it were involved in Nazi crimes. In East Berlin a 'Dora Working Group' was formed to coordinate support, including a representative of the Stasi and two graduate students of Bartel who joined Kaul's law office.[32]

Kaul's participation in the trial thus had nothing to do with attacking von Braun, but he eventually found a way to embarrass the rocket engineer with marginally more impact in the West than Mader or the movie. Soon after the trial opened in fall 1967, the lawyer filed a motion with the three-judge panel to call as witnesses von Braun, former Armaments Minister Albert Speer (1905–81) and three key Speer deputies. The primary objective was to demonstrate the role of the 'monopolies' in the V-2 program, including the use of slave labor, but that was all too transparent, and the chief judge eventually rejected the motion as having no demonstrated relevance to the guilt of those charged. Kaul also tried to drag the president of the Federal Republic, Heinrich Lübke (1894–1972), into the proceedings, as he had been a leading manager in Speer's Peenemünde construction group, which had used forced labor. The DDR had launched its campaign against Lübke back in 1964 (although Mader had already attacked him in *Geheimnis* in 1963), the last and largest one against a West German politician. The attacks eventually wore down Lübke, who resigned from office a few months early in mid-1969. Kaul had no luck with the president, however, but finally succeeded in getting Speer called as a witness in October 1968, because he might shed light on Bischoff and Sander's role in prisoner executions resulting from sabotage in the factory. Speer's testimony caused a minor press sensation in Germany and led directly to the court's decision in early November to grant Kaul's motions to call Wernher von Braun and his former military chief, General Walter Dornberger, as witnesses too.[33]

Von Braun was very unpleasantly surprised when, in the midst of the preparations for Apollo 8's historic first human trip around the moon, he received an airmail letter dated 6 November 1968 from the chief judge, offering him dates to testify immediately before Christmas or after New Year. If he could not come to Essen, his testimony could be taken through the help of an American court. But that meant von Braun might have to speak publicly about the Mittelwerk issue, which he had so far successfully avoided despite two or three scares in the mid-1960s, including the one about Mader. The NASA General Counsel at the time, Paul Dembling (1929–2011), relates that von Braun was definitely 'troubled' by this letter, 'certainly didn't want to go back to Germany,' and was 'afraid they were going to do something to him.' The Marshall Director was particularly worried about the impact on US public opinion regarding the postwar use of ex-Nazis. Concern about the fall-out for NASA's programs was also on his mind. When von Braun answered on 22 November, he declared that he could not come to Germany because of his obligations to the US space program. Moreover, he had nothing to do with running the Mittelwerk or the Mittelbau-Dora camp, only visited the former on several occasions and had little to offer as a witness. If they still thought he was useful, however, the court should contact his center's Chief Counsel.[34]

Negotiations began. Dembling recalls getting an angry call from someone in Essen about von Braun's absence. In conjunction with the State Department, NASA then proposed that testimony be taken at the West German consulate in New Orleans. That site was chosen, according to Arthur Konopka, the Headquarters lawyer assigned to the case, precisely because it was off the US media's beaten track. Despite their efforts, on 4 January 1969 a wire service report from Essen revealed that the court had called von Braun as a witness, an item that appeared in newspapers across the United States. Two days later, NASA offered 6 February in New Orleans as a meeting date, later postponed to 7 February because of a conflict in von Braun's schedule. A week beforehand, the State Department conveniently denied Kaul a visa, thus keeping the East German from joining the chief judge and the defense attorneys on the trip, but NASA was still nervous about bad publicity.[35]

By the time Konopka accompanied von Braun to the consulate, however, he felt that the Marshall Director was no longer worried, but clearly did not like answering questions about prisoner mistreatment, feeling that it was not his responsibility. Von Braun's 7 February testimony, which was not made available to the press, shows a very clear memory of the Mittelwerk and of the key people involved, but not only did he deny any personal involvement, he also denied ever having received a report of prisoner sabotage – although von Braun cleverly phrased it as an *official* report of sabotage so as to leave the false impression that he had hardly heard of sabotage at all. Afterward, von Braun gave a short statement on the consulate steps in which he declared he had 'nothing to hide, and I am not implicated.' The West German press featured his assertion of a 'clear conscience.' Most of the American media either ignored the statement or remained in ignorance, as neither NASA nor the State Department had informed them of the testimony. But in answering questions on the steps, von Braun lied: he denied there had been any concentration camp prisoners in Peenemünde – a story that indeed did not come out for decades. Afterward, according to Dembling, he was pleased that the matter had turned out so well – certainly they had controlled the US publicity problem. Ten months later, von Braun wrote Dornberger: 'In regard to the testimony, fortunately I too have heard nothing more.' The retired general and aerospace executive had been questioned after him in Mexico, where he had begun wintering.[36]

In Germany, one surprising aspect of the press coverage was how much it was a Western phenomenon and how little attention the Eastern newspapers paid to it – yet another example of the disconnected media discourses in East and West. The SED's official organ, *Neues Deutschland*, ran an article on 1 February 1969, 'SS Leader Wernher von Braun Will Testify as Witness,' but it was a small item on page three, and there is no evidence of further stories in that newspaper. Despite Kaul's close connections to the party leadership, and his occasional requests to run certain stories in the press to bolster his position, it is apparent that the DDR's ruling elite simply did not see much propaganda value in playing up the Essen trial or von Braun's testimony. When Julius Mader wrote a member of the SED Central Committee in late March asking that *Die gefrorenen Blitze* be broadcast on DDR television in July 1969 as part of a massive campaign against the Apollo 11 moon landing, with one of the aims being the 'unmasking of the Nazi von Braun, who is celebrated as a Prometheus in the USA and West Germany,' he got nowhere. Presumably there were many other issues the SED leadership thought of greater importance.[37]

This disinterest was likely connected to an important transition in inter-German relations. The DDR dropped the Lübke campaign in fall 1968 as soon as he announced he would leave office, and the tactic faded away. With the election of Willy Brandt's Social-Democratic-led government in fall 1969 came *Ostpolitik* (Eastern policy) and an easing of tensions, leading to the four-power Berlin treaty of 1971 and the mutual recognition of the two German states. The bitter Cold War rivalry continued, but in less overt forms. Mader published occasional press attacks on von Braun in the early 1970s, but after the American lunar success and the fading of the US-Soviet Space Race, his fixation must have seemed increasingly pointless to Eastern editors. It is telling that, when Wernher von Braun died prematurely from cancer at age 65 in June 1977, the DDR official news agency issued only a short, one-paragraph announcement, shorn of any propaganda content other than a perfunctory mention of his Nazi career.[38]

Kaul's success in calling von Braun as a witness thus only had an impact in the Federal Republic, as a part of the Essen trial's influence on the memory of the V-2 program, reinforced by a generational change in West German attitudes to the Nazi period. A subtle but noteworthy sign was the relative honesty about the horrors of Mittelbau-Dora that journalist Bernd Ruland inserted into his hagiographic, authorized von Braun biography that appeared shortly after Apollo 11. In fall 1969, Speer published his memoirs with further, if self-justifying, information about Dora; in 1970–71 came the first Western scholarship about the camp. The rocket engineer picked up on the shifting attitudes; in late 1971 he told the leader of the West German ex-Peenemünders that their plans for publicly celebrating the thirtieth anniversary of the first successful V-2 launch on 3 October 1942 were ill advised. It was a wisdom notably lacking 20 years later, in a unified Germany, when the old rocketeers and their new allies, ex-East German Air Force officers at Peenemünde, blundered into an international controversy over their celebratory plans for the fiftieth.[39]

By then, attitudes to von Braun and the V-2 had been fundamentally altered, especially in the United States. Thanks to changing public knowledge of the Holocaust, in 1979 Congress authorized the creation of an Office of Special Investigations, a 'Nazi-hunting' unit in the Justice Department. One of its early investigations was the connection between the German-American rocketeers and Mittelbau-Dora, leading to a case against one of von Braun's closest associates. In 1983 Arthur Rudolph (1906–96), who had been production manager in the underground plant, signed an agreement to leave the United States for Germany and denounce his citizenship rather than contest a denaturalization hearing. He did so in spring 1984. When the Justice Department issued a press release in October, it set off a worldwide echo, and opened the doors to revelations about von Braun's record as well. Investigative journalists used the relatively new Freedom of Information Act to get copies of classified documents from his Army security file – notably regarding his SS membership and his relationship with Dora, but also the US government's behind-the-scenes battles in the late 1940s over his immigration status. Until then the Western media, especially in the United States, had studiously ignored, or simply remained ignorant of, information that Mader had often published two decades before. In the aftermath, von Braun's posthumous reputation was greatly damaged.[40]

IV Divided discourses

Why, then, was the East German campaign against Wernher von Braun – if it merits that description – essentially a failure in the West? The key DDR actors in this affair, namely Mader, Veiczi, Thürk, Wolf and Kaul, clearly would like to have destroyed his reputation on *both* sides of the Cold War divide. Given the glaring omissions for the Nazi years in von Braun's quasi-official biography, not to mention the outright falsifications in *I Aim at the Stars* and other popular representations, one can understand that their outrage was not simply motivated by ideology. But whatever hopes they may have had for their impact in the West were largely frustrated by the depth of the chasm between the public and media discourses on the two sides and by limitations, both national and specific to the case, that hampered their ability to bridge that chasm.

Beyond the bitterness of the German divide, the manifest bankruptcy of a regime that had to build a wall to prevent its own population from running away meant that any propaganda coming from the East in the 1960s was almost automatically dismissed in the West. Only against a few Western politicians, notably Lübke, did the DDR score some successes when it could produce credible Nazi documents and invested much effort into bringing that information to the attention of the world press. In von Braun's case, Mader had the support of the Stasi, but he had no major party/state campaign behind him, plus he had very few documents from the Stasi archives that substantiated von Braun's links to the SS and to Dora's horrors. That was in part because of luck – the Western Allies captured the bulk of the Party and SS central membership files in 1945 – and partly due to a deliberate policy of secrecy on the part of the United States government, which kept damaging information about von Braun's past classified in order to protect one of its key technical assets. The Western media shared that motivation with their governments; in the United States and West Germany particularly, von Braun was a hero, and in the latter also a symbol of the alliance with the United States.

The film-makers of *Die gefrorenen Blitze* faced another East German limitation: fear of a libel lawsuit making it impossible to export the movie to the Federal Republic. Behind the Ministry of Culture's decision lay an unsolvable economic problem for a Soviet-bloc economy, the shortage of convertible Western currency and the resulting scramble to earn more money to carry out necessary trade and activities outside the Warsaw Pact. Exporting DEFA films was one way. The studio's requirement for foreign earnings stoked the ambitions of Veiczi, Thürk and Wolf in their desire to shoot a Second World War epic that Western audiences might watch. Yet fear of hard-currency losses in fighting a lawsuit, as had happened to earlier DEFA films, ultimately won out over the desire to make money. When the studio eventually did sell the film in France, it found it could only do so by gutting *Die gefrorenen Blitze* of anti-von Braun content. It would not be surprising if similar legal considerations had earlier hindered the publication of *Geheimnis von Huntsville* in the West.

In the end, only Kaul's motion to call Wernher von Braun as a witness had any impact at all outside the Soviet bloc, and then only in the Federal Republic, where war crimes trials contributed to a gradual shift in public attitudes toward responsibility for the Third Reich. It is noteworthy that in the West Germany of the 1970s there was more discussion of Mittelbau-Dora than in the United States, although von Braun's heroic reputation was still eroded only around the edges. But to conclude by emphasizing the failure of Mader, Veiczi, Thürk, Wolf and Kaul would in many ways leave a false impression: their primary efforts were to bolster 'antifascist' public opinion in the East by instilling distrust and fear of the Western powers who had made an ex-Nazi weapons designer into a hero. Their attacks on von Braun were perhaps less an orchestrated campaign of the East German state than a loose collective effort of well-known, even famous East Germans with significant autonomy to pursue the objectives of the SED Party state in their own way, but they succeeded in creating, or at least greatly fortifying, an alternate discourse about the rocket engineer in the Warsaw Pact, one that aired many of the scandals of his past 20 years before almost anyone did so in the West.

Notes

 Portions are modified from Michael J. Neufeld, *Von Braun: Dreamer of Space, Engineer of War*, New York: Alfred A. Knopf, 2007.

1. Julius Mader, *Geheimnis von Huntsville: Die wahre Karriere des Raketenbarons Wernher von Braun*, Berlin (Ost): Deutscher Militärverlag, 1963 (2nd edn 1965, 3rd edn 1967). Wernher von Braun (hereinafter WvB) to Lehmann, 11 December 1962, in file 417–10, US Space and Rocket Center, Huntsville, Alabama, Wernher von Braun Papers (hereinafter WvBP-H); and WvB to Weiss-Vogtmann, 14 January 1963, in 418–2. The series was in *Forum: Organ des Zentralrats der FDJ* 16 (1962), *wissenschaftliche Beilagen* to issues 36–49 (microfilm copy, Yale University Library). See also Julius Mader, 'Die Karriere des Wernher von Braun,' *DDR in Wort und Bild* (July 1963), 32–3, copy in National Air and Space Museum, WvB bio. file.
2. A partial exception is Horst Körner, *Stärker als die Schwerkraft: Vom Werden und von den Zielen der Raumfahrt*, Leipzig: Urania, 1960; see also his note to Mader in *Geheimnis* (1963 edn), 378–9.
3. See Neufeld, *Von Braun*, 252–78, 284–90, 323–7, 346–53, 408–10.
4. Idem, 'Creating a Memory of the German Rocket Program for the Cold War,' in Steven J. Dick, ed., *Remembering the Space Age*, Washington, DC: NASA, 2008, 71–87.
5. On the history of the V-2 program and Dora, see Michael J. Neufeld, *The Rocket and the Reich: Peenemünde and the Coming of the Ballistic Missile Era*, New York: Free Press, 1995; and Jens-Christian Wagner, *Produktion des Todes: Das KZ Mittelbau-Dora*, Göttingen: Wallstein, 2001. For the best example of the earlier apologist literature, see Frederick I. Ordway and Mitchell R. Sharpe, *The Rocket Team*, New York: Thomas Y. Crowell, 1979. For the muckraking literature after 1984, consult Linda Hunt, 'US Coverup of Nazi Scientists,' *Bulletin of the Atomic Scientists* 41.4 (April 1985), 16–24; idem, *Secret Agenda: The United States Government, Nazi Scientists and Project Paperclip, 1945 to 1990*, New York: St. Martin's Press, 1991; and Tom Bower, *The Paperclip Conspiracy: The Battle for the Secrets and Spoils of Nazi Germany*, London: Michael Joseph, 1987. Erik Bergaust's hagiographical *Wernher von Braun*, Washington, DC: National Space Institute, 1975, may be contrasted with Rainer Eisfeld's critical *Mondsüchtig: Wernher von Braun und die Geburt der Raumfahrt aus dem Geist der Barbarei*, Reinbek: Rowohlt, 1996, and Neufeld, *Von Braun*.
6. Mader's primary Stasi file is at the Berlin central archive of *Der Bundesbeauftragte für die Unterlagen des Staatssicherheitsdienstes der ehemaligen Deutschen Demokratischen Republik* (abbreviated BStU), his 'cadre file' (*Kaderakte*) is MfS-ZAIG 25335/90. Mader's career is thoroughly evaluated by Paul Maddrell's 'What We Have Discovered About the Cold War is What We Already Knew: Julius Mader and the Western

Secret Services During the Cold War,' *Cold War History* 5.2 (May 2005), 235–58. Maddrell focuses on Mader's primary work as a propagandist against Western intelligence agencies.

7. Mader file, MfS-ZAIG 25335/90, BStU; Maddrell, 'What We Have Discovered,' 236–40.
8. Halle and Gleißner 'Beurteilung,' 18 December 1961; and Mielke order 207/62, 1 April 1962, signed by Mader 21 April 1962, 15, 46–8, MfS-ZAIG 25335/90, BStU. All translations are mine.
9. Mader to WvB, 8 July and reply, 21 July 1961; and WvB to Magnus von Braun Sr., 29 October and reply, 7 November 1961, in Nr. 85, N1085 (Nachlass Magnus von Braun Sr.), Bundesarchiv Koblenz. WvB's letter to Mader is reprinted in the foreword to *Geheimnis*, 12–13. WvB's Mader file has disappeared from the WvB Papers in Huntsville. Its existence is revealed in the list of personal and 'sensitive' files attached to Ruth von Saurma's note to WvB, 22 August 1973, in file 607–15, WvBP-H. Most files on this list relating to Mittelbau-Dora are missing too.
10. WvB to Lehmann, 11 December 1962 (and WvB-edited R.v. Saurma draft 27 November), and Lehmann reply, 16 January 1963, in file 417–10; and WvB to Weiss-Vogtmann, 14 January 1963, in 418–2, WvBP-H; Mader, *Geheimnis*, 5–39.
11. Ibid., 72–96, 344–6, 350–1.
12. Ibid., 69, 92–5; WvB SS-Reitersturm-Stammrollenblatt, 28 February 1934, and SS officer list, 1943, former MfS Nazi archive, now Bundesarchiv Berlin-Lichterfelde (hereinafter BArch Berlin) SS file card for WvB, former Berlin Document Center records (BArch Berlin and microfilm, National Archives College Park, hereinafter NACP); WvB 'Affidavit,' 18 June 1947, Accession 70A4398, RG330, NACP. For further information, see Michael J. Neufeld, 'Wernher von Braun, the SS, and Concentration Camp Labor: Questions of Moral, Political, and Criminal Responsibility,' *German Studies Review* 25.1 (February 2002), 57–78; and idem, *Von Braun*, 63–4, 120–2.
13. Mader, *Geheimnis*, 285–325. The definitive work is Wagner, *Produktion des Todes*. See also André Sellier, *A History of the Dora Camp*, Chicago: Ivan Dee, 2003.
14. Webb to WvB, 20 December 1963, copies in HRC 2563, NASA History Division, and in Box 3.16, Ms. 147, Dryden Papers, MSE Library Special Collections, Johns Hopkins University.
15. 'Thought Police,' *Washington Post* (6 August 1964); Dieter K. Huzel, *Peenemünde to Canaveral*, Englewood Cliffs: Prentice-Hall, 1962; James McGovern, *Crossbow and Overcast*, New York: William Morrow, 1964; Ernst Klee and Otto Merk, *The Birth of the Missile: The Secrets of Peenemünde*, New York: Dutton, 1965 (first published as *Damals in Peenemünde: An der Geburtsstätte der Weltraumfahrt*, Oldenburg: Stalling, 1963); David Irving, *The Mare's Nest*, Boston: Little, Brown, 1965.
16. Bibliographic research by Professor Mark Kulikowski, SUNY Oswego, given in two e-mails to me, 16 and 19 November 2006, and in mailed copies of East-European bibliographies. Mader, *Geheimnis* (3rd German edn 1967), rear book flap. Interestingly, this edition (I do not have the second) has a different book jacket, one less inflammatory than the SS drawing. I do not have the first edition book jacket, but it is pictured in an advance book announcement, 'Raketenboy Nummer 1,' copies of in WvB's FBI File 105–130306, 434–5, in RG 65, NACP,

and in folder III./153, Samuel Goudsmit Papers, Niels Bohr Library, American Institute of Physics, College Park, Maryland.

17. Michael Lemke, 'Kampagnen gegen Bonn: Die Systemkrise der DDR und die West-Propaganda der SED 1960–1963,' *Vierteljahrshefte für Zeitgeschichte* 41.2 (April 1993), 153–74, and 'Instrumentalisierter Antifaschismus und SED-Kampagnenpolitik im deutschen Sonderkonflikt 1960–1968,' in Jürgen Danyel, ed., *Die geteilte Vergangenheit*, Berlin: Akademie, 1995, 61–86; Hubertus Knabe, 'Die missbrauchte Vergangenheit: Die Instrumentalisierung des Nationalsozialismus durch SED und Staatssicherheitdienst,' in Manfred Agethen, Eckhard Jesse and Ehrhart Neubert, eds, *Der missbrauchte Antifaschismus*, Freiburg im Breisgau: Herder, 2002, 248–67.
18. I searched electronic finding aids of the Bundesarchiv's SAPMO (*Stiftung Archiv der Parteien und Massenorganisationen der ehemaligen DDR*) records of the 'Büro Alfred Norden' in record group DY 30, BArch Berlin, and found only one reference to Mader, not related to von Braun.
19. Maddrell, 'What We Have Discovered,' 237, 240 (quote).
20. The records of the Humboldt University 'Forschungsgemeinschaft Dora' ended up in the papers of Professor Dr. Friedrich Karl Kaul; see Nr. 268–85, N2503, BArch Berlin; for group origins, see files 275 and 280; for correspondence with Nordhausen, 279. On the latter, Jens-Christian Wagner kindly gave me his unpublished lecture 'Remembering the Nazi Camps in East and West Germany: The Case of the Mittelbau-Dora Camp.' On the German contribution to the Soviet program, see Matthias Uhl, *Stalins V-2: Der Technologietransfer der deutschen Fernlenkwaffentechnik in die UdSSR und der Aufbau der sowjetischen Raketenindustrie 1945 bis 1959*, Bonn: Bernard & Graefe, 2001; Asif A. Siddiqi, *Challenge to Apollo: The Soviet Union and the Space Race, 1945–1974*, Washington, DC: NASA, 2000; and Boris E. Chertok, *Rockets and People*, vol. 1, Washington, DC: NASA, 2005.
21. On the film, see Thomas Heimann and Burghard Ciesla, '*Die gefrorenen Blitze*: Wahrheit und Dichtung. FilmGeschichte einer "Wunderwaffe,"' *Apropos: Film. Das Jahrbuch der DEFA-Stiftung* (2002), 158–80. Archival records are found in DR 117 (DEFA files), BArch Berlin, and in Carl von Ossietzky Universität Oldenburg's Mediathek, Bestand Dieter Wolf (hereinafter UOM/BDW), particularly the Produktionsakt (production file) for *Die gefrorenen Blitze* (hereinafter *DgB*). The quote comes from the latter file, Wolf to Sorin (Kiev), 27 June 1966.
22. Quoted in Heimann and Ciesla, '*DgB*,' 161 ('die Legende um den faschistischen Raketenbaron Wernher von Braun zerschlagen'). They do not footnote it and I have not found it in original documents, but do not doubt its veracity. The WvB 'Legende' is mentioned frequently in the movie-related documents, esp. in 1966–67.
23. '1. Konzeption J. Veiczi,' 1963–64, in Produktionsakt *DgB*, UOM/BDW, 2; Wolf, 'Stellungnahme zum Fahrplan "Gefrorenen Blitze" von H. Thürk u. J. Veiczi,' stamped 28 February 1964, in vorl. BA(I)3671, DR 117, BArch Berlin, 2–3; 'Autor, Regisseur und ein neuer Titel: Gespräch mit Harry Thürk and János Veiczi,' *Neues Deutschland* (21 November 1964), Beilage Nr. 47, 'Die gebildete Nation.'
24. *DgB* Rohdrehbuch, 19 June 1964, vorl. BA(I)3564 (Part 1) and BA(I)3565 (Part 2), DR 117, BArch Berlin, quotes from Part 1, 24, and moon morphing, 36; Ruth Kraft, *Insel ohne Leuchtfeuer* [1959], Berlin: Vision, 1994. Interesting

light on the film's prehistory is cast by a mid-1966 controversy between Kraft, Mader and Wolf; see Wolf to Kraft, 12 May, Mader to Kraft, 19 May, and Kraft to Wolf and Veiczi, 7 July 1966, in Produktionsakt *DgB*, UOM/BDW.

25. *DgB* Rohdrehbuch, 19 June 1964, vorl. BA(I)3565, DR 117, BArch Berlin, quote from Part 2, scene 181; Gröttrup entry in address list, 1964, in Produktionsakt *DgB*, UOM/BDW; Heimann and Ciesla, '*DgB*,' 163, 168–9.
26. Wolf minutes of meeting with Kaul on 10 June 1965, film summary, 1966, and film summary, 25 October 1966, in Produktionsakt *DgB*, UOM/BDW; 1965 scripts, vorl. BA(I)469–72, DR 117, BArch Berlin; Heimann and Ciesla, '*DgB*,' 178–9.
27. Heimann and Ciesla, '*DgB*,' 164–77; Schneider to Wolf, 20 March 1965, in Produktionsakt *DgB*, UOM/BDW.
28. Jerzy Bossak 'Bemerkungen für Regisseur Veiczi,' ca. May 1965, Goßens/Defense Min. letter, 7 July 1965, and other correspondence in Produktionsakt *DgB*, UOM/BDW; minutes of meeting, Veiczi and Wolf with Dora student group, 24 October 1964, in Nr. 275, N2503, BArch Berlin; 'Sitzung des Filmbeirates,' Min. of Culture, 16 March 1967, in Nr. 133, DR 1-Z, BArch Berlin.
29. Heimann and Ciesla, '*DgB*,' 159, 164, 177–80; copies of press clippings on *DgB*, 1967, in Universität Oldenburg's Mediathek, HFF 'Konrad Wolf' – Pressedokumentation; DEFA statement of attendance and sales, 14 August 1967, in Produktionsakt *DgB*, UOM/BDW (listing total earnings, including foreign rights, of about 1.6 million marks).
30. Wolf, 'Stellungnahme zum Rohschnitt,' 29 November 1966, and Hauptdirektor Bruk, VEB DEFA, 'Stellungnahme,' 10 February 1967, in Produktionsakt *DgB*, UOM/BDW; DDR press clippings cited in previous note; Staats to Wagner/HV Film, 22 February 1967 (also in Produktionsakt *DgB*, UOM/BDW), and May–June 1967 correspondence on foreign export, in Nr. 133, DR 1-Z, BArch Berlin; Heimann and Ciesla, '*DgB*,' 179. On West German hero worship, see Neufeld, *Von Braun*, 323–4, 349, 409–10.
31. Heimann and Ciesla, '*DgB*,' 179-80 (quote, 'two [...],' 179); 'Französische Minister sahen DEFA-Premiere,' *Neues Deutschland* (15 June 1969), clipping in Produktionsakt *DgB*, UOM/BDW; and Wolf memo, 14 July 1969 (quote), in latter. A veiled reference to the French broadcast issue appears in Sigismund von Braun to WvB, 18 March 1969, in file 403-10, WvBP-H. Von Braun's file on the issue, entitled 'Notes on East German film "Die gefrorenen Blitze, 1969,"' probably by Sigismund, is noted in Ruth von Saurma's list to WvB, 22 August 1973, in file 607–15, but the file disappeared like the Mader file, hence it is impossible to say when WvB first heard of the movie.
32. Kaul's Dora trial files are in Nr. 239 to 247, N2503, BArch Berlin, and newspaper clippings in Nr. 414; trial records in Ger. Rep. 299, Nordrhein-Westfälisches Hauptstaatsarchiv Düsseldorf, Zweigarchiv Schloß Kalkum (hereinafter NWHSA/ZSK). For the origins of Kaul's participation and the Dora working group, see Kaul's meeting with Foth and Fassunge of the Oberste Staatsanwaltschaft of the DDR on 2 February 1967 (quote on neo-Nazism), Kaul to Norden, 3 February 1967, and Noack report on meeting, 8 March 1967, in Nr. 239, N2503, BArch Berlin. On the East German recognition issue's centrality, see the excellent study by Georg Wamhof, *Geschichtspolitik und NS-Strafverfahren: Der Essener Dora-Prozeß im deutsch-deutschen Systemkonflikt*, MA thesis, Georg-August-Universität Göttingen, 2001.

33. Abschrift of Kaul motion, 4 December 1967, and Kaul to Pötschke/ZK SED, 11 November 1968, in Nr. 243, N2503, BArch Berlin, decision 29 b Ks 9/66, ca. May 1968, and Kaul motion, 11 June 1968 (WvB and others), in Nr. 242 and Kaul 'Betr.: "DORA"-Prozeß,' 3 January 1969, Nr. 244; Wamhof, *Geschichtspolitik*, 125–46. On the Lübke campaign, see Lemke, 'Instrumentalisierter Antifaschismus'; and Hubertus Knabe, *Die unterwanderte Republik: Stasi im Westen*, Berlin: Propyläen, 1999, 121–52, 461–70; Mader, *Geheimnis*, 103–5, 122, 206–9, 313–14, 402, 407.
34. Hueckel to WvB, 6 November 1968, and reply, 22 November 1968, in Ger. Rep. 299/160, NWHSA/ZSK; Dembling phone interview, 29 July 2004. After Mader, the most important previous scare was in 1965–66, when *Paris Match* received protest letters from French Dora survivors over a flattering profile of von Braun and his center. See Neufeld, *Von Braun*, 408–9, and the sources cited, notably von Saurma to WvB through Slattery, 2 May 1966, in file 227–8, WvBP-H.
35. Dembling phone interview, 29 July 2004; Konopka phone interview, 5 October 2004; 'Betty' to Paine, 3 January, re: Dembling call, and Dembling to Paine, 7 February 1969, in Box 32, Paine Papers, Library of Congress Manuscript Division; 'Von Braun Evidence in Nazi Trial Sought,' *New York Times* (4 January 1969), 24; Guilian to Hueckel, 6 and 21 January, Pauli telegram to Hueckel?, 28 January, and Kaul to Hueckel, 31 January 1969, in Ger. Rep. 299/160, NWHSA/ZSK. Kaul anticipated rejection and planned to exploit it propagandistically: 'Betr.: "DORA"-Prozeß.'
36. WvB deposition, 7 February 1969, in Ger. Rep. 299/160, 69–80, NWHSA/ZSK, and West German newspaper clippings, 8–10 February 1969, in Ger. Rep. 299/104; 'War Crimes Trial Hears Dr. Von Braun,' *Chicago Tribune* (8 February 1969), ('nothing'); WvB to WD, 8 December 1969, in WvBP-H, 423–4 (quote). No February 1969 testimony stories were found in the *New York Times, Washington Post* or four other American newspapers available at the Smithsonian on Proquest. But the West German press did respond; see the clippings in Ger. Rep. 299/104, NWHSA/ZSK, and Nr. 414, N2503, BArch Berlin.
37. 'SS-Führer Wernher v. Braun wird als Zeuge vernommen,' *Neues Deutschland* (Republik-Ausgabe) (1 February 1969), in Nr. 414, N2503, BArch Berlin (a similar article ran in the *Berliner Zeitung*); Kaul to Pötschke/ZK SED, 12 December 1968 and 4 February 1969, in Nr. 244, N2503, BArch Berlin; Mader to Lamberz/ZK SED, 25 March 1969 (quote), in UOM/BDW, Produktionsakt *DgB*.
38. J. Mader, 'USA-Prometheus mißbrauchte Arbeitssklaven,' noted in Abt. Agitation document, 12 January 1972, in Teil 2, 549, MfS-ZAIG 10227, BStU, and 'Strippel belastet W. von Braun,' 26 July 1973, in Teil 1, 224–5 of same; 'Wernher von Braun gestorben,' *Neues Deutschland* (18/19 June 1977), in Nr. 414, N2503, BArch Berlin (the same ADN story ran in the *Berliner Zeitung*).
39. Bernd Ruland, *Wernher von Braun: Mein Leben für die Raumfahrt*, Offenburg: Burda, 1969, 233–7; Albert Speer, *Erinnerungen*, Frankfurt am Main: Ullstein, 1969, translated as *Inside the Third Reich*, New York: Avon, 1970; Manfred Bornemann and Martin Broszat, 'Das KL Dora-Mittelbau,' in idem, ed., *Studien zur Geschichte der Konzentrationslager*, Stuttgart: Deutsche Verlags-Anstalt, 1970, 154–98; and Manfred Bornemann, *Geheimprojekt Mittelbau*, Munich: J.F. Lehmanns, 1971; WvB to Grösser, 18 November 1971, in file 427–7, WvBP-H.

On the 1992 controversy, see Johannes Erichsen and Bernhard M. Hoppe, eds, *Peenemünde: Mythos und Geschichte der Rakete 1923–1989*, Berlin: Nicolai, 2004, 10–12. Grösser and WvB also corresponded about a two-part television play broadcast on the ZDF network in August 1971, but it is not clear how critical that film was.

40. Neufeld, *Von Braun*, 474–5; 'German-Born NASA Expert Quits U.S. To Avoid a War Crimes Suit,' *New York Times* (18 October 1984), A12; 'Road to Departure of Ex-Nazi Engineer,' *Washington Post* (4 November 1984), 1; Hunt, 'US Coverup of Nazi Scientists,' and *Secret Agenda*; Bower, *Paperclip Conspiracy*.

CHAPTER 7

Transcendence of Gravity: Arthur C. Clarke and the Apocalypse of Weightlessness

Thore Bjørnvig

I Mythological fuel for rockets of reality

Arthur C. Clarke (1917–2008) was one of the most influential advocates in the twentieth century for the exploration and colonization of space (Figure 7.1). He was a major figure in the British space lobby organization known as the British Interplanetary Society (BIS), serving as chairman in 1946–47 and again from 1950–53. His popular science book on spaceflight, *The Exploration of Space*, made it to the Book of the Month Club in the United States in 1951 and thus secured him a wide readership not only in Britain but also in the United States. As a science-fiction writer, he was not only one of the defining authors of the genre but also one of the first British science-fiction writers who reached a mainstream readership. The movie *2001: A Space Odyssey* (1968), co-authored with film director Stanley Kubrick (1928–99), and the accompanying novel, became a defining moment in the history and popularity of science fiction.[1]

Clarke stood out as one of the foremost spokesmen for the promise of progress offered by science and technology. He was in the same league as other major space advocates such as German-American rocket scientist Wernher von Braun (1912–77), and American astronomer Carl Sagan (1934–96). At the same time, Clarke was also a fervent spokesman for skepticism, rationalism and natural science. Yet, despite the fact that Clarke was an avid critic of religion, religious ideas and longings ran deep in his thinking about the human future in space. This religious side of Clarke must be studied seriously and in detail not only in order to give his work the critical attention it deserves, but

Thore Bjørnvig (✉)
Copenhagen, Denmark
e-mail: thorebjoernvig@gmail.com

Alexander C.T. Geppert (ed.), *Imagining Outer Space*
European Astroculture, vol. 1
https://doi.org/10.1057/978-1-349-95339-4_7

Figure 7.1 Photograph of Arthur C. Clarke (1917–2008), one of the most influential writers in the genres of science fiction and popular science, ca. 1965.
Source: Courtesy of Hulton Archive/Getty Images.

also in order to understand how and why religion and science work together in what De Witt Douglas Kilgore has labeled 'astrofuturism,' of which Clarke's writings are an example.[2]

The role religion plays in a techno-scientific milieu such as the pro-space movement and its astrofuturistic literary outlet is too easily neglected because of the propensity to see science and religion as incompatible categories each of which deals with completely different aspects of existence. In this view, religion deals solely with the 'why' of existence, whereas science deals with the 'how.' Thus, the pro-space movement with its insistence on scientific and technological solutions to problems seems at first glance to represent the very essence of the scientific, anti-religious mind. This picture, however, is much too simplistic. Scientists need meaning and purpose as much as anybody else. The function the idea of progress fulfills for scientifically minded people is similar to the function faith fulfills for a religious person. The teleology inherent in the idea of progress is rooted in Christian apocalyptic and eschatological thought and plays a central role in the pro-space movement. This makes good sense – if pro-space advocates only dealt with questions of how to get into space, they would have great difficulty motivating the public and politicians – not to forget technicians, engineers and astronauts.[3]

There may be many economic and military reasons for going into space, but when space enthusiasts such as Clarke or Carl Sagan tried to convince the public of the importance of exploring and colonizing space their appeals often went beyond such reasons. In fact, they fulfilled an important role. NASA needs independent advocates, or what Piers Bizony calls 'deniable assets,' to show and talk about 'the religious and mystical desires implicit in cosmic exploration' in order to win the support of the public and, in consequence, policy-makers. NASA, Bizony argues, cannot itself talk about spiritual motivations of going into space because of its position as 'a government entity funded by tax-payers.' Clarke fulfilled the role of space advocate described by Bizony, aware as he was that it was not only LOX/kerosene that would get rockets into space – of equal importance was the mythological fuel.[4]

The reasons for going into space put forward by pro-space advocates such as Clarke present themselves to the cultural historian (or, as in my case, the historian of religion) like any other cultural expression dealing with meaning. They are historically contingent and often draw on powerful idea complexes that already exist in a given culture – what H. Porter Abbott has called 'cultural masterplots.' These are stories that are widespread in a culture and through time have accumulated enough power to form an important frame of reference on which people feed cognitively and emotionally. Drawing on such masterplots makes for a very powerful discursive strategy. In a slightly different terminology, masterplots can also be labeled 'myths.' Idea complexes become available to us through narrative, and when such narratives, and the ideologies they harbor, are naturalized, they become myths. That is, they become an unquestioned cognitive matrix through which we organize and interpret reality. What we deem essential and non-essential is influenced by the myths we live by. They set up ideals, present metaphors, produce values, help give a sense of purpose, and influence the choices we make. Myths create a horizon of meaning that functions precisely because the outlook and values it produces have become invisible. They can become part of the implicit religion surrounding instances of what William A. Stahl has termed 'technological mysticism.' Such myths must be scrutinized with what Kilgore has adequately called a 'cold eye' – without doing this, cultural and historical studies will too easily end up reproducing the underlying naturalized belief system instead of critically examining it. This goes for the pro-space movement as well.[5]

In *Astrofuturism* Kilgore describes a 'language of aspiration' on which the prospace movement thrives and in which several narratives, such as the American Dream, utopianism, transcendence, heroism and the liberal promise of a better world play major roles. This chapter argues that a central narrative is lacking in this catalogue without which one cannot fully understand the writings of Arthur C. Clarke: apocalypse. Kilgore's recurrent emphasis on utopianism and transcendence is highly relevant in this context, but not enough to account for the religious – and more specifically apocalyptic – themes present in Clarke's *oeuvre*. My specific contribution to the study of the religious aspects of Clarke focuses on the connection between apocalyptic thought and imaginations about the transcendence of gravity.[6]

II Science, fiction and apocalyptic myth

To clarify the role apocalypse plays in the writings of Arthur C. Clarke, Frederick Kreuziger's studies of apocalypse and science fiction are helpful. According to Kreuziger, 'apocalyptic' can be defined according to the function it fulfills: to give hope to a marginalized group of believers who find themselves thrust into disillusion. In traditional apocalypse the disillusion stems from the promises made by God – whether to Jews or to Christians – that remain unfulfilled. In science fiction it is the promise of science and technology to humanity that remains unfulfilled. Science and technology is to the disillusioned reader of science fiction as God is to the disillusioned Jew or Christian. Both had promised salvation and both disappointed – and in each instance an apocalyptic literature developed to give new hope to the disenchanted believers.[7]

The disillusion felt by space enthusiasts and science-fiction readers, to whom the future has become something imminent, almost tangible, stems from the fact that even though many fantastic things may happen as a result of science and technology, there is still death, sickness, war, poverty – and we still have not established a permanent presence in space, or taken to the stars. Thus, Kreuziger calls science fiction a 'secular apocalyptic literature' that responds to a historical crisis and copes with this crisis by providing its followers with myths and symbols. According to Kreuziger, the 'future as promise' is central to the apocalyptic genre and one of its main objectives is to give hope to believers. The apocalyptic game views the present time as both a time of crisis and expectation, it judges and consoles, and shines a light in dark times by making promises of a new and better world to replace the old.[8]

A brief sketch of the apocalyptic narrative as it originates in the Bible is appropriate here. In Pauline Christianity, Christ is interpreted as a second Adam. The first Adam, tempted by Eve, eats the forbidden fruit and is therefore to blame for the suffering and toilsome way in which humans have to live – 'By the Sweat of Thy Brow.' This is the mythological background of original sin. The second Adam is Christ who at last puts everything right by sacrificing himself on behalf of humanity, thereby – in principal – freeing all from the consequences of the acts of the first Adam. But apparently not everything went well; otherwise the Apocalypse of John would not be a very meaningful ending of the New Testament. Christ must return, after having been crucified, resurrected and gone to heaven, before everything *finally* will be put right. This was seen by John in an apocalypse (an 'unveiling,' or 'unfolding,' typically translated as 'revelation') where the many visions culminate in the return of Christ, the shackling of Satan and the establishment of the 1,000-year reign of Christ on earth. After the 1,000 years have elapsed, Satan 'shall be loosed out of his prison' and together with the nations he has deceived, he lays siege upon 'the camp of the saints' – but Satan is 'cast into the lake of fire and brimstone,' his army destroyed by fire, and finally John sees 'a new heaven and a new earth,' a new Jerusalem descends from heaven and a loud voice explains that 'God shall wipe away all tears from their eyes; and there shall be no more death, neither sorrow, nor crying, neither shall there be any more pain: for the former things are passed away.'[9]

This narrative has permeated Western civilization and become a masterplot. It has influenced the Western concept of history and connected it with a teleological sense of necessity, even inevitability. The apocalyptic culmination of history was foreseen by John and according to this scheme history moves towards a new earth, a new Jerusalem and a heavenly reign. By activating apocalyptic narratives and metaphors Clarke mobilizes this deeply ingrained teleological concept of history. The link to the idea of progress is clear.

III The apocalyptic teleology of Clarke's writings

It is not difficult to find traces of an apocalyptically inspired teleological understanding of history in Clarke's work and its reception. In the 1962 essay 'Aladdin's Lamp' Clarke even played openly on the myth of Adam and Eve's expulsion from Paradise, when, in Clarke's words, 'the gates of Eden clanged shut with such depressing finality.' Since then, said Clarke, humankind has been engaged in a never-ending struggle to obtain the necessities of life. But there is a happy end to this story. Salvation by means of technology lies ahead – the construction of the 'Replicator,' a machine able to reproduce any material thing in whatever quantity necessary, may result in 'the lifting of the curse of Adam.'[10]

In *The Coming of the Space Age* from 1967, a collection of writings anticipating the Space Age edited by Clarke, an essay by J.B.S. Haldane (1892–1964) called 'The Last Judgment' is reprinted. Haldane exchanged letters and ideas with Clarke, and the essay in question throws more light on the connections between apocalypse, space and the teleological necessity connected with space. In his essay Haldane explained that higher religion teaches that individuals only lead good lives if they conform to a greater plan. If this is true, Haldane concluded, it is our duty to realize the scope of such a plan – with the afterthought that this is so whether the plan originates with mankind or God. He continued, in a way that leaves no doubt concerning the apocalyptic content of his message and thus merits quotation at length: 'Man's little world will end. The human mind can already envisage that end. If humanity can enlarge the scope of its will as it has enlarged the reach of its intellect, it will escape that end. If not, then judgement will have gone out against it, and man and all his works will perish eternally.'[11]

There is judgment – but, seemingly, the outcome is uncertain. Yet, there is little doubt as to what direction Haldane (and Clarke) want history to take. The future is mobilized in order to influence the choices made in the present. The possible end of the world hinges on mankind's ability to conform to a greater plan. Part and parcel of this plan is humanity's future in outer space. If we do not follow this plan it will result in the end of the human race. Conforming to the plan will ensure our survival and give us purpose.

Thus, a strong necessity is conferred on the prospect of conquering space – a necessity that is also, in a sense, inevitability. This inevitability is created by the power of the apocalyptic narrative. The importance of an overarching masterplot, or myth, for creating purpose and meaning for a community was also stressed in Clarke's 1953 novel *Prelude to Space*. Here, a historian, Dirk, who has the task of documenting the work of a space organization called 'Interplanetary' in general, and more specifically its launch of the first manned rocket to the moon, says of his time studying Interplanetary that 'he had been amongst men whose lives had a purpose which they knew was greater than themselves.' This 'purpose' is exactly the kind of purpose that Haldane saw as necessary for the survival of the human species and was that which Clarke's apocalyptic narrative supplied to the space community.[12]

In the same novel Clarke also underscores the importance of creating imaginative narratives in order to motivate people and gather support for the space-cause among the public. Dirk has a conversation with two scientists from Interplanetary – Collins and Professor Maxton. Collins mirthfully tells Dirk that Maxton once wrote for '*Stupendous Stories*' – a fictional pulp science-fiction magazine probably meant to refer to *Astounding Stories.* Maxton, though, is by no means ashamed of this. Rather, he states that 'someone had to write about space-travel before people would believe it was possible.' Collins interjects that actually that kind of science fiction had quite the opposite effect and made everyone think that it was just kids' stuff. Maxton agrees that maybe that was how it was earlier, in the 1940s, but then adds: 'They read about it – and when they grew up they made it happen.' Indeed, Clarke has influenced and given hope to generations of scientists, engineers, science-fiction writers – and astronauts.[13]

Clarke has often been hailed as the 'prophet of the Space Age.' His statements and ideas have been treated as almost oracular and even though he pointed out that the future cannot be predicted, he was, nevertheless, obsessed by the possibility of doing exactly this. His autobiography *Astounding Days* presents many examples of 'curious coincidences' and 'uncanny predictions' that are suggestive of something more than mere chance. Clarke's attitude towards the paranormal gradually changed from a measure of credulity to outspoken skepticism. Yet, relatively late in his life he still speculated that the explanation of certain unique and mysterious mental abilities, such as performing advanced mathematical calculations, can be found in the possible existence of some sort of 'universal field of information' and that, as a consequence, memory may be something that actually exists outside the brain. From here it is only a small step to suppose that this mysterious field may also hold information about the future. In some of his works of fiction, Clarke clearly speculated that this might be the case.[14]

Clarke often portrayed the road to the stars as inevitable, to the degree that we may be driven into space by 'some mysterious force,' and he described soaring into space as a movement from the past into the future, from a world (and world view) that is stagnated and decadent into a new world (view) of infinite promise and possibility. In *Prelude to Space*, Clarke wrote that during the countdown of the first spaceship to fly to the moon 'an age was dying and a new one was being born.' *Prelude to Space* was written while Clarke was still active in the BIS and in many ways expressed the longings of this wild bunch of space enthusiasts. In a memoir of his time with the BIS, Clarke quotes T.E. Lawrence's *The Seven Pillars of Wisdom*, 'It felt like morning, and the freshness of the world-to-be intoxicated us.' All this fits well with the apocalyptic expectation of the arrival of a 'new heaven and a new earth.'[15]

In the epilogue to *Prelude to Space*, a literally millenarian piece of prose is presented, saturated with poetic affirmation of apocalyptic promise. It is close to midnight and people expectantly await the stroke of the hour that will propel them into the new millennium, full of wonder, without barriers or frontiers, whether of the mind, nations, or space; finally emancipated from the suffocating restraints of a dying century and entering the Second Renaissance – which for the chiliastic children of the Enlightenment is rather like the Second Coming of Christ. Clarke set the temporal scene purposefully. Often the apocalyptic imagination revolves around significant calendar events, and this is particularly the case when a shift occurs between two millennia. It is no accident that Clarke's best known novel is called *2001*, which is the year in which the new millennium really began. Thus, there are a number of reasons why it is fruitful to read Clarke's work in light of apocalyptic thought. In the following section, I shall concentrate on the connections between gravity, apocalypse and original sin.[16]

IV Gravity and apocalyptic thought

Clarke often used language reminiscent of traditional apocalypse to describe gravity. Consequently, apocalypse as a comparative type can shed light on how Clarke imbued the physical phenomenon of gravity with meaning and, consequently, what role it played among space enthusiasts. Heavenly sojourn is one of the defining characteristics of the apocalyptic genre. The receiver of apocalyptic visions often travels upwards into heaven, or through several heavens, and receives knowledge regarding the world and the future, mediated by some supernatural agent. The journey into another, supernatural world easily maps onto the journey into space. This motif feeds into a multitude of science-fiction works and it also seems to feed into the way astronauts experience and talk about their journeys into space. Indeed, by the very power of the apocalyptic narrative they – happily or inadvertently – become part of, they are almost forced to return to earth with revelations. Likewise, aliens in science fiction and the search for extraterrestrial intelligence (SETI) often fill in the role of supernatural agents. Though Clarke does not explicitly refer to a connection between gravity, apocalypse and the Christian concept of original sin, it nevertheless implicitly structures his metaphors and narratives concerning gravity and weightlessness.[17]

In Clarke's 1962 essay 'Beyond Gravity,' the 'mysterious' force of gravity was portrayed as a malevolent force that controls our lives from beginning to end. According to Clarke gravity was to blame for the – sometimes deadly – injuries we receive when we fall to the ground. We are, however, aware of this our 'earth-bound slavery' and that is why we have always looked longingly at birds and clouds and why we have imagined gods as living in the skies. Combining statements like these with allusions to the loss of Paradise experienced by Adam and Eve mentioned above, a picture begins to emerge: according to Clarke, gravity has about the same effect that original sin in a Christian view of existence has on the lives of human beings, making them follow earthly desires, with a number of dire consequences, rather than lead lives of heavenly purity. In his essay, Clarke continued by saying that we have been dreaming of weightlessness and of becoming a 'heavenly being' – indeed this very phrase 'implies a freedom from gravity.' Psychologists, Clarke said, have tried to explain such dreams as reminiscences from our arboreal past. But Clarke added yet another idea: perhaps the dream of levitation is not a primitive leftover from our past but rather 'a premonition from the future.' From such a perspective, it is our preordained destiny to reach the stars – and along the way we have, since time immemorial, been guided by premonitions. Forced by the teleological narrative of Paradise – Fall – Redemption, Clarke was bound to see the future as a divine promise which we, through the revelations of science fiction, are able to catch in glimpses.[18]

The inevitability inherent in this conception of time and history is sometimes portrayed as evolutionary, even though the processes of evolution ought to be governed by chance and natural selection. A favorite idea of astrofuturistic writing is that since we have originated in the sea, we long to return to that other greater sea, space – and by doing so we will make an evolutionary leap that compares to when our ancestors in the form of fish left the sea. Taken together the leap onto land and the leap into space initiate a radical break with the past and a move towards a promising new realm of experience and opportunity.[19]

How the metaphoric projections of apocalypse onto space travel are structured becomes clearer if we apply George Lakoff and Mark Johnson's conceptual metaphor theory. According to the *Encarta World English Dictionary*, 'fall' means 'to come down freely from a higher to a lower position, moved by the force of gravity.' This surely facilitates the projection of the Biblical Fall onto the 'fall' of our antediluvian ancestors into the ever-pulling clutch of gravity. The project is further aided by the primary orientational metaphors 'Happy Is Up, Sad Is Down.' These metaphors structure linguistic expressions such as: 'my spirit is soaring'; 'I'm on top of the world'; 'I feel down'; and 'I'm depressed.' It is grounded in human experience on a basic level: we are upright beings in a three-dimensional world influenced by gravity. When people get sick or die, we usually find them in a horizontal position close to the ground. This may help explain why all that is good and benevolent has a tendency to be placed above earth (in the heavens) and all that is bad on the ground, or beneath it (hell), or at least these projections are facilitated by such orientational metaphors. Thus, it is easy to feel that a rocket going upwards must be going towards something good. Building partly on the primary metaphor 'Happy Is Up,' the apocalyptic narrative of the Christian Bible is mapped onto the narrative of the conquest of space.[20]

According to Clarke, colonization of space will mean an escape from history. Sometimes the way Clarke described history and tradition is strongly reminiscent of the way he describes gravity, as when he wrote about 'the crushing burden of tradition.' Just like gravity, tradition threatens to crush humanity – really, they are two kinds of the same evil. The radical break with the past initiated by spaceflight mirrors the apocalyptic break with history. But the two evolutionary leaps mentioned above have different qualities. The leap of our fish ancestors was an evolutionary revolution and immensely important, but in Clarke's overarching apocalyptic narrative, more like a necessary evil that made the development of intelligence and technology possible. Without a period on dry land, under the influence of gravity, there would be no technology and thus no spaceships to take us into the weightlessness of space – just as without original sin there could be no redemption. Existence in the sea is paradisiacal because the effects of gravity are annulled by the buoyancy of water. Moving onto land is a 'fall' because existence on land means existence under the influence of gravity.[21]

Dry land is, in Clarke's rendering, a horrible, desert-like place under a merciless sun and existence there a constant struggle against gravity. We do not belong on land, but in the sea. In fact, according to Clarke, we are only able to leave the sea because our skin functions as a space suit filled with water. Elsewhere, Clarke compared the origin of life in the ocean with the origin of the individual in the 'mini-ocean' of the womb, pointed out the severe physical dangers of gravity and stated that, on land, we are but 'exiles – refugees in transit camp' bereft of the freedom offered by the sea that we have left and still without the freedom that awaits us in space. Thus, the view of land-existence as something foreign to humans and a sorry interlude between sea and space is underscored. Clarke described the sea – at least as it is during the day – as an 'underwater Eden that knew nothing of sin or death,' thus making the Biblical reference explicit. We left the freedom of the Edenic sea to spend our lives under the punishing force of gravity. Only in space awaits redemptive weightlessness.[22]

In fact, Clarke's predilection for scuba diving, a major reason that he moved to Sri Lanka in 1956, was sparked by the realization that it offered a way to achieve 'one of the most magical aspects of spaceflight – weightlessness.' In several places Clarke highlighted the shared properties of scuba diving and spaceflight. To Clarke, scuba diving offered an opportunity to experience the blessings of space weightlessness in what was really a surrogate for space: the ocean. But the ocean is not just the ocean. In mystical traditions, especially in Sufi mysticism, the ocean is a symbol of the ultimate, mystical ground of existence. The French mystic and writer Romain Rolland (1866–1944) coined the term 'oceanic feeling' denoting mystical experiences in a letter to German psychologist Sigmund Freud (1856–1939), who incorporated the term into his psychology. The haunting metaphor of the 'cosmic ocean' as something we long to return to because it mirrors our origins in the sea, was often used by Carl Sagan, and has long since become a stable in the rhetoric of the astrofuturist community. The concept of space as an ocean draws some of its evocative power from descriptions of mystical experiences that echo the apocalyptic vision in so far as they offer glimpses of a transcendent, heavenly state of being.[23]

A negative stance towards worldly, earthly existence is found in many religious aspirations, such as in the Buddhist goal of transcending *Samsara*, the Gnostic urge to escape the material world, or the Scientological wish to liberate the inner Thetan. A common denominator is the idea that freedom and salvation lies elsewhere, or at least, as in mysticism, in a different state of being. Despite his critical stance towards 'religion' and his dedication to natural science and technology, Clarke openly admitted to the religious underpinnings of his urge to escape earth. The sudden feelings of estrangement that haunt poets and mystics, and make them ask existential questions are, in fact, said Clarke, accurate premonitions. Humankind does not belong on earth and is on the move to somewhere else – space.[24]

Thus, the cross-domain conceptual mappings between, on the one hand, ocean, land and space and, on the other, Paradise, Fall and Redemption that emerge from Clarke's work invite the reader to think of gravity in certain ways. Paradise maps onto primordial existence of our ancestors in the sea; the fall caused by eating the forbidden fruit maps onto the first fish crawling onto land; existence under the yoke of original sin maps onto existence under gravity; the Ascension and Second Coming of Christ maps onto the ascension of the spaceship into space. Some of these conceptual mappings, such as eating the forbidden fruit and the first fish on land, for instance, are not directly evident from Clarke's text itself but follow from the logic of the apocalyptic narrative structure that underlies it. These mappings draw on various primary and conventional metaphors to form novel and highly complex linguistic metaphors, and all point to the fact that the urge to transcend the force of gravity is by no means just an urge to go beyond this force in a concrete, physical sense. It is also an urge of a religious nature, at least as Clarke described it. Seen in this light it is no coincidence that Clarke, in the short story 'The Road to the Sea,' called a spaceship drive that finally takes humanity to the stars 'the Transcendental Drive,' nor is it a coincidence that he made a fictional character divulge that part of humanity has founded a colony on a habitable world, which they call 'Eden.' By leaving earth's gravity field the apocalyptic transformation is effectuated and humankind is reinstalled in Paradise.[25]

All in all, focusing on the religious aspects and allusions in Clarke's writings shows a world view that, with its many religious underpinnings, seems to stand in paradoxical opposition to the otherwise naturalistic world view heralded by this 'prophet of the Space Age.' But looking deeper into Clarke's discourse, one can see that the scientific project of progress and enlightenment is inextricably bound up with religious ideas, metaphors and visions of a mystical and apocalyptic nature. This claim can be further substantiated by focusing on another image used to visualize gravity, namely that of a well, pit or crater. In a seemingly straightforward description of the problems involved in surmounting gravity, Clarke compared going into space to going up a hill. He offered what he called a 'mental picture' describing how gravity weakens the further from earth you get: that of a crater, a pit, or a mountain turned upside down. From a 'gravitational point of view,' said Clarke, humanity lives at the bottom of 'a gigantic funnel, 4,000 miles deep.' This description is accompanied by an illustration that has become commonplace in books about astronomy: a gridded square with a funnel projecting beneath it, at the bottom of which is a small representation of some astronomical body (Figure 7.2).[26]

Despite its rational, mathematical origin, the image of gravity as a well, pit or crater, nevertheless, resonates with a premodern, Christian conception of the world according to which we spend our lives on the bottom of the 'world' from where we can only barely make out the heavens, or as Clarke had it, 'the eternal sunlight beyond the reach of storms.' In addition, Clarke often portrayed space – the heavens – as the abode of superhuman beings, whether biological or mechanical, compared to whom we are low the ladder of cosmological development. This idea resonates strangely with a medieval religious view of the heavens and it seems to point to the fact that even though proponents of a modern, naturalistic world view directly oppose religious conceptions, they nevertheless sometimes subscribe to ideas that are hardly separable from the thing they oppose. In several places Clarke described extraterrestrial beings as being able to master gravity, thus connecting the ability to control gravity with the powers of superhuman beings. The model of gravity as a pit may be a mathematically correct model and may well

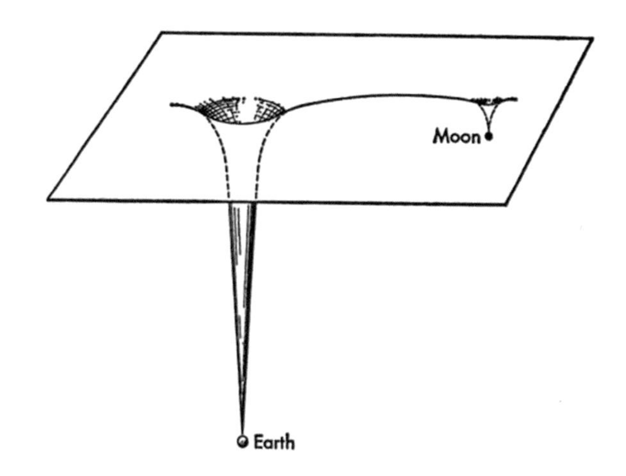

Figure 7.2 A depiction of a 'gravity well' from Clarke's *The Promise of Space*, first published in 1968.
Source: Arthur C. Clarke, *The Promise of Space*, London: Hodder & Stoughton, 1970, 70.

function as a pedagogic tool facilitating the understanding of the effects of gravity. But it also subtly expresses certain ways of feeling about it. It is from out of this deep pit that the protagonists of science fiction and the astronauts of the real world travel on their way to meet the heavenly mediators of esoteric knowledge, and to gain the privileged position of being able to see earth from above which has usually been reserved for gods, and the recipients of apocalyptic revelation.[27]

V The discarded body and the end of man

Clarke did not always present weightlessness as the great redeemer. He was well aware that zero-g presented many problems to astronauts and sometimes he described it not as a blessing but an obstacle – yet still as a vehicle of transcendence. According to Clarke, the problems of extreme forms of pressure, temperature and gravity in space would spark the evolution of mechanical, or artificial, intelligence on behalf of organically developed intelligence. Indeed, Clarke argued, only in space would intelligence realize its full potential in a mechanized, artificial form. In Clarke's essay 'The Obsolescence of Man' humanity's role in the wider scope of things is only that of a chrysalis and the body something ultimately to be discarded. Soon we will have fulfilled our real purpose, namely creating artificial intelligence (AI), or what Clarke called '*M. sapiens*'; that is, *machina sapiens*. One day, *machina sapiens* would outmatch *homo sapiens*, just as homo sapiens outmatched the Neanderthal, ultimately leading to the extinction of our species. We do not have anything to fear, though, for according to Clarke, the advent of machine intelligence might very well usher in a golden age for un-mechanized mankind who will revel in the fantastic powers of their new AI cousins, even though this period may be brief. So enthralled was Clarke by the prospect of AI that to him the idea that AI should become the enemy of humanity is one that merely thrives in pulp sci-fi and comics. From what Clarke said it is clear that to think such lowly thoughts is a matter of humankind projecting instincts developed in our violent past into a future where such instincts will no longer prevail.[28]

In fact – and this makes perfect sense in Clarke's overarching project of transcendence – we should be content to give way to German philosopher Friedrich Nietzsche's (1844–1900) Superman in the form of the machine. For what are we, Clarke asked, quoting Nietzsche's *Thus Spake Zarathustra*, but 'a rope stretched between the animal and the superman – a rope across the abyss.' As such we will have served 'a noble purpose' as parents to our AI successors. The posthuman dream of conquering and transcending the flesh is clearly religiously motivated, as Clarke was well aware. Though intelligence is born of biological life it may, at a later state, leave it behind. Even matter itself, says Clarke, may one day be transcended – just as mystics have suggested. By openly referring to mysticism and the ascetic longing to transcend the material world, Clarke himself put his dreams of the prospects of technology in a religious context. In this Clarke was not alone.[29]

The material world and the body as evils to be conquered is summarized in the Christian formula 'the world, the flesh, and the Devil.' The British scientist J.D. Bernal (1901–71) used this formula as the title of a book in which 'world' becomes nature and the earth, 'flesh' the body's biological foundation, and 'Devil' superstition and irrationality. *The World, the Flesh and the Devil* prophesized a future in which the scientific elite will migrate into space, just as Clarke thought they would. In space the elite will transform themselves into machines and connect into a network that will ultimately become incorporeal by converting itself into light. Clarke was full of praise for Bernal's book, was clearly heir to many of his ideas, and shared with him the idea that the conditions of space will spark the development that leads to a mechanized existence.[30]

Thus, in this scheme yet another level of exaltedness is superimposed on the already exalted state of weightlessness. Weightlessness is no longer a goal in itself, but rather a subtler kind of evil to be transcended. To be able to fully roam the galaxy humankind must abandon biology, in the end perhaps even matter. We must reach an entirely new mode of existence – 'Beyond the Infinite,' as the last part of the movie *2001: A Space Odyssey* is called – and as *2001* suggests, what 'we' will ultimately become are what the spaceship's computer HAL and the human crew encounter in the form of the movie's alien protagonists, the black monoliths. As is made clear at the beginning of the movie, where a monolith seems to ignite a chain of events that leads to our far ancestors' development of technology, it was the monoliths that made us what 'we' – they – are.[31]

In short, in the narrative chalked up by Bernal and Clarke we first transcend gravity, then biology, perhaps even matter itself, and finally we become something so utterly omnipotent and alien that we tremble before it as we tremble before the mystery of God, whether in the form of the monoliths of *2001*, or the group mind of *Childhood's End*. Again, Clarke did not shrink from embedding this narrative in a context that is outspokenly religious. Perhaps, said Clarke, the theologians were not so far off the mark after all concerning humanity's role in relation to God. A minor correction is needed, though. Humanity is not here to worship God but to create him. This, Clarke insisted, should make the debate about whether God is dead obsolete. As co-creator of the movie *2001*, Clarke was well aware that he was in the business of myth-making.[32]

VI Prelapsarian props: myth and motivation in the pro-space movement

What is the function of myth in the community of astrofuturists and why did the great advocate of science and technology – who was also at times a critic of 'religion' – time and again lapse into both implicit and explicit religious discourse? Clarke's writings set forth a narrative in which gravity – and overcoming it – plays a central role. It contains sets of metaphors that combine evolutionary ideas with elements of apocalyptic thought and it serves a dual function. On the one hand, it functions as a common mythology of space enthusiasts which strengthens the community, gives hope and invests pro-space activists with a meaningful narrative offering salvation. On the other hand, the same myth works as a tool of propaganda, which serves to convert both the public and politicians and make them realize that mankind has only one possible future: to take into space.[33]

In Clarke's own understanding, his prophecies and predictions about the future were based on extrapolation from already known facts and guided by logic and rationality. But as Kreuziger has shown, the 'future as promise' in science fiction certainly cannot be accounted for by saying that it is mere 'extrapolation.' This is so because extrapolation, as a rational process guided by historical evidence and what exists here and now cannot move beyond the context set up by these parameters; that is, it cannot say anything meaningful about the future as something that fulfills a promise. Since hard science-fiction's sources are science, logic and rationality, how could anything like a promise emerge from it? Once again, Clarke gave an answer. He insisted that we need logic to predict the future – but added that, as a sort of corrective that occasionally defies logic, 'faith and imagination' are also needed. In the popular science writing and science fiction of Arthur C. Clarke we are not dealing with mere extrapolation but also with myth, propaganda and religion. As the sources used for this chapter show, it is mainly the Clarke of the 1950s and 1960s, and thus the early Clarke – contemporary with the triumphs of the Soviet and American space programs – that has been scrutinized, with his 'optimistic teleology' at full steam. But as Kilgore implies, the wish for salvation, as well as the frustration with the aims and methods of the actual space programs, remain intact up through the various generations of astrofuturist writers and thus the relevance of the analysis of apocalyptic elements remains constant today.[34]

Clarke clearly expected that science and technology would bring about the apocalypse of his dreams. Earlier he was inclined to accept other mediums of a more 'spiritual' kind such as paranormal powers, but later in his life he put this more or less behind him. Time and again Clarke lashed out against the belief in God in particular, and religion in general. At the same time, Clarke subscribed to a variety of religious ideas, seemingly only regretting that religion's way of realizing its beliefs was misguided. As Clarke himself argued in the essay 'Science and Spirituality,' it would be science and technology, not 'religion,' that would put us in touch with divine beings and thereby, at least potentially, deliver us. Clarke added that this was 'a strange thought,' and thus admitted to the inherently paradoxical nature of this notion.[35]

Yet, there is a solution to this paradox – to accept that science and religion are far from always at odds but often work synergistically together. Using a masterplot of such cultural power and spread as apocalypse increases its chances to win new adherents, especially among groups that are already heavily Christian, such as found in, for instance, the United States, where Clarke traveled and lectured extensively before and during the Apollo program. As Kilgore has noted, 'science must engage familiar aesthetic conventions and legitimating narratives to make its knowledge attractive to the lay public.' This may help explain why religious ideas thrive among people who generally adhere to science and technology.[36]

In conclusion, the effect Clarke wished to have on his readers and listeners is the effect a visit to the moon has on a character in one of his short stories, namely the understanding that science, technology and space exploration 'had a poetry and a magic of its own.' For it is magic and poetry that will give hope to the disillusioned space enthusiasts and fuel the motivations of voters, politicians, technicians and astronauts alike – and, in a figurative sense, the rockets flying into space, propelling mankind into a prelapsarian state beyond the gravitational pull of original sin.[37]

If you believe.

Notes

1. On Clarke as chairman, see John Clute and Peter Nichols, *The Encyclopedia of Science Fiction* [1993], London: Orbit, 1999, 229. For *The Exploration of Space* as Book of the Month, see Neil McAleer, *Arthur C. Clarke: The Authorized Biography*, Chicago: Contemporary Books, 1992, 75–7. On Clarke as one of the first to reach a mainstream audience, see Terry Prachett, 'Pratchett Tribute to Arthur C. Clarke,' *BBC News*, http://news.bbc.co.uk/2/hi/uk_news/7304329.stm (accessed 1 October 2017).
2. De Witt Douglas Kilgore, *Astrofuturism: Science, Race, and Visions of Utopia in Space*, Philadelphia: University of Pennsylvania Press, 2003. Clarke criticized religion both inside and outside fiction. For examples of the former, see, for instance, Arthur C. Clarke, *The Fountains of Paradise*, London: Gollancz, 1979, 88, and *The City and the Stars* [1956], London: Gollancz, 2004, 131; for examples of the latter, see Matt Cherry, 'God, Science, and Delusion: A Chat With Arthur C. Clarke,' *Free Inquiry Magazine* 19.2 (Spring 1999), https://www.secularhumanism.org/index.php/articles/2690 (accessed 1 October 2017); and Arthur C. Clarke, 'A Quick Chat With Arthur C. Clarke,' *BBC Focus* 184 (December 2007), 42–3.
3. I use the term 'pro-space movement' to indicate, as does Taylor E. Dark III, people who are united through their belief in the idea of progress and the necessity and unquestionable benefits of exploring and colonizing outer space. However, I use the term in a temporally broader sense than Dark, covering such people as those mentioned above from the beginning of the twentieth century to the present, as opposed to Dark who speaks mainly of a pro-space movement from the 1970s onwards; see Taylor E. Dark III, 'Reclaiming the Future: Space Advocacy and the Idea of Progress,' in Steven J. Dick and Roger D. Launius, eds, *Societal Impact of Spaceflight*, Washington, DC: NASA, 2007, 555–71. On religion as answering 'why'-questions, see Dennis Ford, *The Search for Meaning: A Short History*, Berkeley: University of California Press, 2007, 90; and on the idea of progress as the faith of scientists and rooted in Christian eschatology, ibid., 96–100. A major motivating factor in, for example, the Apollo project was national pride

and Cold War competition with the USSR – but Clarke said next to nothing about these factors, and the astronauts often referred to the spiritual and global importance of the Apollo missions as something transcending national interests. Two important studies deal with the latter, Denis Cosgrove, *Apollo's Eye: A Cartographic Genealogy of the Earth in the Western Imagination*, Baltimore: Johns Hopkins University Press, 2001; and Robert Poole, *Earthrise: How Man First Saw the Earth*, New Haven: Yale University Press, 2008.

4. Piers Bizony, 'Viewpoint: The Bigger Pictures,' *ASK: The NASA Source for Project Management and Engineering Excellence* 33 (Winter 2009), 19–22; http://askmagazine.nasa.gov/pdf/pdf_whole/NASA_APPEL_ASK_33_Winter_2009.pdf (accessed 1 October 2011). For more on the same subject, see Svetlana Shkolyar, 'Conquering Space By Capturing Imaginations,' ibid., 42–6; and, with an emphasis on religious-like experiences, Wendell Mendell, 'Space Activism as an Epiphanic Belief System,' in Dick and Launius, *Societal Impact*, 573–83. Carl Sagan worked as adviser for NASA but still fulfills Bizony's definition of a 'deniable asset.' Regarding mythological fuel for rockets, American film director and avid space advocate James Cameron has made almost the same point; see 'Transcript of Closing Keynote Address by James Cameron at the AIAA's First Space Exploration Conference,' 9 February 2005, http://www.spaceref.com/news/viewsr.html?pid=15381 (accessed 1 October 2017).
5. For the term 'masterplot,' see H. Porter Abbott, *The Cambridge Introduction to Narrative*, Cambridge: Cambridge University Press, 2003. Abbot defines masterplots as 'stories that we tell over and over in myriad forms and that connect vitally with our deepest values, wishes and fears.' He adds: 'It is tempting to see these masterplots as a kind of cultural glue that holds societies together'; ibid., 42, 44. On myth as a cognitive matrix, see Mary Midgley, *The Myths We Live By*, London: Routledge, 2004, 1–6. See also Peter L. Berger, *The Social Reality of Religion*, Harmondsworth: Penguin, 1973, for a general social constructionist theory of society and religion. For more on the connection between metaphor and the construction of society, see Thore Bjørnvig, 'Metaphors and Asceticism: Asceticism as an Antidote to Symbolic Thinking,' *Method and Theory in the Study of Religion: Journal for the North American Association for the Study of Religion* 19.1–2 (2007), 72–120, esp. 78–87. Concerning myth as naturalization of ideology, see Russell T. McCutcheon, 'Myth,' in Willi Braun and Russell T. McCutcheon, eds, *Guide to the Study of Religion*, London: Cassell, 2000, 190–208. For the concepts of 'implicit religion' and 'technological mysticism,' see William A. Stahl, *God and the Chip: Religion and the Culture of Technology*, Waterloo: Wilfrid Laurier University Press, 2001, 1–34; for Kilgore's 'cold eye,' see *Astrofuturism*, 29.
6. For Kilgore's 'language of aspiration,' see ibid., 16, and for his 'narratives,' ibid., 17–21. I call them 'narratives' because each of them can be seen to represent a certain type of story. Others have explored religion in Clarke's writings; see, for instance, Clute and Nichols, *The Encyclopedia of Science Fiction*, 230; and Eric S. Rabkin, *Arthur C. Clarke* [1979], Rockville: Wildside Press, 2006, esp. 27–35. In general, this chapter is indebted to several studies dealing with religious aspects of technology and, in some instances, outer space: Erik Davis, *Techgnosis: Myth, Magic, and Mysticism in the Age of Information*, New York: Three Rivers, 1998; David F. Noble, *The Religion of Technology: The Divinity of Man and the Spirit of Invention*, New York: Alfred A. Knopf, 1997; Mary Midgley, *Science as Salvation: A Modern Myth and Its Meaning* [1985], London: Routledge, 1992, and idem,

Evolution as a Religion: Strange Hopes and Stranger Fears, London: Routledge, 2002; and Marina Benjamin, *Rocket Dreams: How the Space Age Shaped Our Vision of a World Beyond*, New York: Free Press, 2003.

7. Frederick A. Kreuziger, *Apocalypse and Science Fiction: A Dialectic of Religious and Secular Soteriologies*, Chico: Scholars Press, 1982; and *The Religion of Science Fiction*, Bowling Green: Bowling Green State University Popular Press, 1986. 'Traditional' apocalypse refers to generic works such as the Apocalypse of John.
8. Kreuziger, *Religion*, 41. By 'secular apocalyptic' Kreuziger does 'not mean that it has no relation whatever to the biblical, religious forms,' but 'that it has a dynamic of its own, that one need not be aware of the roots and interconnections to appreciate how its mythology, cosmology, and future history have evolved'; ibid., 2–3. For Kreuziger on 'future as promise,' see *Apocalypse*, 49.
9. Quotes from *The Holy Bible: King James Version,* Standard Text Version, Cambridge: Cambridge University Press, n.d., Genesis 3, 19; Revelation 20, 7, 9–10 and 21, 1, 4. Regarding Christ as the second Adam, see Paul's letter to the Romans, esp. Romans 5; and in general Carsten Breengaard, *Kristenforfølgelser og kristendom*, Copenhagen: Forlaget ANIS, 1992, 12–22.
10. Arthur C. Clarke, 'Aladdin's Lamp' [1962], in *Profiles of the Future*, London: Pan Books, 1983, 171–8, here 171 and 177. A similar idea is found in a short story where the 'coming of metal brains' produces a society where 'the curse of Adam is lifted forever'; see Arthur C. Clarke, 'The Lion of Comarre,' in *The Collected Stories*, ed. Malcolm Edwards and Maureen Kincaid Speller, London: Gollancz, 2004, 119–54, here 126. Furthermore, in the July 1930 issue of *Astounding Stories of Super-Science*, there was a story that Clarke himself mentioned in his autobiography, 'The Power and the Glory,' in which a 'young scientist succeeds in releasing atomic energy [...] and calls his old professor to come and witness his triumph,' which, he is sure, means nothing less than 'a new heaven and a new earth – the liberation of mankind from the curse of Adam'; see Arthur C. Clarke, *Astounding Days: A Science Fictional Autobiography*, London: Gollancz, 1990, 54.
11. On friendship and the exchange of letters between Clarke and Haldane, see McAleer, *Arthur C. Clarke*, 331; J.B.S. Haldane, 'The Last Judgement,' in Arthur C. Clarke, ed., *The Coming of the Space Age: Famous Accounts of Man's Probing of the Universe*, London: Panther Books, 1970, 326–7.
12. On discursive uses of the future to mobilize social power in the present, see Bruce Lincoln, *Discourse and the Construction of Society*, New York: Oxford University Press, 1989, 38–50. On Haldane and humanity being predestined to go into space, see McAleer, *Arthur C. Clarke*, 331. Quote from Arthur C. Clarke, *Prelude to Space* [1953], London: The New English Library, 1968, 79. In *Prelude to Space* there is much talk of an organization called 'Interplanetary,' and it is safe to say that this is but a slightly fictionalized portrait of the British Interplanetary Society.
13. Clarke, *Prelude*, 122–3. As astronaut Joe Allen wrote to Clarke: 'When I was a boy, you infected me with [...] the space bug'; see Clarke, *Astounding Days*, 129. Steven J. Dick has pointed out connections between science fiction and SETI, and his observations can easily be transferred to science fiction and space exploration. See Steven J. Dick, *The Biological Universe: The Twentieth-Century Extraterrestrial Life Debate and the Limits of Science*, Cambridge: Cambridge University Press, 1996, 222–66; and in particular his contribution, Chapter 2 in this volume.

14. For Clarke as prophet, see John Reddy, 'Arthur Clarke: Prophet of the Space Age,' *Reader's Digest* 94.564 (April 1969), 134–40; and, for instance, the blurb on Arthur C. Clarke, *Greetings, Carbon-Based Bipeds! A Vision of the Twentieth Century as it Happened*, London: HarperCollins, 2000. On the future as unpredictable, see Clarke, *Astounding Days*, 182. Clarke mused over his frequent use of the phrase 'curious coincidence,' ibid., 194; as to 'uncanny predictions' see ibid., 34. On Clarke's changing attitude towards the paranormal and his suggestion of the existence of a universal field of information, see ibid., 185–6. Fictional works that present similar ideas are, for instance, Arthur C. Clarke, *Childhood's End* [1954], New York: Random House, 1990, 200–1, and *Imperial Earth* [1975], London: Gollancz, 2001, 279. Rabkin, who has a keen eye for the 'Biblical resonances' of Clarke's work, says of Clarke's understanding of time and causality: 'In God's eye, all things happen at once. The Old Testament prefigures the New, and in the eternity of myth, the flow of history becomes insignificant.' See Rabkin, *Arthur C. Clarke*, 27–9.
15. The quote 'some mysterious force' can be found in Arthur C. Clarke, 'Space Flight and the Spirit of Man,' in *Voices From the Sky*, London: Mayflower Books, 1969, 11–18, here 15; and the quote 'an age was dying [...]' in idem, *Prelude*, 153. T.H. Lawrence in Arthur C. Clarke, 'Memoirs of an Armchair Astronaut (Retired)' [1963], in *Voices*, 141–52, here 147. Lack of space precludes comments on Clarke's *Childhood*, which is the eschatological work par excellence in this context.
16. Clarke's millenarian piece is found in Clarke, *Prelude*, 158–9. For apocalypse and significant calendar events, see Frank Kermode, *The Sense of an Ending: Studies in the Theory of Fiction*, Oxford: Oxford University Press, 1968, 9–15. Clarke himself points out that the new millennium begins in 2001, and not 2000; Clarke, 'Out of the Cradle, Endlessly Orbiting [...],' in *Collected Stories*, 697–701, here 697.
17. For heavenly sojourn as central to apocalypse, see John J. Collins, 'Introduction: Towards the Morphology of a Genre,' *Semeia* 14 (1979), 1–19, esp. 9; and in general, Martha Himmelfarb, *Ascent to Heaven in Jewish and Christian Apocalypses*, Oxford: Oxford University Press, 1993. To give an example of apocalyptic thought influencing the way astronauts talk about their experiences in space, American astronaut Walter Schirra (1923–2007) said that weightlessness was connected with a feeling of pride and solitude and a 'freedom from everything that's dirty, sticky.' You become more energetic and work better 'without difficulty as if the biblical curse *in the sweat of thy face and in sorrow* no longer exists. As if you've been born again'; quoted in Benjamin, *Rocket Dreams*, 22 (emphasis in original), who quotes from Oriana Fallaci, *If the Sun Dies*, New York: Atheneum, 1965. Schirra was Christian and an active member of the Episcopal Church; see Noble, *Religion*, 138. On Clarke, space and rebirth, see also Howard E. McCurdy, *Space and the American Imagination*, Washington, DC: Smithsonian Institution Press, 1997, 103. On aliens inserted into Collins's definition of apocalypse, see Thore Bjørnvig, *Science, Apocalyptic, and the Quest for Meaning in the SETI Movement: An Examination of the Interfaces between Science Fiction, Religion, Science and the Search for Extraterrestrial Intelligence (SETI)*, MA thesis, University of Copenhagen, 2005, 86–101. It should be noted that John J. Collins finds the 'analogy' between ancient apocalypses and science fiction 'limited, as science fiction is not presented as revelation and lacks the religious and instructional dimensions of the ancient apocalypses'; see Collins, 'Apocalypse: An Overview,'

in Lindsay Jones, ed., *The Encyclopedia of Religion*, 2nd edn, vol. 1, Detroit: Macmillan Reference, 2005, 409–14, here 413.

18. Arthur C. Clarke, 'Beyond Gravity,' in *Profiles*, 58–72. A similar idea can be found in a short story where the reduced gravity of the moon gives a 'freedom that before the coming of spaceflight men only knew in dreams'; Arthur C. Clarke, 'Venture to the Moon,' in *Collected Stories*, 530–49, here 538. In fact, Neil Armstrong had recurrent dreams of weightlessness during his childhood, see Tom D. Crouch, *Aiming for the Stars: The Dreamers and Doers of the Space Age*, Washington, DC: Smithsonian Institution Press, 1999, 221–2.
19. For evolutionary leaps onto land and into space, see Clarke, 'Spaceflight and the Spirit of Man,' 15, and 'Rocket to the Renaissance' in *Profiles*, 94–109, here 107. In fact, it was Clarke who invented the analogy between the evolution of fish and the exploration of space; see McAleer, *Arthur C. Clarke*, 136. Frank White has worked this kind of evolutionary speculation into a religious system, including visualization exercises; see his *The Overview Effect: Space Exploration and Human Evolution*, Boston: Houghton Mifflin, 1987. As to a promise of a new realm of experience and opportunity, see Arthur C. Clarke, 'Across the Sea of Stars,' in *Report on Planet Three and Other Speculations* [1972], London: Pan Books, 1984, 124–9, here 125.
20. For the conceptual metaphor theory, see George Lakoff and Mark Johnson, *Metaphors We Live By*, Chicago: University of Chicago Press, 1981; eidem, *Philosophy in the Flesh: The Embodied Mind and Its Challenge to Western Thought*, New York: Basic Books, 1999. 'Orientational' is a term used by Lakoff and Johnson, referring to metaphorical concepts drawing on orientations in a three-dimensional space, for instance, 'up' and 'down.' Lakoff and Johnson have a special notation system for metaphors, writing them either in upper case, or, as I have done here, capitalized. For orientational metaphors, see Lakoff and Johnson, *Metaphors*, 14–21, and for primary metaphors, eidem, *Philosophy*, 45–59.
21. For the 'crushing burden of tradition,' see Arthur C. Clarke, 'Earthlight,' *Collected Stories*, 332–70, here 347. Kilgore also deals with the utopian longing for a break with terrestrial history; see, for instance, *Astrofuturism*, 1. In Clarke, *The Songs of Distant Earth*, New York: Random, 1987, 251, fire is seen as a prerequisite for the development of technology and, as fire is not possible under water, a period of existence on land is necessary for a species to develop technology. On land-existence and intelligence, see idem, 'Rocket to the Renaissance,' *Profiles*, 107.
22. For land-existence as desert-like, see Clarke, 'Space Flight and the Spirit of Man,' 11, and as battle against gravity, ibid., 14. On skin as space suit, ibid. For the comparison between our species' origin in the sea and the individual's in the womb, physical dangers of gravity, and that we are but exiles bereft of the freedom of both sea and space, see Clarke, 'NASA Sutra: Eros in Orbit,' in *Greetings*, 428–33, here 433. In another essay the image of the 'transit camp' is repeated with the addition that we are 'waiting for our visas [for space] to come through'; see Clarke, 'Across the Sea of Stars,' 125. For sea as an Eden, idem, *The Lost Worlds of 2001*, London: Sidgwick & Jackson, 1972, 95. It should be noted, though, that Clarke also described land-existence in more positive terms, calling it, for instance, 'a brief resting place between the sea of salt where we were born, and the sea of stars,' and portrayed the sea negatively as containing 'a meaningless cycle of birth and death' and the return to it, in evolutionistic terms,

an atavistic occurrence. See Clarke, 'Rocket to the Renaissance,' 107, and 'The Obsolescence of Man,' in *Profiles*, 228–43, here 239. But the gist of the story remains: gravity-ruled existence on land is unhappy and but a stage to be transcended. For a passionate and burlesque description of a 'gravity hater' by one of Clarke's heroes, see Konstantin Tsiolkovsky, *The Call of the Cosmos*, Moscow: Foreign Languages Publishing House, 1960, 80. Interestingly, the French captain and underwater explorer Jacques Cousteau (1910–97) – whom Clarke knew personally (see, for example, McAleer, *Arthur C. Clarke*, 139) – believed that to conquer gravity meant to conquer death; see Walter A. McDougall, 'A Melancholic Space Age Anniversary,' in Steven J. Dick, ed., *Remembering the Space Age: Proceedings of the Fiftieth Anniversary Conference*, Washington, DC: NASA, 2008, 389–95, here 393.

23. On Clarke moving to Sri Lanka, see McAleer, *Arthur C. Clarke*, 115–23. For weightlessness as magical aspect of spaceflight, see Clarke quoted in ibid., 105; see also Clarke, 'Space Flight and the Spirit of Man,' 13. For further examples of comparisons between scuba diving and spaceflight, see idem, 'Which Way is Up?,' in *Report on Planet Three*, 225–35, here 226, and *Astounding Days*, 114. As Eric S. Rabkin has rightly said of Clarke, he 'seems to want the ability to dissolve himself into something great and powerful that is of the order of importance of the universe as a whole.' This possibility of transcendence is offered both by sea and by space, and Clarke 'never gets onto the ground, physically, in a happy way. It's always in the sky or in the ocean'; Rabkin quoted in McAleer, *Arthur C. Clarke*, 384. For Sufi and the ocean as metaphor for mystic experience, see Farid-ed-din Attar and Djalal-ed-din Rumi, in Vilhelm Grønbech and Aage Marcus, eds, *Mystik og mystikere*, Copenhagen: Gyldendalske Boghandel, Nordisk Forlag, 1930, 70–3. On the meaning and history of the term 'oceanic feeling,' see William B. Parsons, 'The Oceanic Feeling Revisited,' *Journal of Religion* 78.4 (October 1998), 501–23; and Caroline Rooney, 'What is the Oceanic?,' *Angelaki: Journal of the Theoretical Humanities* 12.2 (August 2007), 19–32. For Sagan's 'cosmic ocean,' see Carl Sagan, *Cosmos: The Story of Cosmic Evolution, Science and Civilization* [1980], London: Abacus, 2003, esp. the first chapter.
24. On the feeling of estrangement of poets and mystics, see Clarke, 'Across the Sea of Stars,' 124.
25. I use Lakoff and Johnson's theory of metaphor in a highly sketchy fashion. Detailed analysis is needed to fully unravel the complex conceptual mappings presented here. For the spaceship drive, see Arthur C. Clarke, 'The Road to the Sea,' in *Collected Stories*, 263–300, here 265. Elsewhere it is called 'the Transfinite Drive'; see Clarke, 'The Star,' in *Collected Stories*, 517–21, here 520. The 'space drive' is often imagined to be an 'anti-gravity' device; see, for instance, idem, *The Promise of Space* [1968], Harmondsworth: Penguin, 1970, 267.
26. Ibid., 65, 68–9.
27. 'Eternal sunlight': Arthur C. Clarke, 'The Other Side of the Sky,' *Collected Stories*, 631–46, here 645; on being low on a universal ladder of development, see Clarke, *Lost*, 198. For extraterrestrials as masters of gravity, ibid., 224, 233. Worth noting is also the fact that in *Fountains of Paradise* the great invention to annul gravity – the space elevator – is called 'a stairway to heaven'; see Clarke, *Fountains*, 56. For a description of a medieval view of the heavens, see C.S. Lewis, *The Discarded Image: An Introduction to Medieval and Renaissance Literature* [1964], Cambridge: Cambridge University Press, 1994, 98–9. Even though

Clarke qualifies the above statements by saying that 'this picture is only a mathematical model,' it nevertheless has become the way in which many people now envision gravity; Clarke, *Promise*, 71. Mette Marie Bryld and Nina Lykke's feminist-deconstructionist analysis of space enthusiasts' understanding of gravity and the gravity well parallels mine; see their *Cosmodolphins: Feminist Cultural Studies of Technology, Animals, and the Sacred*, London: Zed Books, 2000, 106–9. Bryld and Lykke even quote Timothy Leary for saying 'the original sin of "Genesis" is gravity; the fall' (107).

28. Clarke, 'The Obsolescence of Man,' in *Profiles*, 232–43.
29. Having just stated that intelligent machines will be peaceful ('The higher the intelligence, the greater the degree of cooperativeness') it is a strange move to evoke Nietzsche, given his great emphasis on the 'will to power' inherent in all life. Though admitting that 'intelligence is developed by struggle and conflict' (ibid., 238), Clarke nevertheless seems to think that intelligence will rise above that which brought it to its victorious position. However that may be, it is no coincidence that a central musical theme of *2001* is Richard Strauss's *Also sprach Zarathustra*. For Clarke on mysticism and transcendence of the material world, see ibid., 237. A highly relevant study in this context, which argues along the same lines as I do, is Robert M. Geraci, 'Apocalyptic AI: Religion and the Promise of Artificial Intelligence,' *Journal of the American Academy of Religion* 76.1 (March 2008), 138–66; and his recent book *Apocalyptic AI: Visions of Heaven in Robotics, Artificial Intelligence, and Virtual Reality*, New York: Oxford University Press, 2010.
30. J.D. Bernal, *The World, the Flesh and the Devil: An Inquiry into the Future of the Three Enemies of the Rational Soul* [1929], London: Jonathan Cape, 1970. Clarke agreed with Bernal on the idea of the elite taking to space when he said that the 'dullards may remain on placid earth' while 'real genius will flourish only in Space'; see Clarke, 'The Obsolescence of Man,' in *Profiles*, 239. For praise of Bernal, see ibid., 241; 'Beyond Centaurus,' in Clarke, *Voices*, 39–50, here 42; and 'The Planets Are Not Enough,' in *Report on Planet Three*, 89–99, here 94. Gregory Benford has noted that Clarke's work, esp. *Childhood's End*, stands in the tradition of Bernal; see McAleer, *Arthur C. Clarke*, 381.
31. This interpretation of *2001* is inspired by Robert Sawyer, 'Artificial Intelligence, Science Fiction, and the Matrix,' in Glenn Yeffeth, ed., *Taking the Red Pill: Science, Religion and Philosophy in the Matrix*, Chichester: Summersdale, 2004, 56–71.
32. Arthur C. Clarke, 'The Mind of the Machine,' in *Report on Planet Three*, 133–45, here 145. For Clarke as aware of producing myth, see 'The Myth of 2001,' in ibid., 253–5.
33. In his book, Geraci reaches the exact same conclusion regarding the apocalyptic content of popular books on robotics and AI by spokesmen for transhumanism: 'Apocalyptic AI [...] is a [...] strategy for the acquisition of cultural prestige, especially as such prestige is measured in financial support.' See Geraci, *Apocalyptic AI: Visions*, 3. In general Geraci's studies complement the content of the present chapter in many intriguing ways, though Geraci does not investigate the space dimension of apocalyptic AI in depth.
34. A recurrent theme that preoccupied Clarke throughout *Astounding Days* was a reassessment of the accuracy of the scientific predictions made in the science-fiction magazine *Astounding Stories*: '*Absurd* science – yes. *False* science – no!'

(emphasis in original); see Clarke, *Astounding Days*, 45. For Kreuziger on extrapolation and 'future as promise,' see *Apocalyptic*, 86. For Clarke's 'optimistic teleology,' see Kilgore, *Astrofuturism*, 112. On the continuity of frustration and wish for salvation in the pro-space movement, see ibid., 132, 144. The nature of the frustration in the pro-space movement makes the theory of apocalyptic presented by Kreuziger retain its relevance, even with the Apollo program at its peak. Geraci also argues that the frustration and alienation resulting in apocalypticism can arise in both weak and powerful groups; see *Apocalyptic AI: Visions*, 14–21, and 'Apocalyptic AI: Religion,' 142, n. 5.

35. On the uses of 'faith and imagination,' see Arthur C. Clarke, 'Hazards of Prophecy: The Failure of Nerve,' in *Profiles*, 15–26, here 26. Especially enlightening regarding Clarke and the paranormal is the prologue to *Childhood's End*, v–viii. On the idea that science and technology will put us in touch with the divine, see Clarke, 'Science and Spirituality,' 154.
36. Clarke on the lecture circuit in the United States, see McAleer, *Arthur C. Clarke*, 126–7, 133–6, and 216–17. Geraci also argues that the lines between science and religion 'are neither clearly nor permanently demarcated,' see Geraci, 'Apocalyptic AI: Religion,' 159. Kilgore, *Astrofuturism*, 8.
37. Arthur C. Clarke, 'Holiday on the Moon,' in *Collected Stories*, 309–31, here 329.

PART III

Visualizing Outer Space

CHAPTER 8

Per Media Ad Astra? Outer Space in West Germany's Media, 1957–87

Bernd Mütter

I Spaceflight and the media

The contribution of mass media to the public image of space travel is eminent. When science editor Werner Stratenschulte (1926–) of German television network Zweites Deutsches Fernsehen (ZDF) reviewed the television reports on NASA's Apollo 8 mission in 1968 he wrote, 'thanks to television the best show of the year has reached its audience.'[1] Stratenschulte's declaration underlines the fact that human activity in outer space was a media event. The topic of people in space was both new and extraordinary since space travel essentially means leaving all basics of terrestrial life behind, and therefore spaceflight events received a high degree of attention from an awed public.[2] At the same time, space events depended on the mass media to be made publicly known ('thanks to television'). In contrast to other events covered by the media, such as wars, natural disasters, economic crises, or sports events, individuals – aside from a minority of space *personae* – could not have any experiences, observations or memories of spaceflight independent of media coverage. The only alternative sources of information and interpretation on outer space were science-fiction literature, cartoons and movies, as well as the popular science books that had earlier caught the public eye. After Sputnik's launch in October 1957, however, nearly all social communication on spaceflight, in terms of public attention, was based on narratives offered by the mass media. Hence, mass-media narratives not only shaped, but also reflected what the public perceived as the realities of spaceflight.

Bernd Mutter (✉)
ARTE, Strasbourg, France
e-mail: muetter@macbay.de

Alexander C.T. Geppert (ed.), *Imagining Outer Space*
European Astroculture, vol. 1
https://doi.org/10.1057/978-1-349-95339-4_8

The meaning of space exploration is not determined by way of technological progress or through technical devices and artifacts.[3] Especially during its earlier stages of development, space technology promised to impact other social fields such as science, technology, the military, economy and ecology. This presumption was based on implicit expectations for the future. Thus, the meaning of space travel, including its purpose, its societal relevance and the prospects of space exploration, was a cultural construct defined through social communication. Even if spaceflight delivered transnational media events, the construction of outer space semantics took place largely within nationally limited spheres, and through these channels, the media set up the cultural framework for the respective national space policies.[4]

This chapter focuses on West German media, in which the impact of space differed significantly from that in the United States or in any other Western European country. On the one hand, private space associations in Germany had already developed popular utopias during the 1920s, and the technical achievements of Wernher von Braun's team led to the development of the V-2 rocket during the war.[5] From the 1950s West Germany had one of the biggest economies in Western Europe with a productive scientific community. In terms of economic and scientific power, West Germany could have contributed substantially to Western European space efforts. On the other hand, when the war was over in 1945, von Braun and many of the V-2 experts were brought to the United States; the defeated Germans were not allowed to develop their own missiles. If West Germany sought to participate in the development of space technology, it had to do so with a non-military nature. West Germany was not able to use the military value of launcher technology for reasoning in favor of a national space program, which was a major argument for other countries, such as France.[6] The discourse on space technology in West Germany relied on future benefit. At the same time, German science fiction and popular science writings on outer space had had their day. In contrast, the writings on American astrofuturism – a tradition of science fiction and science fact, with writers who described the future of mankind in outer space as a real and concrete possibility – became more and more popular in West Germany, even if many of them were based on the American frontier narrative (although many of the most prominent authors such as Wernher von Braun or Willy Ley were German by birth).[7]

This leads to three different questions: How did national distinctions influence the image of spaceflight and the imagination of outer space in mass media? What impact did American astrofuturism have on West German mass-media reports? And what effect did the strategies and criteria of the media system itself have for the image of spaceflight? The first source for the present study is the coverage of space travel in the *Frankfurter Allgemeine Zeitung* (FAZ), a West German national daily center-right newspaper. Here, it is particularly important to focus on newspaper coverage between June 1957 through March 1962; that is, from the coverage of the launch of the first artificial satellite, Sputnik, until the foundation of ELDO, the European Launcher Development Organization.[8] In this period, West Germany's space policy shifted from total abstention from spaceflight to participation in supra-national organized activity. During these five years, there were a total of 1,031 articles on space topics published in the FAZ. As it focused mostly on the political implications of spaceflight, the FAZ thus based its reports and commentaries on explicit arguments.

Space travel's inherent visuality was the primary reason why television was considered the most appropriate medium for covering advances in spaceflight.[9] Television was, from the 1960s, the leading medium in West Germany.[10] Therefore, ZDF's television science show *Aus Forschung und Technik* serves as a second source, especially between 1964 and 1987. Unlike West Germany's first television network Arbeitsgemeinschaft der öffentlich-rechtlichen Rundfunkanstalten der Bundesrepublik Deutschland (ARD), which was composed of eight and later nine local stations delivering different series, shows and formats to the joint television program, ZDF, the central national public broadcaster on air since 1963, had a permanent place for reports on space technology and outer space: its monthly scientific program *Aus Forschung und Technik* (Of Research and Technology), introduced in fall 1963 and headed by Heinrich Schiemann (1916–2002), an aeronautics engineer, who came to ZDF from Nordwestdeutscher Rundfunk/Norddeutscher Rundfunk in 1962.[11] For the first time *Aus Forschung und Technik* reported on space travel in March 1964, during a period of growing public excitement about spaceflight. Until 1987, when ZDF replaced this show with *Abenteuer Forschung*, 47 out of 237 episodes – almost 20 percent – featured spaceflight topics. As a long-standing series, it offers the opportunity to analyze both continuities and discontinuities in the way television presented space travel. The end of *Aus Forschung und Technik* in 1987 coincided with two major events affecting Western space policy. The Space Shuttle *Challenger* disaster occurred in 1986 and prompted the grounding of all Space Shuttles for 32 months, leading to a delay for many Western space projects. By 1989–91, the end of the Cold War and the fall of the Soviet Union concluded the East-West space rivalry and resulted in a downswing in long-range space projects, which affected ZDF's science shows, too. From 1988 to 2003, when the *Abenteuer Forschung* series ended, only 17 out of 195 episodes – less than 10 percent – included space topics. When the production of *Aus Forschung und Technik* was stopped, the great time of space television on ZDF had come to an end.

II Germany's crux: the military dimension of space

The launch of Sputnik, the first man-made satellite, on 4 October 1957 not only started the Space Race, but it also proved the beginning of enormous media hype and unprecedented public attention towards spaceflight. After this first Soviet success, the predominant interpretation of the Space Race in the US media was equivalent to that of a substitute for war, rooted in the military relevance of launcher technology.[12] Shortly after Sputnik, the West German media interpreted spaceflight in a very similar way. Heinz Gartmann (1917–60), a former engineer and one of the most prolific space authors in post-war Germany, wrote on 7 October 1957: 'The nation that is able to deploy a certain payload (in this case a satellite) [...] into earth orbit, will be capable of transporting any other load (or perhaps an atomic bomb) with a similar rocket from its territory to any point on the earth's surface.'[13] In this light, Sputnik appeared to be a new step forward in the superpowers' race for nuclear superiority. West Germany had a more or less passive part in the race and seemed to be a US ally only in the sense that it had to defend the frontline of the Cold War in Central Europe through conventional means. Thus, in West Germany, Sputnik was a question of a defense spending increase rather than of national space projects.[14] In this military interpretation, the satellite was considered an immediate threat. At the same time there were accounts that considered Sputnik an indication of a new era. Jürgen Tern, FAZ editor for politics, wrote on 12 October 1957: 'The existence of the red moon intensifies the danger. Its very emergence provokes emulation and outdoing. [...] Columbus's voyage caused the other powers of his era to follow him into new spheres.'[15] Tern introduced two motifs to FAZ's description of spaceflight that would later emerge time and again: by comparing spaceflight to Columbus's exploration of the New World, he justified his prediction of growing international tension with the colonial wars in the Americas. This was a new version of the frontier narrative that defined the urge for exploration as an element of human nature; that is, as an anthropological constant. In this regard, Tern stood in the tradition of American astrofuturism.[16] However, he created a Europeanized version of the frontier narrative that emphasized the historical experience of conflicts in the wake of exploration, rather than on pioneering spirit and anticipation of wealth and freedom, thus mirroring the then-current West German security concerns at the border to the Eastern bloc. Moreover, Tern turned to an analysis of Sputnik's future consequences. The satellite was not considered as having an immediate societal impact, but it instead indicated future problems. Focusing on the future introduced a new understanding of spaceflight, which was based on the anticipated future implications of the new technology. Astrofuturist writings primarily influenced these anticipations, which described spaceflight as a long-term project with impacts in the far future. Many science-fiction authors had imagined military conflicts in outer space.[17] Not long after the launch of Sputnik, however, the military interpretation changed. On 15 October 1957, the FAZ had reported that, according to Wernher von Braun, a space station could execute nuclear attacks with high accuracy. In later reports, any offensive capabilities were suppressed. The FAZ called the first experimental reconnaissance satellites 'eyes in outer space,' emphasizing their defensive rather than their offensive value.[18]

Even more important than the semantic modification of the military interpretation of spaceflight was its loss of relevance. Since the most important code within the media system is the distinction of 'attention' and 'non-attention,' the relevance of interpretations can be evaluated at the level of frequency of occurrence.[19] The number of references indicates media attention towards certain interpretations and, therefore, their relevance. The more frequently an interpretation was published in the FAZ, the more important it was for the shaping of the image of space travel in West Germany. In order to obtain relevant statistics, it makes sense not to analyze this on a monthly basis, but rather to consider longer periods. The time frame from October 1957 to March 1962 is therefore divided into five periods that coincide with the major events of space travel. In each period the articles are classified under the headings of Military, Prestige, Science/Technology, Economy, Evolution and Peacekeeping (Figure 8.1). The proportion of articles mentioning each of these interpretations shows their relevance. This method indicates a substantial shift in the meaning of spaceflight.

During the first months, then, the military interpretation of spaceflight was dominant. The scientific and technological interpretation ranked second, the evolution of mankind in outer space third. Less than one year after the launch of Sputnik the military interpretation saw a dramatic downswing. It was only graded fourth during the second period, between August 1958 and February 1960, while the interpretation of spaceflight as a factor of national prestige became more important for a short time. However, it never became formative in FAZ coverage, whereas the US media interpreted space travel

	Military	**Prestige**	**Scientific/ Technological**	**Economic**	**Evolution**	**Peacekeeping**
Oct. 1957–July 1958	26.2%	9.4%	22.3%	5.5%	8.6%	2.0%
Aug. 1958–Feb. 1960	7.1%	10.0%	26.0%	7.8%	10.3%	3.2%
March 1960–Jan. 1961	7.2%	5.0%	22.1%	19.3%	5.0%	3.9%
Feb. 1961–Sept. 1961	4.7%	9.3%	21.8%	13.5%	8.8%	4.2%
Oct. 1961–March 1962	4.1%	2.4%	22.8%	15.5%	6.5%	2.4%

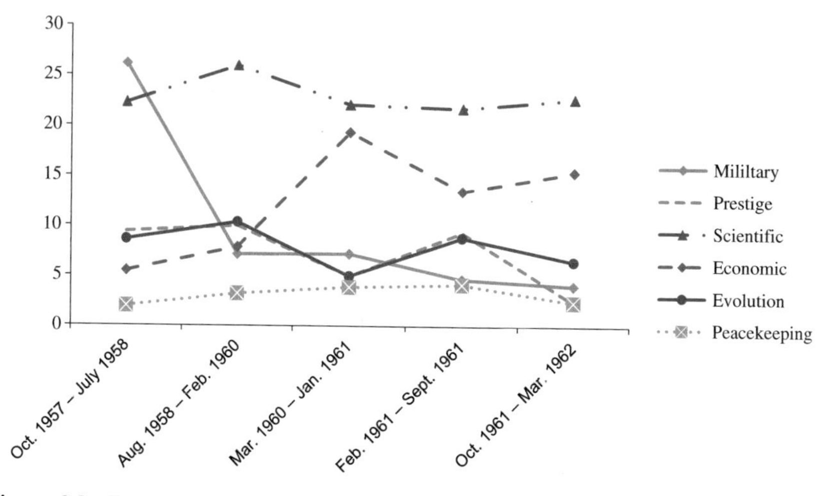

Figure 8.1 Frequency of interpretations of outer space in FAZ articles between October 1957 and March 1962.[20]

even during the Gemini program 1965–66 as a means of gaining superiority in terms of prestige or military power.[21] Under different national political circumstances and cultures, the same technical artifacts evoked different interpretations of space travel. The interpretation of space travel as an evolutionary step for humanity seemed to lose its importance as well. In fact, this interpretation can be found only when impressive spaceflight projects occurred, for instance the first satellite in orbit, the first photographs of the far side of the moon made by Luna 3, and the first men in space, Yury Gagarin and Alan Shepard. Whenever a new chapter in the history of space travel was opened by a spectacular 'first,' the idea of spaceflight as a step in human evolution was picked up again in the FAZ. Nevertheless, the most important interpretation of spaceflight apart from scientific and technological progress became its economic dimension, ranked second from March 1960 onwards. The relevance of each interpretation measured by frequency of occurrence reflected West Germany's position within the international space community. In 1957 the FAZ predominantly interpreted Sputnik as a threat to the nuclear balance and the West German federal government regarded outer space as a Cold War battlefield for the superpowers. Only five years later, in 1962, West Germany joined ELDO, and the FAZ interpreted spaceflight predominantly as a factor of scientific and technological progress and economic profit. Within these five years, space travel had lost its military dimension almost entirely, while alternative interpretations became dominant that were very compatible with, and relevant to, the identity and the goals of West Germany's postwar politics and society.

The television show *Aus Forschung und Technik*, starting its space coverage in 1964, demonstrates that dealing with the military or strategic dimension of outer space remained unpopular for decades. Until the West Europeans launched their own successful space projects in the mid-1970s, the interpretation of space exploration as a field of conflict and armament was largely ignored.[22] The episode *Vom Feuerwerkskörper zur Mondkapsel* (From Fireworks to the Moon Capsule), airing on 14 September 1965, constitutes an exception to this rule. In the end, author Peter G. Westphal, ZDF science editor and colleague of Heinrich Schiemann, expressed his hope that space technology would never serve 'purposes of warfare, but always only peace.' It does not come as a surprise that Westphal presented a negative view on the military dimension of space. Yet most remarkable is that in this statement the military purposes are still a matter of the future, even if reconnaissance and military communication satellites were a reality in 1965. Thus, Westphal's desire remained an abstract appeal, which upheld the idea of peaceful space exploration while ignoring the actual situation.

The military side of spaceflight remained a taboo for many years in the television show *Aus Forschung und Technik*. As a matter of fact, NASA itself addressed the military purposes of spaceflight only very cautiously, at most. Being both patriotic and enthusiastic for spaceflight, most American space journalists did not question NASA's reluctance to discuss the military relevance of space exploration.[23] Nor did the authors of *Aus Forschung und*

Technik develop a critical attitude, a situation that did not change until NASA and the Soviets carried out the Apollo-Soyuz test project in July 1975. At first glance, it may seem surprising that the military value of space travel was only frankly mentioned for the first time in the context of a project that should symbolically support a policy of *détente* and peace. In a review on recent plans for a Soviet space station, Schiemann pondered openly about its military benefit. In the same episode, he repudiated the concept of space exploration as a substitute for war: 'Were they [the landings on the moon] really a victory over the Soviets in the race to the moon? [...] Today it seems that the Soviets may have never [...] worked towards landing on the moon.'[24] This quote reflected a significant shift in understanding space travel. In the first decade of *Aus Forschung und Technik*, spaceflight appeared to follow a technologically determined path; all national projects shared the same goals, milestones and development. In the 1970s, this idea was substituted by the insight that each space-faring nation followed its individual path. Thereby the military taboo in Germany was broken, because if the development path was no longer considered universal, a non-military alternative for West Europeans/Germans was conceivable.

In March 1983, President Ronald Reagan (1911–2004) publicly announced the Strategic Defense Initiative (SDI), a program for the development of a ground-and space-based anti-ballistic missile system challenging the idea of mutual assured destruction. Following Reagan's announcement, the militarization of space became an intensively debated topic in the media for many years.[25] Joachim Bublath (1943–), Schiemann's successor since 1981, explicitly criticized the plans for SDI on various occasions. For instance, in an episode aired on 19 November 1987 he said, 'The shield in space is an illusion from this [technical] point of view.'[26] With this blunt statement Bublath not only retained the negative attitude towards the military dimension of outer space but also showed that he had developed an independent and critical stance, his viewpoint being that of an expert inspecting official plans.

The objection to the military use of outer space was a constant factor in the West German media. The military value of spaceflight was not compatible with West German postwar identity because of West Germany's complex attitude to the military and the nation. Nevertheless, the way of dealing with the military aspects varied. In the wake of Sputnik, there was a short time during which outer space exploration was regarded as a cause for increased defense spending. Thereafter, attitudes towards the military use of spaceflight changed from concealment in the Apollo era, to open criticism at the time of SDI. Considering space travel as a utopian conflict-free enterprise proved a typical aspect of the imagination of spaceflight. However, this view was much older than spaceflight itself. It can already be found in the writings of early science-fiction writers.[27] West Germany's unique position in international and space affairs made this account especially interesting for the West German media – and this was not only for political, but media reasons as well.

III Including the West Germans: universalist versus nationalist interpretations of spaceflight

Only after the reading of spaceflight shifted from military to scientific, technological, and economic interpretations, did the FAZ publish the first newspaper articles calling for West German space efforts. However, these interpretations were ambiguous. On the one hand, space travel was predominantly described as a scientific project belonging to all mankind.[28] Of course, this interpretation was often used in the aftermath of highly symbolic spectacular 'space firsts,' especially when these were of Soviet origin. This universalist interpretation involved West German readers and viewers in spaceflight since it appeared to affect humanity as a whole. After the flight by Soviet cosmonaut Yury Gagarin in April of 1961 – the first man in space – Nikolas Benckiser (1903–87), FAZ editor for politics, called the flight a 'triumph for mankind' and argued, 'in moments like this, when a man's mind pushed forward in areas that have been closed for thousands of years, there arises a fine awareness of solidarity.'[29] Spaceflight appeared not just as universal, but even universalizing. This sense of global fellowship reflects the internationalist ideas and activities that were popular among European science-fiction writers and widespread within the rocket societies of the postwar era such as the British Interplanetary Society or the West German Gesellschaft für Weltraumfahrt.[30]

On the other hand, space travel was considered a symptom of technological progress. In this case, spaceflight was not a universal project of all mankind, but rather a selfish project of each nation. The technological and economic interpretations coalesced because economic growth was regarded as the outcome of technological advancement. Between March 1960 and January 1961, this nationalist interpretation saw a dramatic upswing in frequency of occurrence, yet could coexist with the universalist interpretation. At the same time, when editor Benckiser argued that space exploration was a universal project, FAZ science editor Kurt Rudzinski demanded a German space effort using the nationalist interpretation: 'Spaceflight is just a symbol,' he wrote in April 1961: 'It stands for the performance and progress of science and technology. Today it secures the freedom of the Western world and will drive our economy and guarantee the existence of our nation tomorrow.'[31] Such an understanding of spaceflight was based on the implicit expectation that space technology would be decisive for future industry. It perfectly matched the postwar *Wirtschaftswunder* mentality and became therefore the basis of West German participation in European space projects.[32] The fact that it contradicted the universalist interpretation was never expressed in the FAZ. To arbitrarily choose between these two interpretations made dealing with space topics easier. The universalist interpretation included Germans in the achievements of other nations (making reports more interesting), whereas the discourse about West Germany's engagement with space exploration was grounded on pragmatic arguments of cost and profit (making the arguments acceptable to the German public).

The television series *Aus Forschung und Technik*, by contrast, depoliticized spaceflight in the early years. Space travel was not considered to be competing with other fields such as welfare and education for public resources. Moreover, negative opinion on spaceflight was never mentioned in the show.[33] And there was only a single episode featuring an interview with a politician, Gerhard Stoltenberg (1928–2001), the federal Minister for Research, on 10 January 1966. As a result, spaceflight as presented in *Aus Forschung und Technik* was purely technocratic and not a matter of political dispute.

The interpretation of spaceflight, universalist or nationalist, depended on the question of whether the project was of German or of American origin. In the 1960s, reports on the Soviet space programs were extremely scarce, since both reliable information and footage were limited in the Western World.[34] The European space efforts could hardly present anything more substantial than abstract blueprints, whereas the US space program offered concrete and detailed information on rockets, vehicles and mission schedules; furthermore, the US space program provided a great deal of film footage to the media. As a result, most of the reports dealt with US space efforts. One example is the episode of 20 March 1964. Although the reports were on NASA's plans for a manned flight to Mars, it was entitled 'How can *we* reach Mars?' 'If possible the Americans want to land on the moon before 1970 [...]. After that, the next destination of spaceflight will certainly be Mars. [...] It is of course impossible to say when man will land on Mars,' Schiemann introduced the topic.[35] Within a few sentences he altered the 'American' moon-flight project to 'man [...] on Mars.' The goals of US space efforts did not appear to be the product of a national space program, but rather as the result of the determined path of technological development. It did not matter which nation; what mattered was the fact that man landed on Mars.

This was the same implicit assumption used by the FAZ, originating in the previously mentioned internationalist ideas of the first astrofuturists. In television it had an analogy in science fiction. The popular first West German television series *Raumpatrouille: Die phantastischen Abenteuer des Raumschiffes Orion* (directed by Michael Braun and Theo Mezger, written by Rolf Honold), aired on ARD in 1966, told the story of the crew of the spaceship 'Orion.' The main plot of *Raumpatrouille* was a war against extraterrestrials. On the show, however, mankind was united under a universal government, nations had been abolished and so the crew of 'Orion' came from different, not exclusively Western countries. In this respect, science fiction and popular science propagated – through the same medium – the same concept of spaceflight as an enterprise of humanity. However, the universalist interpretation helped to achieve television's aim to present new and spectacular topics in *Aus Forschung und Technik*. Only if the audience regarded NASA's projects as affecting their lives, did space topics have a chance of gaining public interest and attention. By providing footage and interviewees to ZDF, NASA's media strategy of open communication to international media paid off. Both German television and the German public accepted the US approach to space exploration as being universalist and serving all mankind. From such a perspective, *Aus Forschung und Technik* contributed to the American-led globalization of the media.[36]

With regard to media attention, it does not come as a surprise that the universalist interpretation was challenged when German and European space projects were launched. On 10 January 1966, *Aus Forschung und Technik* reported on 'rocket development in West Germany.' West Germany's contribution to the Europa rocket was nationalized and called 'the third stage, the *German* stage.' The project simply gained attention because it was German, neither did the rocket achieve any relevant 'first' nor deliver any technology unequaled by other nations.[37] For the same reason *Aus Forschung und Technik* reported on Spacelab in July 1974. Spacelab was a research laboratory designed by the European Space Research Organisation (ESRO), one of the precursor organizations of ESA, to be used with the American Space Shuttle. One of two Spacelabs was given to NASA free of charge in exchange for flight opportunities for European astronauts. Schiemann's co-presenter Franz Buob emphasized: 'Thus something happened which hardly anyone dared to hope until recently, that Europe was to undertake manned spaceflight within a few years.' Buob did not refer to the scientific relevance of Spacelab (which was presumed by the German federal government and disputed by some scientists), but to the fact that European astronauts were about to enter earth's orbit.[38] In fact, Spacelab was a laboratory module depending on NASA's launch and re-entry capabilities. The Spacelab project did not include any technological development related to these life support systems so that even after its completion, Europe would not be able to undertake manned spaceflight independently. Hence Buob overestimated the actual value of Spacelab, but his interpretation demonstrated that manned spaceflight was an important factor for the national prestige of a space nation. Since scientists had always disputed the need for man in space, the public's preference for manned space projects reflected the influence of astrofuturist ideas that never stopped on the level of unmanned space exploration.

At the same time, whenever spectacular long-range concepts were topics in *Aus Forschung und Technik*, spaceflight was still universalized. Most of the visionary concepts publicly discussed in the 1970s were ideas that responded to the planet's problems of limited resources and overpopulation, raised by the Club of Rome's influential book *The Limits to Growth*.[39] These ideas were mainly formulated by second-generation astrofuturists, who had – in contrast to the generation of Wernher von Braun and others – no management positions in official space agencies, such that their success depended completely on the strength of their arguments.[40] However, *Aus Forschung und Technik* reported extensively on Krafft A. Ehricke's study of possible industrial involvement on the moon, Gerald K. O'Neill's space colonies, and Peter E. Glaser's plans for solar power satellites in outer space.[41] Their visionary concepts were presented as evolutionary leaps of mankind as a whole, rather than potential projects for single nations. For example, Schiemann introduced Glaser's solar power satellites in the episode of 26 November 1979 by saying: 'Let's discuss a couple of things that concern our daily life.'[42] By this Schiemann made the plans relevant to the audience and adopted Glaser's assumption that spaceflight was a universal solution for tangible and international problems. While rockets and space vehicles were the property of nations, the grand designs of Glaser, O'Neill and Ehricke belonged to humanity. In other words, existing technical artifacts were national, while visions of the future in outer space were universal. Whenever astrofuturist visions became a topic in *Aus Forschung und Technik*, they were universalized. Until the landing on the moon, the first generation of astrofuturists were in the focus of *Aus Forschung und Technik*, and their project of manned exploration of the moon and Mars appeared to be universal. When their ideas faded in the wake of the Apollo disillusion, the technology of this era of spaceflight was nationalized, and the visions and speculations of the next generation of astrofuturists were universalized. Their plans never materialized into technologies, however, so nothing remained that could have been nationalized.

By the 1980s, the heyday of influential new space visions had passed. The result was that on the television shows, spaceflight was no longer described in a universalist way. This was emphasized by the fact that many episodes featured reports on different nations' individual projects. On 14 July 1980, *Aus Forschung und Technik* covered the plans for the American Space Shuttle but also took a glance at the European Ariane program. On 19 November 1987, Joachim Bublath contrasted the drive of Soviet space ambitions – as being 'part of an ideology' – with those in Japan: 'The Japanese are considering where to best invest their money, into space research or into research on earth.'[43] This is one of the few statements addressing the fact of the public funding of spaceflight. Since the idea of determined progress in space was no longer undisputed, the question to what end spaceflight should be pursued and how funds were to be allocated depended on individual decisions of each space nation. Thus, spaceflight was re-politicized in the 1980s. Between 1964 and 1987 *Aus Forschung und Technik* had slowly shifted its understanding of spaceflight from an enthusiastic account, defining space travel as apolitical, universal and technically necessary, towards a critical viewpoint which considered space travel as costly, undetermined and contingent on national political debates. Over the course of little more than two decades, spaceflight had ceased to deliver sensational news. In the early 1970s, the access to information on the Soviet space program for West German television journalists had been improved as a result of political *détente*, for instance the *Ostpolitik* of West Germany's chancellor Willy Brandt (1913–92). In the late 1970s and the early 1980s, Europe and the ESA had developed considerable space capabilities worth reporting. *The Limits to Growth*, published in 1972, and the growing ecology and anti-nuclear movement had disputed the sustainability of space technology and set different topics on the agenda for the future.

IV Technocracy and astrofuturism

In the 1960s and 1970s, West German television producers assumed that they were analyzing, if not shaping, tomorrow's society.[44] In *Aus Forschung und Technik* this orientation towards the future was a typical and recurring theme. Many reports dealt with upcoming technology. Reviews of completed projects were rare and short. In the series, spaceflight was described by explaining the technical devices, the flight maneuvers and the astronauts' life on board the spaceships.[45] Whenever the word 'future' was mentioned, it was identified with the future of spacecraft, rockets and technology. Thus, technology appeared as the central means of shaping the future. This linking of the future with space technology persisted throughout the 1970s and 1980s and remained a constant of spaceflight. Regardless of the other changes in the show during the 1970s and 1980s: the transition from first- to second-generation astrofuturism; the shifting focus from interplanetary manned exploration towards earth-centered space technology solving environmental problems; and the change of the presenter's attitude from an affirmative and sometimes enthusiastic position to a more independent and critical view.

In contrast to American media, where the astronauts were the most important protagonists of spaceflight, these heroes of the Space Age were hardly ever mentioned in West German media such as *Aus Forschung und Technik*.[46] There were no interviews, family portraits or personal stories. The true protagonists were the engineers and researchers such as Wernher von Braun, Kurt Debus, Ernst Stuhlinger, Krafft Ehricke and, later, Gerald K. O'Neill. In the pre-Apollo years, they were frequently interviewed and they themselves the main subject ('Why do *you* do it this way?'). From 1964 to 1969, *Aus Forschung und Technik* featured altogether 14 interviews with American space experts, 12 of which were of German origin. Undoubtedly, interviews with German-Americans had many advantages for German television. Voice-over translation was not necessary, and a German audience might consider their former fellow countrymen more interesting. They embodied the German roots of US rocket technology, particularly as the history of the V-2 was not yet considered controversial. Moreover, these interviews had yet another implication: For the viewers of *Aus Forschung und Technik*, space travel appeared as a matter of engineering. The engineers explained not only the spacecraft and their maneuvers, but also their plans and visions. This broader approach delivered astrofuturist ideas of shaping the future that did not appear as 'dream but, with current science and technology, an immediate possibility' directly on the screens.[47]

Such an effect was reinforced by two dominant modes of imaging future projects: models and animations. The 'modeling method' means that an expert holds exemplars of technical artifacts, rockets and other devices in his hands and demonstrates the course of a spaceflight with their help (Figure 8.2). In *Aus Forschung und Technik*, this method was predominant until the landing on the moon. Original technical devices, rockets in particular, were usually shot with a wide-angle lens showing a worm's-eye-view or panning from bottom to top. Whenever rocket engines, test stands or other huge objects were shown, humans

Figure 8.2 TV presenter Heinrich Schiemann (1916–2002) with a small model of the Apollo Lunar Excursion Module. In early episodes of *Aus Forschung und Technik*, spaceflight was reduced to technical devices, flight paths and maneuvers, thus letting the presenter effectively take over control, if in model scale only.
Source: Courtesy of ZDF, ca. 1969.

were usually included in the picture as well, allowing a demonstration of the enormous outer dimensions of the devices. Such an impression was emphasized when the technology was referred to as the 'pyramids of Alabama' or as coming from a 'world of giants.'[48] Whereas next to the original device men appeared as being miniature (or insignificant), the person (or more precisely: the expert) had full control over his inventions in the 'model flights.' In the hands of the designing and controlling expert, giant technical devices appeared small and easy to handle. Such devices were virtual tools, technical extensions of man's extremities.

In the model flights, only technical artifacts such as rocket stages, or command and landing modules were used. Presenters carried out launches, flight maneuvers, rendezvous maneuvers and landings. There was no earth, no other planets or celestial bodies, no distances given, and never anybody aboard the spaceship. Thus, space technology and its devices, processes and procedures revolved literally, as well as metaphorically, around each other. No experiments were carried out in these model flights: artifacts cruised through space without serving any apparent purpose. As their scale was too small to register on any measuring instruments, the missions' scientific tasks were ignored. Regarding their appearance, in particular their shape, color, inscriptions, and flags, the devices were modeled as accurately as possible. In this respect, *Aus Forschung und Technik* proved self-referential. The audience was able to recognize the devices during upcoming live broadcasts. The model-flight approach reached its zenith when both national German television stations used it during their Apollo 11 live shows. In this case, the models were to scale. They were used to re-enact what was happening on the moon and how the astronauts had left their landing module until Neil Armstrong could activate the outboard television camera. The actors in the studio acted on the presenters' command. Thus television had literally become a creator of reality, with the model-flight sessions shifting from illustrating future plans to visualizing an endeavor of which no live footage existed. Notably, the model-flight approach lost importance in the late 1970s and 1980s when most of the space projects took place in lower earth orbits, such that their flight maneuvers, space rendezvous and so on were far less spectacular than those of the Apollo and Skylab era.

Simultaneously to the decline of the model flight, a second method of portraying the future space projects arose: hand-drawn or computer-generated animations, which experienced a reverse development from models. The first space animation to be shown on German television demonstrated how Sputnik was launched into its orbit.[49] This animation made visible something that had occurred but was visually inaccessible, since no images from an onboard camera were available. When the same method was used to visualize future plans, the future appeared as something already existent, just not yet visible. Again, the focus was on technical devices and their maneuvers. Contrary to the hand-model method, animation appeared more authentic since it was similar to film. The animations in *Aus Forschung und Technik* did not explain the procedures in an abstract way or in the form of a diagram, but in continuously animated films. In animations, these planned space projects appeared indubitable, like a real event captured on film.

The animations in *Aus Forschung und Technik*, partly originating from space organizations, were similar in style within the different episodes and to the space programs of the competing television network, ARD.[50] Thus, a visual code was maintained between different shows and media. The American prototype for these German animations was the 1955–57 *Tomorrowland* trilogy on space exploration by Wernher von Braun and Walt Disney.[51] The imagination of astrofuturists and space artists influenced television's image of spaceflight from the start. The idea of presenting the future through animation film was avant-garde in itself. The Disney trilogy set the benchmark for visualizing spaceflight even for West German television. In the 1980s, computer animation became increasingly more important. It was not yet as photo-realistic as today's animations, but its shortcomings (pixilation, flickering, etc.) constituted a futuristic style of its own. Hence the artistic style of the images supported the claim that spaceflight itself was avant-garde.

Both visualizing methods showed a technocratic conception of the future as an orderly course of planned procedures. Philosopher Günther Anders (1902–92) wrote in 1970 that 'these images cannot be wrong [...] unless the real events fail.'[52] If the images were not congruent to reality, something had gone wrong in reality, not in the creation of the images; an impression that was intensified by the narrator's choice of words. Only very few expressions were used, such as 'it is planned/intended that.' Usually, the procedures were explained using future or present tense, allowing past tense to refer back in time. For this reason, the projected future did not appear uncertain or open-ended, but rather predetermined and set just like past events. There were no uncertainties, unforeseen complications or accidents. Everything had to go according to a preconceived plan in which eventualities were already considered. In an interview of 20 March 1964, former rocket pioneer and then director of NASA's Kennedy Space Center Kurt Debus (1908–83) extensively explained the operation of the rescue rocket. When three astronauts died in the so-called Apollo 1 accident on 27 January 1967, the event was not even mentioned in *Aus Forschung und Technik*. Neither was the earlier statement on the rescue systems called into question – which is consistent with the finding that the show's account was anything but critical in its early years.

During the 1980s, *Aus Forschung und Technik*'s way of dealing with predictions and speculations began to change. Due to Joachim Bublath's encouragement of critical science journalism, now even failure was considered a possible outcome of a vision or a space mission. In the 2 December 1985 episode, Bublath referred to plans to grow plants in space: 'Decades or centuries will pass before one will have entire fields in space. Maybe it can never be achieved.'[53] Whereas in the 1960s all plans and visions seemed to be realized over a span of 20 to 30 years, in the 1980s the future of space travel was a matter of centuries. For the first time, the possibility was mentioned that speculation could turn out to be wrong. Not until this time did *Aus Forschung und Technik* present visions of a future in space as a product of the imagination, as speculation that could come true, but could just as easily not.

Remarkably, the producers of the series imposed a strict barrier between science fiction and science fact. They constantly dissociated themselves from science fiction. As a public television network, ZDF was obliged to promote public education, and *Aus Forschung und Technik* was one of its educational programs. The show tried to provide a 'deeper insight' into science to the audience, which was considered possible only through a purely positivistic approach to empirical science.[54] For ZDF's producers, science and popular culture were irreconcilable. Although engineers such as Wernher von Braun used science fiction as an instrument with which to communicate their ideas, such a close link between science fiction and science fact was hardly ever mentioned on German television. Even though the show frequently referred to speculations, science fiction had negative connotations.[55] Arguments for dissociating oneself from science fiction always followed the same pattern: First it was admitted that the reported future might 'look,' 'sound' or 'seem' 'fantastic,' 'utopian' or 'daring'; then a reason was given why this prognosis might turn out to come true despite the immediate opposite impression. If space technology appeared as the means of making fantasy become reality, television played the role of the prophet announcing the future. Studies of space companies or of government administrations were used to give credibility to the predictions, so that the future appeared to depend on the decisions of engineers rather than on an open political process. This, again, showed a technocratic understanding of a future in space, which was popular among astrofuturists.[56]

Another topic proving the close links between the imagination of space in science fiction and the image of space presented by the show was Mars. There were more reports on the Red Planet than on any other planet. Whereas in the 1960s the central focus was on the idea and the technology needed for a manned exploration to Mars, in the 1970s reports dealt with the question of extraterrestrial life on Mars. One example is Peter G. Westphal's documentary 'Life on the Planets' in *Aus Forschung und Technik*'s 19 July 1976 episode. 'That sounds fantastic [...], but it is perfectly possible,' the presenter commented on speculations regarding the existence of extraterrestrial beings.[57] Unless proven otherwise, an assumption was considered valid, even though there was no evidence to support it. The mere possibility justified its visualization. This type of argument referred to the open horizons of science fiction and astrofuturism rather than the hard facts of empirical science. Science fact could not be separated from popular culture.

V Via media to the stars?

The media presentation of space travel in West Germany was shaped by West Germany's unique situation in international and space affairs, general media tendencies and strategies of gaining attention, and astrofuturist ideas. West Germany's political position largely determined the meaning of spaceflight as communicated in the mass media. In the wake of Germany's complex attitude towards the military since the Second World War, any military dimension of outer space was underemphasized, ignored, rejected or criticized. In addition, the idea that space travel might affect national prestige never became dominant. In West Germany, outer space was not seen as either a battlefield or a substitute for war. The astrofuturist idea of the universality of spaceflight proved popular, but in the FAZ this approach was challenged by the nationalist interpretation of technological progress and economic profit. In television, the universalist interpretation was dominant during the 1960s and even supported by science-fiction programs such as *Raumpatrouille Orion*. Whereas the FAZ described spaceflight as a political topic, ZDF's *Aus Forschung und Technik* depoliticized space travel until the 1980s. Thus, television 'manufactured consent' (Noam Chomsky) on spaceflight in West Germany. Certainly its universalization helped television's goal of evoking involvement in the German audience and, therefore, benefited from television's general strategies. When German or European space plans became reality in the 1970s, they were never universalized, but always nationalized or Europeanized. However, this was not tantamount to a re-politicization of spaceflight in *Aus Forschung und Technik*, which would occur only when Joachim Bublath succeeded Heinrich Schiemann as head of *Aus Forschung und Technik*. The SDI proposals cultivated a critical view on space technology in the West German public, and the astrofuturist visions of the second generation failed to produce any technological outcome.

While the textual (written or spoken) interpretations of spaceflight were much influenced by West Germany's unique position towards outer space, this was not the case with visual presentation. The underlying technocratic understanding of spaceflight was the legacy of Wernher von Braun's astrofuturism and his first space show in television, the Disney trilogy *Tomorrowland*. The media's preference for manned spaceflight, for Mars and the interest in extraterrestrials were the outcome of science fiction and astrofuturist literature. However, with topics like these, the FAZ and *Aus Forschung und Technik* satisfied the audience's demand for space visions. From a dramaturgical viewpoint, spaceflight in its technocratic reality was not very exciting. When everything is planned precisely and nothing can go wrong, there is no room for drama. As sociologist William Sims Bainbridge put it: 'The dream of space travel is glorious, the contemporary reality is dismal.'[58] This expected public reaction caused special attention to be paid to more spectacular and long-range prospects.

Aus Forschung und Technik tested the borders of television and of the popular science genre. In animations and model-flights the producers showed what they considered to be the future (Figure 8.3). And for this future they created the necessary images. During the first few minutes after Apollo 11 had landed on the moon, there were no live images available. Both German television stations reenacted the procedures with actors.[59] They created their own images of a present event. Thus, signifier and signified lost their usually formative link for television's live coverage of real events. As early as 1956, philosopher Günther Anders argued that television in general feigned objectivity and that the televisual experience substituted created images for direct experience, calling the re-enacted television images 'images of images of images,' which nevertheless were 'authentic' insofar as in spaceflight reality became the reproduction of prior simulations.[60]

Figure 8.3 In July 1969, *Aus Forschung und Technik* simulated the moon landing with a life-sized model of the Apollo Lunar Module, thus creating its own images and dissolving the link between signifier and signified.
Source: Courtesy of ZDF, ca. 1969.

In the 1980s, *Aus Forschung und Technik* illustrated speculations that were explicitly regarded as improbable. Television brought the stars to its audience in ways that the reality of spaceflight could no longer bring them. In this, it assumed the function of science-fiction literature. Even more, it is astonishing that the show dissociated itself constantly and explicitly from science fiction, thus reflecting a positivistic understanding of science and a deep reservation about popular culture. Obviously, in West Germany the ideas of astrofuturism evoked attraction to such a high degree that it was easier to declare them science fact than to discuss strictly the basic issue of the link between popular culture, science and technology.

Notes

1. Werner Stratenschulte, 'Apollos Mondfahrt – die Show des Jahres,' *ZDF-Jahrbuch* (1968), 35: 'Daß die Show des Jahres ihr Publikum tatsächlich erreichte, ist ein Verdienst des Fernsehens.' The author thanks ZDF, Mainz, for funding this research with a Karl-Holzamer-Stipendium and for the permission to use and reprint archival materials. Special thanks go to Axel Bundenthal, HA Archiv/Bibliothek/Dokumentation, and Umberto Biagioni, Abteilung Bilderdienst, for their help.
2. Georg Ruhrmann, 'Ereignis, Nachricht, Rezipient,' in Klaus Merten, Siegfried J. Schmidt and Siegfried Weischenberg, eds, *Die Wirklichkeit der Medien*, Opladen: Westdeutscher Verlag, 1994, 238–44.
3. Johannes Weyer, *Akteurstrategien und strukturelle Eigendynamiken: Raumfahrt in Westdeutschland 1945–1965*, Göttingen: Schwartz, 1993, 10–48.
4. Horst Röper, 'Das Mediensystem der Bundesrepublik Deutschland,' in Merten, Schmidt and Weischenberg, *Wirklichkeit der Medien*, 506–43.
5. Frank H. Winter, *Prelude to the Space Age: The Rocket Societies, 1924–1940*, Washington, DC: Smithsonian Institution Press, 1983, 35–53.
6. Weyer, *Akteurstrategien*, 56; Walter McDougall, 'Space-Age Europe: Gaullism, Euro-Gaullism, and the American Dilemma,' *Technology and Culture* 26.1 (January 1985), 179–203, here 184; idem, *...The Heavens and the Earth: A Political History of the Space Age*, New York: Basic Books, 1985, 424.
7. De Witt Douglas Kilgore, *Astrofuturism: Science, Race, and Visions of Utopia in Space*, Philadelphia: University of Pennsylvania Press, 2003, 52–81.
8. Weyer, *Akteurstrategien*, 252; Niklas Reinke, *Geschichte der deutschen Raumfahrtpolitik: Konzepte, Einflußfaktoren und Interdependenzen 1923–2002*, Munich: Oldenbourg, 2004, 79–93.
9. In 1965 the critic of West Germany's leading television guide and several letters to the editor criticized the television coverage on Gemini 5 for being weak (*dürftig*) since it had been less extensive than radio coverage. See 'Das Wort hat: Der Kritiker,' *Hörzu* (11 September 1965), 31; R. Werner, 'Astronauten,' ibid. (25 September 1965), 57; Achim W., 'Weltraumflug,' ibid. (11 September 1965), 73. When ZDF broadcasted the missions of Apollo 8 live and *in extenso*, they regarded this as a 'Höhepunkt in der jungen Geschichte des Fernsehens' (milestone in the short history of television) and themselves as 'großes Schaufenster, vor dem sich die Menschheit drängte' (a big showcase for mankind); see *ZDF-Jahrbuch* (1968), 35. Concerning Apollo 11 the yearbook stated: 'Durch das Fernsehen wurde das historische Ereignis der Mondlandung gleichzeitig zu einem weltweiten Erlebnis' (Through television the historic event became a global experience); ibid., 1969, 61.
10. Monika Elsner, Hans Ulrich Gumbrecht, Thomas Müller and Peter M. Spangenberg, 'Zur Kulturgeschichte der Medien,' in Merten, Schmidt and Weischenberg, *Wirklichkeit der Medien*, 163–87, here 181. In 1964 more than half of West German households had access to television; see *ZDF-Jahrbuch* (1962–64), 195. However, ZDF was using the UHF band for broadcasting, a frequency range different from that of the first West German television network, ARD, so that some of the early television sets could not receive ZDF. In 1963 43 percent of all television households were able to watch ZDF: by 1969, this number had increased to 96 percent. See *ZDF-Jahrbuch* (1969), 184.
11. 'Heinrich Schiemann gestorben,' *FAZ* (12 November 2002), 41.

12. Karsten Werth, *Ersatzkrieg im Weltraum: Das US-Raumfahrtprogramm in der Öffentlichkeit der 1960er Jahre*, Frankfurt am Main: Campus, 2006, 53–5.
13. Heinz Gartmann, 'Der Satellit im Weltraum – ein roter Stern,' *FAZ* (7 October 1957), 2: 'Die Nation, die eine Nutzlast (in diesem Fall den Satelliten) [...] in Umlauf um die Erde zu bringen vermag, kann mit einer ähnlichen mehrstufigen Rakete eine andere Nutzlast (dann vielleicht eine Atombombe) von ihrem Territorium aus nach jedem Punkt der Erdoberfläche befördern.'
14. Alfred Rapp, head of the FAZ office in Bonn, wrote in a commentary: 'In the new session the parliament will come to the hard realization that defending people is expensive. Sputnik is to influence all these considerations.' ([Der neue Bundestag] 'wird vor der schweren Erkenntnis stehen, daß der Schild für ein Volk teuer ist. Sputnik wird in allen diesen Überlegungen eine Rolle spielen'); see 'Bonn und Sputnik,' *FAZ* (15 October 1957), 2.
15. Jürgen Tern, 'Raketen-Diplomatie,' *FAZ* (12 October 1957), 1: 'Die Existenz des Roten Mondes steigert die Gefahr. Sein bloßes Auftauchen fordert zur Nachfolge und zum Überbieten heraus. [...] Die Fahrt des Kolumbus brachte die anderen großen Mächte seines Zeitalters auf die gleiche Reise in die neuen Sphären.'
16. Alexander C.T. Geppert, 'Flights of Fancy: Outer Space and the European Imagination, 1923–1969,' in Steven J. Dick and Roger D. Launius, eds, *Societal Impact of Spaceflight*, Washington, DC: NASA, 2007, 585–99, here 596; Kilgore, *Astrofuturism*, 74 and 87–90.
17. Ulrich Suerbaum, Ulrich Broich and Raimund Borgmeier, *Science Fiction: Theorie und Geschichte, Themen und Typen, Form und Weltbild*, Stuttgart: Reclam, 1981, 167.
18. 'Weltraumstation in 1700 Kilometer Höhe,' *FAZ* (15 October 1957), 5; Jan Reifenberg, 'Amerika jubelt über das "Auge im Weltraum,"' *FAZ* (22 December 1958), 3.
19. Jürgen Gerhards, 'Politische Öffentlichkeit: Ein system- und akteurstheoretischer Bestimmungsversuch,' in Friedhelm Neidhardt, ed., *Öffentlichkeit, öffentliche Meinung, soziale Bewegungen*, Opladen: Westdeutscher Verlag, 1994, 88–91.
20. The total is less than 100 percent because the remainder did not contain any of these interpretations.
21. Werth, *Ersatzkrieg*, 279–80, 283.
22. It is possible that there were other television programs dealing with military aspects. However, since *Aus Forschung und Technik* was aimed at 'the interested core audience' ('ihr interessiertes Stammpublikum') and the show never addressed an episode to the military dimension of spaceflight, it is unlikely that this ever became a relevant interpretation in the West German media; see *ZDF-Jahrbuch* (1966), 50.
23. Werth, *Ersatzkrieg*, 123, 194.
24. 'Waren sie [die Mondlandungen] auch wirklich ein Sieg über die Sowjets in einem jahrelangen Wettlauf zum Mond? [...] Es sieht heute eigentlich eher so aus, als ob die Sowjets vielleicht nie, spätestens aber seit Mitte der 60er Jahre gar nicht mehr darauf hingearbeitet hätten, auf dem Mond zu landen.'
25. 'President Reagan's Address to the Nation on Defense and National Security,' 23 March 1983, http://www.reagan.utexas.edu/archives/speeches/1983/32383d.htm (accessed 1 October 2017).
26. 'Der Schutzschild im Weltraum ist schon aus dieser [technischen] Sicht Illusion.'
27. Suerbaum, Broich and Borgmaier, *Science Fiction*, 168; Kilgore, *Astrofuturism*, 40.
28. Karl Korn, 'Wir sind dabei gewesen,' *FAZ* (7 October 1957), 1: 'It [Sputnik] has been invented by the human mind, not by one power bloc in rivalry with the

other' ('Erfunden hat ihn der Menschengeist, nicht ein Machtblock in der Rivalität zum andern').

29. Nikolas Benckiser, 'Sieg für die Menschheit,' *FAZ* (13 April 1961), 1: 'In solchen Augenblicken, in denen dem menschlichen Geist ein Vorstoß in Bereiche gelungen ist, die ihm durch Jahrtausende verschlossen waren, stellt sich wie von selbst ein schönes Bewußtsein der Solidarität ein.'
30. Geppert, 'Flights of Fancy,' 595; Reinke, *Raumfahrtpolitik*, 38.
31. Kurt Rudzinski, 'Raumflug mit Konsequenzen,' *FAZ* (14 April 1961), 2: 'Der Raumflug ist nur ein Symbol. Er steht für die wissenschaftlich-technische Leistung und ihr Fortschreiten schlechthin, von dem schon heute die Freiheit des Westens abhängt und morgen die Leistung unserer Wirtschaft und damit die Existenz unseres Volkes.'
32. Helmuth Trischler, *Luft- und Raumfahrtforschung in Deutschland 1900–1970: Politische Geschichte einer Wissenschaft*, Frankfurt am Main: Campus, 1992, 377; Reinke, *Raumfahrtpolitik*, 79.
33. This is all the more astonishing since one of the most prominent opponents to spaceflight was of German origin and gave television interviews readily: physicist and Nobel prizewinner Max Born (1882–70) who was interviewed in Rüdiger Proske's television feature *Auf der Suche nach der Welt von morgen*, episode 1 (ARD, 8 June 1961).
34. The documentary *Russlands Weg zum Mond* (*Russia's Way to the Moon*) by Rainer M. Wallisfurth and Günter Siefarth (ARD, 10 January 1968) and the relevant papers in WDR, Historisches Archiv, Cologne, Sig. 11254, show how difficult gathering reliable information on Soviet space projects was.
35. 'Die Amerikaner wollen also möglichst noch vor 1970 auf dem Mond landen [...]. Das nächste Ziel der Weltraumfahrt wäre dann gewiss der Mars [...]. Natürlich kann man nicht sagen, wann der Mensch auf dem Mars landen wird.'
36. James Schwoch, *Global TV: New Media and the Cold War, 1946–69*, Urbana: University of Illinois Press, 2009, 154.
37. Concerning technological performance the Europa rocket was similar to the US Atlas-Agena rocket which had its maiden flight in February 1960; see Reinke, *Raumfahrtpolitik*, 80.
38. 'Damit ist etwas geschehen, was vor kurzer Zeit noch kaum jemand zu hoffen wagte, daß nämlich Europa in wenigen Jahren bemannte Raumfahrt treiben wird.'
39. Donella H. Meadows, Dennis L. Meadows, Jørgen Randers, William W. Behrens and the Club of Rome, *The Limits to Growth: A Report for the Club of Rome's Project on the Predicament of Mankind*, New York: Universe Books, 1972.
40. Kilgore, *Astrofuturism*, 151.
41. Krafft A. Ehricke (1917–84) worked as a rocket designer for different companies and was one of the most influential space philosophers during the Apollo and post-Apollo years. In the 1970s, he presented concepts for industrialization and commercialization of space, in particular of the moon. Gerard K. O'Neill (1927–92) was a high energy particle physicist at Princeton University. In the years following 1969, he worked on the concept of self-supporting habitats in space. Peter E. Glaser (1923–2014) was employed by Arthur D. Little, Inc., Cambridge, MA. In 1968 he presented the idea of solar power satellites to supply solar power from space for use on the earth.
42. 'Wir wollen jetzt noch ein paar Dinge behandeln, die unser praktisches Leben betreffen.'

43. 'Die Japaner überlegen sich recht genau, wo sie das Geld investieren, in die Forschung im Weltraum oder in die Forschung am Boden.'
44. Dieter Stolte, head of programming at ZDF from 1967 through 1973, emphasized topics 'relevant for the public's consciousness and the resulting actions and patterns of behavior' ('die für deren [der Öffentlichkeit] Bewußtseinsbildung und daraus resultierende Aktionen und Verhaltensweisen von Bedeutung sind'); see *ZDF-Jahrbuch* (1969), 8. The ARD had great success with its long-run show *Auf der Suche nach der Welt von morgen* (*Searching for the World of Tomorrow*) by Rüdiger Proske which presented future society in a variety of different fields, for instance work, traffic and space.
45. Examples are the episodes of 9 March 1964 and of 14 November 1966.
46. Werth, *Ersatzkrieg*, 129–31, 223, 284.
47. Kilgore, *Astrofuturism*, 65.
48. 'Pyramiden von Alabama,' *Aus Forschung und Technik* (9 March 1964); 'Welt von Riesen,' ibid. (14 November 1966).
49. The animation was part of a report to the Astronautical Congress 1957, held in Barcelona and aired in *Tagesschau* on 10 October 1957.
50. Examples include elaborate shows such as *Auf der Suche nach der Welt von morgen* by Rüdiger Proske, episode 1 *Der Schritt ins Dunkel*, 8 June 1961; episode 9 *Zum Mond und weiter*, 8 February 1966 (part 1), and on 10 February 1966 (part 2); and *Auf dem Wege zum Mond*, by Rüdiger Proske and Max Rehbein, aired on 7 November 1968.
51. Kilgore, *Astrofuturism*, 57–60.
52. Günther Anders, *Der Blick vom Mond: Reflexionen über Weltraumflüge*, Munich: C.H. Beck 1970, 113–14.
53. 'Ehe man nun ganze Felder im Weltall haben wird, werden Jahrzehnte, Jahrhunderte vergehen. Vielleicht wird man es nie erreichen.'
54. Wolfgang Brobeil, 'Kultur und Bildung,' *ZDF-Jahrbuch* (1967), 34.
55. Kilgore, *Astrofuturism*, 64. For example, in the 20 March 1964 episode, Schiemann added frankly after he had described the idea of a Phobos base: 'All this, ladies and gentlemen, may seem fantastic to you, but please take into consideration that big American companies invest a lot of money in these studies' ('Das alles, meine Damen und Herren, mag Ihnen ja nun sehr fantastisch vorkommen. Aber bitte bedenken Sie, es sind große amerikanische Firmen, die viel Geld in diese Studien investieren').
56. Kilgore, *Astrofuturism*, 89.
57. 'Das klingt sicher sehr fantastisch [...]. Aber sie sind durchaus möglich.'
58. William Sims Bainbridge, *The Spaceflight Revolution: A Sociological Study*, New York: John Wiley, 1976, 13.
59. Andreas Rosenfelder, 'Medien auf dem Mond: Zur Reichweite des Weltraumfernsehens,' in Irmela Schneider, Torsten Hahn and Christina Bartz, eds, *Medienkultur der 60er Jahre: Diskursgeschichte der Medien nach 1945*, Wiesbaden: Westdeutscher Verlag, 2003, 17–33, here 28–9. Elmar Zenner, 'Per Tele live auf den Mond,' in Annette Deeken, ed., *Fernsehklassiker*, Alfeld: Coppi, 1998, 121–39, here 132.
60. Anders, *Blick vom Mond*, 114–15. Rüdiger Zill, 'Im Wendekreis des Sputnik: Technikdiskurse in der Bundesrepublik Deutschland der 50er Jahre,' in Irmela Schneider and Peter M. Spangenberg, eds, *Medienkultur der 50er Jahre: Diskursgeschichte der Medien nach 1945*, Wiesbaden: Westdeutscher Verlag, 2002, 25–49, here 45.

CHAPTER 9

Balloons on the Moon: Visions of Space Travel in Francophone Comic Strips

Guillaume de Syon

Reflecting on the 1969 Apollo 11 lunar landing, French astronaut Patrick Baudry (1946–), then a fighter-pilot candidate, admitted years later that he and his classmates had felt a sense of *déjà-vu*. After all, Tintin, a world-famous comic strip character, had been there 15 years earlier.[1] Though anecdotal, the comment nonetheless points to an important element of European astroculture – the comic book – where 'bubbles' or 'balloons' filled with text intertwine with pictures to move the story forward. The two volumes that cover Tintin's lunar odyssey, along with other episodes from francophone comic strips, became a classic of the genre, reprinted and discussed years later. Because of the combined importance of the Francophone comic strip in twentieth-century European popular culture, and the fact that its golden age coincides with the beginnings of the Space Age and the associated Cold War Space Race, it is fascinating to consider the factors that account for the public interest in space-themed comics. To do so, an examination of the space episodes that were published in the *Tintin*, *Buck Danny* and *Dan Cooper* series will show not only the commonality of themes, but the seemingly contradictory aspirations the theme of 'space' elicited.

Though all are great classics, the series under consideration differ notably from one another. Aside from *Tintin*, they include *Buck Danny*, a series that depicts the adventures of an American navy pilot; and *Dan Cooper*, another aviation strip, this time focusing on a Canadian pilot. All three series enjoyed a good measure of success and remain in print, with *Tintin* far exceeding the others' success. Each series was also translated into other languages. Though one cannot interpret directly the reaction of readers,[2] the widespread sales of

Guillaume de Syon (✉)
Albright College, Reading, PA, USA
e-mail: gdesyon@alb.edu

Alexander C.T. Geppert (ed.), *Imagining Outer Space*
European Astroculture, vol. 1
https://doi.org/10.1057/978-1-349-95339-4_9

such comics suggest a popularity that cannot be overlooked, especially when it comes to genre-defining icons of the 'Franco-Belgian' school of comic artists. Their successes also depended on factors unrelated to space travel, and while they intended to display the possibilities of a fruitful technologically oriented future, space travel would be just one facet. To consider such icons of European popular culture also helps respond to recent calls for a better understanding of the cultural history of the space program.[3] The importance of NASA in European popular perceptions of space should never be discounted, especially during the Cold War. The European understanding of satellites and astronauts was, of course, informed by the Cold War, but also by film, novels and the comic strip tradition.[4]

The comic tradition under investigation evolved within the new wave of mid-twentieth-century European mass culture from a standpoint of 'leisure-time' consumer goods directed at youth into classics of the 'ninth art.' This shift is a characteristic example of popular culture that, through mass production and circulation, becomes a common culture.[5] Consequently, a visual dimension began to appear and was expressed first in colorful weekly magazines and new genres of movies, but also through comic books. Whereas Roland Barthes (1915–80) gained fame by analyzing single images in the 1950s, the comics themselves relied on a series of images, which both intrigued critics and also opened new avenues for artists to depict adventurous themes such as space. Though a similar phenomenon existed in the United States, a special European comic identity was beginning to crystallize at that time.

I A primer on comics and space

To understand these strips' importance further, it is necessary to first briefly situate the comic tradition of postwar Europe, its antecedents, and what political factors played a role in their conception. The European comic tradition derived partly from political caricatures as well as from illustrated stories that first appeared in the nineteenth century.[6] The new art, well developed in France and Belgium, was considerably influenced by the United States during the interwar years when classics like *Mandrake* and *Tarzan* were serialized in Europe and registered a stunning success.[7] The Second World War interrupted the supply of American comics, and paper restrictions prevented European series from appearing. But by 1945, American 'funnies' began to reappear, first as small gifts from G.I.s to children and then, more formally, in magazines in Belgium and France.[8]

By 1946, in remembering the prewar American wave and eager to carve out their own niche, several European publishers began targeting the new youth market. Not only were the profit margins of American series clearly a concern, but these series were, in the opinion of European publishers, culturally alien at times, because they failed to account for local tastes.[9] It is in this context that the so-called Franco-Belgian school of cartooning, already in gestation, exploded on the European scene. Its success was the result of a complex mix of market creation and market demand, as much as it depended on new talent at the level of story-writing and comic artistry.[10] It includes the three series this article analyzes.

A reading of these strips focusing acutely on syntax and paradigm is beyond the scope of this chapter, partly because of the widely differing styles each artist adopted. Comparisons are, nonetheless, possible because the heyday of such comics matches roughly the contours of the Space Race between the superpowers. It also constitutes a paradox whereby space became a topic of interest to European cartoonists even though Europeans were not involved in manned spaceflight for decades. In so doing, artists emulated some of their predecessors who had taken to utopian projections of future technological advances. For example, Alain Saint-Ogan (1895–1974), a pioneering cartoonist whose *Zig et Puce* characters experienced, among many adventures, a trip into the future aboard a fantastical spaceship, may have also sought to choose topics deemed 'safe' as far as censorship went.[11]

The measure of popularity of comics in Europe is based on a temporal and political corollary. After the Second World War, comics became a major source of entertainment in youth culture, for television did not emerge as a widespread household feature for another two decades. Serialized in weekly comics magazines, such strips acquired the moniker 'p'ti-Mickeys,' thus referring to Disney's character, which had its own, widely successful weekly. The pejorative name pointed to another issue, namely whether comics had a negative influence on childhood.

In the early 1950s, Francophone popular culture had been experiencing a substantial exposure to foreign comics and foreign science fiction. Some came in the form of novels and movies. The curtailing of the latter through French government action – akin to the US Comic Code Authority – had a substantial influence on what publishers could do. And since it required that all comic strips, serialized and otherwise, be approved prior to sales on the biggest Francophone market, its influence extended beyond the French border. Created to enforce a 1949 law on youth publications, the 'Surveillance Commission' has since been decried as a censorship organization. It did indeed err through the bias of its members, suppressing pro-American comic strips and openly favoring others.[12] However, the Commission's call for the removal of political commentary from comics (an order intended, among others, for the *Buck Danny* comic strip) did have a positive impact, as this also included the removal of racist comments and drawings.[13] But the concern for the psychological well-being of children also extended to what science fiction could do. In that respect, shielding youth from 'nefarious influences' became a key element of early comic strips depicting science-fiction scenarios. As stated in the Commission's 1950 report:

> Such speculation is allowable only in as far as it can continue to claim a scientific base. Whereas Jules Verne projected the future development of science solely on the basis of existent knowledge, the authors of child literature ignore too often scientific information and improvise deliberately any means necessary to fantastic tales without linking these to any

> scientific category. [...] It therefore appears that one should exercise extreme caution in science fiction and try to extract from the science side as much authenticity as possible rather than widen deliberately the field of fiction.[14]

In so doing, the Commission encouraged the use of comics for educational purposes. Its statement also implied that depictions of Western technology should affirm Western values by supporting positive outcomes. It would follow logically that space travel would fit the bill, much in the fashion of the early writings on rockets of science-fiction writer Arthur C. Clarke, but without the associated social commentary.[15] Whereas early novels of moon exploration, including the classic *Cyrano de Bergerac*, had become successful by emulating the principle of *Gulliver's Travels*, the three strips we shall consider would limit themselves to relying on technology for the purpose of adventure, where discovery does not involve any questioning of the inner self.

This does not mean that no other strips took on 'hard' science fiction or some kind of reality-based space theme. In Germany, for example, Hansrudi Wäscher (1928–2016) produced a cheap and very popular strip, *Nick, der Weltraumfahrer* (Nick the Space Traveler, 1958–63), which was commemorated in April 1993 when a German astronaut flew three original drawings aboard the STS-55 shuttle mission. But the series' success included the necessary shift towards a wider science fiction that combined aliens with the likes of American-style super heroes.[16] Across the border in France, a series called *Meteor* filled the same purpose. Closer to the notion of 'hard' science fiction, artist Raymond Chiavarino (1927–2005) serialized a strip that ran for three years in the French press, *Pour la conquête de l'espace* (To Conquer Space, 1960–63). None of these, however, gained a substantial readership, though they likely contributed in their own way to the visual culture of space travel. It is not possible to know how far influence spread, however, as these series all disappeared, eclipsed by other events as well as the growing success of the two *Tintin* albums.

II V-2 realism and the shift in comic strip representations of space

The episodes from *Tintin*, *Buck Danny* and *Dan Cooper* that include space dimensions generally fall in line with an emphasis on 'realistic' science fiction. As Wolfgang Höhne has noted in the context of realism as applied to comics, this was a new phenomenon that followed the Second World War, either displacing the caricatural side of comics or mixing with it. The shift, he points out, was provoked by the V-1 missile and the V-2 rocket, the latter becoming the first man-made object to actually penetrate space.[17] Fantasy had overtaken reality, though popular literature tended to conflate the two machines despite their glaring differences.

The postwar era in Europe involved intense popular concern with atomic power. Consequently, early scenarios depicting rockets in a realistic way usually associated the booster with war rather than exploration.[18] Cartoonist Jacques Martin (1921–2010), a contemporary of Hergé, readily acknowledged that Wernher von Braun's brainchild inspired his scenario in *La grande menace* about a nuclear conflict using missiles (before such things became reality in the 1950s).[19] Similarly, Edgar P. Jacobs's (1904–87) first volumes in the series *Blake and Mortimer* cast a series of rocket attacks and counter-strikes that depicted aircraft heavily inspired by the Nazi missile. Though all stopped short of discussing actual space travel, they became instant successes; the postwar rebirth that science fiction experienced in turn fed the rebirth of comics.

The marketing and consumption of the Franco-Belgian strips relied on artistic innovation, a need to emulate as well as separate from the American comic strip tradition, and an ability to develop storylines that used multiple tropes, most notably science fiction. As these comic series developed, they moved beyond the 'one-dimensionality' critics assigned to mass culture and became a complex media capable of appealing to children, their parents, but also to intellectuals and new generations of comic artists.

All three series are considered part of the second generation of European comic strips that matured in the decades following the Second World War. Their track loosely follows that of the early space program and helps visualize the coming to terms of popular culture with new modes of technological life. Like the postcard which a half-century earlier helped people come to terms with the automobile and the airplane, comic art reveals facets of the machine and its impact on the human imagination.[20]

III The double pioneer: Hergé opens the moon ball

Georges Rémi (1907–83), who published under the pseudonym Hergé, is considered one of the grandmasters of modern comic strips worldwide. The author himself was an extremely complex and conflicted figure, sometimes considered far more interesting than his hero, Tintin.[21] Hergé's genius, aside from his unique approach to drawing, also consisted of researching scenery that would allow for any scenario to 'feel' real. A scene taking Tintin to Geneva had to include, for example, precisely drawn features of the Geneva airport and of the Hotel Cornavin in the same city.[22]

In the case of his double feature *Objectif lune* (*Destination Moon*, 1952) and *On a marché sur la lune* (*Explorers on the Moon*, 1954), a peculiar combination of inspiration and research made the Tintin adventure possible. The storyline itself, though very much a creation of Hergé (he favored a trip to the moon, thinking Mars was too grandiose an idea), bears very strong parallels to Robert Heinlein's *Destination Moon*, which was made into a feature film by the same name in 1950, though Hergé never mentioned whether he read Heinlein.[23] The double album also reflects an important stage in Hergé's work, as it inaugurated the third and final stage of the author's production (Figure 9.1).

The premise – using a single-stage nuclear powered rocket to reach the moon and return safely to earth – reflects the state of knowledge available to

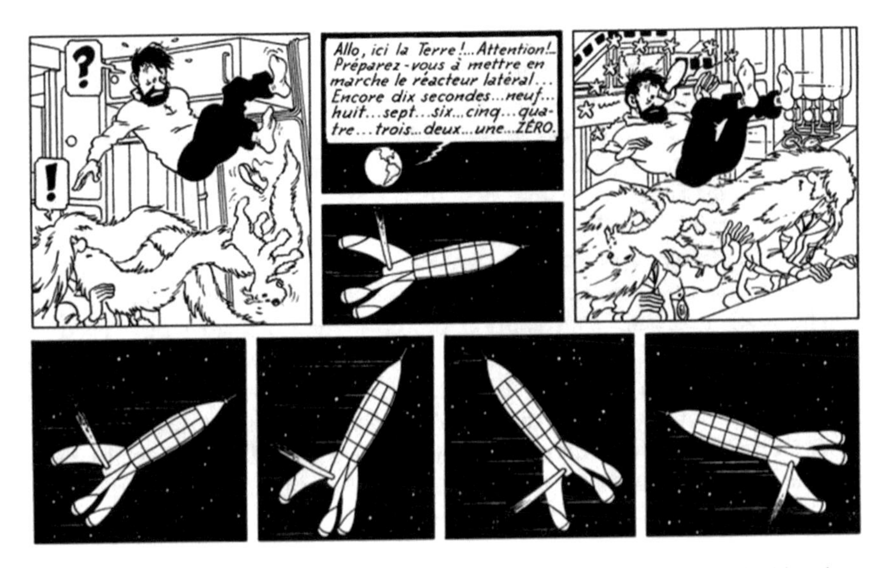

Figure 9.1 Georges Rémi aka Hergé (1907–83) did not hesitate to make liberal use of humor in his comic strips to draw in his readers, as when the rocket retrofires in preparation for its moon landing.
Source: *On à marché sur la lune*, 1954, 17, top. Courtesy of Hergé/Moulinsart.

the public at that time. Hergé appears to have made use of the material coming from publications by Wernher von Braun and Hermann Oberth, often summarized in the work of Francophone writers when they had not been translated.[24] The sounding rocket 'XFLR-6,' shown in the first volume, is a spitting image of a V-2, though the checkered red paint was borrowed from the French Véronique sounding rockets. The actual design may likely have derived from the work of comic strip artists Jacques Martin and Edgar P. Jacobs (both collaborators with Hergé), who were more attuned to technological innovation in the series they developed on their own. Yet Hergé maintained full control over the storyline and the drawing, insisting on a space booster that would be realistic.[25]

As for the main rocket, Hergé relied on the work of science writer Alexandre Ananoff (1910–92), who approved of a specially built model to help Hergé's assistants draw it realistically. Ananoff was a public persona in 1950s France, very much sought after for public scientific conferences, who had presided over the first international astronautical congress held there. In so doing, Ananoff had contributed to creating a space culture of sorts, though one that was limited in its impact.[26] By following scientific advice, Hergé sought to remain within the confines of realistic science fiction, a tradition traceable to Jules Verne's classics. While it helped the comic strip master avoid such pitfalls as 'monstrous animals, incredible beings and two-headed men,' he did take certain liberties attributable to artistic license for the sake of his story. The pressure suit helmets, for example, were all Plexiglas, against the advice of his sources. Also, a falling meteorite barely stunned the heroes on the moon instead of instantly killing them. Finally, whenever the storyline called for scientific explanations, Hergé was careful to balance it with some form of slapstick humor to keep younger readers entertained.[27] His general insistence on realism, however, did help him deal directly with the concerns of the aforementioned French censorship commission, though it took the unlikely form of an argument with the publisher.

The two-album series was serialized through the *Journal de Tintin*, one of the longest and most laborious of the whole series. Begun in 1948, it was set aside, then first appeared on 30 March 1950 for six months, then disappeared for two years before resuming anew. It was not until 1954 that the series finally concluded.[28] As if to confirm the fears of 'too much imagination,' when Hergé announced the title of *Explorers on the Moon* (the French title *On a marché sur la lune* suggests the feat is a done deed), the editors of the *Journal de Tintin* fought tooth and nail to have the title changed to a conditional tense. They lost.[29] Their reasons for fighting Hergé were, according to the latter, the lack of realism of such a title. In 1954 any suggestion of space travel seemed unlikely, much in the same manner as it had seemed impossible to the American public.[30]

In their view, space was simply another entertaining dimension, and their logic was defensible. The scenario of the moon trip involves tropes we are now familiar with throughout science fiction: oxygen running out, clandestine passenger attempting to hijack the rocket, and the expected sacrifice of one crew member to save the others. At the time, these could be found in Fritz Lang's popular 1929 movie *Frau im Mond* (Woman in the Moon), which may have influenced Hergé. But it pushed the boundaries of veracity.

Perhaps the editors were right. Much to Hergé's disappointment, the two-volume set of *Tintin* on the moon literally failed to take off.[31] It would be another three years before the set actually attracted renewed interest. At the same time, however, the double album remains to this day one that few critics have specifically analyzed using literary tools.[32] It stands on its own, almost in the manner of earth's natural satellite. An added paradox, which critics of the science-fiction genre have noted, is the challenge Hergé faced in developing a storyline within a limited physical space. Simply put, most of Tintin's adventures involve travel and follow the traditional boundaries of an epic hero with his sidekicks. As such, it requires interaction with other beings in order to move the plot along. How does one apply this to space, where no one answers? The boundaries of human interaction are limited to the rocket realm. It was precisely the elements of adventure, such as the attempted hijacking combined with the danger of space (asteroid near-collision, meteorite crash, etc.) that allowed the author to circumvent such challenges but also to elicit new interest once reality joined his fiction.

The flight of Sputnik 1 in October 1957, though avidly followed in Europe, ranked more as a form of high-tech entertainment than anything else. However, the beginning of the Space Race actually rekindled the interest in Hergé's two books who had since gone back to earthbound stories for his Tintin character. The French weekly *Paris Match*, in particular, published a multi-page spread in 1958 casting Tintin (and his 'father') into moon pioneers.[33] This association of symbols of popular culture with the moon remained strong until the end of the Apollo moon missions. By then, Hergé had done several commissions illustrating spaceflights for the same weekly. The two-volume set was now among the bestsellers of his whole series. Hergé did not accept the media claim that he was the precursor of space travel, arguing that Jules Verne alone deserved such an honor; nonetheless, he drew a humoristic greeting welcoming Neil Armstrong to the moon that appeared in *Paris Match* and which he reportedly sent to Armstrong personally.[34]

This was the last of Tintin's involvement with space, as Hergé moved to seek inspiration elsewhere. Asked later on to judge his work on the two volumes, he commented that it had been one of the hardest endeavors because of the unknown dimensions in which his characters functioned. Adventure certainly was present thanks to the inherent threats to the hero and his friends, but space was a fantasy that, interestingly, did not lend itself to the kinds of character exploration Hergé favored.[35]

IV The other realist wave: *Buck Danny*

Whereas *Tintin*, though realistic, maintained a comic tradition associated with its pre-Second World War origins, a different kind of comic strip that borrowed from the realistic American tradition developed soon after the war. Based in aviation, *Buck Danny* concerned itself in matters pertaining to space. Loosely inspired by American Milton Canniff's *Terry and the Pirates*, the strip follows the adventure of a US Navy flyer from Pearl Harbor to the present day. Though based in fantasy (Danny is the only Navy pilot in history able to transfer to the Air Force and back whenever it is convenient),[36] the realistic approach adopted by authors Jean-Michel Charlier (1924–89) and Victor Hubinon (1924–79) made the series a hit among Francophone readers eager to learn more about American military pilots at a time when the US military presence in Europe was particularly strong and aviation publications were fairly expensive, especially for younger readers. The series also incurred the wrath of the censorship commission on multiple occasions. Not only did this require changes in storylines, but the authors had to eliminate any hint at a discussion of contemporary political events. The result was a series based entirely on adventures in flying with little space left for character development or stimulation of the imagination.[37] Writer Jean-Michel Charlier scrupulously assembled documentation: during a research trip, he met astronaut John Glenn (1921–2016) and witnessed an X-15 flight piloted by Neil Armstrong (1930–2012).[38] Consequently, Charlier and Hubinon incorporated several aspects of the US space program into four stories, including two that spread over two volumes. All published in the late 1950s and early 1960s, they reflected a positive image of the American space program, though the view was certainly limited by the parameters set for the series.

The first set of stories inspired loosely by the American Project Paperclip involved the tracking down and rescue of a 'Professor von Brantz' – clearly a fictional depiction of Wernher von Braun – to work on the US space program. The actual space potential was incidental to the story, which tended towards a typical race to the finish against unidentified enemy agents.[39] The spying feature, however, allowed the authors to avoid any political commentary. Much in the manner of James Bond's 'Specter' organization, the unidentified spying organization was one way to skirt the French youth commission. The resulting stories, though repetitive, nonetheless proved entertaining, though they took place in the margins of space. One, *X-15*, focused on the test plane by the same name and sabotaging attempts. The realism of the story (despite the fictitious elements) owed much to Charlier's observation of the rituals of an X-15 flight, including the pilot's family pre-flight visit.[40] Naturally, to ensure a moral end that matched the younger readership (and the wishes of the French censorship organization), no ambiguities existed among the 'good' characters; the spies were eventually caught and the project saved. This redemptive dimension matched the space program's portrayal in the popular press and was found even more strongly in the next two-volume episode.

Opération Mercury and *Les Voleurs de satellite* (Satellite Thieves, 1964) concerned a fictitious Captain Dayton's launch aboard a Mercury capsule, and the sabotage of the recovery operation after landing. Complete with a portrayal of a concerned President Kennedy, the thrill of the storyline revolved around the resistance Danny, his companions and the astronaut each raised when taken prisoner. The Mercury capsule flight in the comic was, unlike the real occurrence, a non-event. The astronaut waiting to be picked up by helicopter (only to see a mysterious submarine emerge from the water) was depicted as damning the water tightness of his 'tin can,' a direct reference to real astronaut Virgil 'Gus' Grissom's (1926–67) close call whose Mercury capsule sank after splashdown in July 1961.[41] In fact, Dayton became a stand-in character always needing to be rescued. Excitement was not about space, but about sabotage of an aircraft carrier seeking to pick him up, the various challenges the heroes faced, and of course the successful capture of the spies.

The same went for *Alerte à Cap Kennedy* (Alert at Cape Kennedy, 1965), an episode featuring the destruction of rockets launching from the Cape. Inspired by a NASA film of early rocket failures, and combining it with the Cuban Missile Crisis, Charlier wrote a scenario involving an independent Latin American state using powerful radio waves to throw rockets off course. Danny and his companions were supposed to train to become astronauts, and thrilling though the launch sequence was, the story once again went into the routine of destroying the spy network. Danny & Co. would never return to space, except for a humoristic stint. In June 1969, *Journal de Spirou*, which ran the serialized version of the strip, published a five-page spoof of the aviators flying NASA interceptor spaceships instead of their usual US Navy airplanes, decades into the future. The time of publication coincided with the brief lull between Apollo 10's return and the launch of Apollo 11, and was considered a humorous acknowledgement of the real adventure. The punch line was a bit edgy, though: it was all a daydream of one of the bored flyers.[42] In fact, the very nature of the series, the restrictions on its storylines and the actual events transpiring in space, suggested this was the only venue Charlier and Hubinon could adopt for their characters. Space was in a category by itself, requiring special technology that, though fascinating, was too remote to further develop realistic storylines around it. Since he followed loosely the development of news events, Charlier may have faced the difficulty of finding new space venues as the Apollo program was winding down. The threat of censorship had already prompted the authors to direct their Southeast Asian-themed volumes away from Vietnam towards imaginary lands filled with drug traffickers. It proved much easier to resurrect such old tropes instead of generating new ideas, even in space.

V Balancing reality and fantasy: the Canadian-Belgian solution

Whereas both *Tintin* and *Buck Danny* faced a clear restriction in the fantasy elements associated with space travel, another comic strip, also about an aviator, saw far greater use of space imagination. Drawn by Belgian artist Albert Weinberg (1922–2011), *Dan Cooper* became a major classic through the weekly *Journal de Tintin* by focusing on an unlikely hero: a Canadian fighter pilot (Figure 9.2).[43]

Cooper's nationality was a clear advantage in the context of the Cold War and possible press censorship in France, where the biggest Francophone market lay. By casting Cooper into a quasi-American thanks to the machines he flew, yet dissociating him from Washington, DC, Weinberg was able to emphasize such themes as international cooperation and thus avoided the issues that had offended censors when it came to portrayals of *Buck Danny*. In so doing, he echoed one of Arthur C. Clarke's fictional characters, who argued that nationalities did not matter 'beyond the stratosphere.'[44] The contrast with Hergé's choice is notable in that context: *Tintin*'s political setting had become almost entirely fictitious after the Second World War; the moon shot originated in imaginary Syldavia. Not only did this limit the risk of offending censors, but to Hergé it was also a means to accentuate the secrecy (and thus the attraction) of the rocket project itself. As one of his biographers later noted, a lift-off from the United States (originally considered) would have taken away the excitement of the unknown, yet promising technology in favor of a showmanship of 'cheerleaders and jazz bands.'[45]

Figure 9.2 The series *Dan Cooper* delved liberally in aviation and space matters, often intermixing the two. In so doing, it tended towards science fiction. The cover shown here represents the serialization phase of the strip. The comic book hero embarks on a new adventure to explore Mars. In later albums the author, Albert Weinberg (1922–2011), returned to more realistic portrayals of the Space Race.

Source: *Tintin* 515 (4 September 1958), cover. Courtesy of Hergé-Moulinsart and Weinberg, Editions du Lombard.

Dan Cooper, on the other hand, dealt with real nations. In press for some four decades, the series underwent several transformations partly in response to changing reader (and publishing) interest after the 1950s. Of the 45 albums published, almost a third dealt thematically with science fiction, in one way or another. Weinberg was no stranger to the genre, having previously illustrated a *Buck Rogers*-inspired series, *Roc Meteor*.[46] In the story, the pilot *cum* astronaut first journeyed to a secret space station directly inspired from the von Braun/Disney/*Saturday Evening Post* illustrations. Such illustrations appeared in weekly magazines but also in popular treatises on future space travel.[47] *Le maître du soleil* (Master of the Sun, 1958) featured an Austrian scientist, Professor Shaffer, who had worked with Hermann Oberth (1894–1989) and eventually built in secret a ring-shaped satellite. The machine, clearly inspired by von Braun's model of a wheel-like space station, became the setting for a spy adventure (Figure 9.3).

The ease with which the hero joined others in orbit reflected the popular belief that stellar navigation was easy. More important was the fact that Europeans were involved in the completion and operation of the space station. Like other adventure strips, the ultimate struggle involved saving the station from sabotage, but at the conclusion of the story Dan Cooper hoped the project would soon become an international space station.[48] This emphasis on technology and peaceful cooperation would become the leitmotiv of all space-related volumes Weinberg would produce. In addition, Weinberg went beyond Hergé and the Charlier-Hubinon team in dallying with fantasy elements. Not only are humanoid extraterrestrials to be seen, but Cooper does make it to a Mars satellite by 1960.

Weinberg developed several storylines featuring extraterrestrial beings. But by stressing that the advent of space was 'here and now,' rather than in the future, Weinberg placed *Dan Cooper* into an openly fantastic routine. One album, *Le mystère des soucoupes volantes* (The Mystery of the Flying Saucers, 1969) openly borrowed from popular culture and showed shadowy creatures emerging from crashed spacecraft. Yet there was no resolution offered to the 'first contact.' The same went for a subsequent story, *SOS dans l'espace* (SOS in Space, 1971), where the hero travels into orbit aboard a Gemini capsule, only to discover an alien space station whose humanoid inhabitants are dying of an unknown illness. While the astronauts are infected, they eventually recover, but the aliens remain a mystery. Both episodes make clear that the author is a prisoner of the realistic framework he established for the series in the 1960s and of his endeavor to seek an alternative to the familiar trope of enemy aliens.

The advent of a real Space Race as well as a marked readership preference for stories that were closer to technological reality prompted Weinberg to revise his storylines in subsequent volumes to offer 'reality-based' fantasies. The risky business of 'too much sci-fi' was proving challenging, but Dan Cooper's Canadian identity balanced this by providing a special protection from political issues. Simply put, Cooper stood halfway between the rigid

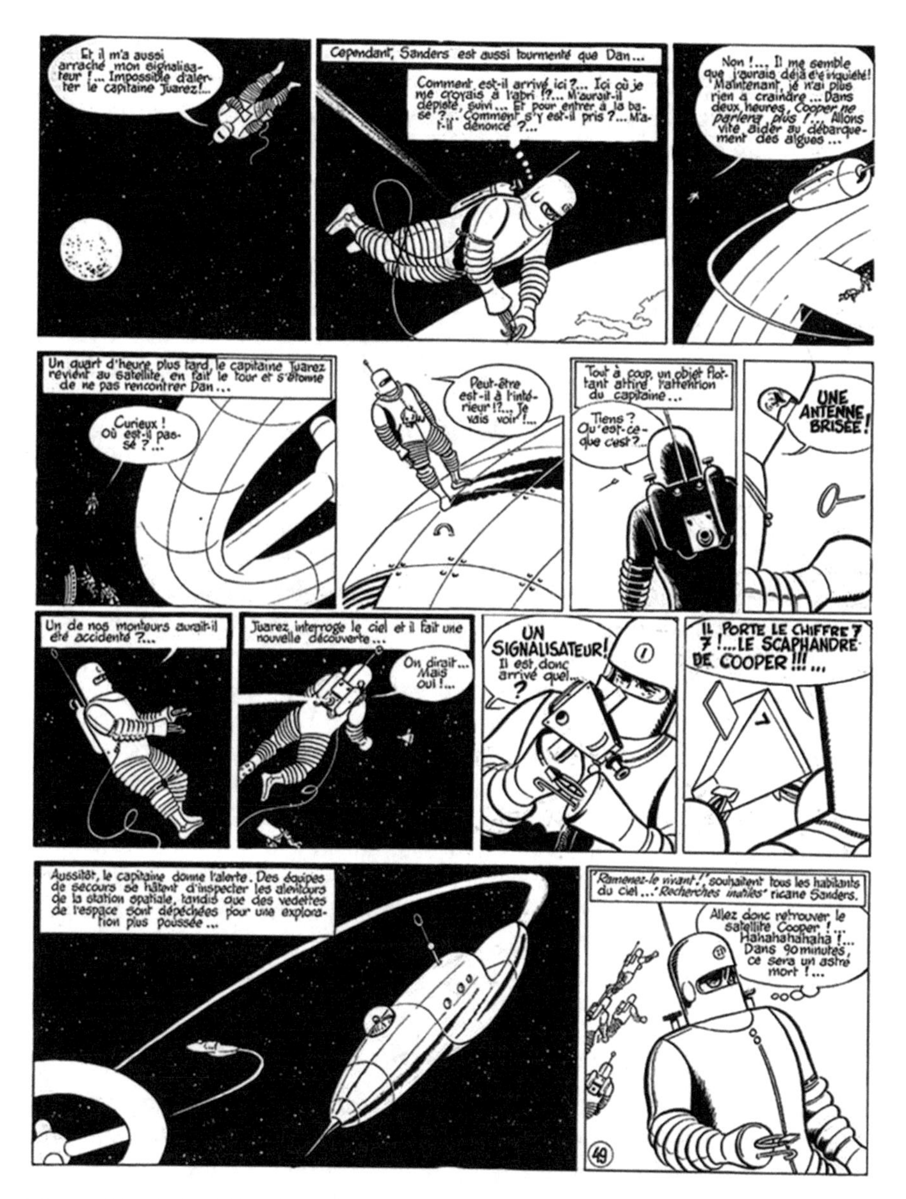

Figure 9.3 Artist Albert Weinberg put his early experience illustrating science fiction series to good use in the early installments of his own *Dan Cooper* series. Here, the Canadian pilot-turned-astronaut journeys from the ringed space station to challenge a saboteur.

Source: Albert Weinberg, *Les Aventures de Dan Cooper: Le maître du soleil*, 1958, plate 49. Courtesy of nufnuf-art collection.

Buck Danny strip and the more colorful *Tanguy et Laverdure*,[49] an aviation strip that gained national renown in France. The citizenship issue allowed Weinberg to cast a European space program which, though fantasy, made for an interesting look away from the classical Space Race. The album *Trois cosmonautes* (Three Cosmonauts, 1966), for example, had Dan Cooper launch into space accompanied in separate boosters by a female cosmonaut and a French astronaut atop 'Europa' boosters in Mercury-like capsules (which were to glide back, much in the manner of a cancelled USAF project). The whole point of the endeavor was to work out an international cooperation deal, which of course was sabotaged by unknown spies.[50] A later episode, *Panique à Cap Kennedy* (Panic at Cape Kennedy, 1970), placed the NASA center under attack. The rockets and gantries were perfect backgrounds for exciting if unlikely intrigues, thanks in part to the realistic drawing style.[51] This episode, however, offered a means of transition into a space adventure based on actual hardware.

Later, however, Weinberg could do little but acknowledge the need for a focus on the superpower space programs. By associating Cooper to a fictitious American astronaut and casting the latter into his sidekick (something he had already done in earlier episodes), the author was able to send him into space several more times. The most memorable of these space stories was *Apollo appelle Soyouz*, which appeared in serialized version in the early 1970s, thus matching the Apollo moon missions. In tune with Weinberg's preference, the superpowers, though competing, ended up cooperating in rescuing a Soyuz orbiter crashed on the moon's surface. The behavior of the characters, however, suggested a kind of friendship of convenience, and a strong sense of moral commitment. Their actions and behavior might well have been set in the desert, or out to sea. Space, once again, was a convenient background, and technology provided the necessary thrill to an otherwise banal story. All this suggested that space would be part of a better future, but without a full-scale science-fiction dimension, there could be no utopian resolution.

VI Moving into a different orbit: sex, smurfs and parsecs

The three sets of space-related comic books briefly described and analyzed above represent the most widely diffused series of the European comic strip tradition. Space did appear as a prop in various other series, but time and again served as a background for entertainment or new social commentary rather than an outright goal.[52] When other series used space as a specific theme, the scientific dimension was usually abandoned in favor of social commentary. For example, the famous 'smurfs' series portraying tiny blue men included an episode *Le Cosmoschtroumpf* that depicts an 'astrosmurf' who meets the 'schlips,' proverbial Martians (Figure 9.4). Technology played a definite role (even though heavily caricatured), but the escapist dimension reflected the traditional tension in science-fiction writing, whereby the

Figure 9.4 Later developments in comic strips involved fantasy works that doubled as social commentaries. The *Astrosmurf*, from the highly successful children's series, humorously questioned both the representation of the Other and the need for a space program.
Source: Peyo, *Le Cosmoschtroumpf et le Schtroumpfeur de pluie* 6 (1970). Courtesy of IMPS.

storyline supported the development of fictitious characters rather than a focus on scientific futures. In fact, the smurf episode is very much a re-enactment of the traditional philosophical tale, one in which wants or needs are sacrificed for the good of the whole.[53] The 'astrosmurf,' fooled into believing he has actually met beings from another world, wants to go back to visit, but decides his own family matters far more. The moral lesson echoes what several aging Apollo astronauts have acknowledged privately in retirement: moving away from earth may be a necessity but also a bit of an aberration.[54] Yet fantasy is usually what readers derive from such stories, and this is especially the case in facets of science-fiction comics that came to fill the niche of travel into outer space.

Inaugurating wilder fantasy was *Barbarella* (1964). Though generally associated with the erotic movie (1968) starring Jane Fonda (1937–), the title refers to the series that inspired the movie. Artist Jean-Claude Forest (1930–98) resurrected the old sci-fi premise that space is no longer mysterious and features an array of alien species congregating and fighting. Though this paralleled the advent of similar assumptions in TV and movies, Forest's innovation was to instrumentalize the spacecraft into tools of sexual liberation for a woman in search of such experimentation.[55] The same author would eventually draw a staider series about time and space travel. Such developments signaled a new liberation in the realm of comic strips that followed 1968, and thus a relaxation of the standards associated with the French Commission that had emphasized the need for educational content in science-fiction comic strips. But *Barbarella* also marked the symbolic decline of the tradition of the grand exploratory adventure associated with the genre.

The slowdown of the space program combined with new directions in science fiction, especially film, helps explain the move away from realistic comic books that focus on space: 1968, as the year Apollo 8 orbited the moon and a generation challenged its elders, offers an interesting point to mark the shift.[56] The new generation of comic strips appearing thereafter, however, did not completely destroy the tradition examined here. In the realm of youth-oriented comics, *Yoko Tsuno*, drawn by Roger Leloup (1933–), who had trained with Hergé, came to bridge readers' wishes for attractive technical innovations with adventures on other worlds. To succeed, however, Leloup completely separated the extraterrestrial adventures from any association with a real space program. In so doing, he contributed to the maturation of a facet of science-fiction cartooning that also witnessed a fragmentation of the comic medium as it grew successfully: instead of great threats and great adventures, inward journeys and escapist dreams became the norm of comic strips.

VII Conclusion

Hergé never returned his hero to space, but the double volume is still considered one of his finest efforts for the successful combination of humor, character development, technical research and artistic mastery.[57] One later episode suggested a flying saucer briefly visiting the earth, but the volume was panned by critics as simply a thrilling adventure with no new character substance. *Buck Danny* remained ensconced in fighter-pilot culture and continued to fly the latest hardware of the US Navy and Air Force. The *Dan Cooper* series emulated *Buck Danny*, though Weinberg did include an SDI-inspired episode during the golden era of the Space Shuttle, but quickly returned to earth adventure.[58] Brief forays were made by others, such as the more comical and child-oriented series *Les 4 as* (Four Aces, 1986), that featured one episode involving the Soviet Space Shuttle *Buran*. Other space adventures, however, focused directly on the more fantastic notions associated with both utopic and dystopic aliens.[59]

Thus, European youth culture did find a certain vision of the future in the reality-based science fiction of classic comic strips, but fantasy remained earth-based: The final frames of *Explorers on the Moon* has one of Tintin's sidekicks argue 'I'll tell you, I've learned just one thing from all this: MAN'S PROPER PLACE [...] IS ON DEAR OLD EARTH.' Hergé repeated this sentiment years later when he summarized the dilemma for any cartoonist seeking to maintain a semi-realistic science-fiction approach: 'No selenites, no monsters, no fabulous surprises! [...] That is why I will no longer draw such [space] albums: what do you expect to see happening on Mars or Venus?'[60] The master of cartooning had seen the future, but he did not think it worked. In a supreme twist of irony, Hergé may have answered his own question to the interviewer through one of his other masterpieces, *Les bijoux de la Castafiore* (The Castafiore Emerald, 1963). There, the artist played on the very challenge of limited physical space to develop a storyline. Replacing the rocket was the traditional castle where Tintin lives. Only there, over 64 pages, absolutely nothing happened other than interaction among the characters. This was very much in line with the reality of a long trip into space that faces no inherent threat, and thus no adventure story. The journey could happen within.

Indeed, the nature of comics meant that, even with restrictions, fantasies were allowed, especially when invoking future space travel. The three series discussed here were the only ones with a wide readership to attempt to remain within the confines of a solid scientific realm, one which ended up paralleling the years of the Space Race. This may be part of the problem cartoonists encountered: any 'hard' science-fiction tradition in their time had little to offer that the real Space Race had not already covered. This dilemma that Hergé so aptly summarized is arguably similar to that which Hollywood faced in its science-fiction offerings: unless the import is to eventually include aliens, 'hard' science fiction becomes as uninteresting to writers and their readers as an advanced materials experiment on the space station.

That is not to say that space comics have no place in current times. In the mid-1980s the French Space Agency (CNES), for example, sponsored several self-contained 'info-strips' that summarized its activities and promised an exciting future aboard the stillborn Hermès mini-shuttle.[61] In fact, the comic albums discussed here filled a role in their era, as reflected in magazine covers, popular references and simply, their longevity. All three still remain in print. But the difficulties of space travel, the evolution of youth cultures, as well as the shift to new entertainment media, means that comics, and especially the classics discussed here, left the realm of excitement associated with pioneers, and matured along with their readership. By focusing on space as a kind of background decor only, these series forced themselves into a kind of geocentric realm that made space an interesting toy. But they later became prisoners of said toy as the real space programs demonstrated the limits of space exploration. The 'balloons' in fact should have been the perfect vehicle to space and other extraterrestrial fantasies, and became thus in different incarnations. In the case of *Tintin*, *Buck Danny* and *Dan Cooper*, this happened too, but in the imagination of youth who were in turn inspired to join a space program.

Notes

1. Thibaut Dary, 'Tintin le précurseur,' *Le Figaro Magazine* (July 2009), 118 (special issue).
2. Though current fan blogs online provide a certain measure of the impact such comic strips had, an unscientific survey of existing discussion rooms showed that it was mostly participants who either loved or hated a specific series or story who got involved in the discussion, and most did not clarify the impact a particular story had on them.
3. For a literature survey see Asif A. Siddiqi, 'American Space History: Legacies, Questions, and Opportunities for Future Research,' in Steven J. Dick and Roger D. Launius, eds, *Critical Issues in the History of Spaceflight*, Washington, DC: NASA, 2006, 433–80.
4. See Alexander C.T. Geppert, 'Flights of Fancy: Outer Space and the European Imagination, 1923–1969,' in Steven J. Dick and Roger Launius, eds, *Societal Impact of Spaceflight*, Washington, DC: NASA, 2007, 585–99.
5. Dominick LaCapra, 'Is Everyone a *Mentalité* Case? Transference and the "Culture" Concept,' *History and Theory* 23.3 (October 1984), 296–311, here 300. On the theoretical relationship of culture to the space program, see also Martin Collins, 'Production and Culture Together: Or, Space History and the Problem of Periodization in the Postwar Era,' in Dick and Launius, *Societal Impact of Spaceflight*, 615–29.
6. Rodolphe Töppfer, *Docteur Festus, Histoire d'Albert, M. Cryptogram* [1840], Paris: Pierre Horay, 1975, 5–10; Claude Moliterni, Philippe Mellot and Michel Denni, *Les Aventures de la BD*, Paris: Gallimard, 1996, 13–19.
7. Pascal Pillegand, Béatrice Le Rider and Martin de Halleux, *100 ans de BD*, Paris: Atlas, 1996, 56–60.
8. Thierry Crépin, *Harro sur le gangster! La moralisation de la presse enfantine 1934–1954*, Paris: CNRS, 2001, 145–6.
9. Pascal Ory, 'Mickey go home! La désaméricanisation de la bande dessinée (1945–1950),' in Thierry Crépin and Thierry Groensteen, eds, *'On tue à chaque page!' La loi de 1949 sur les publications destinées à la jeunesse*, Paris: Editions du Temps, 1999, 74. Richard Kuisel, *Seducing the French*, Berkeley: University of California Press, 1992, 88–9, 114.
10. For an English summary of this facet of comics, see Matthew Screech, *Masters of the Ninth Art: Bandes Dessinées and Franco-Belgian Identity*, Liverpool: Liverpool University Press, 2005.
11. Alain Saint-Ogan, *Zig et Puce au XXIè siècle* [*Zig and Puce in the 21st Century*], 1935 (repr. Grenoble: Glénat, 1997).
12. See Crépin and Groensteen, *'On tue à chaque page!'*
13. Crépin, *Harro sur le gangster*, 311.
14. Quoted in Pierre Assouline, *Hergé*, Paris: Gallimard, 1998, 510–11.

15. De Witt Douglas Kilgore, *Astrofuturism: Science, Race, and Visions of Utopia*, Philadelphia: University of Pennsylvania Press, 2003, 115–20.
16. See Ole Frahm, 'Different Drafts of a "Future Horizon": *Weird Science* versus *Nick der Weltraumfahrer*,' in Norbert Finzsch and Hermann Wellenreuther, eds, *Visions of the Future in Germany and America*, Oxford: Berg, 2001, 471–85. I am grateful to Ralf Bülow for sharing material on this series.
17. Wolfgang Höhne, *Technikdarstellung im Comic: Der Comic als Spiegel technischer Wünsche und Utopien der modernen Industriegesellschaft*, PhD thesis, Universität Karlsruhe, 2003.
18. Michel Porret, 'La "grande menace": L'apocalypse des armes de destruction massive dans la bande dessinée francophone après la Seconde Guerre mondiale,' in idem, ed., *Objectif bulles: Bande dessinée et histoire*, Geneva: Georg, 2009, 203–31.
19. Hugues Dayez, *Le Duel Tintin-Spirou*, Brussels: Editions Luc Pire, 1997, 68–70; available at http://bibliotheque.livrel.eu/duel_tintin_spirou/duel.pdf (accessed 1 October 2017).
20. The literature on postcards, visual culture and iconography is extensive. See, for example, Gordon Fyfe and John Law, eds, *Picturing Power: Visual Depictions and Social Relations*, London: Routledge, 1988. See also Thomas Brune, 'Bürgerlicher Humor in technischer Welt: Eisenbahn und Automobil in den "Fliegenden Blättern" zwischen 1844 und 1914,' in Utz Jeggle, ed., *Tübinger Beiträge zur Volkskultur*, Tübingen: Tübinger Vereinigung für Volkskunde, 1986, 263–84; and Alf Lüdtke, 'Ikonen des Fortschritts: Eine Skizze zu Bild-Symbolen und politischer Orientierung in den 1920er und 1930er Jahren in Deutschland,' in idem, Inge Marßolek and Adelheid von Saldern, eds, *Amerikanisierung: Traum und Alptraum im Deutschland des 20. Jahrhunderts*, Stuttgart: Franz Steiner, 1996, 199–210. On postcards, see Joëlle de Syon, *Les Premiers fous du volant*, Geneva: Slatkine, 1987; Horst Hille and Frank Hille, *Technikmotive auf alten Ansichtskarten*, Leipzig: VEB Fachbuchverlag, 1989; and Ludwig Hoerner, 'Zur Geschichte der fotografischen Ansichtspostkarte,' *Fotogeschichte* 7.26 (1987), 29–44.
21. In comic art studies, the subfield of *Tintinologie* now includes a substantial body of literature focusing on Hergé alone. See the sources used here in addition to the bibliography available at http://www.tintin.com (accessed 1 October 2017).
22. Hergé, *Tintin: L'affaire Tournesol*, Brussels: Casterman, 1956.
23. Assouline, *Hergé*, 505. Numa Sadoul, *Tintin et moi*, Paris: Flammarion, 2003, 195. If Heinlein inspired Hergé, it would have been solely through his writings, as Hergé rarely went to the movies and did not own a television till late in his life.
24. http://tintin.francetv.fr/uk (accessed 1 October 2017), research file 'Ice on the Moon.' Post-Second World War francophone science writers interested in space exploration included journalist Albert Ducroq, Bernard Heuvelmans (author of *L'Homme parmi les étoiles*, Paris: Gérard Delforge, 1944, which Hergé read), and Alexandre Ananoff, author of *L'Astronautique*, Paris: Librairie Arthème Fayard, 1950.
25. Jacques Martin drew *Lefranc*, the adventures of a journalist; E.P. Jacobs drew the adventures of *Blake and Mortimer*. See Thierry Smolderen and Pierre Sterckx, *Hergé, portrait biographique*, Tournai: Casterman, 1988, 207, 226, 243.
26. Sadoul, *Tintin et moi*, 195; Albert Ducrocq, public address, 30 September 2001, http://www.astrosurf.com/planete-mars/news/2003/0607ducrocq.html (accessed 1 October 2011).

27. Sadoul, *Tintin et moi*, 196; *Destination Moon*, plates 7–8, 36–7; *Explorers*, plates 26–7. Benoit Peeters, *Le Monde d'Hergé*, Tournai: Casterman, 1983, 140.
28. Assouline, *Hergé*, 509. Hergé had suffered from acute depression, which slowed his work down considerably.
29. Sadoul, *Tintin et moi*, 215; Smolderen and Sterckx, *Hergé*, 207. Hergé had even proposed a wildly fantasist title in early drafts, 'le plan mamouth' (The Mamoth Plan).
30. Howard McCurdy, *Space and the American Imagination*, Washington, DC: Smithsonian Institution Press, 1997, 28–32.
31. Assouline, *Hergé*, 416.
32. See, for example, *Schtroumpf: Les cahiers de la bande dessinée* 14–15 (1978), 6–22.
33. *Paris Match* 493 (20 September 1958).
34. Stephanie Paine, 'Welcome to the Moon, Mr Armstrong,' *New Scientist* 182.2441 (3 April 2004), 48–9.
35. 'Tintinologists' are quick to note that, for the first time, some of the characters display ambiguity, as in the case of an engineer turned bad who sacrifices himself to save the returning crew.
36. This 'incident' is due to the fact that Charlier had limited access to documents in the 1950s when the series really took off. See Guy Vidal, *Jean-Michel Charlier: Un réacteur sous la plume*, Paris: Dargaud, 1995, 14.
37. See Guillaume de Syon, '"Don't Read those 'Toons!" French Comics, Government Censorship, and Perceptions of American Military Aviation,' *Contemporary French Civilization* 28.2 (Summer 2004), 274–92.
38. Jean-Marc Charlier and Victor Hubinon, *Tout Buck Danny 8: Pilotes de prototypes*, Marcinelle: Dupuis, 1986, 50, 96.
39. *Buck Danny* albums no. 22, *Top Secret*, Marcinelle: Dupuis, 1958; and no. 23, *Mission vers la vallée perdue* (Mission over Lost Valley), Marcinelle: Dupuis, 1958.
40. Charlier and Hubinon, *Tout Buck Danny 8*, 50, 96.
41. *Buck Danny*, album *Les Voleurs de satellite*, Marcinelle: Dupuis, 1964.
42. Charlier and Hubinon, *Tout Buck Danny 7: Vols vers l'inconnu*, Marcinelle: Dupuis, 1983, 188–92.
43. Weinberg was a contemporary of Charlier and Hubinon, and was involved in some of the early drawings of *Buck Danny*, see Vidal, *Jean-Michel Charlier*, 14.
44. Kilgore, *Astrofuturism*, 120.
45. Smolderen and Sterckx, *Hergé*, 207.
46. The series ran from 1949 to 1956 in the weekly *Héroïc-Albums.*
47. McCurdy, *Space and the American Imagination*, 37–41, 65–6; see also futurologist Georges H. Gallet's heavy use of such pictures (in black and white) in his *A l'Assault de l'espace*, Paris: Pensée moderne, 1956.
48. Albert Weinberg, *Les Aventures de Dan Cooper: Le maître du soleil*, Brussels: Editions du Lombard, 1958.
49. *Tanguy et Laverdure* was a very successful aviation strip of the 1970s, first drawn by Albert Uderzo (Astérix's co-creator) and later taken up by Jijé in a different, yet fairly effective style. Because the two aviators' contact with space remains limited to anecdotal events (the retrieval of a test capsule serves as background to a story of friendship, treason and redemption), it will not be discussed here. See also Vidal, *Jean-Michel Charlier*, 31–2.
50. Albert Weinberg, *Trois cosmonautes*, Paris: Dargaud, 1966.
51. Idem, *Panique à Cap Kennedy*, Paris: Dargaud, 1970.

52. Peyo, *Le Cosmoschtroumpf*, Marcinelle: Dupuis, 1976. The long-running youth comic strip *Les 4 as* (Four Aces) took on space travel in a slapstick fashion by focusing on a trip aboard a Soviet shuttle. François Craenhals, Jacques Debruyne and Georges Chaulet, *Les 4 as et la navette spatiale*, Brussels: Casterman, 1989.
53. Alain Corbellari, 'Du Moyen Âge au conte philosophique: avec *Johan et Pirlouit* et *Les Schtroumpfs*,' in Porret, *Objectif bulles*, 84–109, here 103.
54. *In the Shadow of the Moon*, directed by David Sington, USA/UK 2007 (Discovery Film).
55. Thierry Groensteen, *Astérix, Barbarella & Cie*, Paris: Somology, 2000, 160, 166.
56. Höhne, *Technikdarstellung im Comic*, 86–7.
57. Claudie Guérin, '5-4-3-2-1-0! Französischsprachige Sachbücher und Comics über die Eroberung der Luft,' in Andreas Bode and Marianne Reetz, eds, *Von Ikarus zur Raumfahrtstation: Luft- und Raumfahrt in der internationalen Jugendliteratur*, Mainz: Stiftung Lesen, 1991, 20–6.
58. Albert Weinberg, *Dan Cooper: Navette spatiale* (Space Shuttle), Brussels: Novedi, 1983.
59. This is particularly the case with the series *Yoko Tsuno*, focusing on a Japanese woman's occasional contact with a humanoid advanced species.
60. Sadoul, *Tintin et moi*, 196; Hergé, *Tintin: Explorers on the Moon*, Brussels: Casterman, 1954, 62.
61. Patrick Clément, *Phil et l'appel de l'espace*, Toulouse: CNES, n.d. (ca. 1985).

CHAPTER 10

A Stumble in the Dark: Contextualizing Gerry and Sylvia Anderson's *Space: 1999*

Henry Keazor

'*Space: 1999* occupies a unique and not altogether happy position in the Valhalla of televised space adventures. It is the program that *Star Trek* fans love to hate, even 20 years after its debut. When it is not being ridiculed or attacked, *1999* is often forgotten or overlooked by science-fiction historians and television critics, probably because it was a British-made product that never aired on the major American television networks. Perhaps *Space: 1999*'s biggest problem is simply that it appeared at the wrong time. It stands sandwiched between *Star Trek* (1966–69) and *Star Wars* (1977), two milestones of the genre,' John Kenneth Muir wrote in the introduction to his book *Exploring 'Space: 1999,'* published in 1997.[1]

Indeed, *Space: 1999* does occupy a 'not altogether happy position' between other TV science-fiction series. Although the show earned a certain amount of praise, is still acknowledged by some critics, and continues to enjoy an ever-increasing international fan cult status due to the fact that it was sold to more than 100 countries, the series, originally conceived and produced by Gerry (1929–2012) and Sylvia Anderson (1927–2016), mostly faced severe criticism.[2] This was partly directed against the atmosphere of the series, which was accused of being somber, dull and obscure, and partly because of the way the stories were told. Critics described the plots of the stories as slow, lifeless and heavy-handed. The direction of the main characters and their stiff, formal and sometimes seemingly sedated and wooden acting skills were also criticized. Maliciously referring to the fact that the Andersons had made their first success on British television with children's puppet series like *Thunderbirds*,

Henry Keazor (✉)
Universität Heidelberg, Heidelberg, Germany
e-mail: h.keazor@zegk.uni-heidelberg.de

Alexander C.T. Geppert (ed.), *Imagining Outer Space*
European Astroculture, vol. 1
https://doi.org/10.1057/978-1-349-95339-4_10

Figure 10.1 A view in Commander John Koenig's office on Moonbase Alpha, showing (under the windows) two 'Throwaway' sofas (1965), four 'Toga' chairs (1968) around a 'Mezzatessera' table (1966), flanked by three 'Giano Vani Route' telephone tables (1966).
Source: *Space: 1999*, episode 20, *Space Brain* (22 January 1976). Courtesy of Granada Ventures Ltd/ITV Global Entertainment Ltd/ITV Studios Global Entertainment.

and with *Space: 1999* aimed at live-action science fiction for adults, one critic wrote about the new series and the two American leading actors, Martin Landau and Barbara Bain: 'Gerry and Sylvia Anderson used to make the space stuff with puppets. With Martin Landau and Barbara Bain, I swear, you won't tell the difference.'[3]

But in order to fully understand the originality of *Space: 1999* and the reasons for its mixed reception, it is necessary to consider the historical context in which the series was produced. *Space: 1999* was not only 'sandwiched between *Star Trek* and *Star Wars*,' as Muir stated, but also bridged two different times and their changing cultural, economic as well as social and scientific conditions, all of which had an impact on the series.[4]

Thus, despite the fact that the first season of the series was produced and shot between 1973 and 1975, and was supposed to show the reality of the then distant year 1999, the viewer continuously encounters examples of typical 1960s design, primarily conceived by Italian and international fashion designers: A 'Mezzatessera' round table conceived in 1966 by Vico Magistretti stands in the Commander's office (Figure 10.1) together with the so-called 'Selene' chair by the same designer. Completing this set of 1960s furniture, the audience sees two 'Throwaway' sofas, designed by the Italian Willie Landels and distributed from 1965 by the Italian designer Aurelio Zanotta, as well as 'Toga' chairs, designed in 1968 by Sergio Mazza and manufactured by the Italian company Artemide, and finally three 'Giano Vano Ruote' telephone tables, designed in 1966 by Emma Gismondi Schweinberger.[5]

The mostly white colors, simple forms and smooth plastic surfaces of the set elements are accompanied by costumes designed by none other than Rudi Gernreich (1922–85), the Austrian-born American fashion designer who rose to fame during the 1960s through his minimalistic clothing creations made of synthetic fabric, often shaped in, or ornamented with, geometric forms. For *Space: 1999* Gernreich conceived a simple and austere unisex-uniform, whose bright, cream color was interrupted only by one colored sleeve (indicating the working area of its wearer), zippers along the sleeve and the trouser-leg, and a yellow plastic belt (Figure 10.2).[6]

The sets and costumes thus showed a bright, smooth, clean and efficient-looking world which – given the heavy amount of plastic used for its fabrication and the strong lights coming from the plastic wall-panels – obviously did not know any shortages of oil or energy: This can be seen as either a nostalgic

Figure 10.2 A look into the Command Center with the Moonbase Alpha crew in their unisex uniforms.

Source: *Space: 1999*, first pilot episode, ***Breakaway*** (4 December 1975). Courtesy of Granada Ventures Ltd/ITV Global Entertainment Ltd/ITV Studios Global Entertainment.

look back or an encouraging glimpse into the future, especially in fall 1973 when the energy and oil crisis in the Western hemisphere forced even England to introduce a three-day work week. As a matter of fact, the shooting of *Space: 1999*, which started in November 1973, was hampered by fuel shortages and the threat of strikes; the crew even had to organize a car-sharing system in order to get to the film studios which were, at the time, under the threat of closure. As a result, the production of *Space: 1999* had to move from the Elstree Studios to Pinewood Studios where all of the sets had to be rebuilt.[7]

I *1999/2001:* seriousness and credibility

The fact that the 1960s are so visually present in the sets and costumes, which were themselves created in the early 1970s in order to show a convincing vision of the year 1999, should make the viewer aware of the fact that different temporal layers in the series are obviously inherent.[8] These visual layers not only had an impact on the visual side of the production, but also on its structure and storylines.

Indeed, the late 1960s are also present in *Space: 1999*, insofar as the program was originally conceived as a spin-off of the then successful British TV-show *UFO*, which was also produced by the Andersons starting in 1969. *UFO* became so popular in the United States that in 1972 the American network giant CBS signaled its interest in buying a further series; the spin-off, however, was expected to be bigger, better and more spectacular than the first season. The main idea here concentrated on the especially popular moon base which, according to the premise set up in *UFO*, is part of S.H.A.D.O. (Supreme Headquarters Alien Defence Organization), a secret defense force fighting off the continuous attacks of invading aliens. These hostile enemies should come up with a plan to blast the moon out of its orbit in order to dispose of its base, which served as a defensive outpost against their attacks. The new series, therefore, was supposed to show the adventures the crew of the moon base experienced during their odyssey through space. But when viewer ratings for *UFO* suddenly dropped in 1972, the project of a new series, now called *UFO: 1999* was cancelled, despite the fact that preproduction was already at an advanced stage and that much money, time and energy had already been invested. Given this investment, CBS nonetheless decided to go ahead with the production of a new series.[9] However, in order not to endanger its possible success by associating it to the now suddenly failing *UFO*, the premise of the new series was changed. In place of an alien force, a disaster on the moon's nuclear waste dump would now serve as a means of catapulting the earth satellite out of its orbit and out into deep space.[10]

But despite the fact that not aliens, but ultimately humanity itself was responsible for the moon's odyssey, *Space: 1999* still carried some of the inherently xenophobic bias of *UFO* in its storylines. Whereas a series such as *Star Trek* showed a universe where alien races had established, for example, a 'United Federation of Planets,' the encounters with aliens of both the crews of S.H.A.D.O in *UFO* as well as of the crew of the Moonbase Alpha in *Space: 1999* had – as some frowning critics noted – mostly negative results.

This stands in contrast to another science-fiction production with which *Space: 1999* is closely linked. The circular arrangement of the moon base, its elevator launch pads, and the modular grid-and-girder structure of the space transporter used in the series – baptized as a nod to the first moon landing module named Eagle – bears a close resemblance to the moon base and its moonbus featured in Stanley Kubrick's 1968 film *2001: A Space Odyssey*.[11] Widely recognized as a milestone of the genre, Kubrick's film was compared to *Space: 1999*, often to the latter's disadvantage. In addition to the fact that both narrate a space odyssey – *2001* focuses more on the aspect of odyssey as a metaphor for a shipwreck, while the TV series relies on the structure of Homer's *Odyssey*, only exchanging the different islands for planets – the productions are also linked by the special-effects designer Brian Johnson (1939–), who created the moon bases and space vehicles in both productions. He was later joined by Martin Bower (1952–) who, unlike Johnson, had not worked on *2001*, but his design of a spacecraft used in the episode *The Alpha Child* was admittedly influenced by the Discovery spaceship from Kubrick's film.[12] And in both cases, the seriousness and credibility of the depicted future purposefully enhanced the more mystical bias of both tales, where each time an alien, but obviously friendly force, seems to steer and direct the events.[13] In *2001*, an alien intelligence lures the humans into deep space in order to establish direct contact and propel humankind on a journey leading to a higher evolution. In *Space: 1999* it is made clear, slowly but surely throughout the first season, that the odyssey of the moon – though seemingly inspired by an accident provoked by humanity's short-sighted storage of nuclear waste on the earth's satellite – follows a destiny and purpose determined millions of years ago. This universal force intervenes and helps the Alphans, the inhabitants of Moonbase Alpha, during their expedition through space.[14]

II *The Infernal Machine:* techno-fascination and techno-skepticism

Mysticism in both productions is accompanied by skepticism towards technology. Certainly in both *2001* and *Space: 1999*, technical progress seen in space stations, moon bases and spaceships is stylized as something fascinating because of its efficiency, power and even beauty. At the same time the limits of technical progress are clearly emphasized. HAL, the computer brain of the spaceship Discovery in *2001*, has to be de-activated in the end because it goes berserk; both the dialogue and the action of *Space: 1999* relate to the shared theme regarding the limits of human knowledge and computer intelligence.[15] Not only is the whole premise of the series based on the assumption of a gigantic failure of human technology, leading to the explosion of the nuclear waste storage facility on the moon, but throughout the first season frequent references are also made to earlier, failed and ill-fated deep-space missions.[16]

This ambiguous attitude towards human technology, seen positively in the stylization of the carefully crafted models and the impressive special effects on the one hand, and negatively in the technology-related disasters on the other, also lies at the heart of earlier attempts by the Andersons to rival Kubrick's sci-fi masterpiece.[17] In 1968 they wrote and produced the film *Doppelgänger*, directed by Robert Parrish (1916–95), where the viewer encounters the same contradictory mix of enthusiasm and skepticism towards technology. Here, too, the details of space exploration technology are presented with great care and in a fascinating light: rockets, ramps, space suits, and even an attentively crafted manned docking maneuver, obviously inspired by the first unmanned docking between Gemini 8 and an Agena target vehicle, achieved in March 1966. Further, the show anticipated the real manned docking of Soyuz 4 and Soyuz 5, accomplished by the Russians in January 1969, nine months before the film's première. It is, however, not by chance that in the end, the whole rocket base with all its technological paraphernalia is destroyed by an automated shuttle during its unfortunate re-entry in one of the famous and typical, gigantic 'Anderson' explosions: already, the obvious fascination for the achievements of technology displayed on the show is counterbalanced by a deep skepticism not only towards the impossibility of controlling technology, but ultimately also towards its purpose. In *Doppelgänger*, a manned flight is secretly planned and realized by both the European Space Exploration Centre (EUROSEC) as well as NASA in order to explore a recently discovered new planet in the solar system, which hitherto had been hidden due to the fact that it is in the same orbit as earth, albeit on the opposite side of the sun. After a successful launch and flight to the planet, the shuttle of the English astronaut and his American colleague is hit by an electrical storm and crashes. While the American is critically injured, the English astronaut, when rescued, finds himself back on earth where his superiors accuse him of having willfully aborted the mission and returned to earth. Slowly, however, it dawns on the astronaut that he has not actually returned, but is on the newly discovered planet which turns out to be a duplicate earth, a mirror image of his own world. In order to prove this to his superiors, he tries to return to his still orbiting spaceship, but the maneuver fails – due to an opposed electrical charge. When crashing his shuttle, he destroys the entire base together with the control station, killing all the people involved in the secret project and leaving behind only EUROSEC director Jason Webb (who, in the end, cannot even be sure that it all was more than just a hallucination). Thus, the plot strongly implies that humankind, when leaving earth and steering into deep space, ultimately only encounters itself, with catastrophic results.[18] Other *Space: 1999* episodes frequently dealt with similar storylines, particularly in *Another Time, Another Place* and in *A Matter of Life and Death*, which also employ motifs of the opposed or plots in which anti-matter is unfolded.

III *A Matter of Life and Death:* the impact of the energy crisis

This skepticism towards the effects and risks, as well as the achievement of technology in general and space exploration in particular, was obviously enhanced at the beginning of the 1970s by the worldwide oil and energy crises, and the immediate necessity of inventing alternatives to traditional energy sources. A scene from the pilot episode of *Space: 1999*, *Breakaway*, further spins the idea that atomic energy would be a major response to this challenge: one of the characters remarks that 'atomic waste disposal is one of the biggest problems of our time.' This acknowledgment leads to establishing nuclear waste disposal areas on the far side of the moon, which in the end threaten to explode, transforming them into (as Commander Koenig puts it), the 'biggest bomb man's ever made.'[19] Ultimately, the explosions do catapult the moon out of its orbit. And when, in another episode (*Another Time, Another Place*) the moon temporarily returns to its orbit, it is the earth of the future which it now circles around, devastated by what seems to be a worldwide nuclear disaster that has covered the planet with radioactive ash.[20]

Given the impact that the energy crisis, with its shortages and subsequent strikes, had on the production of the series, it is little wonder that this situation left its mark on the plots of many episodes, in which the preciousness of energy repeatedly characterizes the storyline.[21] In fact, the title of one of the episodes, *A Matter of Life and Death*, can in some ways stand as a headline for the whole *Space: 1999* series which in its plots mostly revolves around the need of energy and the ensuing questions of death, survival and procreation.[22] Interestingly, the topic of procreation in these stories is strongly linked to the idea of death as a necessity for understanding life. The concept of immortality, tackled twice by two episodes in the first season (*Death's Other Dominion* and *The End of Eternity*), is in both cases rejected. In *Death's Other Dominion*, survivors of a former Uranus expedition of 1986, stranded on the ice-planet Ultima Thule, have been granted immortality by their surroundings, but with the price of being chained to the planet and reproductively sterile. Their leader, Dr. Cabot Rowland, wants to discover the secret of their own immortality in order to 'step forward into the greatest scientific adventure in the whole history of man.' 'Unencumbered by death, we shall leap from planet to planet, from solar system to solar system, from galaxy to galaxy!,' he proclaims. But his experiments with some of his crew members have only led to their mental degradation, leaving them as 'living dead,' 'vegetables who sacrificed their minds to science.' It is due to the mortal Alphans, who serve as a positive mirror, that the immortal inhabitants of Ultima Thule eventually turn their backs on the search for the secret and engage instead in trying to heal the demented. The question asked by Commander Koenig at the end of the episode – 'Is it death that gives a meaning to life, in the end?' – is then indirectly answered in *The End of Eternity* where the Alphans have to fight a character called 'Balor.' Balor himself is a criminal who has been exiled and jailed by a civilization of immortal aliens because, as his character claims, he was trying to bring back meaning and purpose to their sterile and apathetic lives by sowing chaos, terror, destruction and torture. As Balor exclaims, 'We tried to instill in the minds of our people the thought that only death gives a purpose to life, that a full response to life can only be measured against a fear of death. But how can you value life, if you do not fear death?'

IV *The Space Ark:* predestined purposes

The Alphans are thus continuously confronted with basic biological notions such as death and life, and connected processes like nourishment, procreation, evolution and even phylogenesis (in one episode they are pushed back to the evolutional stage of Cro-Magnon). All this makes clear that they are taking with them the basic elements of the natural human life they seemingly had to leave behind on earth when thrust on board their artificially created base on the sterile moon and into their deep-space odyssey. But in fact, it is repeatedly hinted at during the development of the series that the positive consequence is that their errant ways allow them to carry human life with them into deep space, dispersing it to dead planets and bringing them, respectively, back into life.[23]

This concept is particularly played out in two further episodes which, despite the fact that they are chronologically quite far from each other, are closely linked. In episode 6, *Another Time, Another Place*, the Alphans are colonizing a future earth, which humanity has devastated by radioactivity, while in the last episode of the first season, *The Testament of Arkadia*, they discover the planet Arkadia, from which the ancestors of humankind once fled after they had destroyed their own world. Taking plants and animals with them, the Arkadians did find earth, inhabited and colonized it, but – as *Another Time, Another Place* depicts – they obviously fell back on their errors, destroying the new planet the same way they destroyed their home planet, Arkadia. In this episode, a man and a woman from the Alpha crew follow an imperative spelled out in the message: 'You who are guided here, make us fertile, help us live again.' Thus, both home planets of humankind, earth and Arkadia, are in the end brought back to life again by the Alphans and the circle is complete. 'The seeds from earth, so carefully stored and nurtured by us, had returned to their place of origin,' Commander Koenig concludes in the *Arkadia* episode, this being a clear sign of the original idea, according to which the entire series should have been titled *The Space Ark*.[24]

In the earlier episode *Mission of the Darians*, this notion of procreation as a biological and universal imperative is rehearsed in the encounter of the Alpha crew with another space ark when they cross paths with a nearly extinguished alien race, traveling from their destroyed home world Daria in a giant space city to a very remote 'virgin planet' in order to restart their civilization with the help of a tersely defended gene bank. A disaster, tellingly very similar to the one which blasted the Alphans out into space (a nuclear reactor exploded), has severely damaged the space city and poisoned all survivors with radiation. Their ruler, Neman, tries to convince the Alphans to join the mission of the Darians by spelling out the parallels between the Alphans, Noah's Ark and themselves. He stresses the fact that these allegories all fulfill the same purpose: to preserve and transport life to another place and to make it flourish again there.

But the Darians are only seemingly defending their civilization when protecting their gene bank, which contains the undamaged genetic material of their race from which, in the future, 'pure, healthy Darians' should stem. In reality, the Darians have already renounced all civilized behavior. They feed on (according to them) 'degenerate creatures, savage, mutant, cannibal' respectively and use them as living organ pools in order to prolong their own sterile lives. Ironically, they defend their horrible deeds with Darwinist elitism: 'to preserve only the fittest.' Thus, it becomes clear that author Johnny Byrne (1935–2008) has given the Darians a name that rings very similar to the notion 'Arian' in order to emphasize the fascistic nature of their thinking and acting.[25] Not to have fought for their survival is ultimately the crime of the Darians for which they are punished: for they have lost respect for life itself and have betrayed the present both in the name of the worship of life's artificially extracted and condensed essence, the gene, and in the name of the future it seemingly stands for. The Alphans, instead, while trying to find a new home for themselves in the first place, are bringing (back) life to hitherto lifeless planets as a positive side-effect. But that this reanimation ultimately follows a predestined purpose is made clear by Arra, the mysterious inhabitant of a huge spaceship the Alphans encounter in the episode *Collision Course.* When Koenig speaks of the explosion that catapulted the moon out of its orbit as an 'accident' nobody could have foreseen, she pities him because he is belittling himself in the scheme of things, and when he asks about the destiny of his crew and himself, she answers: 'You shall continue on; your odyssey shall know no end; you will prosper and increase in new worlds, new galaxies; you will populate the deepest reaches of space.'

In light of this prediction, the title of the first episode, *Breakaway*, which depicts the explosion that launches the moon into deep space, also takes on a positive aspect. It hints at the fact that the moon breaks free and away from earth in order to go on its destined odyssey that will ultimately guide it to the dead cradle(s) of humankind, with the purpose of reviving them, thus also answering Koenig's closing doubts in *The Testament of Arkadia*: 'We still wander the emptiness of space [...]. We must keep faith and believe that for us, for all of mankind, there is a purpose.'[26]

The idea of a migration into space, according to which humankind should venture into deep space, and thus colonize it (in order to assure the survival of the human race in case of a global disaster), had already been put forward in 1929 in *The Aims of Astronautics.* Written by the Russian Soviet rocket scientist Konstantin Tsiolkovsky (1857–1935), who, for this purpose, had also proposed 'generation starships' such as the one built by the Darians, the theory was then taken up with a different bias by the American physicist Gerard K. O'Neill (1927–92).[27] O'Neill not only organized a small conference on space colonization in May of 1974, but also published a highly influential article, entitled 'The Colonization of Space,' in *Physics Today* in September 1974, exactly when author Johnny Byrne was preparing the script for *Mission of the Darians.*[28]

V *Earthbound:* science fiction and contemporary issues

It is no surprise that the writers of *Space: 1999* should have taken notice of such developments in the context of space exploration, given that they continuously paid attention to the past and present of space exploration during the production of the series. But like their general attitude towards technical progress, their feelings regarding space exploration were also, clearly, ambiguous. On the one hand, photos from real space and moon explorations appear repeatedly, giving the series a touch of authenticity (or, perhaps, simply as a nod towards the accomplishment of NASA).[29] For instance, a photo of the Skylab 2 mission, taken by the departing crew during their final inspection flight on 22 June 1973, is visible on the board of the scientist Professor Victor Bergman (Figure 10.3) in *Ring around the Moon*, filmed in March 1974. This image, however, takes on a slightly different meaning because of the context of the story told in the episode.[30]

On the other hand, referring to the launch of probes during the production of the series, *Space: 1999* deals twice with the consequences of such

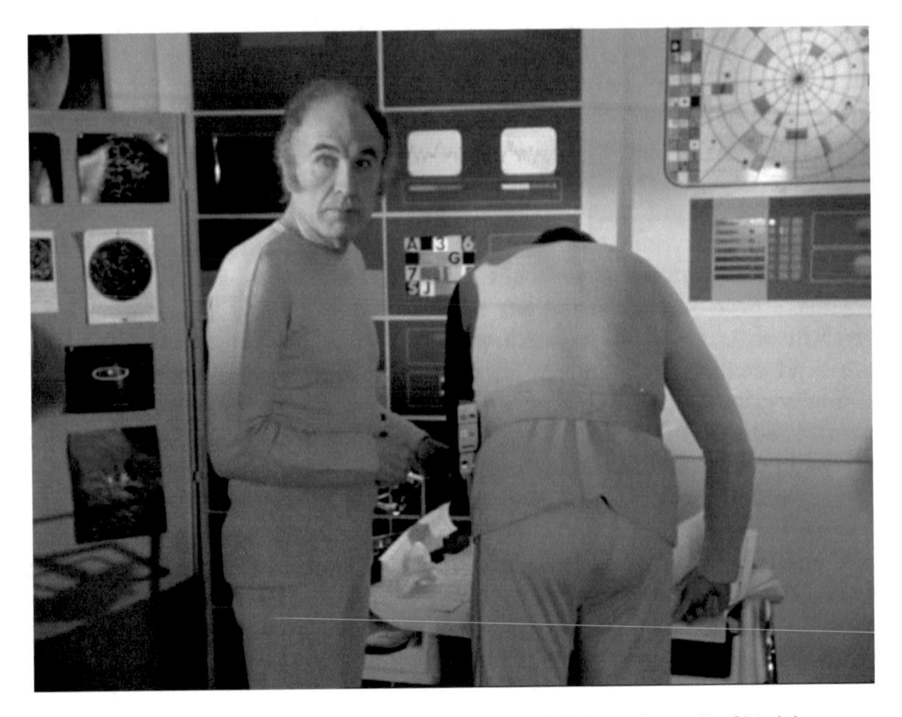

Figure 10.3 Professor Bergman with a photo of Skylab 2 on the wall of his laboratory.
Source: *Space: 1999*, episode 4, *Ring around the Moon*, 15 January 1976. Courtesy of Granada Ventures Ltd/ITV Global Entertainment Ltd/ITV Studios Global Entertainment.

an exploration by encompassing a different perspective each time. In *Ring around the Moon*, the Alphans have to face the consequences of an alien probe which recklessly, and without any regard concerning possible victims, collects its data despite the fact that the planet from whence the probe had been sent no longer exists. In the context of a story about an alien deep-space probe out of control, the photo of the mutilated Skylab – with one remaining solar panel and a rigged parasol solar shield to replace its missing micrometeoroid shield – could serve as a discrete, yet still visible admonition to think about the probes humanity has launched. Further, it points to the obvious consequences they could have while being out of reach or control, a connection underlined when Professor Bergman compares the threatening probe to similar efforts of humankind: 'They're a sort of reconnaissance team, sent out into space to gather information and then transmit it back to Triton. That's all. We did it ourselves with our deep-space probe ships.'

The Alphans later have to deal with exactly one of these 'deep-space probe ships' from earth in *Voyager's Return*, filmed in August 1974, an episode which goes further than *Ring around the Moon* in questioning the consequences our own probes might have. In this episode the fictitious probe, 'Voyager One,' turns out to be equipped with the deadly, so-called Queller Drive, 'an atomic engine used for going across space at incredible speed. [...] Incredible speed produced by fast neutrons spewed out into space, annihilating everything in their path. You'd survive better standing smack in the middle of a nuclear explosion.' When crossing the path of the moon, the probe threatens to destroy the base. However, its engine is switched to a harmless landing device based on ordinary chemical rockets, thanks to the German scientist Ernst Queller, founder of the Queller Drive, who reveals that he has been among the crew along under the assumed identity of Ernst Linden. Queller took this name out of shame after a horrible accident when the twin probe 'Voyager Two' switched to the Queller Drive too early during its launch and killed an entire community of 200 people. But immediately after the danger of 'Voyager One' has been avoided, another threat is revealed: During its exploration the probe has obviously played havoc with the two planets of an alien civilization, the Federated Worlds of Sidon, whose survivors have sent Chief Justifier Aarchon, together with a fleet, on the probe's trail in order to find the planet from where the deadly device has been launched and to destroy it in revenge. Given that the Alphans are former inhabitants of planet Earth, they are also held responsible by Aarchon: 'Once we have located planet Earth, your colony here will be extinguished.' Thus, the scientist Ernst Queller is again confronted with the terrible result of his invention, which he had originally conceived as a means to help 'further the boundaries of man's knowledge.' As Queller pleads, 'My purpose was to unite a divided world, to reach out in the name of science and humanity, to illuminate the mysteries of space, to seek out other worlds [...] and offer the hand of friendship.'

In 1982, the author of the episode, Johnny Byrne, confirmed that Ernst Queller's character, including his name and the elements of his invention, are partly indebted to the 'Father of the H-Bomb,' the Hungarian-American scientist Edward Teller (1908–2003); hence, the reference to 'You'd survive better standing smack in the middle of a nuclear explosion' and the good intentions put forward in defense, and partly to the rocket engineer Wernher von Braun (1912–77), who also tried to defend his involvement with the Nazi government by referring to his noble goals of traveling to the stars, an achievement that would unite a world divided by the Cold War.[31] The reference to von Braun is even emphasized by the fact that his alien victims are coming from a planet called 'Sidon,' which – if taken literally – refers to the port in Southern Lebanon, one of the chief cities of the Phoenician empire in 1200–650 BC, but which – with the central letter 'd' cut out – reads like 'Sion.'[32]

But the plot of *Voyager's Return* not only looks back at the people involved in pushing forward the technical progress of science, but it also casts a glance at the then still ongoing launch of probes, while fantasizing about the possible consequences of such activities. In retrospect, it is quite telling that author Johnny Byrne got his chronology mixed up when reversing the relations between fact and fiction while claiming that the plot struck him as 'quite impressive in a way because the Voyager probes were out there at that time.'[33] As a matter of fact, the filming of the episode took place in 1974 while the launch of the real Voyager probes (as in the series there were 'Voyager' 1 and 2, though they were actually named only in 1977) happened in August and September 1977.[34] Byrne's 'Voyager' pair is thus likely to have been inspired by the earlier twin probes Pioneer 10 and Pioneer 11, launched in March 1972 and April 1973, which indeed 'were out there at that time.'

VI *The Troubled Spirit:* mirror images

At the same time, *Voyager's Return* demonstrates that the above-stated xenophobic depiction of aliens in *Space: 1999* must also be put into perspective. The aliens are not just evil, they have a motive for seeking revenge on the humans. Ultimately, the Sidons's mistake is not that they are hunting down the civilization responsible for sending the deadly probe, but that they are unwilling to reason and revise their judgment even when they learn that behind the genocide stood a tragic technical failure for which the Alphans cannot be made responsible. By insisting on his judgment, Aarchon is guilty of the same 'pride and arrogance' of which Queller in the end did blame himself, showing that in *Space: 1999* human and alien beings do serve as a sort of a mirror of each other. They share the same needs, as seen in the episode *Force of Life*, where the alien creature craves precious energy, just as the Alphans had, and aliens react to the weaknesses and mistakes of humankind (such as in *Voyager's Return*). Moreover, the alien surroundings occasionally function as a passive, reflecting screen which simply confronts the Alphans with themselves. This occurs, for example, in an episode entitled *War Games*, in which inherent human fear leads the Alphans to imagine themselves in vivid hallucinations of war, death and destruction. In *Full Circle*, the Alphans are reduced to the state of Cro-Magnon man and, deprived of the capacity to speak, are forced to live out the emotions of frustration, which results from their misunderstandings.[35] The humans thus encounter mirror images of their own problems (shortage of energy, war), the echoes of their deeds (carelessly sent probes) and of their shortcomings, which partly reflect the typical challenges and questions of the early 1970s, many of which we still face today.

VII *The Space Brain:* cosmic intelligence

Some of the answers indicated do, however, also point back towards the early 1970s. In *Space: 1999*, reference is frequently made to a mysterious 'cosmic intelligence' which, through all seemingly accidental and fortuitous events, actually assures the Alphans that they all serve a purpose and contribute to the fulfillment of a bigger scheme. Behind this concept clearly stands Edgar Mitchell's (1930–2016) idea of an intelligence in the universe, which he experienced as a sort of 'epiphany' during his return from the moon as astronaut on Apollo 14 in 1971.[36] Leaving NASA a year later, Mitchell then founded the Institute of Noetic Sciences (IONS) in January 1973, aiming at a reconciliation of science with religion. According to Mitchell, consciousness itself is the point at which science and religion meet. It is also this connection which connects the cosmic consciousness with the individual consciousness: the key to the universe is contained in our minds and vice versa. According to Mitchell, the kind of epiphany he experienced during his flight is 'a latent event in every individual.'[37]

Space: 1999 shows traces of this concept of 'cosmic intelligence' or 'universal consciousness,' developed and publicized by Mitchell ever since his return to earth, particularly in the wake of Richard Maurice Bucke's book *Cosmic Consciousness: A Study in the Evolution of the Human Mind*, published in 1901. Not unlike Mitchell's description of his 'epiphany,' Bucke described and examined his experience of the universe as a living presence.

In the episode *The Black Sun*, while traveling through a black hole and expecting nothing other than annihilation, the Alphans do encounter this 'cosmic consciousness.' It seemingly not only protects them against the immense forces of the black hole, but also reunites them against all odds with a survival ship they have sent in the opposite direction in order to save at least six crew members. 'Something brought us home,' one of the passengers of the survival ship concludes, after reuniting with friends and colleagues. Already this 'sense of order' is akin to Mitchell's concept of the 'Quantum Hologram' which, as a proof of the existence of 'cosmic consciousness,' keeps related particles connected even if separated through an immense distance.[38] But the idea that such a 'consciousness' can also reveal itself to anybody when a certain 'syntony' is established, is depicted in *Space: 1999* when both Koenig and Bergman begin to wonder how and why they survived the journey through the black hole.[39] Their dawning realization is prepared by a dreamlike, illuminating experience during their passage though the black hole, where their dialogue also echoes parts of Mitchell's 'epiphany.' While the void suddenly seems alive for the Apollo 14 astronaut, Koenig and Bergman perceive the whole universe as 'living thought' where 'every star is just a cell in the brain of the universe.'[40] Just as Mitchell had tried reconciling science with religion, Professor Bergman points out that 'the line between science and mysticism is just a line.'

All this demonstrates that 'noetically' gained knowledge and science, according to this view, are by no means opponents – on the contrary. Again in concordance with Mitchell's philosophy that 'true spirituality' and 'true science [...] are looking for the same thing,' science in *Space: 1999* can actually guide to the point where the 'cosmic conscience' can then be encountered and experienced.[41]

VIII *The Age of Extremes:* optimism and crisis

In his book *Age of Extremes: The Short Twentieth Century, 1914–1991*, English historian Eric Hobsbawm divides the era from the beginning of the First World War to the fall of the Iron Curtain into three periods. The second, called 'The Golden Age,' represents the greatest period of economic expansion, running from the end of the Second World War to exactly the end of the year 1972, followed by an era of 'Landslide' marked by 'Crisis Decades.' As Andrew Smith has observed, this Golden Age and its fallout coincides precisely with the conclusion of the Apollo program which ended with Apollo 17 in December 1972. In Smith's view, this fact confirms that, with this last manned moon landing, an era of 'upheaval, [...] optimism and energy' ended, one that – economically, socially and psychologically – had made the Apollo program possible, which in turn had marked and colored this 'Golden Age.' This age was followed by a generation which – due to upcoming financial, technical and political crises – was far more skeptical, disappointed and pessimistic.[42]

Space: 1999, conceived and produced by Gerry and Sylvia Anderson during this watershed period between 1972 and 1973, can be described as being also 'sandwiched' between these two eras of 'upheaval, [...] optimism and energy' on the one hand, and 'landslides' and 'crisis' on the other.[43] Given the historical nature of the period in which the show was created and filmed, it is not surprising that the series would be imprinted with an ambiguous attitude towards space exploration and technology. As stated above, the attentive and detailed creation of the sets for a future moon base, perfected with lovingly crafted models and ambitious special effects, shows untroubled and unclouded enthusiasm and fascination for technology. Importantly, it testifies, through the bright colors of sets and costumes, a certain level of optimism. However, the dark, elegiac, and sometimes bleak and mysterious stories told in these sets, are deeply rooted in a profound skepticism regarding technology and its aftermath. Ultimately, it expresses a troubled consciousness of the limitations of human knowledge and an outright pessimism towards the possibilities of controlling human behavior and destiny.

The creators of *Space: 1999* were perhaps particularly appropriate for communicating these mixed feelings, given that they had already shown similar tendencies in their first feature film *Doppelgänger*, released in 1969. The film, confirming Andrew Smith's view 'that Apollo was an emanation of popular culture before it was anything else,' on the one hand serves the viewer with glimpses into the spectacular preparation and training of the hero astronauts, but on the other hand also comes up with sometimes rather disheartening insights into the political and private background of space exploration.[44] Thus, as early as 1969, *Doppelgänger* depicted NASA as an organization whose decision to fund a projected flight to a newly discovered planet is not triggered by scientific idealism or curiosity (on this basis the funding was initially denied in the film narrative), but only by the fear that a political force other than the United States could be the first to explore and lay claim to the planet. And despite the fact that the phenomenon of marital breakdowns and increasing divorce rates among astronauts was considered shameful, and hence censured and concealed by NASA in the 1960s, the film – perhaps because it was a British production – shows the failure of the American astronaut's marriage.[45] Nevertheless, the final message of the film – namely that humankind ultimately only encounters itself in space – is still in line with *2001*. Similarly, author Arthur C. Clarke in 'Space Flight and the Spirit of Man' wrote in 1961 of the general feeling of humanity: 'We are in the grip of some mysterious force or Zeitgeist that is driving us out to the planets, whether we wish to go or not.'[46] In *2001* as well as in *Doppelgänger*, humanity decides to follow this siren call out of curiosity and the spirit of adventure; but in *Space: 1999*, the words 'whether we wish to go or not' do get another meaning, because here the crew of Moonbase Alpha actually only involuntary veers away from earth and into deep space. Unlike *Star Trek*, the breakaway of the first episode is not a planned and controlled start of a five-year mission with precise duties and projects, but an endless odyssey defined by fear, anxiety and nostalgia. The adventurers are only partly comforted by the thought that they, too, are unwillingly on a mission that has been mapped out, though without their knowledge, by a mysterious 'cosmic intelligence.'[47]

It is, therefore, little wonder that Koenig bitterly rephrases Neil Armstrong's first words when stepping out onto the earth's satellite. While reflecting over the mission of a projected landing on a planet passing the solar system, and while facing threatening technological failures which will then lead to the disastrous explosion which ultimately kicks the moon into deep space, Koenig says to himself: 'A giant leap for mankind. It's beginning to look like a stumble in the dark.' It is a dark faintly illuminated by often discomforting mirror images from earth and humankind. It is thus too early to mourn together with Andrew Smith for the loss he attributes to the Apollo program when arguing that 'in the Sixties, some feared that Apollo would destroy the moon's value as the most elastic symbol or metaphor since God.' Perhaps exactly because they did send the moon and its moon-base crew into deep space and thus out of the reach of humanity's grasp, the producers of *Space: 1999* were able, even after the Apollo program, to use 'space and the moon' again as 'canopies across which you could spread the psyche.'[48]

Notes

1. John Kenneth Muir, *Exploring 'Space: 1999': An Episode Guide and Complete History*, London: Jefferson, 1997, 1.
2. Gerry and Sylvia Anderson split up after the first year of *Space: 1999*, and Sylvia Anderson left the production of the series. In this chapter only the episodes from the first season will be discussed because, with the arrival of Fred Freiberger (1915–2003) as the new producer in 1975, the series took a very different approach, which – with its inflationary, action-based stories, often involving silly monsters – was heavily modeled on the expectations of the US market and included a more colorful and eye-catching visual style. Because the second year tried to be very 'up to date,' even concerning its music, it has aged much more than episodes from the first season, which were meant to be a relatively timeless representation of the future.
3. Quoted from the film *Space: 1999 Documentary*, produced and directed by Tim Mallett and Glenn Pearce, UK 1996 (Fanderson). This reproach might also have been due to the title sequence of the show where Landau and Bain were introduced in a very stiff and almost puppet-like way.
4. *Star Wars* director George Lucas (1944–) was so impressed by the special effects in *Space: 1999* that he visited the studios during the filming of the episode *War Games* in 1974. From this episode he also borrowed the sight of a giant spaceship flying overhead the camera as shown in the opening of *Star Wars*; see http://catacombs.space1999.net/main/epguide/t17wg.html (accessed 1 October 2017); and Jonathan W. Rinzler, *The Making of Star Wars*, London: Random House, 2008, 102–3, 412. Rinzler reports that, following his visit, Lucas changed the already approved design of the Millennium Falcon because it looked too similar to a spaceship shown in the episode. This might seem strange given that *Space: 1999* and *Star Wars* have very different approaches towards science fiction and concerning their distribution, with *Star Wars* belonging more to the science fantasy genre and being presented as a cinema blockbuster, but it shows how much Lucas was impressed by *Space: 1999*'s special effects.
5. See http://catacombs.space1999.net/main/cguide/umcoffice.html (accessed 1 October 2017).
6. Gernreich had already developed such a design in 1972; see Brigitte Felderer, ed., *Rudi Gernreich: Fashion will Go Out of Fashion*, Cologne: Du Mont, 2000, 156. The whole uniform obviously echoed Gernreich's famous as well as notorious 'UNISEX Project' which in the 1960s challenged basic social assumptions regarding gender rules, limits and taboos. See, for example, Peggy Moffit and William Claxton, eds, *The Rudi Gernreich Book*, Cologne: Taschen, 1999, 184.
7. Pierre Fageolle, *Cosmos: 1999. L'Epopée de la blancheur*, Pézilla-la-Rivière: DLM, 1996, 54.
8. The same holds true concerning some of the architecture shown in *Space: 1999.* In the episode *Missing Link* the buildings of the alien dwellings on Zenno clearly rely, for example, on designs published by the Japanese architect Arata Isozaki

in *L'Architecture d'aujourd'hui* 117 (November 1964/January 1965), xxv. It is obvious that all these 1960s decorations were also chosen for their 'futuristic' look, but it would be too easy to accept this as an already sufficient explanation for such a choice, given that for the earlier Anderson series *UFO* other items were used in order to depict the design of a differently looking future. The mere reference to the 'futuristic' look chosen for *Space: 1999* is insufficient, and deeper analysis must rather ask: what type of imaginary future was the series supposed to depict? *Space: 1999* actor Prentis Hancock commented that 'in some ways it was a '60s series, in the '70s.' See Robert E. Wood, *Destination Moonbase Alpha: The Unofficial and Unauthorised Guide to SPACE: 1999*, Prestatyn: Telos, 2010, 279 and 71, for a similar statement.

9. The project probably triggered the (today almost forgotten) TV series *Moonbase 3*, also a co-production of a British company (here the BBC) and American companies (Twentieth Century Fox and the ABC network). The series, created by the *Doctor Who* producer Barry Letts (1925–2009) and script editor Terrance Dicks (1935–), was commissioned in December 1972 by the BBC and aired in Britain in the fall of 1973. It only ran for six episodes and turned out to be a commercial and critical failure. Unlike *Space: 1999* it was produced on a much more modest scale and had quite talkative scripts; similar to the Anderson series, it tried to depict a technically realistic vision of a future moon base and was also criticized as being too slow in its narration. See Steve Rogers, *Moonbase 3: The Pictorial Compendium*, London: The Mausoleum Club, 2004.
10. *Space: 1999 Documentary*; Fageolle, *Cosmos: 1999*, 20–1; Muir, *Exploring*, 6–7. This supports analysis of the moon's role in contemporary science fiction. For example, Michael Salewski, *Zeitgeist und Zeitmaschine: Science Fiction und Geschichte*, Munich: Deutscher Taschenbuch Verlag, 1986, 99, argues that actual human exploration of the moon has 'caught up' with science fiction, and science-fiction writers no longer know what to do with this celestial body – this is why the moon in the Anderson series *Space: 1999* served as a 'footboard' and catapult for the proper journey into space.
11. See also the parallels given with the dates, both revolving around the year 2000. Another, rather apocryphal and perhaps tongue-in-cheek reference to the first manned moon landing might be seen in the name given to the former commander of the moon base whom the new commander Koenig is about to supersede. His name, 'Gorski,' and the words he says to Koenig ('If you want to talk things over before I leave, I shall be in my quarters. Good luck.') might refer to the phrase Neil Armstrong is (as we know today: wrongly) supposed to have said during the first moon landing: 'Good luck, Mr Gorsky,' thus addressing a former neighbour from his boyhood whose wife was incredulous about Armstrong one day landing on the moon, and thus allegedly promised her husband a special erotic gratification in case Armstrong should succeed; see for this 'urban legend,' Andrew Smith, *Moon Dust*, London: Bloomsbury, 2005, 27.
12. Fageolle, *Cosmos: 1999*, 49. Some particular scenes from *Space: 1999* were also visually influenced by *2001*. A production sketch for the encounter with an alien probe in the episode *Ring around the Moon*, showing close similarities to the final traveling sequence in *2001*, was probably not realized in the end because the probe is too negative and subordinated to associate it with the alien intelligence hinted upon in *2001*. In the pilot episode *Breakaway*, however, some realized visuals are clearly indebted to scenes from *2001*; see Fageolle, *Cosmos: 1999*, 54.

13. It was perhaps also this indebtedness to *2001* which earned *Space: 1999* its hostile reception from the fans of *Star Trek* since the seriousness of Kubrick's intended 'proverbial good science fiction movie' was often put forward against Gene Roddenberry's (1921–91) TV series which, in that confrontation, obviously appeared as 'cheap' and 'rather childish.' For Kubrick's intention, see Arthur C. Clarke, *The Lost Worlds of 2001*, London: Sidgwick & Jackson, 1972, 17.
14. See especially the episodes *Black Sun* or *Collision Course*.
15. See the quotations below from the episode *Earthbound*.
16. See, for instance, the failed expedition to planet Meta in *Breakaway*; the doomed 'Astro-Seven'-mission of Dr. Russell's husband whose ship was incinerated while locked into an orbit around Jupiter in *Matter of Life and Death*; the failed Uranus-expedition in *Death's other Dominion*; and the disastrous mission of the manned probe to the planet Ultra in *Dragon's Domain*.
17. The goal was to transfer the cinematic style and the quality of the special effects from *2001* into TV format. The budget for the series granted in order to achieve this – 300,000 dollars per episode – was quite exceptional for a TV series if, for example, compared to the above-mentioned, rather modest *Moonbase 3* (see note 9 above) – and the result impressive; Fageolle, *Cosmos: 1999*, 22.
18. See also the film's trailer which asks: 'Apollo 11 has conquered the moon! Where to NOW in space?' The answer implied by the film is this: back to earth and back to ourselves. Fageolle, *Cosmos: 1999*, 9, finds it revealing that the moon's movement towards the exterior in *Space: 1999* is actually answered at the same time by a movement towards the interior, thus making the explorer always encounter himself.
19. In the opening titles the nuclear waste area is described as being on the 'dark side of the moon,' a (common) error given that there is no 'dark' but just a 'far' side, as Isaac Asimov already noted in: 'Is "Space 1999" More Fi Than Sci?,' *New York Times* (28 September 1975), Sect. 2, 1.
20. This returns, slightly varied, in the second season with the episode *Journey to Where*, in which the earth has been devastated by human industrial pollution.
21. From the 24 episodes of the first season, eight deal directly with the imperative necessity of energy for survival. See, for this, episodes such as *Earthbound* (where a deserting crew member blackmails the commander by threatening to destroy a central unit from the power station); *The Testament of Arkadia* (where an unknown force drains the energy from the base's generators – something then repeated in the episode *Black Sun*, where a black hole slowly drains the base of its energy); *Ring around the Moon* (where the threat of an alien probe is illustrated by the fact that it forces a crew member to transmit vital information about the life-support system of the base); *The Infernal Machine* (where the title character, a giant robot, blackmails the humans to get energy he desperately needs); or *Force of Life* (where one of the crew members is possessed by an alien entity which needs energy and thus sucks it out from every source it can get, be it light, body heat or – finally – the generators of the moon base).
22. In the above-mentioned episode *Force of Life*, it is suggested that, ultimately, the alien entity which took possession of the body of the crew member did so because it needed to absorb energy in order to procreate. It is certainly not just a mere coincidence that in the following episode, *Alpha Child*, the birth of a human offspring, the first baby born on the moon base, is a key element in the narrative. The title *A Matter of Life and Death* refers to the classic British movie made by Michael Powell and Emeric Pressburger in 1946, which also deals with the frontiers between life and death.

23. In the above-mentioned episode *Force of Life*, for instance, the Alphans are involuntarily functioning as obstetricians for the alien birth, while in *The Guardian of Piri* they wrestle the dead planet Piri out of the control of a computer that had made the planet sterile because it subjected everything to the deadly principle of perfection.
24. Fageolle, *Cosmos: 1999*, 43.
25. Ibid., 15–16 and http://moonbase99.space1999.net/mission.htm (accessed 1 October 2017).
26. This even goes so far as the 'reviving' of old, and therefore dead, myths from earth: At the end of the episode *Dragon's Domain*, Dr. Russell says to Commander Koenig: 'If we ever do find a new place to live, and if we succeed, we're going to need a whole new mythology,' because the deadly fight between a crew member and an alien monster, reminiscent of Saint George and the dragon, has thus 'put new life into an old myth.'
27. Marina Benjamin, *Rocket Dreams: How the Space Age Shaped Our Vision of a World Beyond*, New York: Free Press, 2003, 142–3. In 1959 Wernher von Braun voiced similar thoughts when justifying the need for space travel by saying that it might be the destiny of humankind to grant immortality not only to itself, but to life in general by transporting the 'spark of life' to other planets; see David F. Noble, *The Religion of Technology: The Divinity of Man and the Spirit of Invention*, New York: Alfred A. Knopf, 1997, 163.
28. Gerard K. O'Neill, 'The Colonization of Space,' *Physics Today* 27.9 (September 1974), 32–40. See also Benjamin, *Rocket Dreams*, 141. That another strong inspiration for the *Darian* episode was furnished by Douglas Trumbull's 1972 film *Silent Running* can be seen by the fact that *Daria* with its cupolas has a similar look to the 'ark' spaceships, depicted by Trumbull and designed to enshrine and preserve plants, already extinguished on earth.
29. In the episode *The Last Sunset*, Commander Koenig discusses the possibilities for setting up a colony elsewhere on the moon while inspecting photos taken by the Lunar Orbiter, launched in 1966; see http://catacombs.space1999.net/main/epguide/t11tls.html (accessed 1 October 2017). The 1969 Apollo 9 mission is particularly present in the series: In the episode *The End of Eternity*, pictures of astronaut Russell Schweickart (1935–) during his extra-vehicular activity and of the Lunar Module frame the doors of a pilot's quarters in order to show his enthusiasm for his work as an astronaut. The photographs reappear during a flashback to the year 1996 in the episode *Dragon's Domain* when they are hanging in Professor Bergman's former laboratory on the moon base.
30. http://catacombs.space1999.net/main/epguide/t04ratm.html (accessed 1 October 2017).
31. http://catacombs.space1999.net/main/models/w2mvoyager1.html (accessed 1 October 2017). On the occasion of the American première of the series on 4 August 1975 – and so still very distant from the later episode *Voyager's Return* – von Braun, in his position as president of the National Space Institute, welcomed and recommended *Space: 1999* in a letter dating from 5 September 1975 and addressed to the broadcasting network's President Abe Mandell. Von Braun praised the series because, while 'freeing the creative imagination – so effectively accomplished in *Space: 1999*,' it 'imaginatively captures the excitement of living in the incredible age of space' and thus 'can only make the public more enthusiastic and concerned with the further exploration of our universe.' Von Braun's letter can be found at http://www.space1999.net/~catacombs/main/pguide/wrefcbraun.html (accessed 1 October 2011).

32. Fageolle, *Cosmos: 1999*, 67.
33. See http://www.space1999.net/catacombs/main/crguide/vcwbvr.html (accessed 1 October 2011).
34. The Voyager mission was first called 'Mariner Jupiter-Saturn 1977 Mission' or 'MJS77' and it was only subsequently renamed 'Voyager' about six months prior to the launch of the two probes in the fall of 1977.
35. In the episode *The Troubled Spirit*, the presence of an alien surrounding does not even matter because it is ultimately through his concentrated fear that a crew member conjures up a revengeful specter which haunts the base.
36. See the interview with Mitchell presented in *In the Shadow of the Moon*, directed by David Sington, USA/UK, 2007 (Discovery Film).
37. See Smith, *Moon Dust*, 52, 58, and especially the lectures given and articles written by Mitchell since 1973, summed up in Edgar Mitchell and Dwight Arnan Williams, *The Way of the Explorer: An Apollo Astronaut's Journey Through the Material and Mystical Worlds*, New York: Putnam, 1996.
38. Ibid., 59.
39. Cf. the following dialogue: Victor: 'John. Have you ever wondered just how and why we've survived?'/ Koenig: 'Not until now.'/ Victor: 'Have you got any answers?'/ Koenig: 'You're not referring to God, are you?'/ Victor: 'Oh, I don't know exactly. I'm a scientist, I don't know anything about God, but, no, a sort of "cosmic intelligence" is what I've got in mind.'
40. This concept of the universe as a brain was later also adapted in the episode *Space Brain* where the moon flies through a giant living entity which builds 'the centre of a whole galaxy [...] maybe even hundreds of galaxies [...] planets [...] stars [...] strange life forms [...] and in the middle of it all [...] is this brain [...].'
41. Smith, *Moon Dust*, 61.
42. Eric Hobsbawm, *Age of Extremes: The Short Twentieth Century, 1914–1991*, London: Michael Joseph, 1994; Smith, *Moon Dust*, 292–3. In this context, see also the contributions to Alexander C.T. Geppert, ed., *Limiting Outer Space: Astroculture After Apollo*, London: Palgrave Macmillan, 2018 (= *European Astroculture*, vol. 2).
43. Preparations for the series began in 1972, the filming of the first episode started in November 1973; Muir, *Exploring*, 6 and 15.
44. Smith, *Moon Dust*, 121.
45. See, for instance, ibid., 252.
46. Arthur C. Clarke, 'Space Flight and the Spirit of Man,' *Astronautics* (October 1961); quoted in idem, *Voices from the Sky: Previews of the Coming Space Age*, New York: Harper & Row, 1965, 107.
47. It is perhaps significant that meanwhile also, the once so optimistic *Star Trek* series themselves have steered into more sceptical and pessimistic waters. After the spin-off series *Star Trek: Deep Space 9* (1993–99) where a rather dark and bleak vision of the possibilities of diplomacy was displayed, with *Star Trek: Voyager* (1995–2001), the producers created a starship which – like Moonbase Alpha – suffers the fate of an odyssey in unknown space, desperately striving to find a way back home to earth.
48. Smith, *Moon Dust*, 168.

PART IV

Encountering Outer Space

CHAPTER 11

Life as We Don't Yet Know It: An Anthropologist's First Contact with the Science of 'Weird Life'

Debbora Battaglia

A striking feature of the idea of the alien is the extent to which galaxies of disciplinary discourse cross-connect, collide or pass through one another under its influence. It follows that as their core elements reconstellate, space opens for their becoming, at least in the moment, productively alien to themselves. But there are risks. This chapter proposes a slowing down of the now common practice of interdisciplinary knowledge sampling and intercourse, seeking beyond 'first contact' an appreciation for deeper interdisciplinary dialog, and conditions for mutually flourishing knowledge ecologies of relatedness.

As a case in point, I consider my own first contact as an anthropologist with astrobiology and with questions concerning 'weird life' – its affectionate term for non-carbon, non-human-like life forms of life as we don't yet know it. At least to me, the attraction is obvious. For a start, their relation as alien discourses iterates both disciplines' iconic focus on gaps in knowledge and cultural obfuscations of alien ways of being and relating across extremes of difference. Both are self-aware of the extent to which they rely on interpretation of data to present 'an opportunity to phenomena that […] would not be "given a chance" [to appear]' if subjected only to the gaze of 'hard' facts.[1] And both are currently extending their 'welcoming apparatuses' to other disciplines, notwithstanding the threat to boundaries that is part and parcel of this kind of outreach.[2]

Debbora Battaglia (✉)
Mount Holyoke College, South Hadley, MA, USA
e-mail: dbattagl@mtholyoke.edu

Alexander C.T. Geppert (ed.), *Imagining Outer Space*
European Astroculture, vol. 1
https://doi.org/10.1057/978-1-349-95339-4_11

Yet, this same invitational attitude can create problems for concepts such as anthropology's famously slippery one of 'culture,' as if this could, like modern facts, be 'torn up by the roots' (as Poovey cites Francis Bacon's image) and transplanted. In contradistinction to 'creolization' which occurs when expert knowledge communities work to translate their different 'languages' (for instance engineering, chemistry, physics) in service of dedicated projects,[3] *warp sampling*, as I think of it – the transfer of terms and constructs from a historical moment into a possible future without pausing to contemplate *in situ* translation in the socio-cultural present – is problematic. Even without the accelerant of new media technologies, transfer too readily taken up as translation, not unlike replication too readily taken up as reproduction, lends itself not only to the production of new essentialisms which serve no one, but as well to confusions of familial resemblances for family relations,[4] and perhaps most seriously, to raising from shallow graves the remains of whole programs better left where they were – images of colonies, for example, calling up images of colonialist agendas just waiting for the alchemical rhetoric that will revive and repurpose them in practice. Outer space and its territories elicit such elisions and temptations of sense unmaking in no uncertain terms: we have only to note a former Apollo astronaut's interest in mining the moon or India's in weaponizing space for protecting its 'assets,' to hear the ring of predatory expansion.[5] All this impacts how nationally supported scientific projects and the positions on future projects to which their research leads them relate to the publics they are charged, as a matter of ethics, to answer to.

In this chapter, I begin by considering *The Limits of Organic Life in Planetary Systems*, a life sciences report by the US National Research Council (NRC) quietly released in 2007. The stated object of the report is to broaden awareness of 'life as we don't yet know it' in hopes of increasing the chances of alien life forms being recognized by robotic probes, and decreasing the chances of damaging life in the process. From my point of view as an ethnographer, the report is remarkably sensitive to problems of first contact with exotic life forms, and to the power asymmetries inscribed in the language of 'discovery' of the 'exotic' – not least in reference to entities which are enframed primarily as resources of use to humans.[6] Along these lines it voices concerns with expansionist programs from earth and with issues of sovereignty – scientific and non-scientific – in outer space. However, the NRC report can also feel too comfortable with a nineteenth-century imperial consciousness in which anthropology was implicated in its disciplinary youth, and from which it has long since distanced itself by means of scathing self-critique.

This discordance moves me to inquire more deeply into the disciplinary confluence of exophile space studies and anthropology by considering images and documents that were circulating in public mediascapes when the NRC report was being drafted, and which effectively iterate critiques of Euro-American colonial era discourse. I take this ex-centric approach partly as an experiment in gathering points-of-view that taken together reveal disciplinary 'blind spots.' Three textartifacts, then: The first is the report, which I found through a brief announcement buried in the *New York Times*. The second is Werner Herzog's 2005 docu-fantasy film, *The Wild Blue Yonder*, featuring an assimilated alien entity marooned on planet Earth and seeking to deliver an urgent message to humankind, that it must cease its quest for a better place to colonize than the planet it has ecologically and socially ruined. The message is, 'there's no place like home' and home can do very well without you. The third is the neo-creationist origin myth of the International Raëlian Movement, a new religious movement which holds that alien scientists, not God, created human beings 'in the beginning,' and are warning us through their prophet that abuses of science and technology could be the end of life on this planet. Both the film and the neo-creationist narrative highlight 'second origin' themes (as distinct from de novo creation) and resist as a dangerous romance colonial expansionism and domination of space (respectively). And both emphasize the value of life on Terra, in contradistinction to the outward-bound vision of exobiology.

The NRC report, the film and the new religious movement's foundational narratives all source to weird life origins, discovery and sampling which vibrantly engage orthodox science, and send cautionary messages to humans in respect of their technoscientific knowledge and its deployment. In the NRC report, science does the talking. In the Herzog film and in the Raëlian Movement's *Messages From the Designers*,[7] socially alien-ated entities talk back from positions of historical distance, marginalization and perceived wisdom. This dialogue is crucial. If we accept that outer space has for publics the aura of a cosmology which is shared by space scientists and engineers, anxieties and moral reasoning held in common deserve a hearing from policy-makers.

I Romancing 'weird life'

> *The human mind finds it difficult to create ideas truly different from what it already knows. Recognizing this difficulty, the committee chose to embrace it.*[8]

The National Research Council report makes arresting reading for anthropologists in part because it is disarmingly self-reflexive, in part because its tone is poetic, even romantic:

> No discovery that we can make in our exploration of the solar system would have greater impact on our view of our position in the cosmos, or be more inspiring, than the discovery of an alien life form, even a primitive one. [...] At the same time, it is clear that nothing would be more tragic in the American exploration of space than to encounter alien life and fail to recognize it.[9]

Reading this passage, it is as if the authors had not only taken to heart Steven J. Dick's writings on the cultural history of the search for extraterrestrial life, but responded to his call to articulate the 'intellectual cultures of natural scientists and social scientists,' with specific reference to anthropology.[10] For one thing, its stance on science as cultural practice refuses to sideline the social as a crucial element of scientific inquiry. Also, by linking its technical data to accessible, emotional, sometimes almost playful language, the report in effect positions itself as a 'skyhook' for connecting its experts to publics who already share their curiosity.[11] For example, it refers to producing a 'hierarchy of weirdness,' by way of asking: 'Is the linear dimensionality of biological molecules essential? Or can a monomer collection or two-dimensional molecules support Darwinian evolution?' And: 'Are Darwinian processes and their inherent struggle to death essential for living systems? Can altruistic processes that do not require death and extinctions and their associated molecular structures support the development of complex life?'[12] This, then, is a kindlier, gentler vision of weird life discovery.

Further, as I was reading through the report, I was arrested by an all too familiar anxiety of ethnocentrism. Since the report takes scientific practice as cultural at the start, and its extremophile projects as *technes* – that is, as artisanal productivities which do not claim to be disinterested – the fundamental problem confronting weird life science – robots aside – is the imperfection of human apparatuses of interpretation. It follows that special attention is given to 'terracentrism,' to 'anthropocentric biochemistry' and to 'geocentric' research agendas.[13] 'The natural tendency toward terracentricity,' we read, 'requires that we make an effort to broaden our ideas of where life is possible and what forms it might take [...] given that we are the life form defining it.' The 'natural' in this context being understood as the culturally natural, the responsibility for methods of discovery is fully its human authors'.[14]

In addition, the report is premised on recognizing the value of a life that is different by degrees but not necessarily other than the life we know and are familiar with: weird, but not strange or monstrous. For example, it opens readers to visions of terran life forms 'with a fundamental difference in the method of reproduction'[15]; to those life forms which thrive in environments inhospitable to terran life, perhaps not requiring water; to those inhabiting 'shadow biospheres' as yet unexplored here on earth: life not yet visible, heard or translated, not yet known in its effects or, more to the point, in terms of the relations it could invite us to entertain.[16] All of this points in the direction of fully honoring not only human actors and systems but 'non-human-like' *actants* or mediators of exchanges, significances in their own terms.[17]

Meanwhile, the relation to anthropology, even apart from science and technology studies, remains closely articulated. For example, the report acknowledges how the 'Darwinian signature' of lineal evolution might have occluded expressions predating Darwin's study and times by several billion years, and asks that we turn our attention to expressive genetic networks and such things as lateral gene transfer which were not on Darwin's map, as it were, for tracking the evolution of emergent life forms.[18] The report goes on to consider alternatives such as random and frequent 'mistakes' of 'more fit' structures, and 'altruistic genetic processes that do not require death and extinctions.' It asks, again self-reflexively, could terran forms of 'weird life' evolve 'de novo or would it have to be seeded from a neighboring planet or moon that during its early history had more suitable conditions for spawning life [panspermia]'? Firmly renouncing any myth of objectivity, it concludes, in my view not altogether accurately: 'Human exploration is concomitant with human contamination.'[19]

Look closely, however, and we also find assumptions that belong more properly in the nineteenth century than in the time of the report, which recommends 'development of a new generation of life-detection experiments that can be conducted *in situ* on planetary surfaces or conducted on samples returned from other solar system bodies.'[20] Such collecting practices go unproblematized, and even if a necessary given from the point of view of contemporary science, might have drawn attention to colonial-era collecting practices. Indeed, this moment in the report put me directly in mind of the directive of Everard im Thurn (1852–1932), the pioneering botanist, anthropologist and president of the Anthropological Institute of Great Britain who, in the late nineteenth century, called for practitioners to abandon their armchairs and their racially motivated biometric projects and laboratories of 'species' profiling, and to repurpose their technologies of knowledge production to field-based study of things and people in their culturally 'natural' settings.[21]

Of course, I do not pretend to offer a systematic reading of the NRC report in this chapter. However, even these brief examples reveal that the position it takes on its project is ambivalent, and its ethical stance unsettled. Apprehension noted, there is no indication of caution that sample collecting should at the very least await environmental impact studies, or solid evidence that so-called forward and back contamination can be technically contained.

Meanwhile, its authors do acknowledge in passing that themes from popular ET/UFO culture are circulating in public spheres and attracting attention to the messages we send to ourselves about ourselves from this realm where hopes and fears for the future are so contagiously expressed. Yet, they stop short of seeking in vernacular messages the moral meta-commentaries which would offer forums and terms of reference for engaging broader publics in debates informed by 'nature culture' literatures emergent in extreme ecologies studies.

It is a logical but enormous step away from exobiological discourse to think that the authors might entertain issues of weird life sentience, on the model of cultures that attribute agency to 'multinatural' formations such as mountains, streams, birds, and so on – not to mention plantlife: a model of agency widely employed in the 'speculative ontologies' of indigenous peoples and others seeking to establish a mutual caretaking relationship with their material world.[22] Yet, while not expecting that the NRC report would seek to entertain weird life's point of view of itself, we do not need to look far to find possible world narratives which speak on their behalf.[23] Displaced to film characters and the messages of alternative science religions, such points of view are out there, waiting to be acknowledged – as I turn now to discuss from two such unconventional perspectives.

II *The Wild Blue Yonder:* an exercise in interdiscursive poetics

Industrialized entertainment becomes the entering wedge of religion. [...] Recreation and religion, and their intertwining, are the DNA of his worlds: the tedium of existence forces us toward 'fun'; fun becomes the basis of our faith.

Adam Gopnik, commenting on the science fiction of Philip K. Dick[24]

In 2005, Werner Herzog's (1942–) feature film, *The Wild Blue Yonder*, was released for distribution in the United Kingdom and Europe. Its working title for US release was *Wake for Galileo* – a reference to the suicide mission of the US space probe that crashed into the icy seas of Europa, a Galilean moon of Jupiter. The film uses documentary footage from Arctic deep-sea exploration and NASA images from orbital space to support a fictional narrative of alien voyaging. Listed in the credits are the astronauts of Space Shuttle STS-43, the mathematicians of NASA/JPL/Caltech, Pasadena, and the actor Brad Dourif, who physically resembles Mr. Herzog, and whose character is a disheveled alien living in a trailer in the North American desert surrounded by trash and dilapidated furniture. Now socialized to earth, Brad Dourif's alien is weird life, talking.

The alien's story guides the film through a sequence of chapters. Urgently, desperate to send a message to earthlings, he tells of the exodus of his fellow aliens from their dying planet – an ice and liquid helium ocean that he calls 'The Wild Blue Yonder'; how, after traveling 'hundreds of years' in search of a habitable alternative, the displaced creatures arrived at earth and eventually settled in the desert. However, they had degenerated significantly en route. Most important, they had forgotten almost everything they once knew about mathematics and science. However, this did not deter them from making great plans to establish a capitol rivaling Washington, DC, and from building a thriving shopping mall, for them an icon of Western culture and economic success. Alas, both projects failed miserably. After spending all their energy trying to adapt to their host culture, the capitol building was never actually built and the shopping mall fell victim to a market liquidity crisis. We see its ruins: a dead mall in the middle of nowhere. These were obviously 'unsuccessful aliens,' as the director describes them in an interview, maladaptive by any standard and a far cry from the powerful entities of popular culture images and ET/UFO religions.

Yet, what truly anguishes our narrator is that humans are repeating this sad odyssey, in the reverse direction. Responding to fears of a lethal microbial infestation, NASA astronauts have been sent on a mission to find an alternative planet that humans might inhabit and colonize. The planet they discover is, of course, the Wild Blue Yonder, a past and future world that is anywhere-but-here-now. The images from Antarctic wonderlands of ice-capped sea are wondrous – a romantic water world not unlike the futurist fantasy scene of deep-sea outer spaces which directors James Cameron and Steven Quale are moved to anthropomorphize, Spielberg style, in a fantasy footnote to their IMAX documentary film *Aliens of the Deep* (2005). It does seem as if scientifically informed contact with actual weird life worlds of whatever sort cannot resist the lure of the aesthetic sublime.[25]

This aesthetic sensibility is worth noting since under its influence persons on earth today get carried away into realms of collective imagination. The aura of the extraterrestrial imparts an excessive charge to weird geologic formations, human-produced earth art (for instance crop circles), celestial phenomena – generating unique occasions of sociality and inspiring stories of alien contact. Indeed, as I was writing this paragraph, a UFO sighting over Texas was attracting CNN coverage of a mile-long field of lights, captured on a mobile phone video. One UFO blogger commented: 'Unless you have been under a rock somewhere lately, you have heard about the recent coverage of multiple UFO sightings over Texas.'[26] 'Everybody' knows that marvelous truths are 'out there' and accessible via their television or cell phone or computer screen.

It is precisely against this effect that Herzog defines the Brad Dourif alien – a character unremarkable in every way and designed to be overlooked; an entity with absolutely no effect on the social despite its urgent appeals to humanity to recognize its authority. In short, it is the *de-exoticized* alien to which we are introduced in Herzog's film; one unable to dazzle us, to blind us to its utterly banal, intransigent situation of social disempowerment. This alien has something important to tell space science, and the message concerns its defeating infatuation with weird life as a way out of its own present world predicament. The solution to our failure to fashion a well-made life on earth will be found, Herzog tells us, not by building a shopping mall, or even a shopping mall in outer space, as envisioned in the film (with a totally straight face) by NASA/Caltech mathematician Martin Lo and his entrancing computer-generated algorithms. Rather, as Bruno Latour reminds us, we must begin our relations in space with 'the question of the right ways to build.'[27] And this will require crediting multiple kinds of actors – human, non-human, institutional, and so forth – for their parts in the bricolage, and in fashioning an inter-entity ethics for intervention and productive exchange.[28]

The obvious moral of the film's 'second origin' theme is that whether we seek life in outer space or on Terra, the Wild Blue Yonder that really matters is whichever planet we call home (as Herzog states in a special features interview). But this is only the most obvious message. The film concludes as the astronauts return from their mission of discovery to find earth totally depopulated – as observed from their sublime perspective in orbit, transformed back into a place of 'pristine beauty, [...] prehistoric again.' Nature having declared independence from culture, we leave the space-faring scientists in orbit, weightless, encapsulated, extraneous, and presumably deciding what next to do.

III Alien designers: second origins and the International Raëlian Movement

Earth can support life today, but prevailing views hold that life could not have originated in an atmosphere that is as oxidizing as Earth's today. If that is true, the surface of Earth would be an environment that is habitable but not able to give rise to life.[29]

A second, 'second origin' myth intersects with extremophile science in different, but equally instructive ways. In late December 2002, the International Raëlian Movement's bishop-chemist, Brigitte Boisselier (1956–), announced to the world press that she had successfully cloned the first human baby, whose name was Eve. The mere claim was enough to ignite global press coverage, offering much of the world first contact with the Raëlian religion. It followed that by any measure the prophet Raël was successful in raising visibility for his movement and likewise an audience for its message that benevolent alien scientists had created human beings in their own image, using advanced cloning technologies. Once humans recognized how science and technology could create but also destroy life on earth – Year 1 of the Raëlian calendar, and Raël's birthday, is the day that Hiroshima was bombed – the ground would be prepared for welcoming our wise and benevolent alien ancestors back to earth, just as Klaatu modeled this in the 1951 Cold War era film *The Day the Earth Stood Still*.

The International Raëlian Movement is the largest ET religion in the world today, claiming a membership in Europe, Asia, Oceania, Africa, and North and South America.[30] Raël, born Claude Vorilhon in 1946 in rural France, describes in his foundational text how he first encountered a diminutive alien who told him the Truth of human creation. Subsequent encounters, including one that transported him to the planet of the alien Designers, revealed the threat of nuclear annihilation. The Makers are watching for signs of this self-destructive streak in humans, their 'supreme artworks.' However, in the event of nuclear holocaust the fallback plan is to clone worthy humans from scanned DNA stored in spacecraft archives, for reproduction on a more hospitable planet after earth is destroyed.[31]

During the period of my fieldwork, Raëlians I knew who were bench scientists included at the time a postdoctoral neurobiology researcher at a major Ivy League university, an internationally established nanotechnologist, and a graduate student of biology at a major Canadian university, to name just a few. None of these would refute the theory of evolution from the moment of creation. However, they shared Raël's objection to our genetic kinship with primates, which had a status for them of weird life forms relative to humans.

The Raëlian cosmological narrative is not strange by origin myth standards, of course. And the Raëlian message for exophile science is not irrelevant in the terms of the NRC report. For one thing, the view from space on human social practices of science both iterates science as culture and presents a counter narrative to terracentrism, as it did in *The Wild Blue Yonder*. However, from this point the possible worlds of Herzog and Raël part company. For Herzog's disempowered mobile-home alien, exchanges with the-powers-that-be have little chance of being heard, much less of positively transforming social futures: humankind has all but trashed the hope of that. By contrast, Raël's vision would liberate humans to a world where relations with Designers are good, perhaps even on a par. If the scientist gods are made to feel welcome on earth, technoscientific knowledge exchanges will, over time, produce a 'geniocracy' of pacifists. This kind of exchange relation is feasible because Raël does not rule out the possibility that human knowledge has evolved beyond that of the Designers; that we can all be 'like gods.'

Meanwhile, if the awakened must wait for an afterlife experience off-planet, it will have been worth it. This will be a world of unmitigated pleasure where submissive biological replicants will respond to every desire of one's cloned person and robots will liberate humans from the 'sweat of labor.'[32] In short, as Carly Marchado observes: 'Raëlian cosmology re-invents the colonial model in a [...] world biotechnologically legitimated by the creation of serfs by their masters, granting the latter biological ownership of this sub-humanity.'[33] Raël doubly inflects a value for exotic life – wherever its home – provided it serves human purposes. This is where the Raëlian message comes up against arguments for banning lateral gene transfer's use for human reproduction. In Europe, for example, the ban against reproductive cloning has its legal foundation in colonial era antislavery legislation on the basis that cloned human subjects could be exploited as property (for instance for use as slaves or in parts for growing organs). By way of various loops, Raël would remind NRC authors and publics of the problems that arise when life, terran or non-terran, is denied a value other than as a commodity. Of course, new religious movements tend to thrive to the extent that they make themselves up as they go along in response to world events and trends – no one claims that they are coherent, or above history. Raël's human cloning as originally advertised on www.clonaid.com was explicitly for producing designer babies for a fee. Thus, Raël's messages partially reinforce exophile science's leanings toward colonial ideation, and also partially resist these.

IV Concluding remarks

As a matter of principle, contemporary anthropology revisits, repurposes and complexifies its own terms of reference, in field and office practice sites never far out of sight of one another.[34] Further, it is now taken as an ethical given that practitioners will approach their research sites as if 'once again making contact [...] for the first time, as if history has given [us] an incredible second chance and [we] were back in the seventeenth, eighteenth, nineteenth, or twentieth centuries and are allowed, in the twenty-first, to introduce [ourselves] properly, [we] who had introduced [ourselves] so badly in previous centuries through the most ruthless imperialism.'[35] Needless to say, perhaps, non-sentient life forms may themselves colonize. But enter the programmatic vision and instrumental technologies of humans, not least technologies of persuasion such as specialized disciplinary languages, and the project of mindful interaction with 'weird life' becomes ethically complicated.

For example, on the Internet recently I came across a talk delivered by the popular (if controversial) astrophysicist and Nobel laureate Freeman Dyson (1923–) in 2003. The title is 'Let's Look for Life in the Outer Solar System,' and it quite shocked me to hear Dyson advocating for the search for life on 'real estate' like Europa's.[36] Rather than searching for life in water as the Galileo probe was designed to do, Dyson reasoned that life at a more evolved stage might already have crawled out of the sea onto Europa's surface where, by a version of 'pitlamping' – a term used by night hunters for shining lights into the eyes of their game and aiming their rifles at the reflective surfaces of their eyes – we might detect it. Bloggers on the www.ted.com web site, where the talk was posted, were quick to note Dyson's terracentrism:

> *JoHaNNaSLiLLaBeBBe*: Wait. [...] wouldn't the Europa organisms absorb rather than reflect sunlight?
>
> *CO2Junkie*: Here's my answer, Johannas: No living things absorb sunlight with 100% efficiency. Also, organisms tend to absorb certain wavelengths (the ones they find most useful) and reflect other wavelengths (the ones they find most harmful). Trick is since we don't know what to expect (they probably don't have cat's eyes) we wouldn't know for sure what pattern of reflectivity to look for other than something 'un-geological.'[37]

But Dyson does not stop with life 'stuck' in its place of origin. He goes on to consider the 'second origin' scenario of life forms conceivably dislodged from their planets, say, knocked off by a collision, along with their physical context, and surviving to colonize another heavenly body – spreading out into space 'like sun-flowers' which in their profusion would be quite easy to detect. Or if we can't find them, we can create our own life forms through genetic engineering 'so that we can colonize Europa with our own creatures, [...] transforming the Universe into something more rich and beautiful than it is today.' More rich and beautiful than what, I might ask, as Mr. Herzog does. And for whom, as the Raëlians do not consider, being anthropocentrically self-absorbed? And '*real estate*'? Dyson's blithe statement that developing the real estate of the cosmos as a human laboratory would be 'a fun thing to do' was really almost too much for this reader. Not to mention that the sunflower scenario calls up the specter of a relation to the ecophagic molecular nanorobots created by humans which, critics warn, could conceivably exceed their designers' control to become 'gray goo' that consumes all matter, terran or otherwise. From a science-fiction point of view none of this is news. But Dyson saying it has earned him a popular rating of five stars out of five on the web site for this particular talk, which was sampled from YouTube and, before that, Princeton University. And it has earned the self-reflexive NRC report on *The Limits of Organic Life in Planetary Systems* kudos for humanitarian vision, by contrast.

Co-authorship of possible worlds by human and non-human actants inevitably produces a more complex 'nature' (as revealed by its cultural construction) and a more complex culture (through its defining process of bricolage) in which failure to acknowledge asymmetries of power can prove disastrous for all concerned, on or off world.[38] Too, co-authorship charges us to refine our understanding of the more complex exchange relations to other disciplines we encounter along the way, either by recognizing our inquiries' affinities or kinship or both, but importantly, without mistaking one for the other. Marilyn Strathern makes the point by reference to Herschel's description of the planets from 1833: 'When we contemplate the constituents of the planetary system from the point of view which this relation affords us, it is no longer mere analogy which strikes us, no longer a general resemblance among [the planets]. The resemblance is now perceived as a true family likeness; they are bound up in one chain – interwoven in one web of mutual relation.' Strathern then proposes that 'Herschel wanted to displace a weak sense of analogy between planetary bodies (they look alike) by a strong sense of the affinity between them (their orbits are related to one another).'[39] But the trope of the familial is merely that: 'All he was insisting on was their necessary or systemic connection' as 'an assemblage of objects' – no family required. Science was 'sampling' an anthropological terminology, but applying it loosely. It was intending a poetical effect, not a disfigurement of the kinship concept. Yet the discordant note, amplified perhaps by the scientific instruments of colonial regimes that, even in Herschel's day, were being tuned to racist projects, rings in the ear more persistently for this.

Overall, it is precisely the fact that exobiology and anthropology share a vulnerability to being implicated in colonizing narratives that they can find common ground for constructing a beautiful diplomatic relationship. This to me is altogether more appealing than finding 'ammunition' in one another's fields to use against each other. Herzog, in his credits for *The Wild Blue Yonder*, thanks NASA for its 'sense of poetry.' I find this as good a way as any to move into an inter-entity alien future that considers the consequences of mistaking epistemological maps for disciplinary fields.

Notes

1. Bruno Latour, 'On Actor-Network Theory: A Few Clarifications,' *Soziale Welt* 47.4 (1996), 369–81, here 368.
2. Derrida refers to 'structures of welcoming' in his published session on hospitality as related to *Acts of Religion*, observing that hospitality, on the one hand, 'presupposes waiting, the horizon of awaiting and the preparation of welcoming; while the opposite is nevertheless true on the other hand: to be hospitable is to let oneself be […] surprised.' This element of surprise is a crucial one for the human/non-human relations of exobiology as a practice of the extraterrestrial. See Jacques Derrida, 'Hospitality,' in Gil Anidjar, ed., *Acts of Religion*, New York: Routledge, 2002, 358–420, here 360.

3. Peter Galison, *Image and Logic: A Material Culture of Microphysics*, Chicago: University of Chicago Press, 1987.
4. See Marilyn Strathern, 'Emergent Relations,' in Mario Biagioli and Peter Galison, eds, *Scientific Authorship: Credit and Intellectual Property in Science*, New York: Routledge, 2003, 165–94.
5. I refer to Harrison Schmitt's lecture for returning to the moon on YouTube, http://www.youtube.com/watch?v=AZnBP2Itkwg. The story was released by the Indo-Asian News Service, New Delhi, India, 12 April 2006: 'India Begins Work on Space Weapons'; www.spacewar.com (both sites accessed 1 October 2017).
6. Heidegger's critique of 'enframement' is relevant here, as a framework for valuing 'nature' only as a resource in service of human consumption, such that other values are excluded from the picture and hence disappeared by our own sociocultural 'blind spots.' See Martin Heidegger, *The Question Concerning Technology*, Honolulu: University of Hawaii Press, 1998.
7. http://www.rael.org (accessed 1 October 2017).
8. Committee on the Limits of Organic Life in Planetary Systems, Committee on the Origins and Evolution of Life, and National Research Council of the National Academies, *The Limits of Organic Life in Planetary Systems*, Washington, DC: National Academies Press, 2007, 9.
9. Ibid., 84.
10. See Steven J. Dick, *Life on Other Worlds: The Twentieth-Century Extraterrestrial Life Debate*, Cambridge: Cambridge University Press, 1998; idem and James Strick, *The Living Universe: NASA and the Development of Astrobiology*, New Brunswick: Rutgers University Press, 2004; Steven J. Dick, 'Anthropology and the Search for Extraterrestrial Life: An Historical View,' *Anthropology Today* 22.2 (April 2006), 3–7.
11. My thanks to James Miller for acquainting me with 'skyhooks': hypothetical structures used for transporting material to and from a planet's surface into orbit, continuously supporting it rather than using rockets, catapults or hypothetical anti-gravity effects. Smaller skyhooks include hypersonic skyhooks, rotating cables in lower orbits whose ends dip repeatedly down close to the planet's surface to snag payloads and lift them up. Large rotating tethers can also be used far from a planet's surface to transfer momentum to and from payloads, changing their orbits without the expenditure of reaction mass.
12. NRC report, *The Limits of Organic Life in Planetary Systems*, 10.
13. Referring to 'terran' life forms, the committee states that this denotes 'a particular set of biological and chemical characteristics that are displayed by all life on earth.' Thus, 'earth life' has the same meaning as 'terran life' when the committee is discussing life on earth, but if life were discovered on Mars or any other non-terrestrial body, it might be found to be terran or non-terran, depending on its characteristics. See ibid., ix.
14. Ibid., 6. See Jacques Derrida, *The Other Heading: Reflections on Today's Europe*, Bloomington: Indiana University Press, 1992. Derrida, again a useful supplemental voice, has written on this point of authorship and authority. He writes: 'When the path is clear and given, when a certain knowledge opens up the way in advance, the decision is already made, it might as well be said that there is none to make; irresponsibly, *and in good conscience*, one simply applies or implements a program. Perhaps, and this would be the objection, one never escapes the program.' Ibid., 41 (emphasis mine).

15. NRC report, *The Limits of Organic Life in Planetary Systems*, x.
16. Carol Cleland coined the term 'shadow biospheres' in her 2005 report 'Philosophical Issues in Astrobiology,' part of a NASA-funded study at the Center for Astrobiology, University of Colorado, Boulder. The report draws on astrobiologist Paul Davies's and Charles H. Lineweaver's hypothesis paper 'Finding a Second Sample of Life on Earth,' *Astrobiology* 5.2 (June 2005), 154–63. See Carol E. Cleland and Shelley D. Copley, 'The Possibility of Alternative Microbial Life on Earth,' *International Journal of Astrobiology* 4.3–4 (October 2005), 165–73; and Carol E. Cleland and Christopher Chyba, 'Does "Life" have a Definition?,' in Woodruff T. Sullivan and John A. Baross, eds, *Planets and Life: The Emerging Science of Astrobiology*, Cambridge: Cambridge University Press, 2007, 119–31.
17. Latour, 'On Actor-Network Theory,' 368.
18. This was interesting to me in light of anthropologist Stefan Helmreich's important essay situating Darwinian evolution in the lineal kinship ideology and technologies of his culture and times. See Stefan Helmreich, 'Trees and Seas of Information: Alien Kinship and the Biopolitics of Gene Transfer in Marine Biology and Biotechnology,' *American Ethnologist* 30.3 (August 2003), 340–58.
19. NRC report, *The Limits of Organic Life in Planetary Systems*, 7, 9, 30–7, 84.
20. Ibid., xi.
21. See Michael Young, *Malinowski's Kiriwina: Fieldwork Photography 1915–1918*, Chicago: University of Chicago Press, 1998. Young discusses that in 1892, Im Thurn addressed his colleagues on the subject of photography, stating that he deplored the 'scientific' anthropometric photography of the day, with its 'lifeless bodies,' and advocated instead that photography document unposed native subjects under 'natural' conditions. Im Thurn urged the use of the camera 'for the accurate record, not of the mere bodies of primitive folk [...] but of these folk regarded as living beings.' Also see Anna Grimshaw, *The Ethnographer's Eye*, Cambridge: Cambridge University Press, 2001. In an interesting turn, a recent *New York Times* Op-Ed piece by Seth Shostak, an astronomer at SETI, takes this logic to its extreme by arguing for 'Boldly Going Nowhere' – sending micro-robots into space which would allow us to explore even deep space without leaving our computer screens. Shostak, surprisingly, sees this scheme as satisfying merely the desire to experience outer space, which such microbots and other technologies could aid us in sensorially simulating, mentioning nothing about how this information might be warped or warping in its effects on any and all concerned. See Seth Shostak, 'Boldly Going Nowhere,' *New York Times* (14 April 2009), A21; and Gonzalo Munévar's contribution, Chapter 14 in this volume.
22. For example, Eduardo Viveiros de Castro, 'Zeno and the Art of Anthropology: Of Lies, Beliefs, Paradoxes, and Other Truths,' *Common Knowledge* 17.1 (Winter 2011), 128–45.
23. For a fascinating meditation on the voice and what media technologies do to it, see Jacques Derrida, 'Above all, no Journalists!,' in Hent de Vries and Samuel Weber, eds, *Religion and Media*, Stanford: Stanford University Press, 2001, 56–93, here 70–1.
24. Adam Gopnik, 'Blows Against the Empire: The Return of Philip K. Dick,' *New Yorker* (20 August 2007), 79–83.
25. Our first evidence of this phenomenon comes from the deep-sea journals of William Beebe (1877–1962), who early in the twentieth century produced a first-person account of his pioneering experiences of life in deep aquatic space, in the

company of his colleague, Otis Barton. Published in 1934 as *Half Mile Down*, the book describes Beebe's sense of wonder at the 'blue-black world' beyond his Bathysphere's fused crystal windows – its weird aquatic entities, its unfathomable beauty. Indeed, noting that 'our knowledge will always be extremely fragmentary, given the natural world's immense and complicated history,' he adds a cautionary aside to science on the necessity of being open to knowledge beyond its own discourse, or risk generating 'a colorless, aridly scientific discipline, devoid of living contact with the humanities.' 'No doubt,' he continues, 'a certain spirit of skeptical inquiry should be cultivated even in freshmen, but surely we should realize, like the amateur, that the organic world is also an inexhaustible source of spiritual and esthetic delight.' See William Beebe, *Half Mile Down*, New York: Harcourt, Brace, 1934, xv. Also, for an ethnographer's account of present-day deep-sea explorers' points of view, see Stefan Helmreich, *Alien Ocean: An Anthropology of the Marine Biology and the Limits of Life*, Berkeley: University of California Press, 2008.

26. http://www.youtube.com/watch?v=oo-dx35bCvc (accessed 1 October 2011).
27. Bruno Latour, *War of the Worlds: What About Peace?*, Chicago: Prickly Paradigm, 2002, 40.
28. Casper Bruun Jensen and Teun Zuiderent-Jerak, 'Unpacking "Intervention" in Science and Technology Studies,' *Science as Culture* 16.3 (September 2007), 227–35.
29. NRC report, *The Limits of Organic Life in Planetary Systems*, 9.
30. See Susan Palmer, *Aliens Adored: Raël's UFO Religion*, New Brunswick: Rutgers University Press, 2004; Debbora Battaglia, 'Insiders' Voices in Outerspaces,' in idem, ed., *E.T. Culture: Anthropology in Outerspaces*, Durham: Duke University Press, 2005, 137; and idem, 'Where do we Find our Monsters?,' in Jeanette Edwards, Penny Harvey and Peter Wade, eds, *Anthropology and Science: Epistemologies in Practice*, Oxford: Berg, 2007, 230–60; Carly Machado, *Imagine if it All Were True: The Raëlian Movement among Truths, Fictions, and Religious Modernity*, PhD thesis, University of the State of Rio de Janeiro, 2006.
31. For a historically rich discussion, see Christopher Roth, 'Ufology as Anthropology: Race, Extraterrestrials, and the Occult,' in Battaglia, *E.T. Culture*, 38–94.
32. Raël, *Intelligent Design: Message from the Designers*, Nova Distribution, 2005; see http://www.rael.org/ (accessed 1 October 2017).
33. Machado, *Imagine if it All Were True*, 6.
34. See Marilyn Strathern, *Property, Substance and Effect: Anthropological Essays on Persons and Things*, London: Athlone, 1999; Karen Sykes, 'My Aim is True: Postnostalgic Reflections on the Future of Anthropological Science,' *American Ethnologist* 30.1 (February 2003), 156–68.
35. Latour, *War of the Worlds*, 40.
36. Freeman Dyson, 'Let's Look for Life in the Outer Solar System,' https://youtu.be/wVGjQSnLg4Y [2003] (accessed 1 October 2017). See also 'Hawking: Humans Must Colonize Other Planets,' http://www.msnbc.msn.com/id/15970232/ns/technology_and_science-science/t/hawking-humans-must-colonize-other-planets# (accessed 1 October 2017).
37. http://www.ted.com (accessed 1 October 2017).
38. Biagioli and Galison, 'Introduction,' in *Scientific Authorship*, 1–12.
39. Strathern, 'Emergent Relations,' 179. Strathern draws from John Herschel's 'Treatise on Astronomy' in Lardener's *Cyclopedia*, as quoted in Gillian Beer, *Darwin's Plots: Evolutionary Narrative in Darwin, George Eliot and Nineteenth Century Fiction*, London: Routledge & Kegan Paul, 1983.

CHAPTER 12

A Ghost in the Machine: How Sociology Tried to Explain (Away) American Flying Saucers and European Ghost Rockets, 1946–47

Pierre Lagrange

In 1947, 'flying saucers' emerged as a subject of public, and sometimes scientific, controversy (Figure 12.1). Since that time, a number of historians and sociologists have tried to understand the emergence of this phenomenon on the margins of science. This essay aims to raise questions about how social scientists have studied the subject.

In their research on the topic, historians and sociologists have asked questions such as: Why do people believe in saucers? How did the contexts of the Cold War and science-fiction culture influence this belief? In the early 1960s, Neil Smelser produced one of the classic sociological explanations of flying saucers: 'Recently anxieties over the potentialities of atomic warfare have led not only to predictions of world destruction, but also to many apparent misperceptions and hallucinations of "flying saucers" believed to be omens of destruction.' In 1999, historian of technology Tom Crouch considered the 'flying saucer craze [one of the] signs that US interest in spaceflight, leavened with a bit of Cold War apprehension, was on the rise.' For him, saucers 'spawned a new generation of science-fiction films and had everyone looking to the skies and wondering what might be out there.' And of course, how can we forget Roland Barthes (1915–80), often considered a founding father of cultural studies, who wrote in his famous *Mythologies*: 'The mystery of flying saucers was at first entirely terrestrial:

Pierre Lagrange (✉)
Paris, France
e-mail: lagrange@agence-martienne.fr

Alexander C.T. Geppert (ed.), *Imagining Outer Space*
European Astroculture, vol. 1
https://doi.org/10.1057/978-1-349-95339-4_12

Figure 12.1 'Origin of the Flying Saucers,' a satirical drawing of extraterrestrials shooting tea pots and saucers at neighboring planets, by Ukrainian-born illustrator Boris Artzybasheff (1899–1965) and published in *Life Magazine* less than a month after the first UFO reports. 'The explanation of the flying disks [...] shows residents of the planet Neptune gleefully bombarding the universe with stacks of crockery fired by atomic saucer-launchers,' the original caption read: 'Neptunians thus far have aimed only saucers at the earth (top) but more favored planets have been shelled with tea-pots and dinner plates.'
Source: *Life Magazine* (21 July 1947), 15.

we suspected that the saucers came from the Soviet netherworld, from this world as devoid of clear intentions as another planet.'[1]

The present chapter will demonstrate that these conclusions fail to take into account a significant subset of the data while also failing to distance themselves adequately from the situation they are supposed to study. More specifically, these authors forget that their explanations in terms of Cold War and science-fiction influences do not come from critical historical or sociological analyses, but rather from what was said by participants in the controversy that raged in 1947 and the years that followed. Academics have overlooked the fact that the actors were already, without waiting for the arrival of scholars, their own sociologists and that they did not need Barthes and social historians to produce explanations in terms of Cold War or science-fiction influences. Therefore the problem for sociologists becomes: how can scholars use Cold War or science-fiction explanations in their critical research, when these theories are part of the picture they should describe or explain?

I Saucer believers or skeptics?

To begin, we should question the idea that saucers started as a collective belief influenced by a heightened fear and credulity sparked by the Cold War context. To do this, we need to identify and analyze the origins of the phenomenon.

On Tuesday, 24 June 1947, Kenneth Arnold (1915–84) – an American businessman and private pilot – reported seeing nine strange-looking aircraft flying at twice the speed of sound in the sky above Mount Rainier, Washington (Figure 12.2). Virtually all UFO researchers and historians see this story as the beginning of the UFO controversy.[2] It was this very sighting that launched the public controversy in the press. The day following his aerial sighting, Arnold went to see journalists from the local newspaper of the little town of Pendleton, Oregon, where he had subsequently landed, and discussed his sighting with them. One of the reporters, Bill Bequette (1917–2011), sent an Associated Press (AP) dispatch in which he described the objects seen by Arnold:

> PENDLETON, Ore., June 25 (AP) – Nine bright saucer-like objects flying at 'incredible' speed at 10,000 feet altitude were reported here today by Kenneth Arnold, Boise, Idaho, a pilot who said he could not hazard a guess as to what they were.
>
> Arnold, a United States Forest Service employee engaged in searching for a missing plane, said he sighted the mysterious objects yesterday at 3 p.m. They were flying between Mount Rainier and Mount Adams, in Washington State, he said, and appeared to weave in and out formation. Arnold said he clocked and estimated their speed at 1,200 miles an hour.
>
> Inquiries at Yakima last night brought only blank stares, he said, but he added he talked today with an unidentified man from Utah, south

Figure 12.2 Portrait of American pilot Kenneth Arnold (1915–84), taken by the photographer of the *Idaho Daily Statesman* at his home in Boise, Idaho.
Source: *Idaho Daily Statesman* (28 June 1947).

> of here, who said he had seen similar objects over the mountains near Ukiah yesterday.
>
> 'It seems impossible,' Arnold said, 'but there it is.'[3]

It is this dispatch that generated the controversy over the existence of 'flying saucers.' After this story was published, hundreds of other observers reported their own sightings.

Is this communication the sign, as historians usually say, of the emergence of a collective belief or widespread public credulity? A close reading of this key document suggests otherwise. This article, in fact, does not even suggest that the saucers might have been Russian flying machines, or Russian propaganda – which would have been the case if people had been influenced by the Cold War context. On the contrary, this initial account emphasized Arnold's uncertainty about what he had seen. The fact that the dispatch noted that '"It seems impossible," Arnold said, "but there it is,"' shows that the story provoked skepticism and not belief. And when we consider the very first article published in Pendelton's *East Oregonian* the same day, whose title reads 'Impossible! Maybe, But Seein' Is Believin', Says Flyer,' this point is reinforced (Figure 12.3).

It is true that Kenneth Arnold went on to discuss his sighting with fellow pilots and journalists because he thought he had seen secret weapons, whether American or Soviet, and demanded that a military and federal investigation be conducted. It is true that there were some fellow pilots who mentioned the possibility that these objects might be American or foreign secret weapons. It is also true that when we read the press accounts published in 1947, we find some authors who mention the possibility that these saucers might be Soviet aircraft. For example, in his column 'The End of the Week' for 27 June 1947, *East Oregonian* editor Nolan Skiff wrote: 'We hope they are ours.' We could conclude, then, that these actors were influenced by the Cold War context. But these people were very few indeed, if we compare them to the number who expressed incredulity in response to the story. From Arnold's very first discussion with his friend Al Baxter at Yakima airport, to his conversations with reporters in Pendleton, the debate focused not on secret weapons but on the fact that Arnold's senses played tricks on him. Even among Arnold's friends, many refused to consider the saucer as a secret-weapon issue. Instead, they cast it as a misperception or a product of

Impossible! Maybe, But Seein' Is Believin', Says Flyer

Kenneth Arnold, with the fire control at Boise and who was flying in southern Washington yesterday afternoon in search of a missing marine plane, stopped here en route to Boise today with an unusual story—which he doesn't expect people to believe but which he declared was true.

He said he sighted nine saucer-like aircraft flying in formation at 3 p. m. yesterday, extremely bright—as if they were nickel plated—and flying at an immense rate of speed. He estimated they were at an altitude between 9,500 and 10,000 feet and clocked them from Mt. Rainier to Mt. Adams, arriving at the amazing speed of about 1200 miles an hour. "It seemed impossible," he said, "but there it is—I must believe my eyes."

He landed at Yakima somewhat later and inquired there, but learned nothing. Talking about it to a man from Ukiah in Pendleton this morning whose name he did not get, he was amazed to learn that the man had sighted the same aerial objects yesterday afternoon from the mountains in the Ukiah section!

He said that in flight they appeared to weave in an out in formation.

Figure 12.3 Caption from the first page of the *East Oregonian* issue of 25 June 1947, with the very first newspaper article on flying saucers.
Source: *East Oregonian* (25 June 1947).

Arnold's imagination, thereby placing the story in the context of a discussion over what is real and not real, a debate that would later develop in the press and among scientists over the real/imaginary and later rational/irrational nature of saucers. In the interview he gave over radio KWRC in Pendleton on 26 June 1947, Arnold recalled: 'He [Al Baxter] told me I guess I'd better change my brain (laughs), but he kind of gave me a mysterious sort of a look that maybe I had seen something he didn't know.'[4] While Arnold referred to the military implications of his sighting, his friends continued to question the reality of the events reported. In 1952 Captain Edward J. Ruppelt (1923–60), head of Project Blue Book (the UFO program of the US Air Force conducted between 1952 and 1970), met with a fighter-bomber pilot who was a reporter in civilian life and had worked on Arnold's story. This reporter, most probably David N. Johnson of the *Idaho Daily Statesman*, told Ruppelt 'that when the story first broke, all the newspaper editors in the area were thoroughly convinced that the incident was a hoax, and that they intended to write the story as such.'[5]

From Yakima to Washington, DC, the majority of the experts expressed incredulity over what had been reported. On 26 June, the press noted the skepticism of military spokespersons in Washington, DC: 'As far as we know, nothing flies that fast except a V-2 rocket, which travels at about 3500 miles an hour – and that's too fast to be seen.' The following day, an AP dispatch summed up the debate: 'It is still uncertain whether the shiny discs [...] were objects or optical illusions.' And an astronomer at the University of Oregon formulated a similar hypothesis: 'I believe this man could have been a victim of "persistent vision" from reflections on the glass of his plane,' he asserted. Most newspapers, then, felt compelled to discuss whether the disks really even existed.[6]

If we want to maintain that saucers were a Cold War belief, we must ignore the views of the vast majority of the actors in the controversy who expressed doubt over the reality of the saucers, or we must reduce our analysis only to the claims of the 'believers.' Conversely, we could argue that these skeptical articles were simply reactions to the huge wave of credulity raised by the appearance of flying saucers. But to understand the problem fully, we must ultimately consider the perceptions of all the key players and not just some of them. And the fact is that from the very beginning, the reality of the saucers was seriously challenged. A careful analysis reveals that the saucer phenomenon was not the product of a wave of believers countered by a few skeptics, but just the opposite: the original story was based on the views of a majority of skeptics critical of a minority belief which the majority deemed preposterous. In fact, nearly everyone wanted to appear a skeptic rather than a believer. The believer is always the other.[7]

Even if it is true that there were people who believed in the reality of the saucers, it is also true that these people did not want to look like believers and should not be portrayed as such. Believers and skeptics are not simply two social categories into which we can distribute the people who participated in the discussion. These categories – just like the categories of deviance or superstition – are constructed in the very course of the discussion by participants who are themselves both skeptics and believers, determined to shield themselves from charges of excessive credulity.[8] We need only turn to the experience of Kenneth Arnold to see how complex this situation was. Kenneth Arnold, the man who started it all, turned skeptic – or, more precisely, referred to the divide between skepticism and credulity as a strategy to protect himself – when he saw the furor unleashed by his sighting. Even though Arnold thought the objects he saw might be secret weapons, the nature of the popular debate sparked by his sighting caused him to reconsider. Arnold later wrote: 'From then on, if I was to go by the number of reports that came in of other sightings and of which I kept close track, I thought it wouldn't be long before there would be one of these things in every garage. In order to stop what I thought was a lot of foolishness and since I couldn't get any work done, I went out to the airport, cranked up my plane, and flew home to Boise.'[9]

The divide between believers and skeptics thus appears to be more complex than scholars have often assumed. If we want to explain how Arnold might have been influenced by an awareness of new weaponry in the Cold War context, we must also explain why, at other moments in the discussion, he chose to voice skepticism about sightings reported by others. And we must also ask why we should explain Arnold's skepticism and not the skepticism of the other participants. These categories of skepticism and belief evolved during the debate; thus, we cannot justify the continued study of *why* people believed in saucers, since these people may have considered themselves skeptics, as was the case with Arnold. In other words, we cannot ask how and why the actors believed this or that because, through their actions, they constructed or deconstructed the very categories of belief and skepticism.

We cannot simply ask, therefore, why Arnold and others were influenced by the Cold War, without asking why others were so preoccupied with – even influenced by – the idea of dividing skepticism and credulity. Another key document comes into play here. Between 25 and 30 July 1947, only a month after the discussion started, the Gallup institute conducted a poll on the topic of flying saucers. One of the questions engaged the nature of saucers: 'What do you think these saucers are?' Far from being under the influence of the Cold War context, respondents offered an array of explanations unrelated to possible Cold War fears. A full 29 percent answered that the saucers were optical illusions or imaginary entities, while 3 percent thought them to be meteorological instruments. Another 2 percent responded that the sightings were caused by anti-aircraft searchlights, and 10 percent believed these sightings were simply a hoax.[10] It is clear that most people were 'skeptical' about the existence of real saucers. Though it is true that 15 percent of the population answered that the saucers might be a secret American weapon, only 1 percent attributed them to the Russians. In that same poll, a question was asked about what interviewees considered the most important problem the country confronted, and we see that it is rather difficult to determine whether the Cold War was central among concerns mentioned by the public. In the poll, only 3 percent mentioned the control of atomic weapons as important; 1 percent mentioned communism; but 21 percent considered that the most important problem was to maintain peace and prevent war, which could be considered a fear connected to the Cold War. If we accept the idea that the Cold War was a matter of concern for a significant portion of the interviewees, it is nevertheless clear that the conflict was not the primary factor influencing their reaction to reports of flying saucers.

In spite of this evidence, historians and sociologists might of course reply that it is the scholar's task to unearth the 'real' context of the discussion and that, as an observer and analyst, he is not obliged to accept the actors' explanations. It is this approach that has, until now, led most historians to contend that the proper context of the discussion is the Cold War atmosphere. But it is crucial that we acknowledge that academics have not paid sufficient attention to the situation as it developed in 1947. Moreover, we must explain a series of problems raised, as we will now see, by another phenomenon that appeared in Europe a year prior to the first purported sightings of flying saucers in the United States.

II Constructing European ghost rockets as a Soviet menace

Beginning in late May 1946, northern Europe – especially Sweden and Norway – became the theater of nearly a thousand sightings of strange phenomena described as rockets of unidentified origin. In this case, however, most people believed these craft were of Soviet provenance. The press, just as they had in response to saucers in the United States, participated in the public debate around these sightings.[11]

The two situations were so similar that most historians, including both professional and amateur UFO historians, have considered these so-called ghost rockets to be some sort of 'pre-Arnoldian' saucers.[12] But these observers fail to consider that the two situations, even if they look very similar from a certain perspective – the two events started from sightings reported by individuals, after which the military started an investigation and the press discussed the sightings – were in fact a far cry from one another. That is, the nature of the collective debate they generated varied greatly. From the very beginning, these two public discussions diverged. While the American saucers debate centered on whether they were real or not, no one expressed any doubt about the reality and materiality of the European ghost rockets. Their reality was considered a matter of fact: they were interpreted from the beginning as a secret Soviet weapon launched over the Baltic Sea to create a public scare about the possibility of a third world war.

The different treatment of the ghost rockets and the saucers is reflected in their respective press accounts. Not only the Swedish newspapers – which could be accused of lacking distance from sightings that occurred in their country – but even the *New York Times* in the United States or *Le Monde* in France, both newspapers that would prove to be so skeptical of the reality of saucers in 1947, took the ghost rockets at face value.

For example, in its edition of 9 August 1946, *Le Monde* reported that 'another of these flying bombs has been seen by Lieutenant Lennart Nackman, from the Swedish territorial army staff.' The paper also noted that 'according to the experts, the hypothesis of meteors is absolutely excluded.' The article went on to mention that Swedish authorities had received thousands of letters that allowed them to reconstruct the trajectory of the rockets. From these letters, officials concluded that the rockets came from the Baltic Sea and were much more powerful than the German V-2. On 13 August, *Le Monde* revealed that one of the rockets fell near a city in Sweden, but that the military kept the name of that city secret so that no information could reach the foreign country that might have been the origin of the launching. A *Le Monde* article dated 16 August 1946 outlined how the debris of the rocket had been obtained and suggested that the materials 'were submitted to intense heat.' Not only were the rockets considered real, but, as we have seen, subsequent popular accounts, from the verdict of the experts to the discovery of debris, suggested that no one doubted their reality. Like *Le Monde*, the articles printed in the *New York Times* represented a collection of articles, often on the front page, that recalled the sightings and the discussion in Sweden without expressing the slightest doubt.[13]

The difference between reactions to the flying saucers and ghost rockets is striking. In 1947 most commentators questioned the reality of flying saucers and attributed their emergence to eyewitnesses' limitations and popular credulity; in 1946, the facts were taken at face value, nobody spoke of belief, and the idea of discussing the facts occurred to no one except perhaps as a figure of speech to show how the facts were unquestionable. Thus, if we want to describe a phenomenon that resembles one like the flying saucers *and* that seems to have been influenced by the Cold War, we must turn to the ghost rockets.

Historians, at least, can argue that if Cold War explanations do not work for saucers, they work perfectly for ghost rockets. But we should now examine why this explanation in terms of Cold War influence, even for 1946, is not pertinent after all. There are two reasons.

First, how can we have two reactions so different when concerns about the Cold War would have been so similar? Both in Europe and the United States, there were fierce debates about secret weapons and the fear of a third world war. In 1946 and in 1947, journals and newspapers on both sides of the Atlantic printed numerous articles claiming that the Soviets were far in advance in the air, that the ocean was no longer a significant buffer between the USSR and the United States, and that it had become possible to fly over the pole. The French science magazine *Science et Vie*, for instance, published papers about the progress of rocket technology. Its February 1946 issue cover, which shows a rocket based on the German anti-aircraft missile Rheintochter R-1, crossing the sky at night, appearing exactly like a ghost rocket would, is indicative of this trend (Figure 12.4).[14] In the United States, articles from the beginning of 1947 – a few weeks before the

Figure 12.4 In 1946 and 1947, journals and newspapers on both sides of the Atlantic printed numerous articles claiming that the Soviets were so technically advanced in the air that the ocean was no longer a significant buffer between the USSR and the United States as missiles could be flown over the pole. The French science magazine *Science et Vie*, for instance, published articles about the progress of rocket technology. Its February 1946 issue cover shows a rocket based on the German anti-aircraft missile, the Rheintochter R-1, crossing the sky at night, appearing exactly like a ghost rocket would.
Source: *Science et Vie* 341 (February 1946), cover.

arrival of flying saucers – dealt with similar issues. For example, a story in *Collier's* magazine addressed the progress of Soviet aeronautics and asked, 'Will Russia Rule the Air?' The article reported that there was no longer any meaningful distance between Soviet and American territories because of the possibility of flying over the Pole. It even mentioned as evidence in this regard the mysterious rockets observed in 1946: 'Scandinavian countries have reported mysterious rockets and lights knifing their skies. They are not rockets of course, which could not be seen. They are Russian stepchildren of the V-1, which devastated London. But instead of a range of 160 miles, these have a range that carries them across the Pole – from Europe to a target area in Siberia.'[15]

Of course we can explain the difference in tone between the discussion on flying saucers and the one on ghost rockets by claiming that the only real Cold War context was in Europe and that the Americans never really believed in the possibility of being overflown by Soviet aircraft or guided missiles. Americans, still sheltered from direct contact, would be less susceptible to Cold War beliefs, while Europeans, within easy striking distance, would harbor more active fears of a Soviet threat. But then we have to explain why the Americans participated in this Cold War escalation by responding to the Soviet menace through the launching of spy flights above the Pole. Sometimes these spy flights even flew over the Soviet border.[16] And we have to explain why the T-2 (Technical Intelligence) Department at Wright Field (later Wright-Patterson Air Force Base), near Dayton, Ohio, was dedicated to the prevention of surprise attacks from the Soviets. And last but not least, we have to explain why, while the public debate on saucers focused on the question of public credulity, the military experts at Wright Field concluded that the saucers were real and not imaginary. Indeed, some of these experts believed that the saucers were real flying machines, possibly of Soviet origin, which was a view not expressed in public, of course, and which historians only discovered years later when the formerly secret documents were released to the public.[17]

Why, then, when we are dealing with such similar contexts, did the two stories develop so differently? The answer is not that Cold War concerns were irrelevant to the 1947 American scene in general but that it does not apply to the particular debate on saucers. Historical explanations must be more subtle and specific. While observers in 1946 chose to focus on Russian technology, participants in the saucer debate in 1947, even if they may have also been interested in discussing Russian technological progress, preferred to concentrate on credulity and skepticism when it came to the particular subject of saucers. There is no single context available in which to interpret the facts, but several.

The second reason for which we cannot accept the Cold War context as the determining factor in shaping this discussion is that it cannot account for what transpired over the long term. After two months, opinions expressed about ghost rockets turned from belief to skepticism. Like the saucers, ghost rockets eventually became imaginary in the collective mind. Suddenly, ghost rockets were no longer Soviet rockets. Suddenly, the actors in the debate found it much simpler to explain ghost rockets as beliefs rather than products of Soviet technology.

The first mention of the fact that the rockets might have been imaginary appeared at the beginning of September 1946. The *New York Times* reported that the Soviet journal *New Times* 'denounced as anti-Soviet slander today allegations that radio-controlled shells fired by Russia were crossing Swedish territory.' The journal reduced the story to one about panic and mirages. We might expect people to doubt explanations furnished by the Soviets as nothing but war propaganda. But from that point on, the ghost rockets were no longer perceived as unquestioned facts. By 17 September, the Soviets were not the only skeptics, and the press mentioned the hypothesis expressed by 'Dr. Manne Siegbahn, a 59-year-old Swedish nuclear physicist and Nobel Prize winner,' which cast 'doubt about recent reports that rocket bombs had been fired over Sweden.'[18] The journal quoted scientists who said: 'There is no clear evidence that any guided missiles have been flying over Sweden. [...] I myself examined one reported to be such a missile and found it was a meteorite. I am very suspicious about the existence of any such thing.' The scientist further 'declared that "hysteria" might have been a factor in reports about the missiles.' From then on, skeptical accounts would proliferate in the journals, and the ghost rockets would lose their status as established facts, becoming yet another example of visions and rumor. Just as their counterparts were in the construction of the early flying saucer storyline, the actors in the ghost rocket narrative were always where we do not expect them to be. One moment they looked like believers, the next moment they became skeptics. It seems that far from being under the influence of the context, participants in this drama could move from one context to another very easily.

This brings us to another view of the notion of context. If we return for a moment to saucers, we can see how the Cold War explanation appeared as an element in the public controversy concerning the existence of saucers. In 1951 journalist and commentator Bob Considine (1906–75) published the result of his investigation in *Cosmopolitan*. He stated that 'pranksters, half-wits, cranks, publicity hounds, and fanatics in general are having the time of their lives playing on the gullibility and *Cold War jitters* of the average citizen. It is their malicious fancy to populate the skies over America with a vessel that just does not exist – the flying saucer.' Introducing the Cold War context in his paper allowed Considine to marginalize the people who reported seeing saucers.[19]

Thus, the conclusions of the historians and sociologists who later identified the Cold War influence on the flying saucer phenomenon were anticipated by the very actors of the saucer controversy. These early observers were in fact their own sociologists and historians. They did not need scholars to propose the importance of the Cold War as a factor in shaping what people saw in the skies of Europe in 1946 and 1947. And scholars are often not very far from being actors in the controversy, rather than the analysts they purport to be.

If we remember that Barthes wrote his paper right after the intense public debate that followed the saucer wave in the fall of 1954, we see that he was hardly a neutral observer of the situation, but rather somebody who plunged directly into it.

III Science-fiction influence?

After turning the usual historical explanation that flying saucers were a Cold War phenomenon upside down, it is necessary to discuss another important contextual explanation that is used to explain the rise of 'belief in flying saucers' in 1947. Namely we need to determine whether the sighting of flying saucers can be attributed to the influence of science fiction on the people who saw or 'believed' in UFOs.

This time, at least, the explanation appears credible on the surface. Since the creation of *Amazing Stories* in 1926, the very first pulp science-fiction magazine, the United States was the place where popular magazines dedicated to science-fiction literature multiplied.[20] In fact, it was there that the very word 'science fiction' (first 'scientifiction') was coined. Therefore, it is not unreasonable to assume, as a result of this influence, that saucers were likely connected in the public mind to the idea of their extraterrestrial origin.

As former NASA chief historian Steven J. Dick has argued, 'this controversy [regarding UFOs] would become intimately associated with the debate over the existence of extraterrestrial intelligence.' Dick further states that 'in 1947 […] it was not long before the extraterrestrial hypothesis (ETH) was put forth as a possible explanation.' But, he adds, at first 'very few people immediately sought an extraterrestrial explanation.' He cites the poll conducted by the Gallup institute in August 1947, discussed above, to show that 'most [of the population] thought they were illusions, hoaxes, secret weapons or other earthly phenomena' and did not mention the possibility that these events were linked to extraterrestrials. UFO historian Jerome Clark also shows that the extraterrestrial hypothesis emerged very early in the public discussion of the flying saucer controversy, if not exactly at its beginning.[21]

Unfortunately, when we look carefully at the sources, the explanations offered by Dick and Clark are both true and false. It is true that the extraterrestrial hypothesis was mentioned very early, in fact, even earlier than indicated by these two authors. A review of the articles published on saucers in the press during the summer of 1947 shows that from the very beginning, notions about saucers and other planets helped shape the popular understanding of the phenomenon.[22] For example, as early as 26 June 1947 – when the very first newspaper articles on the subject were printed – the *La Grande Evening Observer*'s title about Kenneth Arnold's sighting read, 'Pilot Sees Planes From Other World.' The *Vancouver Sun*, quoting an International News Service dispatch, explained that Kenneth Arnold reported seeing 'nine saucer-shaped Martian planes.'[23]

But it would be wrong to assume from these references to Mars and to 'other worlds' that they reflected any particular 'belief': was this extraterrestrial hypothesis mentioned to enhance the public's belief or discourage it? The explanations offered by Dick and Clark, then, neglect to address two important questions: Was the outer-space origin a reference to the scientific discussion about life on other worlds or to the Martian invasions of science fiction? In 1947, in addition to the continuation of the old debate about Martian life and the canals, there was a new wave of discussions among scientists on the possibility that life existed beyond the earth, and the public was certainly privy to this information.[24] But it was not *that* extraterrestrial debate that was mentioned in the public coverage of flying saucers; instead, these references typically invoked popular science-fiction culture that referred to 'men from Mars.'

One example will suffice to show with which sort of alien beings flying saucers were associated. Like most of the people who commented on the news, historian Marjorie Hope Nicolson (1894–1981) recalled in her famous *Voyages to the Moon*, published in 1948, that on 11 July 1947, 'when I turned on the radio [I heard] the most recent chapter in the "Strange Saga of the Flying Saucers" that is amusing or terrifying us today':

> I heard over the air – as I have been expecting to hear for some days – that the latest theory about the apparitions is that they come not from Russia but from Mars! And then, as I opened a new box of breakfast food, my eye fell upon the picture of a bold mariner in ultra-modern flying dress, about to take off from the earth to Saturn, complete with a spectacular ring. I stopped to read the captions in Brobdingnagian letters:
>
> BEYOND ROCKET POWER!
> BEYOND THE ATOMIC BOMB!
> BEYOND THE FUTURE!
> BIG NEWS!
> BUCK ROGERS IS ON THE AIR![25]

This anecdote is revealing. Nicolson did not associate the Martian origin of saucers with a scientific debate on life on Mars but with *Buck Rogers* – with pulp culture and comics. Like Barthes in 1955, Nicolson, who through her allusion to *Buck Rogers* served to marginalize the credibility of UFO sightings, was far from being a neutral observer.

Moreover, Dick and Clark forget to pose a second important question. If pulp culture influenced the controversy, did it do so in the sense we usually think – to influence people to believe in the interplanetary origin of saucers, to introduce this hypothesis as a serious solution to saucers? One might still contend that there was a difference between the skeptical comments by the press and scholars, and the beliefs of the public. The answer to this question is a firm no. Instead of pushing people to believe in the extraterrestrial origin of saucers, the influence of science fiction tended to produce just the reverse: it discouraged them from seriously entertaining this idea. It contributed to the debunking of the Martian saucers by making them appear non-scientific. The interplanetary hypothesis was mentioned not because people believed it to be true, but because they wanted to show how this idea was not serious. Invoking the 'men from Mars' storyline was one more method to debunk the reality of saucers. If people referred to 'men from Mars' and not to life on other planets, it is because, in the public mind, saucers were connected not to scientifically sound discussion but rather to the worlds of pulps and comics. It was to show how this idea was silly.

We must remember that pulp magazines were considered by literate people as lowbrow; indeed, as an inferior product of popular culture. Even the people who read them had the feeling they were outside 'real' literary culture.[26] Illustrators were not inordinately proud of illustrating these magazines. And their authors had only one wish: to be published in the slick magazines.[27] Kenneth Arnold, who is supposed to represent the 'popular culture of flying saucer belief,' expressed strong criticism of pulp culture, which he viewed as producing the 'type of publications that I not only never read but had always thought a gross waste of time for anyone to read.' Like Arnold, intellectuals were busy denouncing this 'popular literature' while examining and psychoanalyzing the comics.[28]

Several other episodes show how the extraterrestrials were mentioned in an effort to marginalize the saucers. Kenneth Arnold described how a woman recognized him in a Pendleton café and rushed out madly, saying: 'There's the man who saw the men from Mars.' During the month of July 1947, many newspapers reprinted a story about the experience a journalist, Hal Boyle, claimed to have had. Boyle said he had been abducted by Martians and sent to their planet aboard a flying saucer. His story had the ring of a joke more than a serious story. On 21 July, *Life* discussed, in a humorous tone, the idea that saucers might be sent by aliens. Next to a drawing by Ukrainian artist Boris Artzybasheff (1899–1965) (Figure 12.1 above), *Life* claimed saucers were just that: saucers, but they were sent by Neptunians who have 'attained a civilization far in advance of that now enjoyed on earth [and] are shelling the universe with crockery.'[29] The association of the saucers with the 'men from Mars' allowed critics to dismiss flying disks as products of over-active imaginations or misunderstandings.

As noted above, during the late 1940s there was already a scientific discussion on extraterrestrial life, and it would have been possible for people to connect saucers to this discussion of life on other worlds. But for scientists, this idea of connecting saucers to one of their subjects of inquiry was ridiculous. They considered there to be a huge divide between the way people were 'fascinated' by men from Mars and bug-eyed monsters from science fiction, and the way they themselves 'studied' the possibility of life on other worlds. Until the end of the nineteenth century, scientists, in particular those connected to physics, had been busy constructing a divide between themselves as professionals on the one hand, and with amateurs and the public on the other.[30] Scientists were interested in extraterrestrial life, but their interest – they thought – was vastly different from the interest they attributed to the press and the public. They did not look at the same extraterrestrials with the same tools. In September 1947, a colloquium on astronomy organized by Gerard P. Kuiper (1905–73), professor of astronomy and director of the US Yerkes and McDonald observatories, was dedicated to the question of planetary atmospheres. 'The question of life on other worlds can be settled only through the study of planetary atmospheres,' read the jacket of his book. For participants the question of extraterrestrial life could not be solved by looking directly at the sky for 'space-ships' but by studying planetary atmospheres. The public and the scientists could not have been much further apart one from another.[31]

Apart from scientists, in 1947 the tendency to connect saucers with scientific quests for life on other worlds was not frequent, to say the least, among journalists or the public. Even the few articles that tried to connect saucers with extraterrestrial life tied their content to occult culture and not to science.[32]

For most people, the idea that saucers could be connected to something more scientific than 'men from Mars' was often almost inconceivable. As we have just seen, most were busy making jokes about 'the others' who were supposed to believe in Martian saucers. And in the previously mentioned Gallup poll, one of the two questions asked discussed the nature of saucers: 'What do you think these saucers are?' For 29 percent, they were optical illusions, imaginary things. Of the 42 percent of people who replied, 33 percent responded that they did not know what saucers were, and 9 percent reportedly gave 'other answers.' The interplanetary hypothesis was not even mentioned in the poll. Is it possible that the interplanetary saucers – that is, saucers seriously thought to be of interplanetary origin – were unreported in the 9 percent of 'other answers'? According to the press dispatch released with the poll, the thesis of spaceships could not be measured. These 9 percent mentioned the end of the world, secondary effects of the atomic bomb, and so on. Therefore, we must conclude that when it comes to saucers in 1947, 'men from Mars' were mentioned very often in explanations, while extraterrestrial life was not.[33]

Of course the situation did evolve – but this absence of any mention of interplanetary saucers is one more piece of evidence that this idea had no success in 1947. Between 1947 and 1950, a subculture of people who took seriously the hypothesis of an interplanetary origin for saucers began to emerge. This new tendency culminated in 1950 with the publication of a famous article – and subsequent book – by Donald Keyhoe (1897–1988), a former military man turned journalist, in the magazine *True*.[34] In most cases UFO historians note that Keyhoe launched the extraterrestrial hypothesis, while other UFO historians have tried to demonstrate that Keyhoe was not, in fact, the first.[35] In their attempts to discover who was the first to mention extraterrestrials, historians often forget the tone with which this ETH was expressed. Scholars also fail to remember the status of the author who advanced that view. Authors who discussed the interplanetary hypothesis *before* Keyhoe often had an audience limited to the pulp magazines or occult fanzines and had no access to magazines of better popular reputation.[36] Later authors forget that between the summer of 1947 and the publication of Keyhoe's paper in 1950, the actors of the debate had gradually moved from the 'men from Mars' thesis to the idea of saucers of interplanetary origin. This point is clearly demonstrated when we look at Keyhoe's book. When he first heard about the interplanetary hypothesis, he could not take it seriously. It reminded him of the 'men from Mars' from his pulp-writing period (in the late 1940s, Keyhoe published several stories in pulp magazines). Describing his earlier experiences, Keyhoe writes: '[F]aced with this evidence of a superior race in the universe, my mind rebelled. For years, I had been accustomed to thinking in comic-strip terms of any possible spacemen – *Buck Rogers* stuff, with weird-looking spaceships and green-faced Martians.' The way he recalls the discussion with a pilot who told him he thought saucers were interplanetary illustrates how, for him, the idea of extraterrestrials was connected to the pulp universe, and therefore could not be taken seriously: 'I'd heard some "men from Mars" opinions about the saucers, but this was an experienced pilot. "You don't believe that?" I said.'[37]

It is only after a discussion with two of *True's* editors, who had come to the conclusion that the saucers were of interplanetary origin, that Keyhoe started to reconsider his views. After discussing the other hypotheses (Russian, misinterpretations, etc.), Keyhoe had the following discussion with editors Ken Purdy and John DuBarry:

> 'You've left out one answer,' said Purdy.
> 'What's that?'
> 'Interplanetary.'
> 'You're kidding!' I said.
> 'I didn't say I believed it,' said Purdy. 'I just say it's possible.'
> DuBarry was watching me. 'I know how you feel. That's how it hit me when Ken first said it.'
> 'I've heard it before,' I said. 'But I never took it seriously.'[38]

The idea of linking flying saucers with men from Mars was present from the very beginning, but connecting them with 'serious' discussions on the possibility of extraterrestrials traveling by spaceship took three years to occur among the public. Therefore, if we want to maintain that science fiction had an influence on the way people discussed saucers, we should also accept that it influenced them in the direction of skepticism and not belief: because the pulp universe was peopled with men from Mars, important segments of the public became more skeptical. And it was these skeptics who associated saucers to 'men from Mars' to show how fatuous the subject was. Instead of being under the influence of science fiction, people used the connection between saucers and the pulps to debunk the idea that saucers were of Martian origin. Historians who have tried for years to establish when the extraterrestrial hypothesis was first mentioned, forgot to question *how*, and to what end, it was introduced. This brings us back to the point we raised in the two preceding sections: the public was not simply influenced by the context, they chose it and even constructed it through their own discussions and actions.

IV A new social history of flying saucers

If we want to understand the construction of phenomena like flying saucers in 1947, we must forget the classical explanations in terms of cultural and political influences. Flying saucers were *not* the result of Cold War fears and anxieties. They were not influenced by science fiction – for the very reason that the notion of influence cannot be applied to people in a one-dimensional sense. Even if we cannot simply do what we want in a given situation, influence works in subtle ways, and people are active in the influences they accept or reject. If a context played a role in the narration of these experiences, it was not the Cold War – rather, it became the idea of separating the naïve from the skeptics, and even the idea of creating the very categories of naïve and skeptical. If science fiction played any role, it was in the direction of marginalizing saucers and not in making people believe in them.

Why, then, have so many scholars consistently considered UFOs to be the result of Cold War apprehensions or science-fiction fantasies? Three reasons can be identified:

First, historians have never investigated how the story started and thus have never realized that the points they were discussing were the very points discussed by the actors who launched the public debate on saucers. All scholars previously mentioned – except Clark and Dick – have never taken the care to go back to the original sources.[39]

Second, they write history and sociology from the point of view of the elite, much as historians were used to doing not so long ago in their disparaging analyses of 'popular culture' and 'popular beliefs.' None of the historians discussed here ever mentions the works produced by social historians or sociologists on the concept of 'popular culture' or the critiques these scholars have leveled against earlier biases. Suddenly, when it comes to saucers, the same sociologists who would have been scandalized to see other subjects being treated dismissively as popular culture, or as beliefs, use these same terms without any care. While historians and social anthropologists have learned how to study 'popular culture' by also studying 'elite culture,' when it comes to saucers and UFOs, they do not even think to apply the very tools they apply to other 'popular subjects.'[40] The reason is simple: if historians and sociologists forget to distance themselves from the context in which saucers emerged, it is because they have no distance. Unlike the medieval peasants or the Bororos, American saucer witnesses belong to the same society as the historian. What happened to saucers is similar to what happened to the *Bibliothèque bleue* and its first scholar, Charles Nisard (1808–90), in the nineteenth century. As Michel de Certeau, Dominique Julia and Jacques Revel have argued, the very first studies of popular culture like the *Bibliothèque bleue* coincided with attempts to eradicate it.[41] Is there a popular culture outside the movement that suppressed it, they asked? Is there a flying saucer belief outside the one denounced by learned members of contemporary society? Perhaps we should reconsider the view expressed by Roland Barthes, quoted above. Like Marjorie Hope Nicolson, who published her book *Voyages to the Moon* a year after the 'saucer scare,' Barthes first published the chapter on 'Martians' that later appeared in his 1957 book in *Les Lettres nouvelles* in 1955, a few months after the flying saucer wave that submerged France in the fall of 1954.[42] Was Barthes an outside observer of the situation or an actor participating in its staging? Many scholars ultimately boarded flying saucers like Barthes; they participated in the public debate more than they distanced themselves from it to explain how both skeptics and believers constructed their views.

Third, not only do historians forget to distance themselves from the elite culture that participated in the marginalization of saucers, they also fail to realize that the categories of 'popular' and 'elite' are social constructions, that these categories do not exist apart from the work of the actors who construct them.[43] In his famous work *Outsiders*, Howard Becker has shown that we cannot simply ask why people are deviant, but that we must understand how the notion of deviance, how the categories of deviance and normality, and the divide between them, are constructed.[44] When historian Jean-Claude Schmitt discusses the subject of superstitions, he does not focus on the question 'why are people superstitious?' or 'what context influenced the superstition of the peasants?' Instead, he explains how the concept of superstition was invented by the spokespersons of clerical culture of the Middle Ages to construct a divide between clerical and peasant cultures, and thus tries to understand how the actors of that society collectively constructed the context that gave birth to the notion of superstition. Schmitt maintains that we cannot take the concept of superstition for granted, and must describe its emergence as an explanation by churchmen for the behavior of the people. In other words, superstition is not the explanation; it is what needs to be explained.[45] If we apply Becker's and Schmitt's method to the 'flying saucer belief,' the problem is no longer to understand 'why people believe in UFOs' but to understand how the categories of belief, credulity, irrationality and those of disbelief, skepticism and rationality are socially constructed. In short, we need to explain how people collectively constructed the divide between belief and skepticism during the public controversy that started in 1947 and what words and concepts they used. Schmitt shows that if historians discuss 'how and why people are superstitious,' they are simply reproducing the controversy started by intellectuals in the Middle Ages.

A number of studies have been devoted to controversies on the margins of science like parapsychology, sea serpents or UFOs. Like the studies on deviance, on popular culture and on superstitions, these works have shown that the sociologist cannot take for granted the idea of a divide between what is scientific and what is not, but instead, he must follow the actors, all the actors, 'believers' and 'skeptics' alike, to see how they construct their worlds.[46]

The contextual explanation appears as a weapon used by some actors to discuss the view of other actors. The Cold War did not subconsciously, or otherwise, cause men and women to see saucers in the night skies; instead, participants in the discussions of these sightings purposefully linked UFOs to irrational anxieties about the Cold War or fantasies loosed by science fiction. Therefore we should not search for the real context and the true reality that might exist outside these accumulations of actions and words. The idea of describing who is under the influence of the context only reveals the researcher's prejudices. Like the sociologists and historians just discussed, we should instead describe how historical actors, rather than being believers, collectively constructed the categories; how the participants collectively defined not only the reality but also the context; how they imagined the nature not only of the material world but also of the sociological world; how they gave form to the reality of saucers and ghost rockets and the qualities of the men and women who reported seeing them.[47] The argument that the flying saucers were a Cold War and science-fiction phenomenon is not the solution to the historical conundrum, but rather one of the results of a collective debate set in 1947.

Notes

1. Neil J. Smelser, *Theory of Collective Behavior*, London: Routledge & Kegan Paul, 1962, 90; Tom D. Crouch, *Aiming for the Stars: The Dreamers and Doers of the Space Age*, Washington, DC: Smithsonian Institution Press, 1999, 119; Roland Barthes, *Mythologies*, Paris: Editions du Seuil, 1957, 42: 'Le mystère des Soucoupes volantes a d'abord été tout terrestre: on supposait que la soucoupe venait de l'inconnu soviétique, de ce monde aussi privé d'intentions claires qu'une autre planète.' I wish to thank most sincerely Alexander Geppert and James I. Miller for their astute comments and English corrections. This chapter is respectfully dedicated to the memory of William C. Bequette who helped start it all.
2. For biographical information on Arnold, see Brad Steiger, ed., *Project Blue Book*, New York: Ballantine Books, 1976, 26–7; and Kenneth Arnold and Ray Palmer, *The Coming of the Saucers*, Boise: privately published, 1952, 5–6; David Jacobs, *The UFO Controversy in America*, Bloomington: Indiana University Press, 1976, 36–7. Most accounts of Arnold's sighting include important errors. In 1988 I conducted an investigation in the Pacific northwest of the United States during which I was able to locate and interview some important actors of that period and to reconstruct exactly how the controversy started.
3. I interviewed Bill Bequette in July 1988 at his home in Kennewick, Washington. See Pierre Lagrange, 'E[ast] O[regonian] Journalist Broke First "Flying Saucer" Story,' *East Oregonian* (24 June 1997), 1, 5; 'A Moment in History: An Interview with Bill Bequette,' *International UFO Reporter* 23.4 (Winter 1998), 15,

20; and 'It Seems Impossible, but There It Is,' in John Spencer and Hilary Evans, eds, *Phenomenon: From Flying Saucers to UFOs – Forty Years of Facts and Research*, London: Futura, 1988, 26–45, here 26.

4. Ted Smith, Interview with Kenneth Arnold, Pendleton, OR, KWRC radio station, 26 June 1947 (this document, that I discovered in 1988, is in the form of a disc record). In his report to the Air Force sent at the beginning of July, Arnold wrote: 'I described what I had seen to my very good friend Al Baxter, who listened patiently and was very courteous but in a joking way didn't believe me.' I am referring here to the report addressed by Arnold to the Wright Field base (Dayton, Ohio) in early July 1947; reproduced in Steiger, *Project Blue Book*, 32.
5. Edward J. Ruppelt, *The Report on Unidentified Flying Objects*, Garden City: Doubleday, 1956, 35. Johnson also added to Ruppelt that 'the more they dug into the facts, however, and into Arnold's reputation, the more it appeared that he was telling the truth' and that as a result 'we all put a lot of faith in his story.' This may be true for the *Idaho Daily Statesman* and of newspapers like the *East Oregonian* who knew Arnold in person, but not for the other journals that remained skeptical.
6. Associated Press, 'Whizzing "Pie-Pan" Plane Report Gets Army Skepticism,' *Oregon Journal* (26 June 1947), 1; Associated Press, 'Picture of "7 Dots" Proves Latest in Flying Disk Case,' ibid. (27 June 1947), 8; Tom Caton, 'Officials Doubt Story of Phantom Air Fleet,' *Oregonian* (27 June 1947), 1.
7. French ethnographer Jeanne Favret-Saada has shown in her investigation on sorcery in the French countryside how people exempted themselves of such 'naïve' beliefs and always sent her in search of 'others' in the next village who were supposed to believe in curses. Jeanne Favret-Saada, *Deadly Words: Witchcraft in the Bocage*, Cambridge: Cambridge University Press, 1981.
8. Howard Becker, *Outsiders*, New York: Free Press, 1963; Jean-Claude Schmitt, 'Les "superstitions,"' in Jacques Le Goff and René Rémond, eds, *Histoire de la France religieuse*, vol. 1: *Des dieux de la Gaule à la papauté d'Avignon*, Paris: Editions du Seuil, 1988, 417–551.
9. Arnold and Palmer, *Coming of the Saucers*, 14.
10. George H. Gallup, *The Gallup Poll Public Opinion, 1935–1971*, vol. 1: *1935–1948*, New York: Random House, 1972, 666. The first question asked: 'Have you read or heard about "flying saucers"?' According to the poll, 90 percent replied in the affirmative.
11. Anders Liljegren and Clas Svahn, 'The Ghost Rockets,' in Hilary Evans and John Spencer, eds, *UFOs, 1947–1987: The 40-Years Search for an Explanation*, London: Fortean Times, 1987, 32–8.
12. See, for example, Loren E. Gross, 'Ghost Rockets of 1946,' in Ron Story, ed., *The UFO Encyclopedia*, London: New English Library, 1980, 147–8.
13. 'Swarm of Mysterious Rockets Is Seen Over Capital of Sweden,' *New York Times* (12 August 1946), 1, 7; 'Swedes Use Radar in Fight on Missiles,' ibid. (13 August 1946), 1; 'Doolittle, Sarnoff Stir Swedish Talk,' ibid. (21 August 1946), 3; 'Doolittle Consulted by Swedes on Bombs,' ibid. (22 August 1947), 2.
14. The cover shown in Figure 12.4 illustrates André Fournier, 'Les derniers types de bombes planantes et volantes télécommandées,' *Science et Vie* 341 (February 1946), 67–76.
15. W.B. Courtney, 'Will Russia Rule the Air?,' *Collier's* (25 January 1947), 12–13, 59, 61; ibid. (1 February 1947), 16, 67, 69; and 'Russia Reaches for the Sky,' ibid. (26 April 1947), 18, 95, 97–9.

16. Paul Lashmar, *Spy Flights of the Cold War*, Annapolis: Naval Institute Press, 1996, 88–9.
17. 'Letter from General N.F. Twining to Commanding General, Army Air Forces, 23 September 1947,' in Edward Uhler Condon, *Scientific Study of Unidentified Flying Objects*, New York: Dutton, 1969, 894–5; Air Intelligence Division Study No. 203, Directorate of Intelligence and Office of Naval Intelligence, *Analysis of Flying Object Incidents in the U.S.*, Air Intelligence Report No. 100–203–79, Washington, DC, 10 December 1948.
18. *New York Times* (4 September 1946), 10; ibid. (17 September 1946), 8.
19. Bob Considine, 'The Disgraceful Flying Saucer Hoax,' *Cosmopolitan* 130.1 (January 1951), 32–3, 100–2.
20. Michael Ashley, *The History of the Science Fiction Magazine*, vol. 1: *1926–1935*, Chicago: Henry Regnery, 1974.
21. Stephen J. Dick, *The Biological Universe: The Twentieth-Century Extraterrestrial Life Debate and the Limits of Science*, Cambridge: Cambridge University Press, 1996, 267, 271–2; Jerome Clark, 'Meeting the Extraterrestrials: How the ETH Was Invented,' in Dennis Stacy and Hilary Evans, eds, *UFOs, 1947–1997: From Arnold to the Abductees: Fifty Years of Flying Saucers*, London: Fortean Times, 1997, 69.
22. Ibid. UFO historian and encyclopedist Jerome Clark discusses the fact that the extraterrestrial hypothesis was mentioned before Keyhoe's article, but he does not mention the local newspaper accounts from 1947.
23. All the newspapers began to print articles on flying disks and flying saucers on 26 June 1947. Only one newspaper, the *East Oregonian* of Pendleton, Oregon, printed an article on 25 June 1947, thus starting the public debate. 'Pilot Sees Planes From Other World,' *La Grande Evening Observer* (26 June 1947), 4. The article does not explore this hypothesis. 'Strange "Planes" Sighted Shooting Across Heavens,' *Vancouver Sun* (26 June 1947); cited in Jan L. Aldrich, *Project 1997: A Preliminary Report on the 1947 UFO Sighting Wave*, n.p.: UFO Research Coalition, 1997, 175–6.
24. Of course, the debate on extraterrestrial life is very old, but the book by Sir Harold Spencer Jones, *Life on Other Worlds* (London: English Universities Press, 1940), is considered the starting point of this new wave of discussions on life elsewhere. See Dick, *Biological Universe*, 160–221; more generally, idem, *Plurality of Worlds: The Origins of the Extraterrestrial Life Debate from Democritus to Kant*, Cambridge: Cambridge University Press, 1984; and Michael J. Crowe, *The Extraterrestrial Life Debate, 1750–1900: The Idea of a Plurality of Worlds from Kant to Lowell*, Cambridge: Cambridge University Press, 1986.
25. Marjorie Hope Nicolson, *Voyages to the Moon*, New York: Macmillan, 1948, 257.
26. See the entry 'Pulp Magazines' in James Gunn, ed., *The New Encyclopedia of Science Fiction*, New York: Viking, 1988, 374–7. In particular, the author of this entry argues: 'A cheap and accessible form of entertainment, the pulps were never highly thought of by most academics, slick-paper critics or various defenders of public morality. The fact that they provided on outlet for SF tended to lower the genre's reputation and establish a stigma that still is being lived down.' To see how the science-fiction subculture organized itself in order to demonstrate the legitimacy of science-fiction literature, see among other works, Harry Warner Jr., *All Our Yesterdays: An Informal History of Science Fiction Fandom in the Forties*, Chicago: Advent, 1969.

27. For example, many science-fiction pulp authors were jealous of authors like Ray Bradbury (1920–2012), who had succeeded in being published outside the pulp magazines, in the slicks.
28. Arnold and Palmer, *Coming of the Saucers*, 20. In 1948 French intellectual Jean-Paul Sartre (1905–80) translated and published in his journal *Les Temps Modernes* a study by Gershon Legman entitled 'The Psychopathology of Comics,' which had previously appeared in the American journal *Neurotica*; see Gershon Legman, 'Psychopathologie des comics,' *Les Temps Modernes* 4.43 (May 1949), 916–33. Previously, Sartre had already published another paper denouncing comics; see David Hare, 'Comics,' ibid. 1.1 (1946), 353–61.
29. 'Harassed Saucer-Sighter Would Like to Escape Fuss,' *Oregonian* (28 June 1947), 1; Hal Boyle, 'Green Martians Kidnaps Boyle! Writer Whisked Away in "Saucer,"' *The Idaho Daily Statesman* (8 July 1947), and 'Boyle Returns From Disc Trip and Promptly Goes on Water Wagon,' ibid. (9 July 1947); 'Speaking of Pictures [...] A Rash of Flying Disks Breaks Out Over the U.S.,' *Life Magazine* (21 July 1947), 14–15. In 1942 the same artist Boris Artzybasheff had illustrated the belief in gremlins for *Life*.
30. See Daniel J. Kevles, *The Physicists: The History of a Scientific Community in Modern America*, New York: Alfred A. Knopf, 1978.
31. Gerard P. Kuiper, ed., *The Atmospheres of the Earth and Planets*, Chicago: University of Chicago Press, 1949.
32. Writer R. DeWitt Miller (1910–58), who had previously authored a series of articles on science enigmas in the journal *Coronet* (these articles were collected later in idem, *Forgotten Mysteries*, Chicago: Cloud, 1947), published a paper that was reprinted in many daily journals; see idem, 'Psychic Expert Says Saucers Are Old Story,' *Idaho Daily Statesman* (9 July 1947). As soon as he learned of the articles on saucers, *Amazing Stories* editor Ray Palmer was among the very first to advocate the interplanetary origin of saucers; see 'The Observatory,' *Amazing Stories* 21.10 (October 1947), 6. This October issue was on sale as early as August. In 1947 an occult subculture, keen to embrace saucers as soon as any arrived, already existed. These occult groups constructed a view of outer space that was very different from the outer space of astronomy and close to the universe of channels that developed in other occult and New Age subcultures. See Lagrange, 'It Seems Impossible, but There It Is,' 40–2.
33. Gallup, *Gallup Poll Public Opinion*, 666; Clark, *UFO Encyclopedia*, 377.
34. Donald E. Keyhoe, 'The Flying Saucers Are Real,' *True* 26.152 (January 1950), 11–13, 83–7, and *Flying Saucers Are Real*, New York: Fawcett, 1950.
35. For example, see the opposition between the views expressed by David Jacobs (*UFO Controversy in America*, 56–7) and by John Keel; see his 'The Man Who Invented Flying Saucers,' *Fortean Times* 41 (Winter 1983), 52–7.
36. This is the case, for example, of Raymond A. Palmer (1910–77), editor of *Amazing Stories*, of R. DeWitt Miller and of Meade Layne (1882–1961), director of the occult-oriented Borderland Science Research Associates, among many others.
37. Keyhoe, *Flying Saucers Are Real*, 62, 50.
38. Ibid., 50–1.
39. See notes 1 and 2 above.
40. See the classic studies by Carlo Ginzburg, *The Cheese and the Worms*, Baltimore: Johns Hopkins University Press, 1976; and Schmitt, 'Les "superstitions."'

41. Michel de Certeau, Dominique Julia and Jacques Revel, 'La Beauté du mort' [1970], in Michel de Certeau, *La Culture au pluriel*, Paris: Christian Bourgois, 1980, 49–80.
42. See James Miller's contribution, Chapter 13 in this volume.
43. Of course this does not mean that they do not exist. Rather, they are constructions, and when they are constructed, they become real and have real consequences.
44. Becker, *Outsiders*.
45. Schmitt, 'Les "superstitions."'
46. See, for example, Roy Wallis, ed., *On the Margins of Science: The Social Construction of Rejected Knowledge*, Keele: University of Keele, 1979; in particular the chapters by Harry Collins and Trevor Pinch, 'The Construction of the Paranormal: Nothing Unscientific Is Happening,' 237–69; and by Ron Westrum, 'Knowledge About Sea Serpents,' 293–314.
47. For a beautiful case study that shows how actors, both human and non-human, construct their social and natural worlds, see Michel Callon, 'Some Elements of a Sociology of Translation: Domestication of the Scallops and the Fishermen of St Brieuc Bay,' in John Law, ed., *Power, Action and Belief: A New Sociology of Knowledge?*, London: Routledge, 1986, 196–223.

CHAPTER 13

Seeing the Future of Civilization in the Skies of Quarouble: UFO Encounters and the Problem of Empire in Postwar France

James I. Miller

On 10 September 1954, a young metalworker named Marius Dewilde (1921–96) stepped outside the abandoned railway station house he occupied in Quarouble, France, and encountered an alien world. Even for a man who would later claim to have worked as a traveling circus performer, fought in the resistance, and voyaged around the world in the French merchant marine before his thirty-third birthday, what happened that September night was truly fantastic. Dewilde, who was initially responding to his dog's furious barking, grabbed his flashlight and headed outside to make sure nothing untoward was afoot. What would prove to be a life-changing experience occurred moments later, when he saw posed on the railroad tracks in front of him a shadowy vessel he first took to be a carelessly placed farmer's cart, a common occurrence in the area. As he moved to investigate, Dewilde observed two small figures skulking in the shadows. Concluding they were smugglers returning to their contraband – also a frequent sight along the *chemin de contrebandiers* that passed nearby – he decided to intervene and capture one of them.[1] Before he could act, however, a beam of light projected from the vehicle paralyzed him, leaving his senses free to process what was happening even as his body refused to respond. In that instant, Dewilde, an unwitting victim of an otherworldly civilization and their advanced technology, became one of the early focal points for observers – the convinced, the curious and the skeptical – interested in the UFO phenomenon in

James I. Miller (✉)
Community College of Rhode Island, Warwick, RI, USA
e-mail: jimiller@ccri.edu

Alexander C.T. Geppert (ed.), *Imagining Outer Space*
European Astroculture, vol. 1
https://doi.org/10.1057/978-1-349-95339-4_13

postwar France. And though Dewilde eventually sought to put his imprimatur on a story that had taken on a life of its own, by 1980, when he published his book dedicated to the sighting, ascribing broader meaning to his encounter and others like it was ultimately left to others.[2]

In fact, Dewilde's story quickly drew the attention of the media, the government, and France's nascent community of ufologists, men and women dedicated to researching UFO phenomena.[3] Within days of his initial experience, Dewilde found himself surrounded by government investigators, the press, and inquisitive onlookers (Figure 13.1).[4] From the outset, both the French state and the media believed Dewilde had witnessed *something*, whether some new instrument in the Cold War arsenal, a meteorite or some other unexplained phenomenon. Moreover, Dewilde's demeanor, his obvious anxiety and his insistent, panicked search for authorities in the wee hours were convincing evidence of his conviction for those who crossed paths with him that night or interviewed him later.[5] Most of the observers who arrived on the scene also attested to the fact that some sort of unusual marks had been left at the site where Dewilde claimed to have seen the mystery craft.

At the same time, neither believers nor doubters have ignored the fact that Dewilde found himself the beneficiary of an intense media spotlight generated by essentially unverifiable, and increasingly grandiose, assertions.[6] The sensational nature of his story brought him momentary notoriety, as members of the local and national press wined and dined the once anonymous man.[7] In the end, as those who have studied the case have concluded, one can never know with any degree of certainty what Dewilde saw that evening; but the story he told and the ways others responded to it brought to light anxieties and concerns about a moment where the boundaries between familiar worlds and their occupants were being redrawn.[8] For, as anthropologist Christopher Ross has argued, 'even if we are being visited by immensely superior aliens, our visitors would be understandable and perceptible only after being fitted into a pre-existing cultural scheme for categorizing beings.'[9] And though Marius Dewilde's experience was among the first to draw widespread attention to UFO sightings, his was not unique in the final months of 1954. On that early fall night, he joined the ranks of hundreds of French men and women who reported witnessing unidentified flying objects or encountering alien beings over the span of two months. Unlike many of these stories, however, over time, Dewilde's became an important case in the debates ascribing meaning to this 1954 rash of sightings.[10]

Among those who eventually sought to explain the stories of Dewilde and others were Aimé Michel (1919–92) and Jacques Vallée (1939–), both of whom would emerge as prominent French ufologists in the 1950s and 1960s. Like many ufologists, Michel was a polymath who rejected the disciplinary rigidity of modern science and believed that the prevailing insistence on scientific rationalism was a detriment to its advancement. Trained in philosophy, he worked at times as a radio engineer, a science writer, and a teacher. Born in a small village in the Alps, Michel was also attached to life in France's rural

RADAR

N° 294 - 26 SEPTEMBRE 1954
Canada 15 cents ★
6 fr. belges - 0 fr. 65 suisses

Hebdomadaire
16 PAGES 30 Francs
Maroc (par avion) 40 fr.

VALENCIENNES

RECONSTITUE LA FANTASMAGORIQUE APPARITION "MARTIENNE" QUI SIDÉRA Marius DEWILDE

Grâce à ses dessinateurs et reporters photographes, RADAR s'est attaché à reconstruire avec une extrême minutie l'hallucinante apparition dont fut témoin M. Dewilde (ci-dessous) garde-barrière dans le Nord. Evénement à peine croyable : il aurait vu atterrir, près de sa maison, une soucoupe volante!

HALLUCINATION ? Vérité ?
Marius Dewilde, ouvrier métallurgiste à Quarouble (Nord), prétend avoir vu une soucoupe volante, posée sur la voie ferrée qui dessert les Houillères nationales, à 6 mètres de sa maison. De petites créatures, vêtues d'une sorte de scaphandre, coiffées de casques en matière translucide, avaient réintégré en courant, le mystérieux engin. Ayant voulu s'en approcher, Marius Dewilde avait été paralysé par un rayon qui, brutalement, avait jailli de l'appareil.

Trois jours plus tôt, deux cultivateurs d'Acheux-en-Amiénois (Somme), Emile Renard et Yves de Gillabox, avaient été les témoins, prétendent-ils, de l'atterrissage d'une autre soucoupe volante dans un champ, à 200 mètres de la route d'Harponville à Contay. L'appareil oscillait à quelques centimètres du sol.

Enfin, dernière nouvelle (en en attendant d'autres !), Antoine Mazaud, fermier à Mourièras (Corrèze), se serait trouvé nez à nez avec un passager d'un « navire spatial » qui lui aurait témoigné des sentiments... pacifiques.

De telles nouvelles — sensationnelles, il faut l'avouer — ne vont pas sans susciter l'incrédulité de beaucoup. Mais pour ceux qui, depuis sept ans, se penchent sur le troublant problème des soucoupes volantes, elles ne font que rééditer certains témoignages recueillis par l'Ouranos, un organisme de recherches qui fixe maintenant à sept le nombre de « navires de l'espace » ayant atterri sur le territoire français.

Mais la France n'a pas le privilège de ces visites surprenantes.

Le 24 juin 1947, un aviateur américain, Arnold Kenneth, apercevait au-dessus du mont Rainier (Etat de Washington) « 9 objets scintillants évoluant à hauteur des pics neigeux ». Ils ressemblaient à des soucoupes surmontées d'une protubérance qui affectait la forme d'une tasse renversée. Chacun d'eux avait à peu près l'envergure d'un quadrimoteur « C 54 ».

A partir de cette date, il ne se passa pas de semaine sans que ces mystérieux engins ne fussent signalés quelque part dans le monde.

L'« U.S. Air-Force », sans prendre tout d'abord position, collecta tous les témoignages et chargea l'A.T.I.C. (Air Technical Intelligence Center) d'ouvrir une enquête. Celle-ci, dénommée « Soucoupe Project », démontra que, sur 270 cas examinés, 60 % pouvaient être expliqués : il s'agissait de ballons-sondes, d'appareils de recherches de rayons cosmiques, de météores, voire d'oiseaux. Mais pour les autres 40 % des cas, le mystère subsistait.

Malgré cette restriction, lorsqu'au mois de décembre 1949, l'enquête « Soucoupe Project » fut close, les conclusions furent que « les soucoupes volantes étaient une plaisanterie ; qu'il s'agissait, en fait : ou bien d'objets connus ; ou bien de mystifications; ou bien d'hystérie collective ».

Cependant, comme le phénomène continuait à se manifester avec une ampleur croissante, l'« U.S. Air-Force » recréa « Soucoupe Project » discrètement sous l'appellation neutre de « Project Bluebook » ou « Commission Grudge », ou « Project Sign ».

SUITE PAGES 2-3

Figure 13.1 In September 1954 *Radar* magazine published one of the first and most widely disseminated renderings of Marius Dewilde being stunned by a mysterious ray during his contact with an alien civilization.
Source: *Radar* 294 (26 September 1954). Courtesy of Mary Evans Picture Library.

hamlets and rued its gradual disappearance following the war. Michel, dismayed by what he believed was a relative lack of interest on the part of 'official science' in the UFO phenomenon in Europe, was also among the first to publish on the UFO phenomenon in France.[11]

In fact, Michel completed his first book, *Lueurs sur les soucoupes volantes*, in the spring of 1954, just months before the fall wave of sightings reported around the globe. In his next work, *Flying Saucers and the Straight-Line Mystery*, published in 1958, he used several cases, including the 'uncanny' and 'weird' story told by Dewilde, to construct his 'straight-line' theory, which identified a rational pattern in a series of sightings occurring around the globe at various times as a means of proving their authenticity.[12] Michel suggested that the apparent absurdity of stories like Dewilde's was less important than the existence of corroborating evidence suggesting an intelligence or organized force behind such patterns. Although he was at first unwilling to proffer any definitive explanation for what was happening in the world's skies, Michel warned that if the 'thousands of identical accounts which day by day reach the files of the commissions of enquiry are true, truly the implication must be that we have a sword of Damocles hanging over our heads,' since, as he warned obscurely, if ominously, 'the destiny of our planet is clearly at stake.'[13]

Michel's younger friend, colleague and frequent collaborator, Jacques Vallée, worked as an astronomer, dabbling in the study of UFOs before leaving France in 1962 and moving to the United States, where he eventually earned a PhD in computer science at Northwestern University. While in Chicago, Vallée also worked with the American astronomer and UFO investigator J. Allen Hynek (1910–86), whose ideas and work opened up new vistas for the young Vallée.[14] In his memoir, *Forbidden Science*, Vallée attributes his early fascination with UFOs to the burst of sensational publicity in 'the Fall of 1954 when there was a deluge of sightings in France.' In his entry of 1 September 1958, Vallée acknowledged that it was the story of 'a railroad worker [in fact a metalworker] named Marius Dewilde,' whose unadorned simplicity had most captivated him.[15] By the end of the 1960s, Vallée would conclude that the seemingly far-fetched observations of individuals like Dewilde, though perhaps attributable to misinterpretations or the limitations of human knowledge, were part of a long history of human contact with alien beings, whether angels, demons or fairies.[16] While Vallée adamantly refused to settle on a single explanation for these phenomena, he too believed, at least during the early years, that witnesses were describing visitors from other planets.[17]

At first glance, the principle concern of observers like Vallée and Michel was to identify *what* people like Dewilde were seeing. For them, this task was critical to advancing the cause of science, which could only be accomplished by encouraging their fellow scientists to question accepted wisdom and explore the new frontiers opening in the skies of the postwar world. But ufologists, at heart, were also determined to explain what these encounters meant for the present and future of humankind, what the arrival of new forms of intelligence signified and why they were appearing then and there. As Vallée concluded in

Passport to Magonia, his study of the long history of paranormal activities, 'We cannot be sure that what we study is something real, because we do not know what reality is; we can only be sure that our study will help us understand more, far more, about ourselves.'[18] What, then, did these men hope human beings would learn from alien contacts like Dewilde's? How were these hopes particular to political conditions in postwar France and Europe? And more specifically, what do the conclusions of France's ufologists about the UFOs in the skies of the world reveal about the marginalizations and disruptions generated by the dramatic changes occurring during a period marked by the processes of modernization and decolonization in postwar France?

By the time of the 1954 wave, France was undergoing a transformation marked by the disruptions of economic modernization, uncertainties about political reconciliation in Europe, and new vectors of imperial power.[19] As a group, the ufologists and others captivated by the UFO phenomenon discussed in this chapter hailed from diverse backgrounds, held differing political views, and practiced different professions. Nevertheless, with the possible exception of Dewilde himself, they were linked in their concern about changing social, political and professional boundaries in the postwar world. All of them questioned the limits and value of technocratic society; all of them shared experiences of social and professional isolation; and all of them were concerned about the effects of imperial wars in the twentieth century.

Centering these concerns is critical to understanding the interpretive framework through which stories like Dewilde's were made sense of through the 1950s, 1960s and 1970s, both by the witnesses themselves and the ufologists who attempted to explain to a broader public what was happening in skies around the globe. For regardless of what Dewilde saw on the railroad tracks in front of his makeshift home in Quarouble, his description of the aliens he encountered and the ways others responded to his claims reveal a good deal about prevailing perceptions of, and concerns about, the remaking of political and social boundaries in France during the 1950s and 1960s.

I A marginalized man's troubled contacts with a 'superior civilization'

That Dewilde, a new arrival to Quarouble in the 1950s, felt alienated there is attested to by witnesses interviewed by computer scientist Claude Gaudeau and psychologist Jean-Louis Gouzien.[20] Like others affected by increasing migration from the countryside to the nation's vital industrial centers and the erosion of rural traditions, men like Dewilde were experiencing new forms of professional, social and cultural dislocation that often seemed beyond their control. Forced from familiar settings, many found themselves uprooted from traditions and practices that had defined their cultural frames of reference and identities. For these men, including Marius Dewilde, defining their place in this new order was as pressing as it was difficult.

When asked later by investigators, many of Quarouble's residents characterized Dewilde as an outsider. Queried about the reliability of Dewilde's testimony, several men and women who knew Dewilde were quick to discredit his story because of his marginal status. One of Dewilde's acquaintances, a local game warden familiar with the case, explained how he never bothered to look into the story because Dewilde was a 'Parisian, not exactly the salt of the earth,' who lived quite literally on the margins of local life in 'a rail attendant's quarters because the line had been abandoned.'[21] In fact, Dewilde lived as an outsider, barely assimilated into local life and excluded from the benefits of a postwar consumer society increasingly important to conferring status and prosperity. His home, which he inhabited illegally, had neither electricity nor running water, a factor all his neighbors were quick to point out in their dismissals of the details and reliability of his story.[22] That Dewilde himself felt alienated by this fact was later evidenced in his memoir, in which he fabricated from whole cloth details about his prosperity and success in Quarouble. When he finally sat down some 20 years after that September evening to 'reveal' the full truth about his adventures in his autobiography, Dewilde claimed that his wife, who had been watching television when the craft landed, came out in time to see it depart.[23]

Dewilde's self-consciousness about his status as an outsider was but one concern that shaped his description of his encounter. For Europeans who had lived through the 1930s, as Dewilde and Michel had, or come of age in the 1950s, as Vallée had, concerns about violent imperial encounters and the massive destruction and loss of life they engendered were not uncommon. To be sure, in the days and months immediately following his experience, Dewilde seemed indifferent to the possible motivations of his alien visitors. He readily discussed their behavior, the nature of their craft, and the material evidence of their visit, but he made no attempt to interpret their mission or the timing of their arrival. If Dewilde initially spent little time contemplating the extraterrestrials' motivations, he also left little record of his concerns about the broader problems he believed threatened France or the rest of the planet in 1954.

As an unemployed, unskilled worker living in an abandoned station house in rural France, Dewilde was not politically engaged.[24] In one early instance where a journalist suspected Dewilde's fantastic story might somehow be related to latent fears about mounting Cold War hostilities, the conversation produced scant evidence that Dewilde was aware of, or worried about, such matters. Indeed, when asked ironically by his interlocutor what he, 'the local star,' thought about the 'antagonism between East and West,' Dewilde, who did not recognize the tensions referred to – or even the word antagonism – merely shrugged, muttering he did 'not give a damn about any antagonist.'[25]

Certain aspects of Dewilde's description of the aliens' physical characteristics, however, suggested he – or Marc Thirouin (1911–72), the ufologist interviewing him – certainly harbored some awareness of France's imperial problems at the time. In 1955 Thirouin, the editor and founder of *Ouranos*, France's first journal on UFOs, spent a month with Dewilde. In the second of two articles he produced from the interview, Thirouin recounted Dewilde's careful description of the extraterrestrials' physiognomy, which was based on a then little-reported second encounter Dewilde claimed to have had on 10 October 1954. According to Dewilde, the beings, whom he this time saw in broad daylight, resembled Asians, 'Mongols; the jaw was fairly strong, the cheekbones high, the hair and eyebrows dark, [...] the skin rather brown: it wasn't that of a white man with a mat complexion, not pink but brown, less "cooked" than one thinks of with the Red-Skins, more like that of Arabs, darker in those places where we have a beard' (Figure 13.2). Finally, with some prodding by his interviewer, who offered his best impressions of Chinese, Vietnamese and Siamese, Dewilde concluded that the language the humanoids used to address him sounded less Asian than 'European, but not, in my opinion, English, German or a Latin language.'[26]

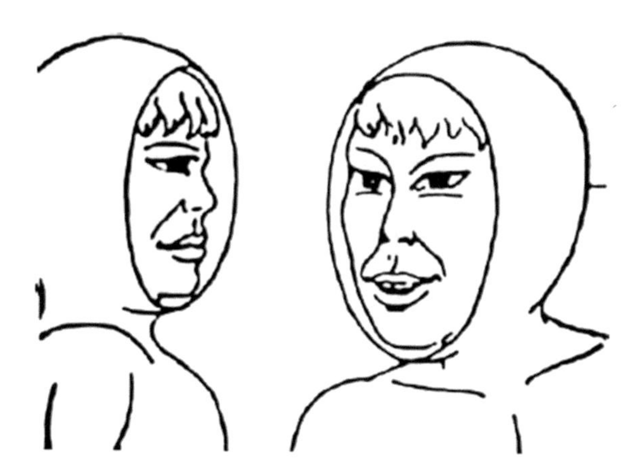

Figure 13.2 This image, which appeared in *Temps X*, a television show featuring Dewilde broadcast by Télévision Française 1 (TF1) on 2 January 1981, appears to be based on Marc Thirouin's characterization of the 'Asiatic' extraterrestrials Marius Dewilde encountered in Quarouble in 1954.
Source: Courtesy of Pathé Actualités Film.

It is impossible to conclude, based on this description alone, that Dewilde himself was in any way concerned about France's recent political problems in Europe or its empire, including the very recent loss of Indochina, rumblings of resistance in Tunisia and Morocco, and the first signs of the problems to come in Algeria. Nevertheless, Marc Thirouin had worked closely with Paul Le Cour (1871–1954), an occultist and the founder of *Atlantis*, a journal of esotericism.[27] Described by a later *Ouranos* publication as a man of 'strong erudition' who blended an interest in the Atlanticist esotericism of Le Cour with a fascination for an 'otherness (*ailleurs*), strange to our world,' Thirouin was certainly aware of the theories of civilizational hierarchy and human origins associated with the works of Annie Besant (1847–1933) and Helena Blavatsky (1831–91), a co-founder of Theosophy.[28] Like Michel and Vallée, Thirouin was convinced that explaining mid-century UFO sightings ranked among the most pressing and momentous challenges humankind had ever faced. Nevertheless, Thirouin was ultimately more interested in verifying Dewilde's evidence and affirming his character than in puzzling over what he admitted was, at best, perplexing behavior on the part of Quarouble's extraterrestrial visitors. In fact, in neither article did Thirouin advance a theory to explain why an amiable crew of humanoids chose twice in the period of a month to pilot its craft onto the abandoned railroad tracks near a metalworker's yard, all the while making no demands, carrying out no apparent mission and expressing no curiosity about the lifestyle or habits of their rural host.[29]

II Roger-Luc Mary interprets Dewilde's contact

It was not until 1980 that Dewilde, who had since 1954 spent countless hours discussing the case with parapsychologists, including Roger-Luc Mary (1930–2002), the co-author of his biography, and UFO aficionados drawn to the occult, would provide any assessment of his visitors' intentions.[30] Dewilde, who now claimed to have had regular contact with the aliens over the years, described their aims as peaceful, even benevolent. In his book, he characterized the beings he came to know as 'good folks' (*gens affables*), beings who were not 'invaders but pacific creatures.'[31] For Dewilde the small, humanoid creatures possessed superior technology and knowledge, but their aims were munificent, as they promised to share their wisdom with humans when the time was ripe. Mary used Dewilde's biography as a call to arms, urging his readers in bold capital letters not to resist the extraterrestrials, who 'possess an enormous science and intend to provide it to those who are worthy of it.' Moreover, he continued, 'in addition to love, they possess wisdom and an immense notion of justice.'[32] In essence, then, Dewilde, with the clear input of Mary, suggested these essentially godlike humanoids were carrying out a new 'civilizing mission,' one where disinterested conquerors delivered technology and compassion from afar as a means of helping humans 'avoid a terrible cataclysm, a cataclysm that men have given birth to and prepared without their knowledge.'[33]

At the same time, Dewilde's book registered a litany of fears regarding the transformation underway in France and Europe. Over the course of some 250 pages that weave together personal testimony, references to subsequent UFO sightings, and Mary's knowledge of esotericism, popular science, parapsychology and popular culture, several clear themes emerge. In fact, the book reveals a profound ambivalence about technological change, the human capacity for violence, the expanding role of the postwar state, and the problems of conquest.[34] In one of the most striking examples of these fears, Dewilde accused the state and anonymous 'European bodies' of kidnapping, imprisoning and torturing him to learn what he knew about the aliens and their advanced technology.[35]

This addition to the story begins with a mysterious object the metalworker claimed to have obtained during the visit of 10 September 1954.[36] Before Dewilde could determine the box's function, however, agents of the French state rushed in and whisked him away, in a manner he likened to the 'dark hours of the occupation, when the Gestapo could emerge from nowhere.'[37] Mary describes how French authorities took Dewilde to an undisclosed hospital site where he was submitted to interrogation tactics shading toward torture. While in custody, Dewilde was, he claimed, injected with truth serum, subjected to electroshock therapy, and held for long periods of time against his will. The scene he describes, in fact, is reminiscent of torture scenes from the occupation or Algerian war.[38] According to Dewilde: 'Sitting, my head, chest, wrists and feet bleeding, I found myself alone, abandoned to myself [...] an intense jolt shot across my body. I bucked in my chair, breaking a strap before I blacked out in total darkness.'[39] At some point during the interrogation Dewilde revealed the presence of the box, thus allowing his captors to seize it and render him powerless to explain its significance or prove its existence.

What also emerged in his biography was a growing disenfranchisement stemming from repeated and troubled contacts with a world where technology and esoteric knowledge simultaneously conferred status and produced isolation. As would Michel and Vallée, Mary expressed ambivalence about materialist solutions to civilizational struggles. Ultimately, Mary implied that the knowledge Dewilde gained was both empowering and devastating; for he found himself increasingly caught between the pincers of authorities in France and those of an alien entity determined to communicate through him. He knew the vast transformative potential of the aliens' technology, which he had seen firsthand as a privileged visitor to their world; yet his inability to embrace the chasteness of their customs, and the unwillingness of the French government to come clean with its findings, meant that this potential might never be realized. In fact, Dewilde speculated, it was his personal failure and the frustration it generated among the extraterrestrials that were somehow responsible for three tragedies that beset him in 1978, including the premature death of his brother and second wife, and the loss of his own right arm, which, he noted, was amputated due to unrelated physical problems.[40]

III Aimé Michel, ufology and benevolent empire

If Dewilde, channeled by Mary, spoke only belatedly about the relationship between alien encounters and the transformation of France and the world after 1945, the same is not true for others who felt compelled to explain experiences like his. Aimé Michel, for example, had already concluded by the 1950s that the challenges facing France, Europe and the world were daunting. In the introduction to the 1954 edition of *Lueurs sur les soucoupes volantes*, Michel contended that, in spite of an almost 'universal blindness, especially in France,' there was no more daunting challenge in the middle of the twentieth century 'for the destiny of humanity' than determining whether flying saucers existed. 'Because,' he concluded, 'if it is true that craft from another world are haunting our skies (*atmosphère*), it is certainly the future of our planet that is at stake.'[41]

The timing of the rash of UFO sightings around the world during a moment of political turmoil intrigued Michel from the beginning. Dismissing the possibility that human science and technology could explain away a wave of sightings so similar in detail, arguing that neither the USSR nor the United States could have produced such vehicles and not used them against the Nazis or kept them secret, Michel concluded that the phenomena were necessarily extraterrestrial in origin. If this were the case, he asked, why was there no long historical record of such encounters? Why were they appearing only then?

Michel, like his colleague Jacques Vallée, assumed that contacts had in all likelihood existed since the emergence of life on earth. Before the modern age, Michel speculated, these contacts would have been understood differently, viewed as visitations from the gods or ghosts. The absence of aliens, then, was due to a failure of the human imagination and intellect alone. How could it be otherwise, he asked, 'if it only requires a million years for the first intelligent race to appear in the galaxy to conquer it completely, how can we escape the conclusion that at least one of these races having appeared near a star very much older than the sun would have already occupied the entire sky for hundreds of millions of centuries?'[42] Why, then, he wondered, were they reappearing so cautiously and irregularly?

Although part of this hesitation could be explained, to his mind, by an inhospitable climate and atmosphere on earth, it was not environment alone that accounted for their caution. Just as important, Michel posited, was the history of human interactions with each other and their environment. Contemptuously dismissing Byron Haskin's 1953 film *War of the Worlds* and its terror-invoking scenario, he reasoned that even if the existence of UFOs was only probable, the civility of the ships' crews, their reluctance to make widespread and sustained contact with humans or meddle in human affairs, was beyond doubt. Michel speculated that the reason UFOs had limited their contacts, in spite of their ample opportunity and means, was obvious: 'fear.' 'Suppose,' Michel reasoned, 'that [...] courteous and respectful beings have kept watch over (*surveillé*) our history for centuries, or even just for several decades: aren't they right to think, given the bloody past of humanity, that their best safeguard is an iron curtain?'[43] Invoking in a single sentence the horrors of Europe's recent past and the threats of its present in a politically polarized world, Michel's proposed explanation for the erratic appearance of these celestial creatures was clearly linked to his concerns about the unabated human propensity for violence and subjugation.

If Michel was unable to explain to his complete satisfaction why the visitors were arriving intermittently and why they chose to appear randomly in rural areas of France, he had no doubt that they had good reason to be cautious and stood to gain little. Given human history and the technological 'backwardness' it demonstrated in comparison to alien spaceships, Michel argued, direct extraterrestrial contact with humans would 'diminish [...] their margin of superiority over us' and teach them nothing. At the same time, he asked, would they have any guarantee, in face of the human proclivity for conquest, that humans would utilize the aliens' technology against them 'with the same discretion?'[44] If he believed that these life forms represented a higher form of consciousness, and brought with them an advanced science and more highly evolved moral precepts, Michel assumed that the impulse behind their expansion was a familiar one. For Michel, expansion was a fundamental drive governing civilizations, wherever they arose, and he supposed that the very arrival of alien life forms was evidence of this fact.

Michel pointed out that recent work in astronomy suggested that if extraterrestrial life existed – and he believed there was a high probability it did – then, given the relative age of our solar system, chances were high these visitors had been present from the beginning of human history. And the reason they came, he averred, was directly related to another statistical probability: all civilizations face mounting pressures for expansion due to a combination of scientific enlightenment, technological development, resource exhaustion and solar decline. It was only logical, Michel reasoned, that if 'the ineluctable evolution of stars, which end by exploding into supernovas, exterminates all life that does not manage to get away, [...] all intelligent life in the universe is obliged by the force of things to abandon its original planet to undertake galactic colonization.' If this was the case, he concluded, 'the galaxy is thus entirely explored, if not colonized for millennia.'[45]

For Michel, the presence of this silent force represented the future of human civilization, once humans were prepared to imagine civilizational contact and expansion without violence and domination. 'Why,' he asked again, 'do they remain silent?' In his view, the reason was again best explained as a shortcoming of the human imagination; for man, he believed, was like the proverbial mouse chewing a book without digesting its message, a being insufficiently evolved to discern meaning right under its nose. In order to see these beings and understand their motivations, to be worthy of their acknowledgement, 'perhaps [...] we must transform ourselves, renounce the old earthly law of violence that makes our species and always guides us.'[46]

If human – and especially French – science was blinded by its refusal to embrace alternative paradigms and eschew rigid institutional orthodoxies, Michel believed, the technology it gave rise to also threatened the richest of its civilizational accomplishments and its ability to think differently. In

1962, Michel, who had grown up in a small mountain village and chosen to remain there throughout much of his life, wrote an essay entitled 'La Fin de la civilisation villageoise' for the journal *Planète*, in which he rued the passing of an age-old world view in the countryside under the assault of technological modernization – 'trains, the radio, television, newspapers, economic and migratory circuits.'[47] Michel opined that modernization had swept aside the ancestral past, so central to the peasant's understanding of life, and left him or her alone to interpret change at a moment when 'the evolution of technology, accelerating without end, will demand more and more of him.'[48] More importantly, though, this process marked the destruction of another way of thinking and being, where 'village time is not constructed according to Cartesian coordinates,' but rather adheres to a flattened, non-linear logic of '*dans le temps*.'[49] If such logic seemed constrictive to the sensibilities of a society shaped by modern science, Michel seemed to suggest, it offered an example of how one could think differently and see what rationalist science could not.[50] In his view, contemporary scientists, obliged by the dogmatism of their methods, would choose to look away from unexplainable phenomena rather than 'spend their time listening to the account of an illiterate peasant who believed he had seen something in the sky.'[51]

To be sure, in the end Michel was not convinced that either scientists with their world view, or peasants with theirs, were any more equipped to understand what was happening before their eyes than a dog, because of evolutionary hindrances, could ever understand the reasoning and consciousness of its human master. Thus, whereas Jacques Vallée, his colleague and admirer, concluded that studies of mystical responses to these phenomena, of paranormal explanations, might yield a more satisfactory understanding of what was happening, Michel was unwilling to draw that conclusion.[52] What he did believe, in the end, was that France, at a time when its very fate seemed challenged, 'is the smartest in the world at the basic level of the average plumber, the farmer, the mailman,' even if it had 'managed to acquire the most pretentious, cynical and rotten elite one can imagine.'[53] Aimé Michel, in other words, was, in ways similar to Dewilde, a man looking toward the skies over the land where his roots were so tenuously embedded, in a bid to overcome his own alienation, but unlike Dewilde, also as a way of coming to terms with the long history of violent European conquest.

IV Jacques Vallée, UFOs and the difficulty of imagining a world without empire

No French ufologist has written more about the phenomenon than Jacques Vallée. Like Dewilde, Vallée had seen a UFO and led a peripatetic life.[54] Like Michel, his friend, Vallée expressed an acute disdain for the study of science as he saw it practiced around him and was deeply troubled by the international politics of the twentieth century. Early in his career Vallée avowed to a friend that, despite his disenchantment, he would continue his study of science, but 'with the knowledge that an appreciation for art, fantasy and sensitivity is not a "negative trait" that I ought to suppress within myself.'[55] Vallée found institutional science, especially French science, stultifying and believed it prevented creative thinking and significant advancement, which only an infusion of ideas from other disciplines could overcome. Shortly after sharing these thoughts with his friend, Vallée happened upon a copy of Aimé Michel's *Mystérieux objets célestes* and recorded his reactions in his journal. Michel's book, Vallée wrote, 'challenged the very depths of my mind.' Vallée believed that for those 'rooted in the ordinary world' the extraordinary ideas of men like Michel about UFOs, who had 'found the immense contour of other shapes, other civilizations,' perhaps did not matter. But he was so taken with the possibilities offered by this new research that he '*passionately want[ed] them to wait for [him]*.'[56]

Like Michel, Vallée was deeply ambivalent about the social and economic transformation of France in the postwar period, simultaneously believing it necessary and unfortunate. On the one hand, Vallée was repelled by what he viewed as the institutional inertia of French science and management, even as he admired what he saw as a similar trait in rural life. On the other hand, he was drawn to American science, which he believed was more dynamic than its French counterpart, even as he was greatly dissatisfied by what he viewed as the shallowness of America, 'this greedy Freudian continent.'[57] As a result of his feeling of marginalization, Vallée bounced back and forth between the United States and France, content with neither, before finally settling in California at the end of the 1960s. But before he established himself in California, he decided, in part because of his profound disappointment in 1967 with official American efforts to take the UFO problem seriously, to pack up and return to France for what would turn out to be the last time.

After having lived in Chicago for nearly five years, Vallée pronounced himself torn between an opportunity to move near the beaches and mountains of southern California, where he could 'take part fully in the novel, exciting frontiers of our creative time,' and the opportunity to savor the 'good ancient books, rain falling over narrow cobblestone-paved streets,' he identified with France. Vallée worried that 'real cataclysms are coming to Europe, to sweep away the sandcastles people have erected to protect their privileges.' In the end, as he prepared to return to France, he seemed ambivalent about these changes, seeing in the collapse of the old 'a genuine deep current that carries the world towards its future, from the convulsions of Mao's China to secret California where everything is crashing through old barriers.'[58]

If Vallée eventually chose to settle in the United States for the professional and material advantages it offered, his decision was also influenced significantly by his profound disillusionment with recent French and European politics. For ufologists with political sympathies to the left, men like Michel and Vallée, their interest in UFOs was shaped by an acute concern about the asymmetrical and often violent social arrangements and civilizational contacts defining recent human history. And while evidence of this is less direct in the published writings of Michel, such concerns enjoy a prominent place in the memoirs of Vallée.

In the 1950s, Vallée was clearly alarmed by the events unfolding in France in particular and the West in general. Although his journals begin in 1958, when he was 19, his concern for what was happening both at home and in Algeria was obvious. In 1958 Vallée looked on in horror as the situation seemed to be spiraling out of control. In just the fifth entry in his diary, he expressed outrage over 'the lies spread by the bureaucracy, the censorship, the denial of tortures committed in North Africa by the French army.' He worried that France was engaged in 'the same kinds of actions that we were taught would forever designate the German Nazis to the shame of the whole world.'[59]

Another incident related to the situation in Algeria, the February 1962 protest at the Charonne subway station, which ended in police violence and the death of nine, produced one of the strongest reactions Vallée had to any event of the 1950s and 1960s. Responding to the incident, Vallée wrote that there was in France 'a palpable, terrible sense of despair, a feeling that reminds me of the dark days of the German occupation.' Vallée, who pronounced himself 'fed up with such violence, fed up with this country and its absurd political intrigues,' asked, how could anybody 'ever build anything of value here?'[60] Even when he looked back on the incident from the perspective of two years and his new life in Chicago, Vallée remained scarred by the incident. 'For me,' he wrote, 'the recent history of France remains summarized in the Charonne massacre: what I saw there after the police riot will always be in my memory.'[61]

By the end of the 1960s, with the Vietnam War looming in the headlines, Vallée worried that his adoptive country was heading down a similar path. He described numerous instances where friends and colleagues, dismissive of France's bid to remain in Indochina, treated his warnings with 'quiet arrogance,' arguing 'you can't compare what happened to the French with what would happen if we went in there.'[62] He later lamented that 'our American friends, even the most enlightened, look at me with some suspicion when I tell them that a war like Vietnam will do more damage to this country than it will to Indochina, no matter who wins in the end.'[63]

Vallée, who left Europe for America because he believed the ruling elite in France stifled innovation and blocked the paths of non-conformers, found other disturbing problems in American society. Certain aspects of life in America reminded him of a 'foreign planet,' like the one he witnessed from the window of his room on the twentieth floor of a hotel room in Philadelphia in 1969. Describing one of the aspects of American society he found most troubling, Vallée noted in his daily journal that much was wrong in a world that 'keeps the blacks among blacks, the whites among whites, the poor among poor, and cops at all the crossroads.' He cautioned that if 'no deeper current comes to provide new vision,' the United States would face a future marked by permanent stratification, where 'people will be locked and entertained within their own little spheres.' Given the history of the previous decades, Vallée asked, 'who could say that this depressing future would be any worse than our past, the odious history of mankind?'[64]

Whether criticizing social injustices in France or the United States or each nation's colonial missteps in the postwar era, Vallée's memoirs suggest a clear link between his early quest to understand UFOs and his critique of the reconstitution of various forms of political, especially imperial, power in the late 1950s and 1960s. Interspersed with his condemnations of French science and the war in Algeria are musings about the meaning of what he believed were increasing human contacts with extraterrestrials. In describing a letter he had written to Aimé Michel on 9 September 1958, Vallée mused that contact with the aliens was far from certain, since, 'in order to educate us they would have to find us worthy of a dialogue with them.'[65] Six years later, writing in Chicago, he returned to this question, wondering whether the return of UFOs could be explained 'by the need for some "unknown superiors" to boost our religious vaccinations.' Perhaps, he continued, some 'super-scientific group of cosmic origin, considering mankind as its own creation and seeking to experiment on us, or to guide us benignly towards galactic status, might behave as the saucer operators do.'[66]

When he found time away from a busy career as a computer scientist and active ufologist, Vallée managed to write novels, too. In his memoirs he discusses the motivation for the plot of one of his novels, *Dark Satellite*. The story is set on Venus in the twenty-first century, where society is organized according to the principles of a system he calls 'Peripherism.' In a world governed by the principles of Peripherism, he notes, 'most of Europe [...] would break up into areas similar to old Provinces, with autonomy for well-defined cultural, linguistic and economic units which would receive key services from the larger world.' Rendering explicit the relevance of this imaginary alien world for his own times, Valleé continues, 'breaking down the major countries into their local components it would be much easier to defuse cultural antagonisms, to force people to be responsible for their own destiny, and to create an enormous federation serving everyone's interests.'[67]

Even by the end of the 1960s, before it was reflected in his published writings, Vallée's optimism about the promise of these contacts had dimmed significantly. Although he continued to believe that 'we are, as a society, developing a great thirst for contact with superior minds that will provide guidance for our poor, harassed, hectic planet,' he had concluded that what was happening, though still very real physical phenomena, were not 'successive waves of visitations from space,' but rather, 'part of a control system.'[68] Vallée's alienation in two countries, his frustration about the opacity and neglect of government UFO research, and his sense that human beings seemed incapable of living in peace, led him to reject Michel's faith that '[w]e are witnessing the first example of contact between two societies in which the inferior group, namely mankind, will not have died as a result.'

For Vallée, Michel's determination to 'break down scientific and official opposition to their reality' meant that he was 'in danger of becoming an evangelist for the Saucer cause rather than an analyst trying to find the truth.'[69] Vallée worried that his friend seemed 'to be hoping for the ultimate landing, a grandiose and apocalyptic manifestation that will prove to the world that he was right.'[70]

It was perhaps Vallée's reaction to the first moon landing that captured his increasing pessimism. He once believed, he wrote near the conclusion of his memoirs, that landing on the moon 'could only be the culminating achievement of a supremely enlightened scientific culture, something that would not be possible until men became immensely wise.' For him, the act itself was 'a signal of cultural eminence, an irreversible break with the ancient world of wars and human misery.' What he feared instead, in the present environment, was 'the entire sky may soon become filled in every direction with spy satellites, flying bombs, orbital barracks and the cosmic latrines of the new secret armies.'[71] What Vallée read in the skies by the end of the 1960s had been profoundly shaped by France's – and his – changing place in the world.

V Failing to conquer the final frontier

In many ways 1954 was a pivotal year in the re-imagining – and even the remake – of imperial space in postwar France. In the realm of national politics, France's geographical boundaries and the limits of its sovereignty were increasingly called into question. Perhaps it is coincidental that a spike in UFO sightings in 1954 occurred at a moment when French social and political space were both undergoing this rapid and unpredictable transformation. For C.G. Jung (1875–1961), the succession of close encounters was occurring because of 'the threatening situation of the world today, when people are beginning to see that everything is at stake.' To his mind, as humans struggled to construct new myths built upon archetypes consistent with a technological age, one became disinclined to accept metaphysical, non-material explanations for incomprehensible occurrences, 'the projection-creating fantasy soars beyond the realm of the earthly organizations and powers into the heavens [...] where the rulers of human fate, the gods, once had their abode in the planets.'[72] Jung also argued that 'Man's living space is, in fact, continually shrinking and for many races the optimum has long been exceeded.' And man, Jung concluded, when thrown together and forced to create meaningful existences under unfamiliar and threatening circumstances, often 'looks for help from extraterrestrial sources since it cannot be found on earth.'[73]

Marius Dewilde, increasingly reclusive in his later years, claimed to have looked into the skies and found that help. Dewilde, who became a latter-day prophet in the hands of Roger-Luc Mary, attested to having seen a world where biology, physics, philosophy and love had conquered what most humans believed was a universal biological impulse: colonization and conquest to counteract death and decay. By the end of his life, an increasingly delusional Dewilde feared that his demise would come not at the hands of extraterrestrials, whom he knew to be benevolent, but through the treachery of shadowy authorities or 'Europeans,' an amorphous group who despised him for telling the truth about the nature of the cosmos and revealing that there was a civilization perhaps thousands of years ahead of their own.

In spite of the increasingly fantastic tale Dewilde told, his initial experience – and the similar experiences of hundreds of others in 1954 and after – suggested to some observers that, hovering on the horizon of a country adjusting to the challenges of technocratic modernization and the reorganization of its empire in a new bi-polar order, was an advanced alien civilization monitoring and perhaps even guiding human affairs. In the views of some of these men, a new sort of civilizational contact, one transcending the long, tempestuous history of imperial relations defining civilizational contacts on earth, was perhaps in the making. Aimé Michel, who feared the increasingly potent capacity of human violence and rued what he saw as the leveling effects of technology, seems to have held onto this hope until the end of his life. For Jacques Vallée, however, who had initially shared Michel's optimism about the arrival of UFOs, clinging to this hope was dangerous. Though he had once believed that these sightings were most likely of crafts piloted by the advanced life forms he so emphatically hoped to encounter as a young man, by the 1970s and 1980s, even as he continued his passionate research of the phenomena, his understanding of what was occurring grew darker. In *Messengers of Deception*, one of many of his books to express this changing belief, Vallée concluded that 'Manipulators,' perhaps human, perhaps something else, were behind the appearance of unexplainable phenomena in the earth's skies. Suspecting that the source was in fact terrestrial in origin, likely linked to an unidentified intelligence community or secret society that sought to spark fear and curiosity among the general population and skepticism among scientists, Vallée posited that these activities, if ignored, could produce a collapse of modern civilization. Vallée contended that the rise of contactee religions, together with an erosion of popular faith in science, could produce a collapse on a par with the fall of the Greco-Roman civilization. In the best case scenario, Vallée averred, the 'increased attention given to UFO activity promotes the concept of political unification of the planet.' In the worst case scenario, Vallée foresaw the erosion of humanity's faith in an increasingly sclerotic science and the concomitant rise of contactee philosophies, which 'often include belief in higher races and in totalitarian systems that would eliminate democracy.' No longer certain of what lay at the last frontier, Vallée concluded: 'One would like to know more, then, about the image of humanity such Manipulators harbor in their own minds – and in their hearts. Assuming, of course, that they do have hearts.'[74]

Notes

1. Although there are some minor deviations, this is the standard narrative provided by Dewilde and those who interviewed him. See Marc Thirouin, 'Objets volants nonidentifiés: Marius Dewilde n'a pas menti: Des Ouraniens lui ont parlé,' *Ouranos* 24.3 (1959), 11–13; and Aimé Michel, *Flying Saucers and the Straight-Line Mystery*, New York: Criterion Books, 1958, 44–7.
2. Dewilde makes these biographical claims in his 1980 book, *Ne résistez pas aux extraterrestres! Le 'Contacté' de Quarouble 26 ans après*, Monaco: Editions du Rocher, co-written with Roger-Luc Mary. Claude Gaudeau and Jean-Louis Gouzien, who interviewed Dewilde and those who knew him, dispute the veracity of much of Dewilde's biography in *Etude d'une population de témoins: Etude de la propagation de l'information, vague de 1954*, Tours: IRAME (Institut de Recherches et d'Applications de Méthodes Psycho-Educatives), 1981.
3. For a discussion of the development of French ufology, see Pierre Lagrange, 'Close Encounters of the French Kind: The Saucerian Construction of "Contacts" and the Controversy over Its Reality in France,' in Diana G. Tumminia, ed., *Alien Worlds: Social and Religious Dimensions of Extraterrestrial Contact*, Syracuse: Syracuse University Press, 2007, 153–90.
4. A substantial collection of contemporary newspaper clippings describing Dewilde's encounter can be found at http://www.ufologie.net/1954/10sep1954quaroublef.htm (accessed 1 October 2011).
5. Aimé Michel describes Dewilde's physical and mental anxiety in *Mystérieux objets célestes*, Paris: Arthaud, 1958. The quote is taken from the English edition of the book, *Flying Saucers and the Straight-Line Mystery*, 67. Marc Thirouin, a ufologist who founded the journal *Ouranos*, concluded after living with Dewilde for several weeks that his story was essentially reliable; see Thirouin, 'Objets volants non-identifiés.'
6. For skeptical reactions to Dewilde's story, see the report compiled by Claude Gaudeau and Jean-Louis Gouzien in *Etude d'une population de témoins*; and numerous articles by Jacques Bonabot published in *Le Bulletin du GESAG* during the 1980s.
7. An example of the more sensational reporting can be found at http://www.ufologie.net/1954/10sep1954quaroublef.htm (accessed 1 October 2011).
8. Swiss psychiatrist Carl Gustav Jung was among the first to suggest these links; see his *Flying Saucers: A Modern Myth of Things Seen in the Skies*, Princeton: Princeton University Press, 1978, 22 (first published as *Ein moderner Mythus: Von Dingen, die am Himmel gesehen werden*, Zurich: Rascher, 1958).
9. Christopher Roth, 'Ufology as Anthropology: Race, Extraterrestrials, and the Occult,' in Debbora Battaglia, ed., *E.T. Culture: Anthropology in Outerspaces*, Durham: Duke University Press, 2005, 38–93, here 71.
10. In fact, it was a number of related observations occurring on that very evening that rendered Dewilde's case so compelling. There were independent reports from other observers corroborating the existence of an unidentified flash of light over the Department of the Nord around 10:30 p.m., when Dewilde's story begins, and a farmer in Mazoud working earlier that evening in his field reported making contact with another unidentified being and his craft. On this point, see Gaudeau and Gouzien, *Etude d'une population de témoins*, 11; and Michel, *Mystérieux objets célestes*, who come to similar conclusions about the material 'facts' in the case.

11. Michel makes this claim in *The Truth About Flying Saucers*, New York: Criterion Books, 1956, 6. The book was originally published as *Lueurs sur les soucoupes volantes*, Paris: Maison Mame, 1954. Unless there is a clear difference in the two editions, subsequent citations will come from the translation.
12. Michel uses these terms in reference to Dewilde's story on pages 222 and 224 of *Flying Saucers and the Straight-Line Mystery*.
13. Idem, *Truth About Flying Saucers*, 8.
14. Vallée famously served as a consultant for *Close Encounters of the Third Kind*, directed and produced by Steven Spielberg, USA 1977 (Columbia Pictures).
15. Jacques Vallée, *Forbidden Science: Journals 1957–1969*, Berkeley: North Atlantic Books, 1992, 15. Vallée also attributed his interest in UFOs to the indifference of his colleagues and establishment scientists to the anomalies they detected in satellite tracking; see an undated interview entitled 'Heretic among Heretics: Jacques Vallée Interview,' http://www.ufoevidence.org/documents/doc839.htm (accessed 1 October 2017).
16. This is the crux of his argument in *Passport to Magonia: From Folklore to Flying Saucers*, Chicago: Henry Regnery, 1969. In another instance, Vallée posited that the beings, exploiting scientific discoveries still inconceivable to humans, were passing from one dimension to another and not traveling through space; see *Dimensions: A Casebook of Alien Contact*, New York: Anomalist Books, 1988, 256.
17. Vallée's opinion changed dramatically over time. In *Messengers of Deception: UFO Contacts and Cults*, Berkeley: And/Or Press, 1979, Vallée argues that some unknown force or group, probably terrestrial in origin, was engaged in a bid to control humanity and undermine its faith in science by producing or exploiting a series of real but unexplainable phenomena.
18. Idem, *Passport to Magonia*, 163.
19. A number of scholars have established the connection between the reaction to postwar technologies, fears of imperialism or imperial collapse, and the construction of French social space after the war. See Richard Kuisel, *Seducing the French: The Dilemma of Americanization*, Berkeley: University of California Press, 1993; Kristin Ross, *Fast Cars, Clean Bodies: Decolonization and the Reordering of French Culture*, Cambridge, MA: MIT Press, 1995; Rosemary Wakeman, *Modernizing the Provincial City: Toulouse 1945–1975*, Cambridge, MA: Harvard University Press, 1998; and Gabrielle Hecht, *The Radiance of France: Nuclear Power and National Identity after World War II*, Cambridge, MA: MIT Press, 1998.
20. See Gaudeau and Gouzien, *Etude d'une population de témoins*.
21. Ibid., 15.
22. For examples, see ibid., 2–10.
23. Gaudeau and Gouzien establish there was no electricity, let alone a television, in the home Dewilde lived in at the time.
24. Dewilde attests to his early political naïveté as a panelist during an episode of 'Aujourd'hui Magazine' in November 1977.
25. Ibid.
26. Thirouin felt compelled to include a disclaimer in his footnote, indicating that these were Dewilde's words, and that of course 'the tint [of an Arab's] skin varied a lot from one individual to another, depending on his racial purity.' See Marc Thirouin, 'Marius Dewilde n'a pas menti,' *Ouranos* 25 (1959), 20–4, here 20 in part II.

27. Le Cour was associated with certain strains of antisemitic, Catholic nationalism during the war. In his *Recherche d'un monde perdu: L'Atlantide et ses traditions*, Paris: Leymarie, 1926, Le Cour discusses the importance of recovering the traditions and knowledge of Atlantis as a way of rebuilding a European world profoundly shaken by 'machinism and the simultaneous loss of the meaning of life,' here 55–7. Echoes of certain of his ideas can be found in the writings of Mary, Michel and Vallée on the UFO phenomenon. For an alternative interpretation of this subject, see Pierre Lagrange, 'Les Controverses sur l'Atlantide (1925–1940): L'archéologie entre vraie et fausse science,' in Claudie Voisenat, ed., *Imaginaires archéologiques*, Paris: Editions de la Maison des sciences de l'homme, 2008, 233–64.
28. For a discussion of the role of race in an American permutation of this group, including the link between the lost continent of Atlantis and the Mongols, see Roth, 'Ufology as Anthropology,' 44–7.
29. Thirouin records this investigation in successive issues of *Ouranos*. See Thirouin, 'Marius Dewilde n'a pas menti,' *Ouranos* 25 (1959), 20–4 for part II, and *Ouranos* 24 (1959), 11–13 for part I.
30. Many of the details of Dewilde's biography seem to mirror those of his co-author, Mary, who was conscripted (and imprisoned, for unspecified charges) during the 1950s before being drawn into ufology. Biographical information about Mary's life can be found at a web site created and maintained by his daughter, see http://rogerlucmary.e-monsite.com/accueil.html (accessed 1 October 2011). Ufologist Jacques Bonabot, among those observers most critical of Dewilde's changing story, describes the influence of Mary on Dewilde in 'Dossier Quarouble 1954,' *Bulletin du GESAG* 82 (1985), 2–7. Gaudeau and Gouzien, 'Etude d'une poplulation de témoins,' also confirm the importance of Mary in Section II. 3.2.6, where they conclude, based on the metalworker's general lack of familiarity with his own biography, that the vivid details of the book were a product of Mary's imagination.
31. Dewilde, *Ne résistez pas aux extra-terrestres*, 67.
32. Ibid., 44. Mary refers specifically to the unleashing of the atomic bombs over Japan as part of this impetus.
33. Ibid., 121.
34. In his study of the racial discourses prominent in shaping the work of certain strains of American ufology, Christopher Roth argues that in many contact stories 'older scientific and theological paradigms [...] mix and combine with images from popular culture and the assumptions and belief structures that make up the world view – including especially racial conceptions – that UFO enthusiasts share with their fellow citizens'; see Roth, 'Ufology as Anthropology,' 40.
35. Although Mary does not discuss any 'European' role in the affair, Dewilde told Gaudeau and Gouzien he feared he 'would be killed by [certain] European organizations because I am still a danger on this earth for a lot of people'; 'Etude d'une poplulation de témoins,' III. 3.1. He does not specify why or who these people were or to what institutions they belonged; nor does he explain why these institutions were 'European' and not French. For various explanations of the emergence of abduction stories in the postwar era, see the contributions in Tumminia, *Alien Worlds*.
36. Although Dewilde did not speak about being submitted to harsh interrogation by French authorities until the 1970s, a special issue of *Ouranos* claimed that

Thirouin was aware of these events earlier; see Thirouin, 'Ouranos aux frontiers de la connaissance,' 108.

37. Dewilde, *Ne résistez pas aux extra-terrestres*, 66.
38. Roger-Luc Mary was born in Algeria and would certainly have been familiar with events there during the most violent years of French rule. Kristin Ross has suggested in *Fast Cars, Clean Bodies* the extent to which torture carried out in Algeria had entered the popular imagination in the 1950s and 1960s as part of the French state's broader modernization project. See idem, *Fast Cars, Clean Bodies: Decolonization and the Reordering of French Culture*, Cambridge, MA: MIT Press, 1995.
39. Dewilde, *Ne résistez pas aux extra-terrestres*, 70.
40. Ibid., 116–17. In his book, Dewilde does not suggest that the aliens were punishing him for his shortcomings. The opposite is true in Gaudeau and Gouzien, *Etude d'une population de témoins*, III. 3.2.
41. Michel, *Lueurs sur les soucoupes volantes*, 9.
42. Idem, 'Les Probabilités d'une vie universelle,' *Question De* 22 (1978), 85–91, here 86.
43. Idem, *Lueurs sur les soucoupes volantes*, 267.
44. Ibid.
45. Michel, 'Les Probabilités,' 90.
46. Ibid., 331.
47. Aimé Michel, 'La Fin de la civilisation villageoise,' *Planète* 7 (1962), 13–23, here 13.
48. Ibid., 23.
49. Ibid., 19.
50. For Michel's discussion of what he understood to be the shortcomings and blindspots of modern science, see his 'Les Tribulations d'un chercheur parallèle,' *Planète* 20 (1965), 31–40.
51. Ibid., 36.
52. Vallée notes this difference of opinion in *Forbidden Science*, 422.
53. Ibid., 371.
54. Ibid., 15.
55. Ibid., 11.
56. Ibid., 14 (emphasis in original). In this case the 'them' denotes the extraterrestrials.
57. Ibid., 380. In another passage of his memoirs that captures his alienation in America, Vallée asks: 'How many Americans my age could understand what that era [the Second World War] meant? […] How many […] have ever seen a horseshoe-maker at work, a water mill in operation, a woman using a washboard in the river?'; see ibid., 411. For studies of the history of French perceptions of America, see Kuisel, *Seducing the French*; and Philippe Roger, *L'Ennemi américain: Généalogie de l'antiaméricanisme français*, Paris: Editions du Seuil, 2002.
58. Ibid., 278–9.
59. Vallée, *Forbidden Science*, 6.
60. Ibid., 53–4.
61. Ibid., 83. Vallée, who was normally more measured in his private musings, went so far as to conclude '[t]he worst hours of Stalin, the worst crimes of the Nazis had a name: Tyranny. The killings at Charonne have no name.'
62. Ibid., 65.
63. Ibid., 310.
64. Ibid., 389.

65. Ibid., 17.
66. Ibid., 80.
67. Ibid., 38–9.
68. Jacques Vallée, *The Invisible College: What a Group of Scientists Has Discovered About UFO Influences on the Human Race*, New York: Dutton, 1975, 195. Vallée first mentioned this concern in 1965, after Aimé Michel suggested it; see Vallée, *Forbidden Science*, 155.
69. Ibid., 321.
70. Ibid., 326.
71. Ibid., 401.
72. Jung, *Flying Saucers*, 14.
73. Ibid., 17.
74. Vallée, *Messengers of Deception*, 217–20.

PART V

Inscribing Outer Space

CHAPTER 14

Self-Reproducing Automata and the Impossibility of SETI

Gonzalo Munévar

The fictionalization of space exploration, from H.G. Wells's *War of the Worlds* to *Star Trek* and beyond, has consistently offered us a vision of a future populated by alien civilizations. The scientific underpinnings for that vision were most notably defended by the American astronomer Carl Sagan (1934–96). That future vision is often supplemented by another: a future in which machines take their place alongside other sentient beings, as famously exemplified in all sorts of novels and films, including *Terminator*, with Austrian-actor-turned-California-Governor Arnold Schwarzenegger in the lead. The scientific underpinnings of that vision owe much to extrapolations of work on computers by the Hungarian mathematician John von Neumann (1903–57). It is not widely realized, however, that von Neumann's alleged proofs for the possibility of self-reproducing automata (SRAs) create a conflict between these two visions of the future; for the application of von Neumann's SRAs to space exploration seems to lend support to a famous objection against the existence of alien civilizations by the Italian physicist Enrico Fermi (1901–54).

This chapter discusses the most critical philosophical and scientific assumptions involved in the proposal to explore space with von Neumann's SRAs. It argues that such a proposal depends on von Neumann's mistaken metaphor of the genome as a computer program, and is thus doomed to fail. It discusses also other suggestions to create exploring SRAs. Although the suggestions are ingenious, they are unlikely to succeed. For the moment at least, Sagan's vision remains to fire our imaginations another day.

Gonzalo Munévar (✉)
Lawrence Technological University, Southfield, MI, USA
e-mail: gmunevar@ltu.edu

Alexander C.T. Geppert (ed.), *Imagining Outer Space*
European Astroculture, vol. 1
https://doi.org/10.1057/978-1-349-95339-4_14

I Sagan and Fermi

Are we alone in the universe? Is it really possible that no sentient being on a faraway planet ever contemplated the stars and felt awe as we do? That only humans ever pondered the nature of the universe, or wondered whether similar beings might be asking similar questions? In the view of Sagan and other scientists it is extremely parochial to suppose that we are alone – one more instance of the syndrome that once made us believe that the earth was the center of the universe.[1] According to those scientists, we have no more justification now to believe that we must be the pinnacle of creation than we once had to believe that the earth was so special.

Thus begins the reasoning that takes them to the conclusion that extraterrestrial intelligence (ETI) is likely to exist, a presupposition without which the search for it (SETI) would make little sense. Ironically, the opposition to SETI is buttressed by the key assumption of the SETI proponents themselves: Carl Sagan's so-called 'Principle of Mediocrity.'[2] The Principle of Mediocrity asserts that the sun is a typical star in having a planet like the earth on which life could arise, that terrestrial life is typical in having produced intelligence, and that human intelligence is typical in giving rise to a technological civilization.

Presumably Copernicus taught us humility when he argued that the earth was not privileged but average, and later astronomy reinforced the lesson by discovering that the sun itself was merely an average star in an average galaxy. By extending the Copernican lesson, the reasoning goes, we should learn to be humble about our own position in the scheme of life. The Principle of Mediocrity thus purports to recognize that humanity and the conditions that have brought it about are average. In their arguments, the opponents of SETI stretch this principle slightly, adding that a technological civilization, should any actually exist, would be typically expansionist. As a result they are able to produce a variety of 'impossibility proofs' against the existence of extraterrestrial intelligence.

If intelligent extraterrestrials do exist, Enrico Fermi once asked, why aren't they here? Some of today's opponents of the search for extraterrestrial intelligence realize that it may be unfeasible, even for advanced civilizations, to travel or to send 'unmanned' probes to the (probably) billions of planets in the galaxy. Their impossibility proof depends instead on a technology they believe is inevitable: self-reproducing machines.[3]

II Objection

A contemporary version of Fermi's argument contends that if the SETI proponents are right, there should be technological civilizations far older, and presumably far more advanced, than ours; many 'typical' stars in our galaxy are billions of years older than the sun, and thus, in some of their planets, intelligence should have sprung long before it did on earth. Now, just as we expanded from our beginnings in Africa, any civilization capable of spaceflight is bound to expand throughout the galaxy. Furthermore, this expansion would take place in a short amount of time: Traveling at one-tenth the speed of light, which is within human reach now, it would take only one million years to cross the galaxy. Thus, the ETIs should be everywhere in the galaxy by now. But they clearly are not here; therefore, there are no extraterrestrials in this galaxy.[4]

III Exploring the galaxy with self-replicating automata?

The opponents of SETI realize that it may be unfeasible, even for advanced civilizations, to send 'unmanned' probes to the billions of planets that may exist in the galaxy. Their impossibility proof depends instead on a technology they believe is inevitable: self-reproducing machines.

John von Neumann supposedly proved that human beings could design a machine capable of making copies of itself.[5] Indeed, NASA scientists have investigated the possibility of using such machines to explore the galaxy.[6] Since we already have a mathematical proof that self-replicating machines can exist, all we need is the talent, effort and money to create the technology. Thus, more advanced civilizations would surely have discovered the equivalent of von Neumann's proof and would have developed the appropriate technology by now. Once these machines arrive at their destinations, they endeavor to make copies of themselves, which then move on to the nearest stars, and so on, setting in motion a geometric progression until they overrun the entire galaxy. But we have no evidence of such machines here; therefore no advanced civilizations exist.

Von Neumann offered not one but five proofs. The first one, however, is the main basis on which all these speculations rest. Von Neumann knew that, through evolution, organisms often produce others more complicated than themselves, and he wondered whether machines could ever do likewise. He then pondered the first step in that direction: Was it possible to program a computer to make a copy of itself? He imagined a robot floating in a vat full of robot parts. The robot could be programmed to pick up a part and identify it. Then the robot, which had a blueprint of itself, would look for the connecting parts, and would then begin putting another robot together the way a child puts together a Mechano set. Once the second robot was assembled, the first would pass on to it the self-replicating program (or set of programs, rather). By breaking down the task of self-replication into small, manageable tasks, von Neumann thought, an automaton could surely copy itself (Figure 14.1). This result led him to remark that there were two kinds of automata: artificial automata, such as computers, and natural automata, such as people and cats. Thus, the implications of this very simple conceptual (rather than mathematical) proof go well beyond the concerns of technology and exploration – they also affect our notion of life.

Figure 14.1 John von Neumann's conceptual proof of the possibility of a self-reproducing automaton: A robot floating in a vat of oil is programmed to assemble parts into a copy of itself; it then passes the program on to the new robot.
Source: Illustration by Nicole Ankeny.

IV Self-reproduction of automata in space

If the European Space Agency were to turn one of von Neumann's presumed SRAs into a starship to explore the galaxy, the SRA would have to stop somewhere to make copies of itself. When an SRA arrives at another world, however, it is not going to land in a vat full of parts. It will have to build factories to build the parts from the raw materials that it will mine. But the factories are themselves made of parts, so it will have to build other factories to build the parts to build the factories. The specialists in the field call this the 'closure problem.' They do not fear an infinite regress, however, for we know that the closure problem can be solved. And we know that it can be solved because our technological civilization solves it: We do send rockets to other worlds.

Such an SRA, however, will be an extremely complex machine, both in its computer programs and in its physical realization. Indeed, to build an SRA, including the starship in which it travels, would demand many of the resources of a technological civilization. Whether *we* can write a program that complex, and whether *we* can assemble a machine that sophisticated is open to question. But let us assume for the sake of argument that we can.

Imagine that one of these extremely complex SRAs arrives at a planetary system. Now, we do not yet know what other planetary systems look like, but some theories and recent observations suggest that they would be collections of relatively small rocky planets and gas giants. If an SRA were to come into *our* solar system, Jupiter and Saturn would not be good places to land, even figuratively, because it is unlikely that the SRA can fashion the needed parts and factories out of the hydrogen, helium and the trace gases that can be found in their atmospheres. The moons of the gas giants are not that much better, for surely a machine that complex may be presumed to need a variety of materials, including metals, for the task of self-replication. Unfortunately, the low density of most of these moons (less than 2 g/cm^3) suggests that they would not be good places to search for the necessary raw materials.[7]

Nevertheless, we know that on rocky planets, like earth, an SRA can find practically everything it needs. But even on rocky planets the SRA's problems are far from over. In astrophysical terms, small differences between the planets may lead to significant differences in density and the chemical composition of the atmosphere. These significant differences will, in turn, make it necessary to adopt different strategies for mining and manufacturing. For example, on earth the best way to treat some particular ore may be to throw it into a pot of boiling water. On Mars the water would evaporate before the ore is settled in. On Venus the pot itself might melt.

This is by no means a small problem. No matter how similar planetary systems may be to one another, we should still expect at least some small astrophysical differences between their rocky planets. Thus, the possible combinations of factors that would affect mining and manufacturing may be practically infinite. In addition, the already extremely complex SRA would need some general-purpose programs so that it can begin the task of making the parts for its progeny. But no one knows how to write such a general-purpose program, and there are reasons for thinking that they cannot be written at all.[8] The biggest stumbling block to artificial intelligence has been precisely the inability to write programs that exhibit a flexible response to run-of-the-mill environments, let alone to the incredible variety demanded of SRAs. Nor is there any assurance that this problem can be overcome in the foreseeable future.[9]

Nevertheless, if rocky planets are too heterogeneous for the SRAs' needs, we may instead settle for a homogeneous environment where they can find all the raw materials in question: the asteroid belt. Whether most planetary systems have asteroid belts is unlikely to be determined for quite some time; but if they do, an SRA would be able to move from asteroid to asteroid, picking up metal ores here, carbon compounds there, mining and processing them all in a rather stable environment. An alternative would be available if other planetary systems were to have the equivalent of our Kuiper Belt, which has bodies as large as Pluto, and whose distance from the system's star, about a light day away, would be convenient not only for replication but for the task of exploration itself.

After assuming all this, we are now able to deal with the fundamental problem of SRAs. As von Neumann himself pointed out, the more complex a computer program is, the more likely it is to have errors. But the SRAs would be far more complex than anything we have ever imagined programming and building. These errors, furthermore, involve principally the task of self-replication. It is not a mere question about bugs in the gigantic program, but about errors of execution in manufacturing and assembling the many components, such as an alloy that is not quite up to strength, or a gear tooth that is slightly short and with a bit of wear such that it will no longer catch another as it must, and so on.

Neither quality control nor error-detecting programs will solve this problem, for it takes only a small percentage of error to bring the task of self-replication to a halt. The SRA is already saddled with a computer program so complex that it seems difficult to imagine that we can debug it completely. But now we must add to it a program to equip the machine with ways of checking the complete specifications for all parts, all fittings and all functions. Nonetheless, even this added complexity does not solve the problem. For a program that can foresee all the possible ways in which something can go wrong, including problems as minute as a piece of dust or a loose screw, begins to look like a general-purpose program – which is one reason why astronauts are so useful in space. In a machine that complex, engaged in the extraordinarily complex task of producing another SRA, things can go wrong in more ways than we can imagine. A program that must deal with so many unknown contingencies is, again, a program that can deal with an open-ended environment. And that is where SRAs come to grief once again.

But is it not the case that living beings, very complex in their own right, also make errors in the copying of the information used in replication? If so, a proponent of SRAs may ask why error brings machine replication to a halt when it does not do so in the replication of living beings. The answer is as follows: In living beings, 'errors' caused by mutations or recombinations serve to provide genetic variation in a population. Although mutations often prove themselves to be deadly, in some cases the genetic variation allows the individual, and eventually the population, to adapt to changes in the environment. For example, scientists develop drugs to identify the HIV virus by its molecular structure and then destroy it. But the HIV virus has a high rate of mutation, which means that the drug will kill most of the viruses that have the targeted structure, but spare those few that have mutated their structure somewhat in the interim. In a short time, the surviving viruses will fully occupy the niche left vacant by the demise of the majority, that is, the host body. As the environment changes, then, some of the members of the population may take advantage of past 'errors,' and the population lives on. This is generally how error can be adaptive for living beings.[10]

In the case of the SRAs, however, we must remember that the asteroid environment was chosen precisely because it would not change from one system to another. In that unchanging environment there is no advantage in error. In SRAs, the bulk of the errors that concern us are precisely in the reproductive part, and thus they are maladaptive. When all is said and done, it seems that an SRA technology is not really even a gleam in a scientist's eye.

Nevertheless, many scientists think that we can point to examples of self-replicating machines: trees, cats and humans. These scientists already assume von Neumann's conclusion, that there are two kinds of automata – natural and artificial. They find that assumption reasonable because they believe that the genetic code is the equivalent of a computer program, and thus they conclude that living things are just the realizations, or executions, of their particular programs. This view is no longer popular amongst biologists, but we still need to see why it fails to help the case of the SRAs.

If we are to use analogies, however, we should stress the following: In the case of the SRAs, the machine must make a copy of itself and then pass on the program. Even if it passes on the program to a unit that is not yet completed, making the copy and passing on the program are separate and largely independent tasks. In the case of living beings, however, it seems more proper to say that they pass on the program first, and then, as result of that, the copy is made. This is not a small difference, for in living beings, relatively simple accomplishments like a fertilized egg can produce very complex organisms – the egg is chemically extremely complex, but simple when compared to the unique human child that will result from it.

V Genomes and computer programs

An even more important point demands consideration. To picture the genetic code as a computer program is just to engage in metaphor, and this metaphor is highly misleading. The 'instructions' of the DNA produce the expected results, of specific proteins, cells, tissues, organs and behavior, only because at every one of those levels such 'instructions' can be expressed in appropriate environments; indeed, it is often the appropriate environment that will trigger the next stage in embryonic development. For example, the normal development of the human embryo requires a certain concentration of sodium, and later, after the human is born, the attention paid to it, even being touched, is not only necessary for its survival but provides the stimulation needed for the central nervous system to grow properly.

In other words, the 'instructions' of the DNA do not have meaning by themselves. This issue is similar to that of the meaning of words in the philosophy of language. It used to be thought that words had intrinsic meaning, but it is now generally accepted that the meaning of a word depends just as much on the context in which it is uttered: the meaning is given by the interaction of word and context, where the context may include a large variety of factors including the relationship of the word to other elements of the sentence, the manner of its utterance, the social conditions that the speaker and the listeners take themselves to be in, and so forth. In embryonic development, the 'program' makes sense – has meaning – only in so far as the maturing organism interacts with a sequence of appropriate environments. Those environments provide the biological contexts in which the 'instructions' of the genetic code are instructions at all.[11]

That sequence is itself the result of natural history, a long series of interactions between the ancestors of that organism and the environments in which they evolved. There is a clear sense, then, in which a living being comes into a world that is already made for it. The world of the SRA, on the other hand, must be largely described in its program from the beginning. The meaning of the program must be made explicit beforehand. The development of a nervous system offers a clear contrast. In dissecting an animal we may find that its nerve cells always exhibit a certain pattern, and may thus imagine that pattern is contained in a blueprint within the DNA. Nevertheless, as the nerve cells grow through, say, a muscle tissue, they do not need to be guided by any such blueprint. They may simply have 'instructions' to grow in the general direction of a chemical marker, until they make contact with a membrane that turns the 'instructions' off. But the developing muscle cells will then constrain the manner in which the nerve cells grow, which will then have to grow around the muscle cells. The final pattern is the result of such contingencies, and there is no need for any blueprint whatsoever. Otherwise we would face an uphill battle trying to explain how as complex an organism as a human being, an organic whole whose individual organs have billions of cells, can simply be the 'hardware' construction of the instructions contained in the fewer than 30,000 genes found in the human genome.[12]

As the German molecular biologist Gunther Stent (1924–2008) has pointed out, a true genomic 'program' would have a structure that 'is isomorphic with, i.e. can be brought into one-to-one correspondence with, the phenomenon.'[13] But one 'of the very few regular phenomena independent of human activity that can be said to have a programmatic component is the formation of proteins.' In this case, a stretch of DNA will be 'isomorphic with the sequence that unfolds at the ribosomal assembly site.'[14] This programmatic element, however, does not go very far, given that 'the subsequent folding of the completed polypeptide chain into its specific tertiary structure lacks programmatic character, since the three-dimensional conformation of the molecule is the automatic consequence of its *contextual situation* and has no isomorphic correspondent in the DNA.'[15]

Stent thus explains how the same genes may produce proteins with different tertiary structures because of contextual factors that are independent of those genes. But different tertiary structures may exhibit different chemical properties that are crucial to further development, and so do different quaternary structures (Figure 14.2). In an organism this contextual situation is actually nested within another contextual situation, which in turn is nested within another contextual situation, and so on.

To see how far this nesting of context goes, consider the following examples. One might think that genes determine the degree of 'maleness' or 'femaleness' in the behavior of animals towards those of the opposite sex,

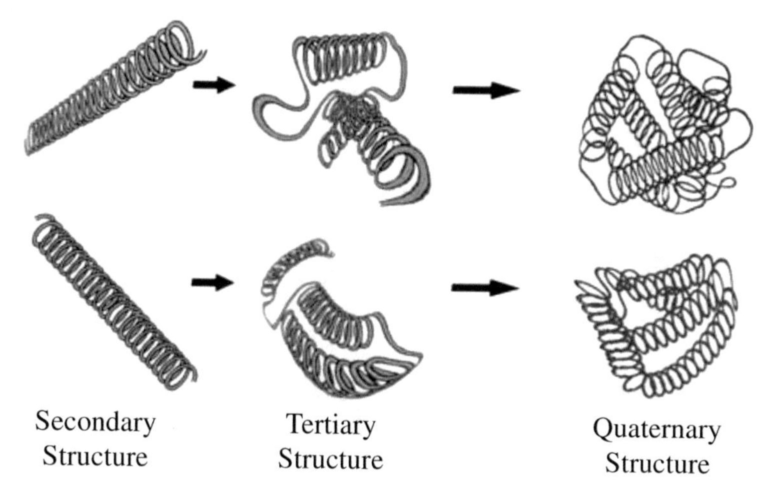

Figure 14.2 The same polypeptide chain (folded as a secondary structure) may fold on itself in a variety of ways to form different proteins, with different chemical properties due to the new and different interactions allowed between amino acids (as they come close together in the tertiary structure). The folding that leads to the tertiary structures depends on the biochemical context. An additional level of complexity, similarly created, is found in quaternary structures.
Source: Illustration by Christina Minta.

including not just sexual preference but also how aggressively the animal seeks copulation. And in some sense the genes do cause such behavior, but only within a context of development that could take place after gestation and even after birth. In the case of rats, the litter in the uterus is made up of males and females. Female rats that lie next to males may try to mount other females later in life.[16] Moreover, when rats are born, the mother will lick the anogenital areas of her pups, but will lick the males longer, guided by their scent. If her olfactory sense is damaged, she will lick all pups equally with the result that the males will be less likely to mount females as adults.[17] Rhesus monkeys raised in isolation from their peer-group also fail to mount females as adults. The reason for this is that in play-acting with other young male monkeys they practice, among other things, the double foot-clasp mount characteristic of their species.[18] Without this social context, the genes for such a basic biological function, that of reproduction, will not be expressed (Figure 14.3).

These considerations show that von Neumann's idea of the genome as a computer program is at best a very loose metaphor. Trees, cats and humans are not natural automata. Like other living beings, they are instead complex creatures whose development depends heavily on context. SRAs, on the other hand, are determined by the programmatic character of their 'blueprint,' and their 'mutations' do not fit the pattern that allows some living beings to have the 'approval' of natural selection bestowed upon them.

Figure 14.3 Genes relevant to developing sexual behavior do not become 'instructions' unless expressed in appropriate contexts, some of which are social. Young male monkeys play-act mounting behavior with others their age. If reared without male playmates, they will fail to mount females later in life. Testosterone does not mediate this social process.
Source: Illustration by Nicole Ankeny.

VI Robots to the rescue

Nonetheless, some space theorists have pushed ahead with plans for interstellar missions using new approaches to SRAs. One researcher, Robert Freitas, has proposed sending to Alpha Centauri a machine about 200 times the mass of the already gigantic Daedalus starship, the British Interplanetary Society's concept of an advanced fusion-propelled ship, which captured the imagination of space enthusiasts in the 1970s and 1980s.[19] Freitas's ship would be populated by specialized robots such as miners and metallurgists, with the ability to divide the tasks of replication into different categories. Freitas would try out this divide-and-conquer strategy in a Jovian-like system.[20] He believes a large-scale mining of a planet-size moon would provide the materials the machine would need to build a factory to replicate itself, at the rate of one replica per 500 years. The ore-processing problems would be overcome by the brute force of a machine that broke up chemical compounds into its component isotopes. Once completed, the new SRAs could be fueled by scooping Helium from the giant planet's atmosphere.

If anything, Freitas's proposal shows the extraordinary complexity of the task. On earth, even far more modest undertakings require decisions by humans, for things are bound to go wrong in unexpected ways. When the complexity of the task increases so dramatically, and when the environment will not be quite like the one envisioned before the ship departs from our solar system, things are even more likely to go wrong.

In response, Freitas has suggested, principally, semi-intelligent trouble-shooting robots to replace those humans. As for the first trait, however, we do not know that we will ever be able to produce intelligent or semi-intelligent robots (what would the latter be anyway, robots with a low IQ?). Even if the strong arguments against this possibility, which are readily available in the literature,[21] were mistaken and we could not rule out true artificial intelligence in principle, we should not conclude therefore that it is feasible. In principle, a woman cannot rule out that she will wake up tomorrow and find herself 30 years younger – perhaps the combined action of 200 key genes caused by an improbable epigenetic event would bring about spontaneously what scientific medicine *might* accomplish in 500 years – but it would be unreasonable for her to expect it. Freitas is just waving his hands.

Computer enthusiasts like to point out that computers have become very 'smart.' Computers certainly compute a lot faster and do more things than they used to. But those abilities have little to do with true intelligence. Brains, real brains that is, do not operate the same way as our admirable desktops and laptops. To arrive at a result, a serial computer may make millions of calculations, one after the other, in a second or so. But our brain must work in a different way, for the speed of transmission in neurons is of the order of 200 meters per second. This means that the decisions we make, often faster than serial computers, cannot typically proceed by serial calculation. We could try building robots that work in ways more akin to the parallel 'processing' that takes place in animal brains. Neural networks along such lines are already able to imitate and sometimes surpass some abilities of human beings in restricted situations, limited to picking up certain aspects of grammar, for example. But even here we encounter two important restrictions.

The first is that a neural network is trained by feeding back to the network the amount of error between its output and a desirable output, for example, telling apart male from female faces. We give it rather a large number of pictures of faces of both sexes, and it has the task of dividing them into male and female faces. By coming up with new ways of 'weighing' features, given the feedback of its error, it improves and eventually it can tackle the training set of pictures with a high degree of accuracy. It can also achieve a decent record of discriminating between male and female faces in new pictures.[22] But humans provide the training set; that is, humans use their knowledge to train the computer to do the correct discrimination. In circumstances that differ from human experience, we can provide much less guidance. And when we send such an extraordinarily complex machine, or set of machines, to deal with new environments, the things that will go wrong will often include what we could not have anticipated. This nearly always happens when we explore new territory.

When the German physicist Wilhelm Conrad Röntgen (1845–1923) entered his darkened cathode-ray lab in the late 1800s, he noticed a strange glow on the wall. His experience told him that such glow should not be there. It had to be energy of some form. But the equations describing the operation of his machine balanced: There should be no extra energy and thus no glow on the wall. The sense that there was something wrong led him to a series of investigations that culminated in his discovery of a new form of electromagnetic radiation, X-rays, and required that a lot of work be redone.[23] Although it would be unfair to demand that robots be as intelligent as a top human scientist, the example, nonetheless, illustrates the ways in which real brains detect and solve problems.

In short, we can train neural networks to make discriminations when we can give them examples of correct and incorrect responses. This would be increasingly difficult to do when we cannot anticipate the sorts of problems that may occur and when the complexity of those problems may be considerable. And in a new planetary system, trying to build a fantastically sophisticated and complicated machine will likely give rise to many new, complex and scientifically challenging difficulties.

Moreover, a characteristic of intelligence, which the Röntgen example illustrates, is the ability to tell relevant from irrelevant, significant from insignificant. This is the second important restriction. Animals achieve this ability, in part, because natural selection has given them some basic emotions that sound an alarm in their brains, or at least tip the balance in pattern recognition, when matters of biological importance come across their gaze, or hearing, or taste, and so on. As Antonio Damasio points out, human beings with lower frontal lobe damage – an area of the brain where, presumably, emotions inform reasoning – may continue to apply the 'rules' of reasoning properly, but are no longer able to make reasonable decisions in aspects of life in which they had been quite competent before their injuries. It is as if the gears in their brains spun aimlessly now.[24] Perhaps someday an equivalent connection to that between emotion and reason can be infused into robots. But right now we do not know that it can, and thus we should not suppose that future research will make good our present waving of hands.

Perhaps Freitas could receive help from self-repairing programs based on genetic algorithms. For example, the 'factory' could produce new robots depending on the nature of the problem. If the standard-issue robots fail to deal with a problem, a super maintenance program could take the computer code for building those robots and introduce, in a virtual environment that represents the archetype of the replica, say, ten copies of that code with a few 'mutations.' The resulting virtual robots would be made to 'perform' in the virtual environment. When one performs better than its rivals – for example, it solves a little more of the problem than the others without causing additional damage – then its mutation is preserved into the next generation: ten new virtual robots with new mutations introduced into them. The idea is that eventually a generation of virtual robots will fit the virtual environment. The 'factory' would then produce actual robots that meet those specifications and send them to fix the problem.

Using genetic algorithms may indeed be a clever way to increase the autonomy of unmanned starships at distances too great for human intervention. But such use would have serious limitations in the case of SRAs. It is one thing to have virtual environments that challenge the competing virtual robots to, say, climb a terrain with simulated large rocks and ravines. It is another to deal with a situation that may require scientific daring, a 'gestalt switch,' or that a robot by itself could not fix without the prior invention of a tool not yet a gleam in any engineer's eye. Of course, even humans may also fail in similar circumstances, when trying to make a replica of a starship under unforeseen circumstances. But the point is that such circumstances are likely to occur when dealing with such an extraordinarily complex machine in unusual circumstances. What are a robot's chances then?

Besides, genetic algorithms may not lead to the perfect match. They are restricted both by the virtual worlds they can produce and by the kinds of code available to them. If you took a gecko into Röntgen's lab and modified it by mutations, there is little guarantee that, even after millions of years, one of its descendants would be able to discover X-rays. Evolution in the direction of higher intelligence does not make sense for all sorts of organisms. It makes sense only for certain types of organisms who can afford a certain type of change in their metabolism.[25] Evolution, real evolution, does not guarantee a perfect match. It does not even guarantee an approximate match. It can only offer chances for improvement when circumstances vary – sometimes. After all, most species that ever lived are now extinct.

Furthermore, evolution occurs not only because organisms change so as to fit the environment. Organisms often change their environment, which change then introduces new factors of selection. In the case of actual, as opposed to virtual, changes in the starship this may indeed introduce organic changes as well. That is, the conditions of interaction between the starship SRA and its environment may well change in unforeseen ways. But we could not then anticipate, and thus program, the changes the computer needs to make in the virtual environment that would be testing the virtual robots. As Pablo Picasso (1881–1973) once said, 'Computers are useless. They can only give you answers.'[26] Of course, the prospect of such organic changes defeats, once again, the choice of a homogenous environment.

VII Nanorobots

A new favorite solution is to appeal to nanorobots (sometimes called nanobots). Some claim that nanorobots will cure cancer and all other sorts of diseases, will create new technologies, make our spaceships a lot smaller, and, best of all, replicate themselves. We should thus send a very small starship full of nanorobots that will make copies of themselves and of the starship. Indeed, the nanorobots would have to reproduce in extraordinary numbers if they are to build any machine large enough to be seen with the naked eye. The idea is to create a so-called assembler that would pick up an individual atom at a time and put it in the right place to make the right kind of molecule. Once assembled, the nano-machines would execute their programs and together produce, say, a starship.

This idea brings up von Neumann's proof again. The little robot picks up the parts (different kinds of atoms) and puts them together in the style of a Mechano set. Of course, the desirable atoms are not going to be floating in the nano equivalent of a vat full of parts. They will be parts of molecules or be found in solution, and so on. We will have to 'mine' them. This would not be easy for a nanorobot, for it would take a considerable amount of energy and finesse. It would require sensors and a way to store and apply energy, and so on. The number of atoms involved to carry out all those functions would make it a very clumsy tool. A good tool would somehow break the bonds the desired atom has created with those other atoms around it and then deposit it into its proper place in the molecule being built by the nano-machine. Perhaps we could solve the problem by some *macro-machine* that would blast the ores into individual elements. But the problem would persist at the point of assembly. I once knew a man who had a watchband accidentally tattooed on his wrist. He was an electronics technician with a penchant for disregarding safety rules. On one occasion he had stuck his hand into a machine with a powerful electromagnet, not bothering to take off his metal-band watch. The powerful magnetic field slammed his wrist against that part of the machine and the current marked him for life. Likewise, the nanorobot is going to interact with the assembly being built, which will not sit there passively like the robot under construction in von Neumann's scenario. Another problem is that the atom being moved will form a bond with the nanobot's 'finger,' and will stick to it.

In the opinion of Richard Smalley (1943–2005), awarded the Nobel Prize in Chemistry in 1996, these problems are insurmountable for nanorobots. Many other distinguished scientists share the same opinion. And, in fact, no one knows how to make the famous 'assembler' proposed by K. Eric Drexler in 1986.[27] Indeed, the problem of stickiness already plagues micro-electro-mechanical systems; it is bound to get much worse at the nanometer level. As if that were not bad enough, the star-trekking nanorobots will be operating in the vacuum of space, where the Casimir effect is likely to play havoc with their operation. The Casimir effect comes about when two thin plates are brought together at very small distances apart in a vacuum. It turns out that the vacuum is not exactly empty but full of virtual particles coming in and out of existence. The narrow space between the plates rules out virtual electrons of certain wavelengths, and this creates an imbalance in the density of virtual electrons within and outside of the plates, which in turn pushes those plates closer together. Nanorobots in space would be plagued by 'stickiness.'

It is not very clear how one would 'program' a nanorobot to perform its tasks: It must not only make a copy of itself but also carry out, in synchronization with many other nanorobots, the construction of much larger, perhaps macroscopic structures. The difficulty does not find parallel in the world of living beings, for they do not reproduce themselves by making their own replicas one atom at a time but rather by serving as templates for the assembling of similar polymers that fold three-dimensionally in useful ways. The action of these polymers then leads to the creation of cells, which in turn form parts of organisms. And as we have seen, such genomes do not really function like computer programs. Thus, using nanorobots does not avoid the previous problem, but instead complicates it.

Insofar as the nanorobot would have a program to accomplish what living beings do without it, such a program would have to go well beyond little tricks like having a molecule 'remember' a shape, that is, revert to a previous alternative configuration under certain conditions. Relatively speaking, assembling a replica and helping in the construction of a starship are complicated tasks for something made up of very few atoms. Nor is it clear how exactly the nanorobot would absorb and transform energy to carry out its tasks. It seems that machines of a larger scale, possibly macro-machines, would have to hold the relevant programs and infuse the necessary energy into their herds of nanorobots. We have already seen that we might need a macro-machine to 'pulverize' ores into individual elements. Moreover, part of the function of an exploring starship is to send information back to its home planet; but electromagnetic transmission at such distance may well require very large structures. All this, in turn, makes it likely that the means of propulsion will also involve large structures. These and other considerations begin to rule out the possibility of building starships the size of a Coke can, let alone smaller.

Of course, today we have already found ways to program computers without having to give them an exhaustive account of every possible action they might take. For example, we can buy relatively inexpensive floor-cleaning robots. They do not have the floor-plan programmed into them, not even the position of the legs of a chair. If the robot encounters an obstacle, it executes a simple maneuver to get around that obstacle. This new approach might be coupled with the ability to cooperate with other robots the way social insects do to create hives, defend them, and so on. The robots in question would cooperate instead to make the replica of a starship.

Although it is true that every long journey begins with one step, not every step leads to a long journey. Learning to swim a lap at the local pool does not warrant optimism about swimming across the English Channel, let alone across the entire ocean. That computers can do a few rather simple things should not allow us to conclude that they can achieve the extraordinary level of sophistication needed to make a working replica of a starship. Maybe they will be able to do so, but at this time this is more a dream than the assurance of a practically inevitable technology that any advanced technological civilization is likely to have mastered. Besides, let us remember that insect hives, like complex organisms, result from the unfolding of biological processes within larger biological contexts that are themselves the product of natural history.

Much of the optimism for these sorts of technological proposals comes from computer simulations. One problem with simulations, apart from the quip that you cannot eat the simulation of a good meal, is that they deal with theoretical scenarios in the sense of 'theoretical' that connotes abstraction from details, which are often unknown. We just do not see how 'in principle' this or that could *not* be done eventually. But God is in the details. In the 1970s, for example, great hopes for the cure of cancer were raised by the development of recombinant DNA. And many specific potential cures were indeed proposed soon after. Theoretically, they should have worked. But they did not. We have learned in the intervening years that the behavior of cancer cells is much more complex than we could have then imagined, for a tumor has many clever ways of protecting itself.

We should also keep in mind, when projecting from small successes to extraordinary feats, that problems of scale can easily arise. If we want to make a dog the size of an elephant, we will have to make it so that it is built like an elephant, with thick bones, and so on.

And even if we perchance figure out how to make new molecules that self-replicate, we do not know that those molecules would work well as nanorobots in a starship. Living cells replicate. DNA replicates. But neither seems a promising candidate for solving the problems that may arise in the building of a starship.[28] Solving those problems would require experience and ingenuity that we cannot program into large computers, and thus what guarantee do we have that the difficulty disappears just because the computers are made extremely small? Those nano-hands seem to be waving extremely fast.[29]

VIII Assumptions

Several of the assumptions in this chapter may be challenged. For example, the starship SRA need not be as complex as envisioned so far: in other words, a machine with millions of parts. A new manufacturing technique called 'quick prototyping' can build rather complicated three-dimensional gadgets in one piece. It works by laying down a substance into the proper shape. When the substance congeals, we have a one-piece stand or handle, and so on. If this could be done with the right kinds of alloys, we might be able to reduce substantially the number of components involved, even if we cannot build the entire hull of the starship in a single block. We may also improve quality control and preserve the structural integrity of the ship by means of electromagnetic identification (ID) tags properly placed to give us a three-dimensional image of the ship, a technology descended from the radio-frequency ID tags of today, which either transmit or respond to radio signals, thus providing information about the location of objects or materials.

Self-healing starships would simplify matters even more. The self-healing is made possible by myriad microcapsules full of a special sealing material. Any crack in the hull, for example, would cut across several of the microcapsules. The sealing material would then be released and fill the crack. These examples give probably just a peek at the extraordinarily clever technologies likely to come the way of spaceship builders in the decades to come. They will surely reduce somewhat the complexity of starships.[30]

Reducing complexity, however, is not the same as eliminating it. The communication, navigation and propulsion systems cannot be of one piece with the hull under any technology so far envisioned. Nor can a similar approach to their construction be undertaken, since each of those systems is made of parts that perform very different functions, undergo different stresses and withstand sometimes drastically different temperatures. Moreover, the quick-prototyping machine that makes the hull will itself have to be carried, and then replicated. This adds complexity. The 'meta machine' that builds it will then have to be replicated as well. And so on.

IX Conclusion

A technology of self-reproducing machines may or may not be feasible, but we do not have sufficient reason for asserting that it is, and even less to conclude that it is practically inevitable. But an impossibility proof requires at least practical inevitability. Therefore the impossibility proof fails.

Even apart from this problem, this impossibility proof would suffer from other serious defects. Unless SRAs were bound to replicate like maggots upon the cadaver of its host, we might well miss them even if they had landed safely on the earth. After all, we sometimes fail to find airplanes and ships in distress even when we are looking desperately for them. A machine that made only a few copies of itself might be unobtrusive, or it might have come a few million years ago, or it might not have arrived yet. And even if it has arrived, as Ronald Bracewell (1921–2007) suggested many years ago, it may be waiting for the right moment to make its presence known.[31] The ability to get a machine to every planetary system eventually is different from the ability to have a machine on every planetary system at any one time.

Besides, if the SRAs were to reproduce without restraint, they would become a rot on the galaxy eventually, and then we could not miss them. But it is not clear why a civilization should wish to create such a nuisance. It is even less clear why a civilization should feel *compelled* to plague other worlds so. And once again, an impossibility proof is not worth much when its conclusions are not inevitable and its crucial assumptions are highly questionable, perhaps even far-fetched.

Nevertheless, the examination of the future conjured up from von Neumann's idea has led to a clearer understanding of life and exploration. And the future painted by Sagan may still guide our way into the cosmos. Not unlike science fiction, these two scientific scenarios force us to contemplate possibilities and encourage us to imagine steps that our descendants might need to take some day.

Notes

1. Carl Sagan, ed., *Communication with Extraterrestrial Intelligence*, Cambridge, MA: MIT Press, 1973. This chapter is based on a section from my book in progress, *The Dimming of Starlight*. An earlier version was published as Chapter 15 of my *Evolution and the Naked Truth*, Aldershot: Ashgate, 1998.
2. Carl Sagan, *Pale Blue Dot: A Vision of the Human Future in Space*, New York: Random House, 1994, 39, 372–3.
3. Fermi asked his now famous question in 1950 during an informal conversation with Edward Teller, Emil Konopinski and Herbert York at Los Alamos National Laboratory. See Eric M. Jones, *'Where is Everybody?' An Account of Fermi's Question*, Los Alamos Technical report LA-10311-MS, March 1985; accessible at http://www.fas.org/sgp/othergov/doe/lanl/la-10311-ms.pdf (accessed 1 October 2017).
4. Authors more contemporary than Fermi have further developed the argument criticized here. See, for example, Frank J. Tipler, 'Extraterrestrial Beings do not Exist,' *Physics Today* 34 (April 1981), 9–38. Sagan twice blocked the publication of this paper. In turn, Senator Proxmire used it to block Sagan's ambitious SETI project Cyclops. For recent commentary (and the reference to Fermi), see Paul Davies, *Are We Alone? Philosophical Implications of the Discovery of Extraterrestrial Life*, New York: Basic Books, 1995, based on his series of lectures at the University of Milan in 1993.
5. John von Neumann, *Theory of Self-Reproducing Automata*, Urbana: University of Illinois Press, 1966.
6. *Advanced Automation for Space Missions*, Washington, DC: NASA, 1982.
7. The exceptions would present a variety of inconveniences: Europa is covered by a layer of ice and presumably an ocean of water. Io has a very unstable surface. And so on.
8. The classic critique is Richard Dreyfus, *What Computers Can't Do*, New York: Harper & Row, 1972.
9. Connectionist approaches, which present a significant alternative to von Neumann's view, are better able to handle context, but this situation would still present a tall order even for optimistic treatments such as Paul M. Churchland's, *A Neurocomputational Perspective: The Nature of Mind and the Structure of Science*, Cambridge, MA: MIT Press, 1989. I find Churchland's view very plausible, and this chapter supports it by undermining the view of mind (and body) put forward by von Neumann.
10. An equivalent situation occurs when living things move into new environments in which their adaptation may leave much to be desired. That would not be the case for the situation of the SRAs.
11. This analogy was inspired by the work of Gunther S. Stent, one of the founders of molecular biology (see below). A recent, updated view can be found in Irun R. Cohen, Henri Atlan and Sol Efroni, 'Genetics as Explanation: Limits to the Human Genome Project,' *Encyclopedia of Life Sciences*, accessible at http://

onlinelibrary.wiley.com/doi/10.1002/9780470015902.a0005881.pub3/full (accessed 1 October 2017).

12. For a very interesting illustration, see Gary Marcus, *The Birth of the Mind: How a Tiny Number of Genes Creates the Complexities of Human Thought*, New York: Basic Books, 2004.
13. Gunther S. Stent, 'Strength and Weakness of the Genetic Approach to the Development of the Nervous System,' *Annual Review of Neuroscience* 4 (March 1981), 163–94, here 187.
14. Ibid., 188.
15. Ibid. (emphasis mine).
16. Simon LeVay, *The Sexual Brain*, Cambridge, MA: MIT Press, 1993, 89.
17. Ibid., 92.
18. Ibid., 93.
19. Robert A. Freitas Jr., 'A Self-Reproducing Interstellar Probe,' *Journal of the British Interplanetary Society* 33.7 (July 1980), 251–64; Alan Bond, ed., *Project Daedalus: The Final Report on the BIS Starship Study*, London: British Interplanetary Society, 1978 (= *Journal of the British Interplanetary Society Supplement*). Daedalus would have been 190 meters long and would have weighed over 50,000 metric tons.
20. We have already found hundreds of extraterrestrial gas giants, some even larger than Jupiter. In some of such planets, we could presumably expect moons even larger than Mercury, similar to Jupiter's.
21. See, for example, the references above to Dreyfus and Churchland.
22. This and many other illustrations are explained by Churchland.
23. This is the interpretation found in Thomas S. Kuhn, *The Structure of Scientific Revolutions*, Chicago: University of Chicago Press, 1970.
24. Antonio Damasio, *Descartes' Error: Emotion, Reason, and the Human Brain*, New York: Putnam, 1994.
25. Munévar, *Evolution and the Naked Truth*, ch. 2.
26. Quoted on many web sites, including www.quotationspage.com/quote/255.html (accessed 1 October 2017).
27. Richard S. Smalley, 'Of Chemistry, Love and Nanobots,' *Scientific American* 285 (September 2001), 76–7; George M. Whitesides, 'The Once and Future Nanomachine,' ibid., 78–83; Steven Ashley, 'Nanobot Construction Crews,' ibid., 84–5. In the same issue Drexler had an opportunity to address the content of the objections, but instead simply suggested that his critics were unqualified and that successful research in the field favored his idea. Ashley's article suggests otherwise. K. Eric Drexler's proposals can be found in his *Engines of Creation: The Coming Era of Nanotechnology*, New York: Anchor Books, 1986, and *Nanosystems: Molecular Machinery, Manufacturing and Computation*, New York: John Wiley, 1992.
28. Under very special laboratory conditions, DNA can be used to make nanomachines, but nothing of the sort relevant to our discussion.
29. None of the foregoing shows, incidentally, that artificial life is impossible. It would be life, even if made from scratch in a laboratory or a factory, as long as such organisms are comprised of cells that replicate, undergo metabolic processes, and so on. Insofar as there is design in them, that design is grafted on to the knowledge we have of natural history, to take advantage of prior interactions with environments or sequences of environments. I am not referring here to the computer field of 'artificial life,' based on von Neumann's other proofs, which has conceptual problems of its own.

30. I wish to thank Ryan Munévar for these and other interesting suggestions and comments.
31. For example, when life in a planetary system begins to use radio and television, the alien probe may save those transmissions for a while, and then begin to broadcast them back to the originating planet, perhaps on a different frequency. This would be a way of making contact, for the inhabitants of the planet would realize that those signals were coming from outer space. In our day and age, however, such rebroadcasts may be mistaken for reruns. Nevertheless, the alien probe may 'edit' *I Love Lucy* to include pictures of its home planet, or come up with another attention grabbing stratagem. For the earliest publications on the subject, see Ronald N. Bracewell, 'Communications from Superior Galactic Communities,' *Nature* 186.4726 (28 May 1960), 670–1; and 'Life in the Galaxy,' in Stuart T. Butler and Harry Messel, eds, *A Journey through Space and the Atom*, Sydney: Nuclear Research Foundation, 1962, 243–8, reprinted in A.C.W. Cameron, ed., *Interstellar Communication*, New York: W.A. Benjamin, 1963, 232–42. See also his *The Galactic Club*, San Francisco: W.H. Freeman, 1975.

CHAPTER 15

Inscribing Scientific Knowledge: Interstellar Communication, NASA's Pioneer Plaque and Contact with Cultures of the Imagination, 1971–72

William R. Macauley

> We believe there is a common language that all technical civilizations, no matter how different, must have. That common language is science and mathematics. The laws of Nature are the same everywhere.
>
> Carl Sagan (1980)

> No one has ever observed a fact, a theory or a machine that could survive *outside* of the networks that gave birth to them.
>
> Bruno Latour (1987)[1]

I Space exploration, universal physical laws and communication with extraterrestrial intelligence (CETI)

Space exploration during the late twentieth century began incorporating 'interstellar messages,' primarily in the form of material artifacts and electromagnetic signals, deliberately created by humans and transmitted from earth, in an effort to establish contact with possible extraterrestrial intelligence in distant star systems.[2] Systematic attempts were made by groups of scientists and their associates to detect incoming interstellar messages from extraterrestrials and, more rarely, send messages from earth to technologically

William R. Macauley (✉)
University of Manchester, Manchester, UK
e-mail: Ray.Macauley@manchester.ac.uk

Alexander C.T. Geppert (ed.), *Imagining Outer Space*
European Astroculture, vol. 1
https://doi.org/10.1057/978-1-349-95339-4_15

advanced civilizations located in astronomically remote planetary systems or traveling through interstellar regions of space. This chapter focuses on a specific interstellar message incorporated on a specially constructed material artifact – NASA's Pioneer plaque – dispatched from earth on board twin Pioneer spacecraft launched in 1972 and 1973 (Figure 15.1).

During the 1970s, American scientists directly involved in research on interstellar communication worked in partnership with engineers, visual artists and others to design messages comprised of representations that were supposed to constitute meaningful and self-explicatory forms of knowledge, considered suitable for initiating communication with extraterrestrials. Interstellar messages carried on physical artifacts such as NASA's Pioneer plaque and Voyager record incorporated science and mathematics primarily in the form of pictures and non-linguistic symbols, explicitly designed to convey factual knowledge about humankind, the planet we inhabit and physical laws that govern, for example, fundamental properties of matter throughout the entire universe.[3]

The late Carl Sagan (1934–96), quoted above and widely recognized as one of the principal historical actors responsible for designing NASA's interstellar messages, often claimed that scientific knowledge and mathematics

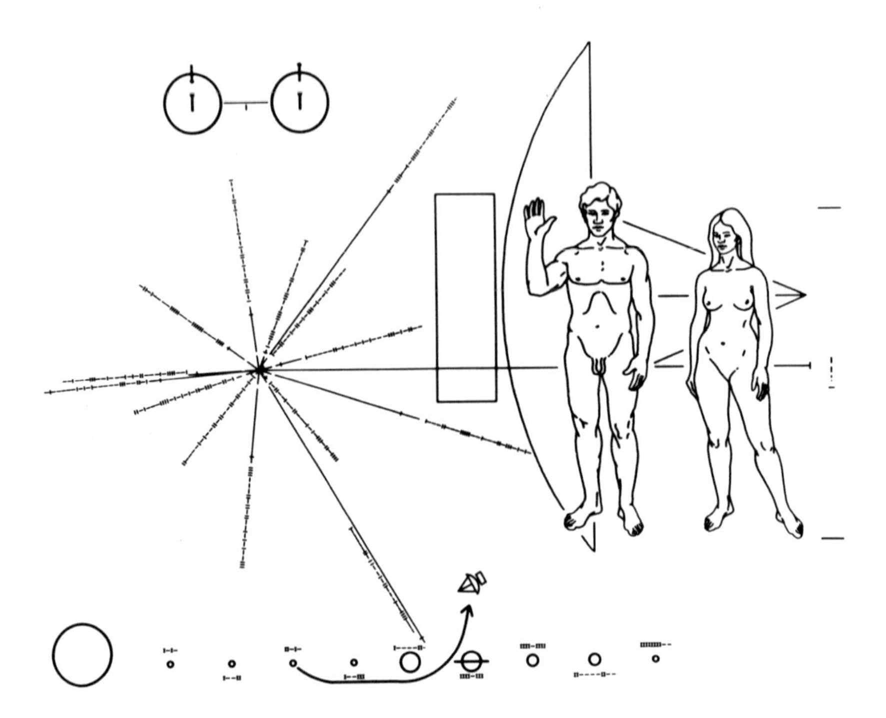

Figure 15.1 Reproduction of the interstellar message inscribed on a 15 × 22.5 cm gold-anodized Pioneer plaque, designed for NASA by Carl Sagan, Linda Salzman Sagan and Frank Drake, 1971–72.
Source: Courtesy of NASA.

would be comprehensible to all advanced civilizations and thus served as valid indicators of social and technological progress throughout the universe. The physical sciences have conventionally been considered by practitioners to be distinct from other forms of empirical endeavor because scientific observations, theories and predictions are for the most part expressed and communicated in the supposedly objective language of mathematics.[4] Many scientists believe that mathematical relationships and numbers correspond to timeless entities and the essential structure of the physical universe; mathematical expressions that somehow reveal these intrinsic properties will be recognized and easily understood by any technologically advanced life form.[5] In contrast, some historians and sociologists such as Bruno Latour have argued that scientific knowledge, mathematics and universal laws of physics – sometimes referred to as 'the laws of Nature' – are created by human actors rather than discovered and as a result bear the hallmark of socially embedded and historically contingent factors.[6] Communication of scientific knowledge and meaning using pictures, diagrams and other forms of visual representation across a range of media is a topic that has been studied closely by historians and other scholars in recent years.[7] Previous research on production and application of visual representation in scientific work and discourse has primarily focused on explaining how and why images are employed as expository devices in scientific journals and educational texts to illustrate and support ideas expressed in the customary medium of writing. There are relatively few academic studies on the history of graphical representations designed to communicate scientific knowledge explicitly and independently of printed text or writing systems that require prior knowledge of spoken language.[8]

The notion of *inscription* as both metaphor and material object is a valuable analytical tool for investigating material practices, discourse, cultural artifacts and social networks that are historically embedded and integral in terms of the production, communication and stabilization of scientific knowledge.[9] Metaphors of inscription and related concepts are applied in this chapter as analytical tools, to highlight and critically examine the materiality of scientific work and rhetorical strategies employed by scientists and others responsible for the design of interstellar messages. Semiotic theory is integral to this research on interstellar communication, because it provides a means of incorporating analytical tools and concepts from a range of disciplines. It thus facilitates an analysis of interstellar messages as social objects or artifacts comprised of signs or units of meaning, designed to communicate supposed universal scientific knowledge through a process of signification or semiosis.[10]

Debates within the scientific community, scientific publications and wider public discourse concerning interstellar messages are attendant to contemporary beliefs, anxieties and multiplicity of meanings regarding complex issues such as the impact of technology on society, intelligent life on other worlds, evolution as a universal principle, and the historical contingency of distinctions between human and non-human. Although images featured on NASA's Pioneer plaque and Voyager record were originally designed as objective and universal in terms of form and meaning, they have been reinterpreted and deployed within a diverse range of narratives, media, networks and social groups to draw attention to or (more frequently) elide the reciprocity of human agency and cultures of the imagination. Indeed, the history of interstellar messages includes excellent historical material for analyzing how visual representations of objects, events and relationships are intimately linked to the imagination as a manifestly creative process. Scientific research and interdisciplinary approaches to the design of interstellar messages are associated with imaginative, as opposed to imaginary, solutions to complex problems concerning communication with non-human entities, including the creation of cultural artifacts that encapsulate human knowledge, experience and desires.

The present chapter argues that interstellar messages designed to initiate contact with extraterrestrials are especially significant in historical and cultural terms because during their design, construction and circulation, scientists give explicit reasons to support sweeping claims regarding the ubiquitous authority of science and mathematics. Further, scientists and other historical actors responsible for the design of interstellar messages such as the Pioneer plaque have employed various forms of discourse and communication media to articulate speculative theories, social agendas and personal beliefs regarding imagined cultures and technologies of the distant future – candid opinions that might otherwise remain tacit or confined within smaller social networks.

II NASA's Pioneer plaque, 1971–72

During the 1970s, NASA launched a series of unmanned spacecraft primarily for exploration of planets and interplanetary scientific experiments on astronomical and astrophysical phenomena within the solar system. A few of these robotic spacecraft, such as Pioneer 10 and 11, were programmed to follow precise flight paths and utilize powerful gravitational forces generated by the outer planets, which propelled them on escape trajectories beyond our solar system, into vast unexplored regions of interstellar space and cruise through the galaxy essentially forever (Figure 15.2).

This prospect was perceived by scientists, science journalists and others as an unprecedented opportunity to design, create and deploy material artifacts containing messages that were representative of all humankind and conveyed scientific knowledge that could be understood by any intelligent extraterrestrial civilization in distant parts of the galaxy or space-faring descendants of humankind capable of interstellar travel in the remote future.[11] Scientists Frank Drake (1930–), Carl Sagan and visual artist Linda Salzman Sagan (1940–) have frequently been identified in scientific publications, scholarly texts and wider cultural discourse as principal historical actors directly responsible for design and production of an interstellar message, commonly referred to as the 'Pioneer plaque,' comprised of images and symbols etched on a metal plate, which was attached to NASA's Pioneer 10 spacecraft and transported into space in March 1972 (Figures 15.3 and 15.4).[12] An identical plaque was attached to NASA's Pioneer 11 launched the following year and, like its sister craft, will eventually be accelerated out of our solar system into interstellar space.[13]

Figure 15.2 Artist's impression of NASA's Pioneer F spacecraft during its Jupiter flyby, December 1973.
Source: Courtesy of NASA.

Figure 15.3 Photograph of the Pioneer F spacecraft at NASA's Kennedy Space Center, February 1972, with plaque attached to struts, shown at center of photograph. The plaque surface containing inscriptions of images and symbols that comprise the interstellar message is faced inwards towards the center-line of the dish antenna to avoid erosion during its journey through interstellar space.
Source: Courtesy of NASA.

Figure 15.4 Photograph showing the position of the interstellar plaque attached to the Pioneer F spacecraft.
Source: Courtesy of NASA.

Four months before the launch of Pioneer 10, in November 1971, expatriate British science correspondent Eric Burgess (1920–2005) arranged a meeting with Carl Sagan at NASA's Jet Propulsion Laboratory (JPL), near Pasadena. At the time, Sagan was on temporary leave from his academic post as director of the Laboratory of Planetary Studies at Cornell University and engaged as science consultant for NASA on the Mariner 9 mission to Mars. Burgess was aware that Sagan had recently attended an international conference on interstellar communication, held in the former Soviet Union and jointly organized by the National Academies of Sciences in both the United States and Soviet Union. During their discussion, Burgess mentioned that he had recently observed Pioneer F undergoing pre-launch tests and suggested that a plaque should be attached to the spacecraft because it would be the first human-made object to travel beyond the solar system.[14] Sagan agreed and subsequently discussed the idea of an interstellar plaque with Hans Mark (1929–), director of the NASA Ames Research Center, Moffett Field, California, where the Pioneer project was managed. After obtaining Mark's support, Sagan arranged meetings with senior scientists and managers at NASA Headquarters in Washington, DC, to seek approval for the proposed message.[15]

By December 1971, Carl Sagan had managed to attain authorization from senior staff at NASA Headquarters and Pioneer mission team at NASA Ames for the attachment of a plaque to the Pioneer F spacecraft and was responsible for designing the interstellar message it would carry. Sagan enlisted the assistance of his colleague Frank Drake, radio astronomer at Cornell's Center for Radiophysics and Space Science, and they decided that the message should consist of a series of drawings engraved on a metal plaque. Drake accepted the task of producing a 'pulsar map' that would indicate the source of the Pioneer craft in terms of the location of its planet of origin, relative to a constellation of 14 pulsating stars (pulsars) in the galaxy. Sagan also enlisted the help of his wife, Linda Salzman Sagan, who had experience as a film-maker and previously studied figure drawing, painting and art history at the School of the Museum of Fine Arts in Boston. She agreed to produce line drawings of human figures, an adult female and male, for the Pioneer plaque. Carl Sagan subsequently explained how severe time constraints meant that the message depicted on the plaque was, to some degree, compromised.[16] In January or early February 1972, the artwork and designs for the Pioneer interstellar message were approved by senior NASA administrators at NASA Headquarters, on the recommendation that changes were made to Linda Salzman Sagan's figure drawings.

Pictures and graphical images designed for interstellar messages are produced through iterative techniques that transform natural phenomena into scientific objects or facts that are amenable to manipulation using mathematical operations. Graphical images such as diagrams of astrophysical phenomena and representational drawings of human figures incorporated on NASA's Pioneer plaque and other interstellar messages are cultural artifacts and manifest forms of knowledge, produced through what ethnographer and sociologist of science and technology Michael Lynch has referred to as 'rendering practices':

> The problem of visibility in science is more than a matter of providing illustrations for publication. Published and unpublished data displays, in the form of graphs, photographs, charts and diagrams, constitute the material form of scientific phenomena. By 'material' I mean sensible, analyzable, measurable, examinable, manipulable and 'intelligible.' Although the procedures for making the object scientifically knowable implicate an independent object, they simultaneously achieve a graphic rendering of the object's materiality.[17]

Further, Lynch has identified specific rendering practices and persuasive graphical techniques such as 'mathematization' that scientists apply in their research and published work to augment or remove visible properties of representations assembled in graphic displays. More specifically, mathematization involves transformation of two-dimensional arrays that contain graphic representations into standardized geometric space, in which natural phenomena are represented as mathematical entities. As such, visual displays in scientific texts are not simply descriptive illustrations; they represent and methodically codify natural objects as inherently mathematical.[18]

III Mapping pulsars and inscribing scientific knowledge

The interstellar message depicted on the Pioneer plaque consists of an assemblage or montage of discrete two-dimensional images and symbols that represent humans, physical objects, and astrophysical phenomena (Figure 15.1). Heterogeneous representational styles and diverse spatial and temporal properties are depicted on the plaque and range in scale from atomic to galactic. Sagan and other scientists claim that comprehension of Pioneer's interstellar message and the intricate interrelationships between the objects that it depicts requires prior scientific knowledge of universal mathematical principles and associated physical phenomena.

The pulsar map featured on the Pioneer plaque consists of a two-dimensional image, as Drake had earlier promised, in the form of a distinctive radial pattern, indicating the sun (at the center of the map) in relation to a constellation of fourteen pulsating stars (pulsars) in the galaxy.[19] In addition, Drake's pulsar map includes temporal attributes in the form of precise numerical values, written in binary notation, for the frequency or period of pulses emitted by each of the 14 selected pulsars, which could potentially be used by recipients of the message to identify the historical epoch in which the Pioneer spacecraft was launched. To facilitate understanding of the physical scales and measurement associated with visual representations featured on the plaque and their referents, the designers included a schematic diagram in the top left on the Pioneer plaque (Figure 15.5), which represents the hyperfine transition of neutral atomic hydrogen in order to specify units of length (21 cm) and time (1420 MHz).[20]

It was assumed by the designers that these particular physical properties of neutral hydrogen atoms – emission of a photon with a wavelength of 21 cm and frequency of 1420 MHz – are abundant throughout the universe and easily detected with radio telescopes or equivalent extraterrestrial technologies. Therefore, predictable quantitative values based on astrophysical observations of the hydrogen emission line would necessarily be recognized by extraterrestrial scientists. Sagan explained that 'this fundamental transition of the most abundant atom in the galaxy should be readily recognizable to the physicists of other civilizations.'[21]

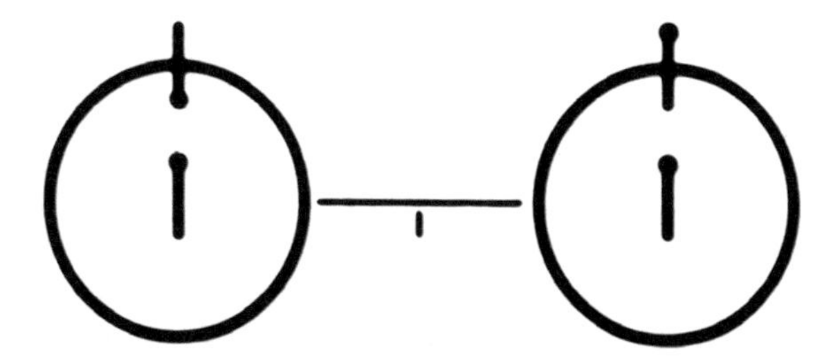

Figure 15.5 Enlargement of schematic diagram on the Pioneer plaque indicating hyperfine transition of neutral atomic hydrogen, 1972. *Source*: Adapted from NASA; see Figure 15.1.

The binary digit for '1' is featured as a small vertical line located at the center of the schematic diagram of hydrogen (see Figure 15.5) in accordance with the stated intention to represent the unitary value for distance (21 cm) and time (1420 MHz). Binary digits featured next to other visual representations on the message in order to specify spatial and temporal quantities that are defined as multiples of the quantitative values associated with the emission spectrum of hydrogen atoms. For example, numerical values for the period of each pulsar (all ~0.1 second) on Drake's map are expressed as multiples of 1420 MHz, written in binary notation at the end of each line. The length of 14 solid lines on the radial map specifies the distance of each pulsar from the origin, with an additional line along the horizontal plane indicating the distance from the Pioneer's launch planet (as specified by our home star) to the galactic center.

The image designed by Drake is organized as a map or navigational tool with each line specifying distance from the sun and binary notation providing periodicity of the cosmic emissions using units based on the emission spectrum of hydrogen. Again, Sagan asserted that:

> With this first unit of space or time specified we now consider the radial pattern at left center. This is in fact a polar coordinate representation of the position of some objects about some origin, with this interpretation being a probable, but not certain, initial hypothesis to scientists elsewhere. The two most likely origins in an astronomical interpretation would be the home star of the launch civilization and the center of the galaxy.[22]

The image deployed a combination of mathematics, geometry and artistic conventions of linear perspective to visualize discrete points within a simulated three-dimensional space, inscribed on the two-dimensional flat surface of the plaque. The map is not an image in the sense of a representational picture, but rather a condensation of multiple inscriptions or cascades (for example, printouts from radio telescopes, mathematical models, tables, illustrations) generated by heterogeneous actors, research sites, communication media and inscription devices. The journal article by Sagan, Salzman Sagan and Drake also inscribed the 14 selected pulsars in a different representational format: a series of numerical values presented in a table that explicitly codified these pulsars as abstract mathematical entities. Presentation of precise numerical values in the table was constitutive of scientific knowledge, concerning the mathematical properties of natural objects referred to as pulsars.[23]

Pulsars and photons emitted by neutral hydrogen atoms have never been observed directly; scientific observations and theories depend on scientific instruments and graphical technologies that mediate and transform astrophysical and quantum events into representational forms that are comprehensible and constitutive of factual knowledge. Indeed, it has become a truism in histories of radio astronomy that the serendipitous 'discovery' of pulsars in 1968, by Jocelyn Bell Burnell (1943–), a junior researcher at the Mullard Radio Observatory in Cambridge, was achieved by sifting through masses of paper charts that featured pen-trace recordings of signals acquired with a radio telescope, to identify and remove 'noise' (or what was referred to at the time as 'scruff') from observational data.[24] Significantly, graphical representations of pulsar-type objects encompassed a diverse range of formats, styles and communication media. The first published account of 'rapidly pulsating radio sources' by Antony Hewish used coordinate graphs to display distinctive temporal properties of these astronomical phenomena.[25]

The image designed by Drake was not based on direct, unmediated observation of natural phenomena, but rather a visual rendering of a long chain of representational forms or inscriptions that mathematize space and objects within it. Repeated claims by the designers regarding the scientific validity, objectivity and universality of the interstellar message elide artistic techniques and imaginative skills required for the design and production of schematic images. For example, the hydrogen diagram depicts quantum events on an atomic scale and the pulsar map portrays a constellation of astronomical objects separated by light years. Arguably, the referent for Drake's map was the table of numerical values featured in the *Nature* article 'A Message from Earth.'[26] NASA's interstellar messages included vivid examples of ways in which natural objects and categories of experience such as perception of time and space were framed as scientific objects that are, in and of themselves, universally comprehensible because of their underlying mathematical structure. Moreover, the image presents time, space and energy as graphical forms that are easily manipulated using arithmetical and geometric rules.[27]

A similar point can be made regarding the historicity of graphical representations depicting neutral hydrogen emission as featured on the Pioneer plaque and other interstellar messages created during the 1970s. Prior to this, completely different visual representations of the same quantum event or astrophysical process were devised and circulated as expository devices in scientific publications during the 1960s, suggesting that despite claims regarding the universality of hydrogen emission spectra there is no consistent representational form or scientifically valid method of deciding what is the most appropriate way of depicting this phenomena.[28] Eric Francoeur has made a similar point with regard to 'graphematic condensation' of knowledge and matter in the field of molecular biology:

> There is no such thing as comparing the structural representation of a molecule to the 'real' thing, since it is through representational work that a molecular structure becomes coherently visible. The realm of molecular structures is thus essentially cultural, i.e., coextensive with the means chemists have given themselves to show, talk about, and work with these structures – means which are, *ceteris paribus*, epistemically equivalent while phenomenologically distinct.[29]

There is a contingent historical relationship between inscriptions on NASA's interstellar messages and physical and quantitative properties of atomic hydrogen. Hydrogen emission spectra was predicted during the 1940s in theoretical studies by astrophysicists such as Hendrik van de Hulst (1918–2000) and Iosif Shklovsky (1916–85), and later confirmed in observational experiments by physicists such as Harold Ewen (1922–2015), Edward Purcell (1912–97) and other researchers associated with the nascent field of radio astronomy during the 1950s. The relevance of quantitative and predictable physical properties associated with the 21 cm hydrogen emission line as a suitable medium for interstellar communication was later analyzed and discussed in a 1959 paper by Cornell physicists, Giuseppe Cocconi (1914–2008) and Philip Morrison (1915–2005). Further, Drake himself used the radio telescope at the US National Radio Astronomy Observatory (NRAO) in 1960 to search for interstellar messages from extraterrestrial intelligence within a narrow band that corresponds to the hydrogen emission line. Drake later explained that his selection of this particular narrowband region was primarily determined by practical and situational constraints, which included the need to calibrate and maintain a prototype narrowband receiver to 21 cm wavelength so other radio astronomers at the NRAO could use it to conduct routine observations of astrophysical phenomena.[30]

Two-dimensional schematic representations of the hyperfine transition of hydrogen and pulsars included on the Pioneer plaque and Voyager record were attempts by scientists and their colleagues to visualize physical phenomena that can only be detected using devices such as radio telescopes or functionally equivalent material technologies. Visual representations of the hyperfine transition of neutral hydrogen and Drake's constellation of 14 selected pulsars are effectively a distillation of previous inscriptions, astrophysical theories and mathematical models. These are not pictorial representations of objects that can be perceived with sense organs or universal knowledge expressed in a form that is by design transparent, intelligible and self-sufficient. 'If scientists were looking at nature, at economies, at stars, at organs, they would not *see* anything,' Bruno Latour has noted: 'Scientists start seeing something once they stop looking at nature and look exclusively and obsessively at prints and flat inscriptions.'[31] Thus, Drake's pulsar map constitutes a practical means of objectifying and mobilizing knowledge claims using graphical techniques. Pulsars were rendered as scientific objects constituted in and through inscriptions that are visible, transportable and have intrinsic mathematical properties. Indeed, the map was designed as a visible record of astronomical objects that conform to universal physical laws, rendering them comprehensible to any technologically advanced civilization.

IV Beyond words: the transformation of human figures

In February 1972, prior to the launch of Pioneer 10, images of the interstellar plaque were widely circulated in scientific journals, newspapers, magazines and other forms of mass media. Frank Drake and Carl Sagan met with journalists and other news producers and described the purpose and meaning of images on the Pioneer plaque. Their motivation to include Linda Salzman Sagan's drawing of a nude man and woman was 'purely scientific' and meant to provide extraterrestrial beings with objective knowledge about humankind such as anatomical differences between male and female and 'diverse racial characteristics.' Lay audiences in the United States and other countries responded in a variety of ways to images of the Pioneer plaque reproduced in newspapers, television and other news media. Mass circulation of the naked figures in national newspapers across the United States and transportation of the plaque into space aboard NASA spacecraft led to public controversy and claims that images engraved on the Pioneer plaque were pornographic, sexist and ethnocentric.[32]

Contemporaneous documents from a number of sources contain contradictory explanations regarding a series of alterations to Linda Salzman Sagan's

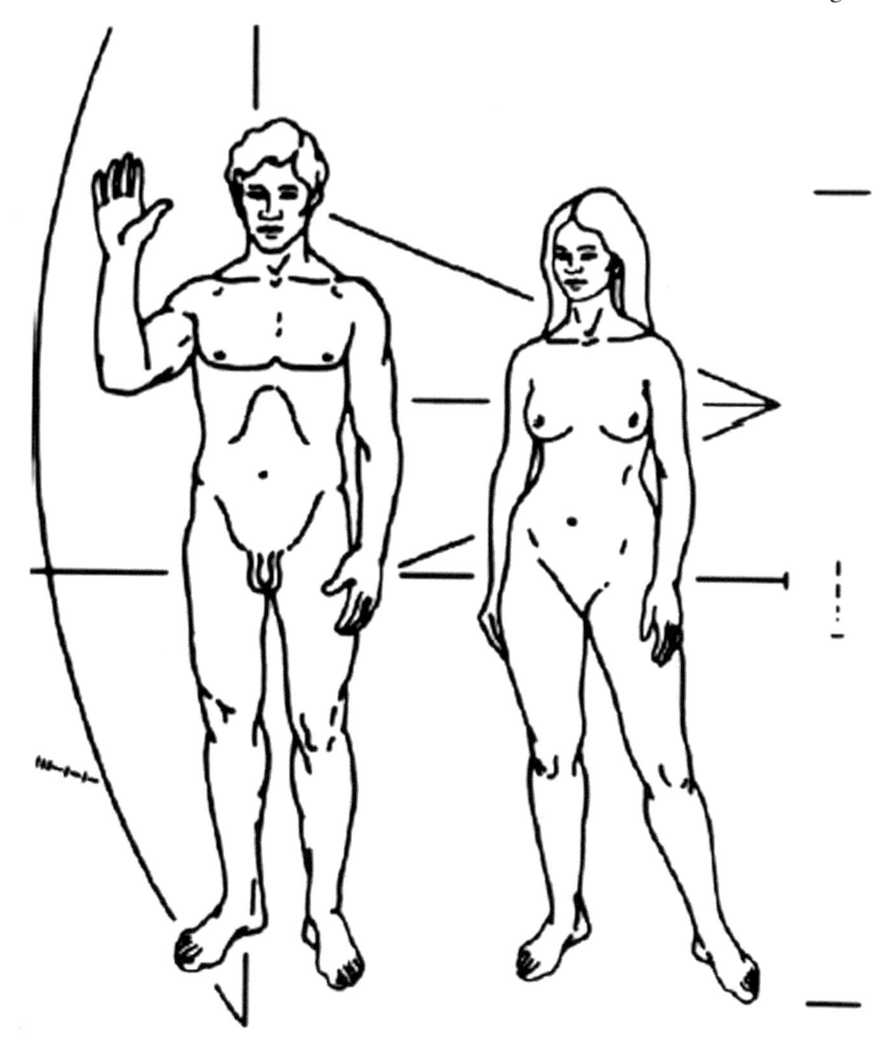

Figure 15.6 Enlargement of detail on the Pioneer plaque showing an illustration of a man and woman, 1972.
Source: Adapted from NASA; see Figure 15.1.

original drawings of human figures during the design phase and actual production of the Pioneer metal plaque, citing a variety of scientific, aesthetic and pragmatic reasons to account for substantive transformation of the original human figures (Figure 15.6). More specifically, there are conflicting accounts about changes in the delineation of female external genitalia and 'ethnic' facial features in both human figures. Inconsistencies with respect to personal recollections of key actors concerning, precisely, why and how the human figures were altered and lack of correspondence between explanations from different sources are instructive in terms of revealing the political, social and unmistakably material practices associated with the collaborative design of NASA's Pioneer plaque and the history of interstellar messages.

Robert Kraemer, director of Planetary Programs on the Pioneer Project Team, was present at the meeting in January 1972 to approve the designs and artwork for the proposed interstellar message, and recalls being rather apprehensive about the response of senior NASA officials to Salzman's drawings. Two decades later he explained:

> Linda was a skilled artist and her naked human figures were very detailed and realistic, as they needed to be. It seems a bit silly today, but at the time I feared that some taxpayers, the true owners of the spacecraft, might label it as pornographic. My boss, John Naugle, had no such fears and approved the design but with one compromise of erasing a short line indicating the woman's vulva.[33]

As far as Kraemer was concerned, it was only after seeing a cartoon of the plaque in the *Los Angeles Times* a few weeks later that he was able to feel less apprehensive about future public response to the pictorial images on the plaque:

> After seeing that cartoon I relaxed, with increased faith in the common sense of the American public. I learned later that there had been one more alteration of Linda's art work. The public affairs people at Ames, thinking the facial features on her figures too ethnic, neutralized them into a racial mixture.[34]

There is no record of Linda Salzman Sagan's response to NASA's recommendations. Her partner and colleague on the plaque design project, Carl Sagan, was inconsistent with respect to published recollections of this incident. A few weeks prior to the launch of Pioneer 10, Carl Sagan acknowledged that the drawing of the female figure was altered in accordance with NASA's recommendation: 'Sagan said the space agency accepted the naked couple, but objected to Linda Sagan's original drawing of the woman as being a bit too explicit. "The final version has been toned down considerably," Sagan said.'[35] In a later response to comments by feminists and other critics regarding the apparent lack of female genitalia in the figure drawing featured on the plaque, Carl Sagan implied that NASA officials turned out to be rather more liberal in their views than he had expected and did not recommend any changes to the original artwork. The decision to exclude details from the representation of the female figure on the finished plaque was, he claimed, based on aesthetic principles and misjudgment on the part of the plaque's designers regarding supposed puritanical attitudes among the NASA hierarchy:

> The decision to omit a very short line in this diagram was partly made because conventional representation in Greek statuary omits it. But there was another reason: Our desire to see the message successfully launched on *Pioneer 10*. In retrospect, we may have judged NASA's scientific-political hierarchy as more puritanical than it is. In the many discussions that I held with such officials, up to the Administrator of the National Aeronautics and Space Administration and the President's Science Adviser, not one Victorian demurrer was ever voiced; and a great deal of helpful encouragement was given.[36]

Official NASA documents and reports by NASA Ames contractors, such as the Bendix Corporation, constitute additional sources of expository narratives concerning the collaborative design and production of the human figures that were physically engraved, along with all the other components of the interstellar message, on the Pioneer plaque. These documents often describe the figure drawings and other visual representations on the plaque by employing technical language and an impersonal mode of description, which overlook the social origins and conditional basis of the pictorial representations and scientific facts inscribed on Pioneer's interstellar message:

> On the plaque a man and a woman stand before an outline of the spacecraft. The man's hand is raised in a gesture of good will. The physical make up of the man and woman are determined from the results of computerized analysis of the average person in our civilization.[37]

Such a textual description suggests that a supposedly generic representation of human physical attributes of an 'average' man and woman depicted on the plaque have been obtained by technological means. Although the document identified the historical actors responsible for the design of the plaque, the description of technological means or inscription devices used to render the human figures do not refer to human operators or agents responsible for their design. The implication is that computerized analysis of quantitative data concerning human physical attributes is sufficient for the construction of an authentic visual representation of humans, supporting attendant claims that the Pioneer interstellar message is fundamentally representative of all mankind; nevertheless, concessions mean it is not.[38]

Technology in the form of inscription devices and graphical techniques is highlighted in other contemporary explanations of how hair and facial characteristics of both the figures featured on the plaque are deliberately comprised of composite anatomical features, believed to be typical of different 'races' in order to create 'de-ethnicized' or, according to Kraemer, 'neutralized,' rendering of the human face. The design and reproduction of these composite details as material visual representations was considered necessary and appropriate for inclusion on the interstellar message contained on the Pioneer plaque, routinely characterized by Carl Sagan and others as representative of all 'mankind.' However, the carefully prepared and collaboratively produced 'panracial' human figures etched onto the 15 by 22.5 cm gold-anodized aluminium plate did not appear as the designers had intended:

> In the original sketches from which the engravings were made, we made a conscious attempt to have the man and woman panracial. The woman was given epicanthian folds and in other ways a partially Asian appearance. The man was given a broad nose, thick lips, and a short 'Afro' haircut. Caucasian features were also present in both. We had hoped to include at least three of the major races of mankind [...]. Somewhere in the transcription from the original sketch drawing to the final engraving the Afro was transmuted into a very non-African Mediterranean-curly haircut.[39]

The juxtaposition of contradictory accounts concerning transformations to the human figures depicted on the plaque are specific examples in which scientific practice and public discourse, ranging from technical documents to news media, purport to provide objective descriptions of the Pioneer interstellar message. Nevertheless, these accounts incorporated speculative theories, rhetorical statements and the personal opinions of scientists and other historical actors. Modification of Linda Salzman Sagan's drawings are literally illustrative of material ways in which artistic or subjective description of, in this case, the human body were altered by omission of visual details and converted through numerical analysis to produce and simultaneously validate what purports to be an objective visual representation of an 'average person.' In short, the narrative accounts described above reveal heterogeneous social processes, inscription technologies and rhetorical strategies for acquisition, production and circulation of a scientifically valid representation or 'fact' regarding human anatomical features considered appropriate for interstellar communication.

Similar rhetorical strategies and application of contemporary inscription devices have been studied by historians Lorraine Daston and Peter Galison in their research on the interplay between social discourse and material aspects of objectivity, encapsulated in scientific atlases from the late nineteenth and early twentieth century. Moral imperatives and mechanical methods were used by scientists during the period to characterize images (reproductions of drawings and photographs, for example) as definitive and objective representations of nature. Similarly, the human figures created for NASA's Pioneer plaque constitute essentialized representations comprised of carefully selected composite visual elements such as facial features, which render 'race' as an essential category or underlying type that can be discerned with the naked eye and incorporated in naturalistic visual representations of the human body that maintain objectivity or 'truth to nature.'[40]

Although the Pioneer plaque was primarily designed as a material artifact for carrying an interstellar message to extraterrestrials in remote star systems, images and written descriptions of the plaque have nevertheless been widely discussed and circulated in scientific texts and news media. The Pioneer interstellar message has been used in a variety ways as an expository tool and repository of universal scientific knowledge, for example, that relies exclusively on visual representations and mathematical symbols to communicate knowledge and meaning. However, a series of transformations applied to Linda Salzman Sagan's figure drawings and repeated claims regarding the correspondence between pictures on the plaque and intrinsic mathematical properties of the physical universe also served a rhetorical function. The interstellar message created for the Pioneer plaque was simultaneously constitutive of scientific practices and the scientific objects it was supposed to render visible.

V Interstellar messages, scientific facts and cultures of the imagination

The history of interstellar communication shows that inscriptions of scientific knowledge and mathematics are produced in and through socially embedded practices and subject to contestation, revision, and transformation. Debates surrounding the Pioneer plaque reveal assumptions held by scientists and co-workers concerning the ubiquity of mathematics and legibility of knowledge rendered as two-dimensional visual representations. Interstellar messages are a manifestation of concerted efforts to administer knowledge and demarcate fact from fiction.

Whilst interstellar communication can be regarded as a niche field or diversion from conventional topics of scientific research; networks, graphical technologies, inscription practices and craft skills applied in the creation of images for interstellar messages are significant components in the production and dissemination of scientific knowledge more generally. Interstellar messages are the product of heterogeneous material practices, discursive forms and imaginative approaches to constructing pictures and mathematical symbols that constitute a reference standard regarded as sufficient for specifying fundamental properties of the universe, humankind and the planet we inhabit as objects of scientific study.

Carl Sagan and others involved in the design of NASA's Pioneer plaque decided to construct an array of images that conveyed scientific facts and objective knowledge, whilst rejecting what they believed to be purely subjective, superficial and parochial expressions of human culture such as language, religious icons and national symbols. Further, interstellar messages designed to initiate contact with extraterrestrials are meant to define and communicate scientific knowledge about fundamental physical properties that apply everywhere in the universe without losing sight of what makes us human. The tension between notions of universal and particular properties is a complex and difficult problem in science as well as other spheres of cultural activity. The history of interstellar messages affords opportunities for critical analysis of the interplay between embodied skills, inscription technologies and human agency in the production of scientific knowledge and the rendering of space exploration as both conceivable and manifest endeavors. Interstellar messages demonstrate how the desire for contact with cultures of the imagination resists categorical distinctions between fact and fiction.

Notes

1. Carl Sagan, *Cosmos: The Story of Cosmic Evolution, Science and Civilisation*, London: Abacus, 1980, 325; Bruno Latour, *Science in Action: How to Follow Scientists and Engineers Through Society*, Cambridge, MA: Harvard University Press, 1987, 248 (emphasis in original). I am grateful to my PhD supervisor, David Kirby, for his constant encouragement and astute observations on earlier versions of this chapter. I would like to thank April Gage, Glenn Bugos and Jack Boyd at the NASA Ames History Office for their hospitality and advice on archival issues. I also wish to thank Hans Mark and Linda Salzman Sagan for their valuable assistance and points of clarification, and Mike Lynch and Alexander Geppert for their support and helpful comments on an earlier version of this chapter.
2. See the collected papers in Carl Sagan, ed., *Communication with Extraterrestrial Intelligence (CETI)*, Cambridge, MA: MIT Press, 1973; Cyril Ponnamperuma and Alastair G.W. Cameron, eds, *Interstellar Communication: Scientific Perspectives*, Boston: Houghton Mifflin, 1974; and Philip Morrison, John Billingham and John Wolfe, eds, *The Search for Extraterrestrial Intelligence (SETI)*, Washington, DC: NASA, 1977, which feature contemporary theories on interstellar messages and specific applications of the term.
3. The most frequently cited texts on the Pioneer plaque and Voyager record are Carl Sagan, Linda Salzman Sagan and Frank D. Drake, 'A Message from Earth,' *Science* 175 (25 February 1972), 881–4; and Carl Sagan, Frank D. Drake, Ann Druyan, Timothy Ferris, Jon Lomberg and Linda Salzman Sagan, *Murmurs of Earth: The Voyager Interstellar Record*, New York: Random House, 1978.
4. See John Heilbron, *The Dilemmas of an Upright Man: Max Planck as Spokesman for German Science*, Berkeley: University of California Press, 1986, 50–4, regarding Max Planck's 'universal constants'; Frank Drake's description of 'quantum laws' in idem and Dava Sobel, *Is Anyone Out There? The Scientific Search for Extraterrestrial Intelligence*, London: Simon & Schuster, 1997, 178; and Carl Sagan's description of 'the laws of Nature' in idem, *Cosmos*, 325.
5. On the alleged universality of science, mathematics and physical laws incorporated in theories and empirical research on interstellar communication, see, for example, Giuseppe Cocconi and Philip Morrison, 'Searching for Interstellar Communications,' *Nature* 184 (19 September 1959), 844–6; Frank D. Drake, 'How Can We Detect Radio Transmissions from Distant Planetary Systems?' [1959], in Alastair G.W. Cameron, ed., *Interstellar Communication: A Collection of Reprints and Original Contributions*, New York: W.A. Benjamin, 1963, 165–75; Frank

D. Drake, *Intelligent Life in Space*, New York: Macmillan, 1963, 110; Iosif Shklovsky and Carl Sagan, *Intelligent Life in the Universe*, San Francisco: Holden Day, 1966, 388; Carl Sagan, *The Cosmic Connection: An Extraterrestrial Perspective*, New York: Doubleday, 1973, 18–20, 30; idem, 'For Future Times and Beings,' in idem et al., *Murmurs*, 1–43, here 13–14; and Bernard M. Oliver, 'The Rationale for a Preferred Frequency Band,' in Morrison, Billingham and Wolf, *SETI*, 63–73. For arguments against the assumption that physical laws are universal, see transcript of discussion between the scientists Francis Crick, Thomas Gold and Vitaly Ginzburg at a conference workshop on communication with extraterrestrial intelligence (CETI) reported in Sagan, *CETI*, 204–6, for example. Ideas and claims concerning putative 'alien mathematics' are critically examined in David Ruelle, 'Conversations on Mathematics with a Visitor from Outer Space,' in Vladimir Arnold, Michael Atiyah, Peter Lax and Barry Mazur, eds, *Mathematics: Frontiers and Perspectives*, Providence: American Mathematical Society, 2000, 251–9; and Edward Regis, ed., *Extraterrestrials: Science and Alien Intelligence*, Cambridge: Cambridge University Press, 1985.

6. For historical research and sociological studies in this area, see Thomas S. Kuhn, *The Structure of Scientific Revolutions*, 2nd edn, Chicago: University of Chicago Press, 1970; Martin J.S. Rudwick, 'The Emergence of a Visual Language for Geological Science, 1760–1840,' *History of Science* 14.3 (September 1976), 149–95; Bruno Latour and Steve Woolgar, *Laboratory Life: The Construction of Scientific Facts*, Princeton: Princeton University Press, 1986; Harold Garfinkel, Michael Lynch and Edward Livingston, 'The Work of a Discovering Science Construed with Materials from the Optically Discovered Pulsar,' *Philosophy of the Social Sciences* 11.2 (June 1981), 131–58; Steven Shapin and Simon Schaffer, *Leviathan and the Air-Pump: Hobbes, Boyle and the Experimental Life*, Princeton: Princeton University Press, 1985; Michael Lynch, 'Discipline and the Material Form of Images: An Analysis of Scientific Visibility,' *Social Studies of Science* 15.1 (February 1985), 37–66; Wiebe Bijker, Thomas Hughes and Trevor Pinch, eds, *The Social Construction of Technological Systems*, Cambridge, MA: MIT Press, 1987; Michael Lynch and Steve Woolgar, eds, *Representation in Scientific Practice*, Cambridge, MA: MIT Press, 1968; Theodore M. Porter, *Trust in Numbers: The Pursuit of Objectivity in Science and Public Life*, Princeton: Princeton University Press, 1995; and Timothy Lenoir, ed., *Inscribing Science: Scientific Texts and the Materiality of Communication*, Stanford: Stanford University Press, 1998. See also Steven Shapin, 'Pump and Circumstance: Robert Boyle's Literary Technology,' *Social Studies of Science* 14.4 (November 1984), 481–520; and idem and Schaffer, *Leviathan*, on the importance of social, material and literary technologies in seventeenth-century scientific practice and historical contingency of scientific facts.
7. See Grant Malcolm, ed., *Multidisciplinary Approaches to Representation and Interpretation*, Amsterdam: Elsevier, 2004; Brian S. Baigrie, ed., *Picturing Knowledge: Historical and Philosophical Problems Concerning the Use of Art in Science*, Toronto: University of Toronto Press, 1996; Gillian Rose, *Visual Methodologies*, London: Sage, 2001; David A. Kirby, 'Science Consultants, Fictional Films, and Scientific Practice,' *Social Studies of Science* 33.2 (April 2003), 231–68; and Edward R. Tufte, *The Visual Display of Quantitative Information*, 2nd edn, Cheshire: Graphics Press, 2001.

8. See Shapin and Schaffer, *Leviathan*; Lynch and Woolgar, *Representation*; Darin J. Arsenault, Laurence D. Smith and Edith A. Beauchamp, 'Visual Inscriptions in the Scientific Hierarchy,' *Science Communication* 27.3 (March 2006), 376–428; and Eric Francoeur, 'Beyond Dematerialization and Inscription: Does the Materiality of Molecular Models Really Matter?,' *HYLE: International Journal for Philosophy of Chemistry* 6.1 (March 2000), 63–84. On the history of non-linguistic interstellar messages, see Douglas A. Vakoch, 'The Conventionality of Pictorial Representation in Interstellar Messages,' *Acta Astronautica* 46.10 (June 2000), 733–6.
9. See Jacques Derrida, *Of Grammatology*, Baltimore: Johns Hopkins University Press, 1974, and *Archive Fever: A Freudian Impression*, Chicago: University of Chicago Press, 1995; Timothy idem, 'Inscription Practices and Materialites of Communication,' in idem, *Inscribing Science*, 1–19; Latour and Woolgar, *Laboratory Life*, 45–53; and Wolff-Michael Roth and Michelle K. McGinn, 'Inscriptions: Toward a Theory of Representing as Social Practice,' *Review of Educational Research* 68.1 (Spring 1998), 35–59.
10. For further discussion of semiotics and semiosis, see James Hoopes, ed., *Pierce on Signs: Writings on Semiotic*, Chapel Hill: University of North Carolina Press, 1991; Roland Barthes, *Image Music Text*, London: Fontana, 1977; Charles Morris, *Signification and Significance*, Cambridge, MA: MIT Press, 1964; Roman Jakobson, *The Framework of Language*, Ann Arbor: University of Michigan, 1980; Mieke Bal and Norman Bryson, 'Semiotics in Art History,' *The Art Bulletin* 73.2 (June 1991), 174–208; and Francoise Bastide, 'The Iconography of Scientific Texts: Principles of Analysis,' in Lynch and Woolgar, *Representation*, 187–230.
11. See Thomas O'Toole, 'Pioneer F Bears a Hello to Space Aliens,' *Washington Post* (25 February 1972), 1; Richard O. Fimmel, William Swindell and Eric Burgess, *Pioneer Odyssey*, Washington, DC: NASA, 1977, available at http://history.nasa.gov/SP-349/sp349.htm (accessed 1 October 2017); Richard O. Fimmel, James Van Allen and Eric Burgess, *Pioneer: First to Jupiter, Saturn and Beyond*, Washington, DC: NASA, 1980; and Kevin J. Kilburn, *Eric Burgess: Manchester's First Rocket Man*, http://www.mikeoates.org/astro-history/burgess.htm (accessed 1 October 2017).
12. See, for instance, Steven J. Dick, *The Biological Universe: The Twentieth-Century Extraterrestrial Life Debate and the Limits of Science*, Cambridge: Cambridge University Press, 1996; Robert S. Kraemer, *Beyond the Moon: A Golden Age of Planetary Exploration, 1971–1978*, Washington, DC: Smithsonian Institution Press, 2000; George Basalla, *Civilized Life in the Universe: Scientists on Intelligent Extraterrestrials*, Oxford: Oxford University Press, 2006; Mark Wolverton, *The Depths of Space: The Story of the Pioneer Probes*, Washington, DC: Joseph Henry Press, 2004; and Gregory L. Matloff, *Deep Space Probes*, 2nd edn, New York: Springer Praxis, 2005.
13. Pioneer 10 was launched from the Kennedy Space Center aka KSC or Cape Canaveral, Florida on 2 March 1972. For further details regarding the chronology of NASA's Pioneer 10 and 11 spacecraft, see, for example, Mark Wade, *Encyclopedia Astronautica*, http://www.astronautix.com/craft/pior1011.htm (accessed 1 October 2017); and NASA's National Space Science Data Center (NSSDC), at their web site, http://nssdc.gsfc.nasa.gov (accessed 1 October 2017). For a chronology of NASA's Pioneer program of unmanned spacecraft and other US and Soviet deep-space missions during the late twentieth century,

see Asif A. Siddiqi, *A Chronology of Deep Space and Planetary Probes 1958–2000*, Washington, DC: NASA, 2002. Individual NASA spacecraft are customarily designated by letter prior to launch and number afterwards; see Kraemer, *Beyond the Moon*, 62. Consequently, references to Pioneer F and G (aka Pioneer 10 and 11) spacecraft can be found in a wide range of documentary material. Further details on the names associated with Pioneer and other deep-space missions in the 1970s are available in official NASA documentation, such as Helen T. Wells, Susan H. Whiteley and Carrie E. Karegeannes, *Origins of NASA Names*, Washington, DC: NASA, 1976; available at http://history.nasa.gov/SP-4402/SP-4402.htm (accessed 1 October 2017). Also, see the compilation of NASA mission reports and other archive material in Robert Godwin and Steve Whitfield, eds, *Deep Space: The NASA Mission Reports*, Ontario: Apogee, 2005.

14. Eric Burgess was a science writer and correspondent for the *Christian Science Monitor*. Before moving to the United States with his family in 1956, Burgess was a keen proponent of spaceflight and worked closely with Arthur C. Clarke and others to reorganize the British Interplanetary Society during the postwar years. For a description of the CETI conference held at the Byurakan Astrophysical Observatory in the former Soviet Union, see 'First Soviet-American Conference on Communication with Extraterrestrial Intelligence (CETI),' *Icarus* 16.2 (April 1972), 412–14; and Sagan, *CETI*.
15. Hans M. Mark, telephone interview with William R. Macauley, 12 November 2008.
16. Sagan, *Cosmic Connection*, 18. Linda Salzman Sagan studied art at the Boston Museum School of Fine Arts while attending a joint academic program for fine art students at Tufts University, near Boston. In June 1968, she graduated from Tufts with a BSc degree in Art Education and interrupted her postgraduate studies to move to Ithaca, New York, with Carl Sagan whom she had married on 6 April 1968. During the early 1970s, Linda Salzman Sagan also produced artwork to illustrate a book written by Carl Sagan, and she is one of the principal actors responsible for another interstellar message, NASA's Voyager record, which was designed and created in 1976–77. See drawing of constellation in Sagan, *Cosmic Connection*, 13; and Linda Salzman Sagan, 'A Voyager's Greetings,' in Sagan et al., *Murmurs*, 123–47. Linda Salzman Sagan email message to William R. Macauley, 27 June 2009.
17. Lynch, 'Discipline,' 43.
18. See idem, 'The Externalized Retina: Selection and Mathematization in the Visual Documentation of Objects in the Life Sciences,' in idem and Woolgar, *Representation*, 153–86.
19. Drake's pulsar map is featured on the Pioneer plaque and on two parts of Voyager's interstellar message – the protective cover and in a picture encoded as a binary signal on the Voyager record. See Sagan 'For Future Times and Beings,' in Sagan et al., *Murmurs*, 1–43.
20. The time interval for hydrogen emission is usually specified in terms of radio frequency (1420 MHz), which is the reciprocal value of a wave period or fraction of time equivalent to less than a billionth of a second. According to the designers of the Pioneer plaque, the precise value for the wave period of neutral hydrogen and the unit of time inscribed on the plaque is $(1.420405752 \times 10^9 \text{ sec}^{-1})^{-1}$; see Sagan, Salzman Sagan and Drake, 'Message,' 882.
21. Ibid., 881.
22. Ibid.

23. Ibid., Table 1, 882.
24. For a short period the 'scruff' was also jokingly referred to as 'LGM-1' (little green men) by Bell and her colleagues. See footnote in Drake and Sobel, *Anybody*, 88. Interestingly, the only recorded instance of a candidate signal transmitted by extraterrestrial intelligence also refers to anomalies discerned in raw data of narrowband radio emissions printed out as matrices of numbers. The incident is widely reported in histories of SETI and described as the 'wow signal' because the scientist, Jerry Ehman, who made the initial observation in 1977, scribbled the word 'Wow!' on the paper printout at Ohio State University Radio Observatory.
25. See figures 1 and 2 in A.J. Hewish, S.J. Bell, J.D.H. Pilkington, P.F. Scott, and R.A. Collins, 'Observation of a Rapidly Pulsating Radio Source,' *Nature* 217 (24 February 1968), 709–13, here 710–11.
26. Sagan, Salzman Sagan and Drake, 'Message,' 882.
27. For example, the relative position of each pulsar is indicated by a radial line, using polar representations; these coordinates can be translated into a Cartesian coordinate system.
28. See figure 4 in Edward Purcell, 'Radioastronomy and Communication through Space,' in Cameron, *Interstellar Communication*, 125.
29. Francoeur, '*Beyond Dematerialisation*,' unpaginated. See also Jane S. Richardson, David C. Richardson, Neil B. Tweedy, Kimberly M. Gernert, Thomas P. Quinn, Michael H. Hecht, Bruce W. Erikson, Yibing Yan, Robert D. McClain, Mary E. Donlan and Mark C. Surles, 'Looking at Proteins: Representations, Folding, Packing, and Design,' *Biophysical Journal* 63.5 (November 1992), 1185–209.
30. Cocconi and Morrison, 'Searching,' 845; and Frank D. Drake, 'Project Ozma' [1962], in Cameron, *Interstellar Communication*, 176–7. See Frank D. Drake interview by David W. Swift, June 1981, Ithaca, in David W. Swift, *SETI Pioneers: Scientists Talk About Their Search for Extraterrestrial Intelligence*, Tucson: University of Arizona Press, 1990, 54–85, here 61–2.
31. Bruno Latour, 'Drawing Things Together,' in Lynch and Woolgar, *Representation*, 19–68, here 39.
32. Sagan, *Cosmic Connection*, 24–32; Wolverton, *Depths of Space*, 71–83.
33. Kraemer, *Beyond the Moon*, 75.
34. Ibid., 76.
35. O'Toole, 'Pioneer F,' 1.
36. Sagan, *Cosmic Connection*, 24.
37. NASA Ames History Office, NASA Ames Research Center, Moffett Field, California. PP03.02, Robert W. Jackson Collection, 3:69. *Pioneer to Jupiter: Second Exploration*, Palo Alto: Bendix Field Engineering Corporation, November 1974, 29.
38. Sagan, Salzman Sagan and Drake, 'Message,' 883.
39. Sagan, *Cosmic Connection*, 26–7.
40. Lorraine Daston and Peter Galison, 'The Image of Objectivity,' *Representations* 40 (Fall 1992), 81–128; see also eidem, *Objectivity*, New York: Zone, 2007.

CHAPTER 16

Alien Spotting: Damien Hirst's Beagle 2 Mars Lander Calibration Target and the Exploitation of Outer Space

Tristan Weddigen

Ah the old questions, the old answers, there's nothing like them!
Damien Hirst, quoting Samuel Beckett's *Endgame*[1]

During the summer recess of 1996, the August issue of *Science* magazine spectacularly revealed the apparent microscopic evidence of fossilized bacteria on the Mars meteorite ALH84001, discovered in Antarctica in 1984 (Figure 16.1).[2] Bill Clinton's public announcement that the meteorite 'speaks of the possibility of life' and that, although 'it promises answers to some of our oldest questions, it poses still others even more fundamental,' fuelled American and worldwide public discussion on extraterrestrial life and future Mars missions – and hence probably alleviated NASA from budgetary restraints.[3] Clinton's declaration that the first Mars mission was 'scheduled to land on Mars on 4 July 1997 – Independence Day' was most likely alluding to Roland Emmerich's namesake science-fiction movie. The president had enjoyed it in a preview screening at the White House – which alien invaders destroy in the movie. The film had been released with great success on 2 July, the day the plot of the movie begins, after it had been publicized with taglines such as 'We've always believed we weren't alone. On July 4, we'll wish we were,' citing and renewing Orson Welles's 1938 mock radio broadcasting of a Martian invasion taken from H.G. Wells's *The War of the Worlds*.[4] The

Tristan Weddigen (✉)
Universität Zürich, Zurich, Switzerland
e-mail: tristan.weddigen@uzh.ch

Alexander C.T. Geppert (ed.), *Imagining Outer Space*
European Astroculture, vol. 1
https://doi.org/10.1057/978-1-349-95339-4_16

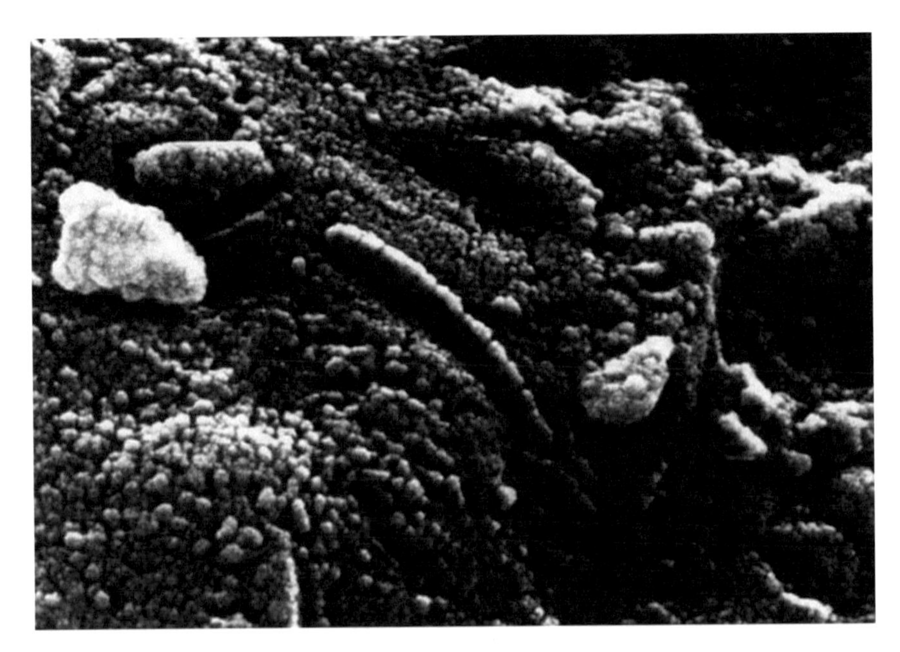

Figure 16.1 High-resolution scanning electron microscope image of the meteorite ALH84001, published most frequently in relation to the possibility of life on Mars, 1996. *Source*: Courtesy of NASA.

conflation of science, fiction, politics, economy, pop culture and globalized mass media makes the rediscovery of Mars in the 1990s a postmodern event.

One year later, in competition with NASA's successful and publicly acclaimed Mars Pathfinder and the succeeding Mars Odyssey and Mars Exploration Rover programs, the European Space Agency (ESA) started its own Mars Express mission to search for 'signatures of life' on the Red Planet.[5] In June 2003, an orbiter and a lander were launched. The lander, called Beagle 2 after Charles Darwin's ship as a tribute to evolutionary theory, had been developed since 1997 by a British public-private consortium under the direction of the principal investigator Colin Pillinger (1943–2014) of the Open University (Figure 16.2). Its explicit purpose was to find answers to the question whether there is or has been life on Mars. Beagle 2 was due to land on Christmas Eve 2003, but has not been traceable since its separation from the orbiter and has therefore been declared missing.[6]

I Blurring boundaries between science, economy, and art

A landing and identification sign that never resounded back to earth had been written into the source code of the lander's software and consisted of a nine-note refrain composed by the celebrated British neo-pop band Blur. The jingle was based on the Fibonacci sequence and was reminiscent of the tune of the popular BBC science-fiction television series of the 1960s, *Doctor Who*.[7]

Figure 16.2 Colin Pillinger and the Beagle 2 model, March 2001.
Source: Courtesy of Beagle 2.

As Pillinger states, because he faced great difficulties in financing the project, the cooperation between the Beagle 2 project and Blur was a marketing strategy to attract the attention of mass media and thus possible public and private sponsorship.[8] He has labeled himself a 'professor of PR' and a 'pop entrepreneur,' creating a 'publicity machine' and even a 'bit of art.'[9]

Blur's drummer, the computer scientist Dave Rowntree, who also collaborated in the digital animation video of Beagle 2, ironically commented on the call sign: 'How do you know what sounds like a friendly, warm greeting and what sounds like a declaration [of war]? But I think Martian bacteria love Britpop.' And: 'We're not expecting to find anything more than bacterial life on Mars, so I imagine they'll like heavy metal,' alluding to the lander's hardware instead of software.[10] Rowntree's statement makes clear that the tune, as a mere programming code imperceptible to both humans and aliens – unlike the Voyager 1 and 2 humanist interstellar messages launched in 1977 – had no serious intention of conveying a universal peace message or to establish a basis for extraterrestrial communication.[11] Even before the launch of Mars Express, scientific evidence did not raise expectations higher than finding bacteria or water, which did not capture people's imagination. Historically, aliens have been the mirror image of humanity's fears and hopes by surpassing us in intelligence or wickedness. In 2003 Martians were imagined as late consumers of mainstream Britpop.[12]

Blur's retro-futurist reference to their childhood's mass media culture such as *Doctor Who* and their own imitation and commercial appropriation of British pop music of the 1960s and 1970s exposes a sincere lack of metaphysical beliefs beyond a capitalist consumerism which recycles and re-exploits historical fetishes and icons. Much like the pop band, Pillinger declared himself inspired by earlier British science-fiction culture of the 1950s such as *Dan Dare*, *Eagle* and *Journey into Space*. The Mars lander itself, before unfolding, appears to be a retro-futuristic citation, for it clearly resembles the type of flying saucer as it was canonized in 1947, playfully transforming the Red Planet into the inhabited Mars we wished it to be.[13] Blur's Mars ringtone was most likely intended to repeat the global media success of the frightening and unforgettable Sputnik beep of 4 October 1957. Instead of the victory of Soviet imperialism, it would announce the cosmic emanation of post-industrial capitalism and global entertainment, the final victory of pop over ideology. Similarly, in 1997, one year after the supposed discovery of fossilized Martian bacteria, the artist Joan Fontcuberta (1955–) had produced the space-historical and post-soviet hoax *Sputnik*, which included a fake meteorite with a microscopic alien message to humanity as an ironic comment on the new retro-futurist US Space Race publicity.[14] Blur's pseudo-acoustic identification signal was clearly addressed to humans alone.

Figure 16.3 Sample of Ferrari *rosso corsa* paint on Mars Express Orbiter, 2002.
Source: Illustration by Medialab. Courtesy of ESA.

The planned landing date of Beagle 2, Christmas Eve – a day even more sensational for the discovery of extraterrestrial life than Independence Day – was probably calculated to back the project's publicity, too. In a post-Cold War and seemingly post-ideological world, the exploration of outer space is not only a political power game and a scientific enterprise but also a matter of economic exploitation.[15] The abuse of Christ's birthday goes beyond commodification of outer space and can be interpreted as a postmodern and ironic reference to the modernist hopes for the epiphany of a higher extraterrestrial intelligence which would act as a material *ersatz* after the proclaimed death of God. The involvement of a pop group with a space agency is not only a business deal, but also a symptom of contemporary culture, which increasingly merges science, economy and art. In contrast, the sample of Ferrari's *rosso corsa* paint attached by ESA to the Mars Express Orbiter demonstrates how uninspiring and meaningless public-private public relations can be (Figure 16.3).[16]

II Spotting aliens

In 1999, after appointing Blur, Pillinger saw Damien Hirst (1965–) on television, one of the most important exponents and curatorial promoters of the successful artists dubbed 'Young British Artists.'[17] They corresponded to Brit-pop in the visual arts of the 1990s and indulged in a close partnership with the art market, and in particular with the influential collector and advertisement manager Charles Saatchi (1943–). The *enfant terrible* Hirst, one of the most successful and wealthiest artists alive, is not only notorious for his dissected sharks, cows and sheep immersed in formaldehyde, but also for his provocatively harmless and meaningless *Spot Paintings* that he has painted since 1990, and that caught Pillinger's attention. Alex James of Blur then contacted Hirst, his friend from their time together at London Goldsmiths College, who had directed Blur's neo-pop music video clip *Country House* in 1995.[18]

Pillinger made a similar arrangement with Hirst, thus generating a great amount of publicity for Beagle 2. In order to spare the spacecraft's mass and power budget, Pillinger asked Hirst not to provide for a dissected animal – 'cows are too heavy' – but to design a calibration target for the several onboard cameras and instruments (Figure 16.4).[19] The small, space-qualified aluminum plate was to resemble Hirst's *Spot Paintings* displaying a series of different benchmarking dots. For example, the Moessbauer spectrometer was meant to analyze iron in Martian soil and needed to be checked against standard samples of different types of iron *in situ*. Accordingly, Pillinger defined nine different hues of the synthetic Mars yellow pigment, named *crocus Martis* in early modern alchemy, varying from yellow to black according to the grade of its oxidation and supposed to be present in Martian soil. A white spot was included for contrasting with the black one, and two additional, technically non-functional colors were added, green and blue, symbolizing planet Earth, thus contradicting Hirst's concept of non-symbolic *Spot Paintings*. Moreover, Pillinger understood the synthetic iron oxides first used in the 1830s as

Figure 16.4 Damien Hirst, Beagle 2 *Spot Painting*, 2002, on gold background. Aluminum plate with indentations cut in the metal to accept colored spots in nine synthetic Mars iron oxides of different shades of yellow, red, orange, 8 × 8 cm, 26.5 grams. Known for his pricey zoological sculptures as well as his *Spot Paintings*, at times incorporated in earthly installations, this leading Young British Artist (YBA) practitioner hoped to install this *Spot Painting* with a useful chronometric function on Mars, as the visual art companion to a sound track to be played upon the Beagle's landing by the Brit indie band Blur. Although the Beagle crash-landed on a Martian plain, Damien Hirst's *Spot Painting* most likely remains partially intact.
Source: Courtesy of Beagle 2; photograph by Mike Levers.

another tribute to Darwin's *Beagle*.[20] Beagle 2 consisted of a disk-shaped vessel containing a swivel arm pointing a number of cameras and instruments first on the calibration target, before it would analyze the Martian soil.

As little of the arrangement of the spots and the inclusion of two colors seem to have been left to the artist's decision, the impression is confirmed that Hirst's art was utilized as a widely recognizable art logo. Again, here, art was a surplus gadget attached to a scientific machine, and the deal between Pillinger and Hirst was based on the mutual benefit of greater notoriety in case life was discovered on Mars on Christmas Eve. According to Pillinger, the Beagle 2 project was clearly not an art performance, but a scientific project aiming at a higher inquiry: 'This collaboration is not about displaying art in space but about finding out if there is life on Mars.'[21]

Still, the calibration target was explicitly staged as a piece of art, including a two-day exhibit at the White Cube gallery in London before being mounted onto the spacecraft. The aluminum target for Mars remains in the possession of Hirst, himself the most important owner of his own works before the spectacular 2008 sell-out at Sotheby's. The calibration target holds the specific status of a so-called multiple, since two spares were produced for Beagle 2, and three are currently owned by Hirst. Although the calibration target is said to be a purely functional work of art as useful as meaningless, its spots inevitably confer meaning: '[B]eing metaphorical is ridiculous, but it's unavoidable.'[22]

III Visual candy

Hirst is not a militant atheist, but a pragmatic nihilist who, under the present circumstances, finds it hard to believe in the existence of any supernatural being or religion: 'Where's God? God's fucked off.'[23] Hirst's pop existentialism is centered on the ideas of life and death. For him, 'art's about life' since there is nothing else to talk about; because death is an 'unacceptable idea,' contradicting our innate desire for eternal life, and can be made bearable only by distraction and amusement, which is art.[24] On the one hand, Hirst cannot believe in art, as art is ultimately powerless to stop death. But on the other hand, foolishly, he feels compelled to make art and prefers theatrical failure to blind belief: 'I love the way that art doesn't really affect the world. Science affects the world much more directly.' With a gesture of self-denigration, Hirst's *oeuvre* persistently flirts with science, the new 'universal panacea' outstripping religion. He seems conscious that some kind of scientific art would fail too, but perhaps more evidently, thus more effectively revealing art's tragic dilemma: 'It's, like, decorative shit in need of a function; looking for a higher meaning for itself.'[25]

At least, the Beagle 2 calibration target had the advantage of being declared a scientifically 'useful' and 'absolutely essential' work of art containing 'a hell of a lot of science,' according to Pillinger.[26] Unquestionably, the plate would have worked perfectly without Hirst's aesthetic arrangement, and art's usefulness was certainly financial and personal. Hirst likes both myths: the modernist one of the 'artist as a scientist' and Andy Warhol's idea of the 'painter as a machine,' because this would free the artist from producing any consistent meaning. The serial and anonymous production of the *Spot Paintings* by assistants applying simple household gloss on canvas was itself, much like Warhol's *Factory*, somehow industrial and explicitly denying traditional ideas of authorship.[27] In this sense, the calibration target, made by scientists and for robot cameras, comes close to a nihilist technological utopia of art made by machines for machines which do not despair at the human condition.

The scientific aspect pervades Hirst's work, and it measures the power of art with scientific solutions to human problems, especially medicine, leading to question both art as pseudo-science and science as a form of art. His *Spot Paintings* follow a simple and potentially endless pattern, and the colors are varied so as to make every painting seemingly unique. In the *Spot Paintings*, the uniqueness, individuality or 'loneliness' of the single colors are harmonized with the uniformity of the shape and pattern of the dots. Hirst claims to have discovered that such grids of colored spots automatically have a harmonious and cheering effect on the viewers: 'No matter how I feel as an artist or a painter, the paintings end up looking happy.' They almost mechanically 'don't go wrong' and help in ignoring the all-pervading existential angst and horror of human existence.[28]

In order to overcome the romantic, emotional Abstract Expressionism he had learned at the art academy, Hirst felt freed from the need to express his own pointless emotions by developing an expressive system which automatically produces happiness in the beholder's mind.[29] Hirst's strategy is heavily indebted to pop art but exacerbates it to pure commodity. For Hirst, the colorism of the *Spot Paintings* conveys an immediate and simple sense of eternal beauty: 'The thing is they're fucking gorgeous; they're fucking delicious; they don't keep still, they'll live for ever. They're absolutely fantastic. They're color. They're as good as flowers, and they're just fucking paintings.' They produce 'the joy of color' without meaning, total superficiality and meaninglessness, which nevertheless has a direct impact: 'The *Spot Paintings* are an unfailing formula for brightening up people's fucking lives.' According to Hirst, they work like 'sweets (Smarties) or drugs,' but in a 'childish' way, like 'visual candy': 'They are what they are, perfectly dumb paintings which feel absolutely right.' Again, the astonishing economic success of the hundreds of *Spot Paintings* produced by his workshop seems to have proven him correct.[30]

Hirst's discovery that the *Spot Paintings* have an immediate uplifting effect on the beholder strengthens his belief that 'art is like medicine – it can heal.' Consequently, in the series of *Spot Paintings*, Hirst searches for 'a scientific approach to painting in a similar way to the drug companies' scientific approach to life.' He therefore called them *The Pharmaceutical Paintings* and named the single paintings in alphabetical order, according to the product names he had found in the catalogue of a biochemical drug company: *Alpha Tocopherol*, *Aminoantipyrine*, *Aminodoexythymidine*, and the like (Figure 16.5). This led Hirst also to display medicines, multi-colored pills and boxes in showcases and cabinets, which is conceptually and visually close to the *Spot Paintings*.[31] Consequently, Hirst's calibration target might still be intended as a simple, chemical message of beauty, happiness and peace to aliens, as ironic and retro-styled as Blur's ringtone based on a never-ending mathematical sequence.

The regular distribution of colored dots on Hirst's paintings refers to the modernist paradigm of the grid as a visually self-referential model of the world and functions like an innocent-eye test for the camera, the humans and

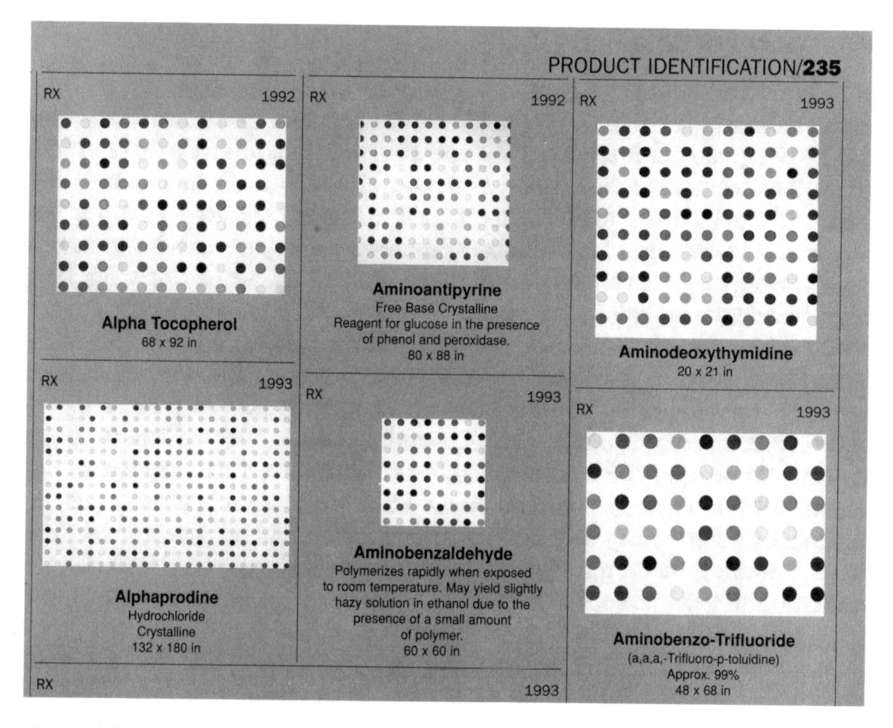

Figure 16.5 Damien Hirst, *Spot Paintings*, detail of chart in the artist's *oeuvre* catalogue.
Source: Damien Hirst, *I Want to Spend the Rest of My Life Everywhere, with Everyone, One to One, Always, Forever, Now*, London: Booth-Clibborn, 1997, 235.

the Martians alike.[32] Interestingly enough, Hirst compared the grid structure of the *Spot Paintings* with the Hasselblad cross-hairs on the Apollo moon photographs: 'If you look at a landscape or one of those photographs they have of the moon, it is a completely unknowable landscape but then NASA always put a grid over it and it gives it a kind of confidence. If you put a grid over something unknowable it is as close as you get to integrating or understanding the image and having power over it. It is the failure of that which I like.'[33] Hirst might be intuitively right: the photographic grids have a scientific look which surpasses their own functionality. As the dot structure of the target is meant to calibrate the Beagle 2 cameras and thereby to reassure humans in their scientific exploration of outer space, art is a means of terraforming Mars and more generally of humanizing the inhumanity of life – an endeavor ultimately condemned to fail. Thus, painting and music seem to be the test cards which reveal what is human, alien or lifeless. 'Where do we come from? What are we? Where are we going?' – these are the fundamental questions both science and art shall ask themselves, Hirst stated by alluding to the title of Paul Gauguin's famous painting, and space exploration offered him a liminal situation for human self-discovery.[34]

IV Art for Martians

Hirst's calibration target stands in a historical tradition of space art, reaching from Kazimir Malevich's (1879–1935) orbital architectures, called *Planites*, to the 1969 Apollo 12 *Moon Museum* and Paul van Hoeydonck's 1971 Apollo 15 *Fallen Astronaut* (see Figures 17.7 and 17.8 below).[35] As much as Pillinger and Blur, Hirst himself owes his ideas about outer space to popular culture and acknowledges his childhood fascination for space travel, but sees in naïve curiosity a positive momentum of science and art. Accordingly, his 1997 *oeuvre* catalogue includes the reproduction of a fake popular astrophysics text book, *Science for All*, illustrating the Great Orion Nebula and the birth of the moon, thus alluding to collective retrograde notions of science as modern myths.[36] But, not being able to ignore a historical series of failed attempts of artists' conquests of outer space – including heroic, utopian, kitsch, ironic and silly strategies – Hirst could hardly avoid making a new statement and comment on humanity's and art's *raison d'être* with his calibration target, the more so by contributing with neo-modern 'high art lite' or 'geometrical decadence,' quoting modernism as an academism of the past. So, ironically, the calibration target becomes a serious message sent from Spaceship Earth, this godforsaken 'Ship of Fools.'[37]

Although Hirst's *Spot Paintings* stand for consumerism, and the calibration target even alludes to the alchemical philosopher's stone whereby art turns matter into money, Pillinger explicitly did not want sponsors displaying their logos on Mars, which might represent a last artificial and inconsistent threshold of sacredness and taste. Still, in a more subtle way, he opted for Blur and Hirst who are widely known brands. Hirst himself sometimes declares that his *Spot Paintings* are 'like a logo' and that the branding of art is unavoidable.[38] Hirst is aware that art cannot evade capitalism and the art market, not even in outer space. Although 'art is about life and the art world is about money,' contemporary life is permeated by commoditization: thus art cannot avoid money, which obfuscates beauty's power to relieve the mind. But art becomes rebellious when it manages to stay art under a regime of commoditization, and when its speculative market value – as in the case of Hirst's – contradicts and transcends its own functionalization.[39]

Concerning the calibration target as a work of art, Hirst has uttered two known assertions: 'I'm sure there'll be a great demand for my work out there – they'll love me!' And: 'If they've got eyes, they'll love it.'[40] As in the case of Blur, such pop star statements are to be taken seriously as part of the work of art's context. Firstly, aliens need human eyes to receive the joy of color and they are even supposed to feel love for art, which they probably do not. This is tantamount to admitting that there is no extraterrestrial life of interest if it is not human-like and thus also attracted to beauty and art. Secondly, Hirst wishes his works to be liked and especially bought by aliens. Therefore, Hirst sees the need to export art with its own existential context, the capitalist art market, which is essential to his own work. In the middle of

Figure 16.6 Simulation of Beagle 2 on the Martian surface, released May 2002.
Source: Courtesy of Beagle 2.

nothingness, on Mars, it is not the context which makes the work of art as such, but rather the other way round: it is the work of art which exports its own context and meaning.[41] Thus, with the calibration target, Hirst states that the commoditization of art is vital to his work, to the Mars Express mission itself, and also to contemporary society as a whole. Capitalism makes humanity, *homo animal oeconomicus.* Thus, in conquering another planet, humans need capitalism as much as oxygen, calibrate the unknown according to it, and spread it like a virus.

Space art is not a means of extraterrestrial, but of human communication and self-assurance, a 'message from earthlings to Mars,' sent back to us by robot cameras.[42] The probe can be read as a scale model of today's society, suggesting that art – although intrinsically nonsensical, powerless and at best uplifting – is a functional decoration, point of reference, symbol, and motor of a technological and capitalist system (Figure 16.6). Essentially an active part of this system, self-conscious space art can effectively reveal the transformation of space exploration into space exploitation and, finally, that human commoditization is expanding well beyond the Blue Planet by means of science and art.

Notes

1. Damien Hirst, *I Want to Spend the Rest of My Life Everywhere, with Everyone, One to One, Always, Forever, Now*, London: Booth-Clibborn, 1997, 6; Samuel Beckett, *Endspiel: Fin de partie. Endgame*, Frankfurt am Main: Suhrkamp, 1974; *Fin de partie*, Paris: Editions de Minuit, 1957, 56. All Internet sources were accessed on 1 October 2011.
2. David S. McKay, 'Search for Past Life on Mars. Possible Relic Biogenic Activity in Martian Meteorite ALH84001,' *Science* 273 (16 August 1996), 924–30. See David R. Williams, *Evidence of Ancient Martian Life in Meteorite ALH84001?*, 9 January 2005, NASA Goddard Space Flight Center, https://nssdc.gsfc.nasa.gov; NASA/JPL *Mars Meteorites*, http://www2.jpl.nasa.gov/snc/index.html.
3. Bill Clinton, 7 August 1996, NASA/JPL, http://www2.jpl.nasa.gov/snc/clinton.html.
4. See audio file of Welles's 1938 broadcast, *Internet Archive*, http://www.archive.org.
5. ESA Mars Express, http://www.esa.int/Our_Activities/Space_Science/Mars_Express.
6. Colin Pillinger, *Beagle: From Darwin's Epic Voyage to the British Mission to Mars*, London: Faber & Faber, 2003. See Beagle 2 web site, http://www.beagle2.com; *The Pleasure Principle: An Interview With Colin Pillinger*, 18 December 2003, ESA web site, http://www.esa.int/Our_Activities/Space_Science/People/The_pleasure_principle_An_interview_with_Colin_Pillinger; Andrew Wilson, ed., *Mars Express: The Scientific Payload*, Noordwijk: ESA, 2004, http://www.esa.int/esapub/sp/sp1240/sp1240web.pdf. For the probable landing site, see http://www.esa.int/Our_Activities/Space_Science/Mars_Express.
7. *Doctor Who* started broadcasting in 1963. See audio file, *Space Place*, http://www.orbit.zkm.de/?q=node/188; Beagle 2 web site, http://www.beagle2.com; BBC's *Doctor Who*, http://www.bbc.co.uk/doctorwho; Blur's song 'Beagle 2' on the 1999 single *No Distance Left to Run*.

8. Pillinger, September 2003, ESA, http://www.esa.int/Our_Activities/Space_Science/Mars_Express; David Whitehouse, 'Beagle Mars Probe Awaits Sponsors,' *BBC News*, http://news.bbc.co.uk/2/hi/science/nature/2950275.stm; Pillinger, 'To Make It in Space You Need Tenacity and a Lot of Stamina,' *RedOrbit* (9 July 2004), http://www.redorbit.com; and archive of Beagle 2, http://www.beagle2.com.
9. Pillinger, 'I'm Searching for Life on Mars,' *BBC News* (24 July 2002), http://news.bbc.co.uk/2/hi/uk_news/1829610.stm: 'The role of a modern scientist in my case is to be a professor of PR as well as professor of planetary sciences. And in the meantime I've become a pop entrepreneur; I do a bit of art with Damien. [...] We had to create a publicity machine to convince people that we had a project worth funding.'
10. Jon Wiederhorn, 'Blur to Play on Mars: Britpop Band Commissioned to Create Track for Beagle 2 Space Lander,' *MTV* (30 January 2002), http://www.mtv.com/news/articles/1452037/01302002/blur.jhtml; Karen Bliss, 'Blur Get Life on Mars: Musical Probe Set for Christmas Landing,' *Rolling Stone* (4 June 2003).
11. See William R. Macauley's contribution, Chapter 15 in this volume.
12. See Michael Schetsche, 'Rücksturz zur Erde? Zur Legitimierung und Legitimität der bemannten Raumfahrt,' in Christoph Heinrich and Markus Heinzelmann, eds, *Rückkehr ins All*, Ostfildern-Ruit: Hatje Cantz, 2005, 24–7, here 26 (exhibition catalogue, Hamburger Kunsthalle, 23 September 2005–12 December 2006).
13. Pillinger, *Pleasure Principle*, and 'Tenacity'; Curtis Peebles, *Watch the Skies! A Chronicle of the Flying Saucer Myth*, Washington, DC: Smithsonian Institution Press, 1994.
14. L. Ishi-Kawa, ed., *Cnymhuk/Sputnik*, Madrid: Fundación Sputnik, 1997, 166–75 (exhibition catalogue, Fundación arte y tecnología, Madrid, 21 May–20 July 1997); see also the Sputnik Foundation's web site, https://www.fundaciontelefonica.com.
15. Melissa Mean and James Wilsdon, *Masters of the Universe: Science, Politics and the New Space Race*, London: Demos, 2004.
16. Hugo Marée, 'Ferrari Red Paint Passes Road Test for Trip to Mars,' ESA, 18 September 2002, http://www.esa.int/ESA.
17. Sarah Kent, *Shark Infested Waters: The Saatchi Collection of British Art in the 90s*, London: Zwemmer, 1994; Norman Rosenthal, ed., *Sensation: Young British Artists from the Saatchi Collection*, London: Thames & Hudson, 1997 (exhibition catalogue, Royal Academy of the Arts, London, 18 September–28 December 1997); Robert Timms, ed., *Young British Art: The Saatchi Decade*, London: Booth-Clibborn Editions, 1999; Rita Hatton and John A. Walker, *Supercollector: A Critique of Charles Saatchi*, London: Ellipsis, 2000; Pillinger, 'I'm Searching for Life on Mars.'
18. Gordon Burn, 'Damien Hirst,' *Parkett* 40.1 (1994), 64–9, and 'Is Mr Death In?,' in Hirst, *I Want to Spend the Rest of My Life*, 7–13, here 11. See also Eduardo Cicelyn, Mario Codognato and Mirta D'Argenzio, eds, *Damien Hirst*, Naples: Electa, 2004 (exhibition catalogue, Museo archeologico nazionale, Naples, 31 October 2004–31 January 2005); Pillinger, 'Briefing Notes on Damien Hirst Beagle 2 Spot Painting,' 2002, Beagle 2, http://www.beagle2.com.
19. Tim Radford, 'Hirst Launches Spots into Space,' *Guardian* (29 November 2002).

20. David Derbyshire, 'Hirst's Art is Out of This World,' *Telegraph* (16 July 2003): 'Pillinger recalls that Hirst was keen to include a blue spot to represent Earth. The pair settled on azurite, a copper carbonate mineral. Another nod to the mother planet came with a blob of Green Earth – a mix of different oxidation states of iron as a hydrated silicate. It also contains potassium needed for age determination.' Kristine von Oehsen, 'Damien Hirst,' in Heinrich and Heinzelmann, *Rückkehr ins All*, 66: 'Wir hatten Gründe für verschiedene Vorgaben bezüglich der Beschaffenheit der Kalibrierungsplatte und gaben diese Erfordernisse an Damien weiter, aber das Design, das Layout und die Wahl der Farben aus der Palette blieben ihm überlassen.'
21. *BBC News* (29 November 2002), http://news.bbc.co.uk/1/hi/entertainment/arts/2522417.stm; 'Damien Hirst Spots on Mars,' 1 June 1999, Beagle 2, http://www.beagle2.com.
22. David Derbyshire, 'Damien Hirst Artwork to Land on Mars,' *Telegraph* (29 November 2002); Hirst, *I Want to Spend the Rest of My Life*, 100; 'Damien Hirst Spots on Mars,' 1 June 1999, Beagle 2, http://www.beagle2.com.
23. Damien Hirst and Gordon Burn, *On the Way to Work*, London: Faber, 2001, 211. See also Hirst, *I Want to Spend the Rest of My Life*, 14, and *The Death of God: Towards a Better Understanding of a Life without God Aboard the Ship of Fools*, London: Other Criteria, 2006, 7 (exhibition catalogue, Galería Hilario Galguera, Mexico City, 2 February–27 August 2006). But see John Gray, 'Damien Hirst: Die Ikone wird in ihre Einzelteile zerlegt,' in Eckhard Schneider, ed., *Re-Object*, 2 vols, Cologne: Walter König, 2007, vol. 1, 96–9 (exhibition catalogue, Kunsthaus, Bregenz, 18 February–13 May 2007).
24. Hirst, *I Want to Spend the Rest of My Life*, 20–2, 82. Gordon Burn, 'Beautiful Inside My Head Forever: Gordon Burn in Conversation with Damien Hirst. Stroud Studio, Friday 27 June 2008,' in Michael Bracewell, ed., *Damien Hirst: Beautiful Inside My Head Forever: Evening Auction. Monday 15 September 2008 at 7 pm*, 5 vols, London: Sotheby's, 2008, vol. 1, 16–25 (auction London, Sotheby's, 27 June 2008).
25. Burn, 'Is Mr Death In?,' 11; Hirst and Burn, *On the Way to Work*, 210; Paul Stolper, ed., *New Religion: Damien Hirst*, London: Other Criteria, 2006, xv (exhibition catalogue, Pail Stolper, London, 13 October–19 November 2005); Gordon Burn, 'The Hay Smells Different to the Lovers Than to the Horses,' in Damien Hirst and Jason Beard, eds, *Theories, Models, Methods, Approaches, Assumptions, Results and Findings*, vol. 1, London: Gagosian Gallery and Science, 2000, 6–21 (exhibition catalogue, Gagosian Gallery, New York, 2000).
26. Eulina Clairmont, 'Hirst Sets the Tone on a Mars Mission,' *Independent* (4 February 2003); Derbyshire, *Hirst*.
27. Derbyshire, *Hirst*; Hirst and Burn, *On the Way to Work*, 210; Gene R. Swenson, 'What is Pop Art? Answers from 8 Painters. Part I,' *Artnews* 62.7 (1963), 24–7, 35, 60–4, here 26; Hirst, *I Want to Spend the Rest of My Life*, 246; Caroline A. Jones, *Machine in the Studio: Constructing the Postwar American Artist*, Chicago: University of Chicago Press, 1996.
28. Hirst, *I Want to Spend the Rest of My Life*, 207, 246; Burn, 'Is Mr Death In?,' 9; Hirst and Burn, *On the Way to Work*, 120.
29. Cicelyn, Codognato and D'Argenzio, *Damien Hirst*, 98.

30. Hirst and Burn, *On the Way to Work*, 90, 119; Hirst, *I Want to Spend the Rest of My Life*, 198, 250, 246; Cicelyn, Codognato and D'Argenzio, *Damien Hirst*, 105.
31. Hirst, *I Want to Spend the Rest of My Life*, 207, 246; Stuart Morgan, 'Damien Hirst: The Butterfly Effect,' in ibid., 68–73, here 68; Burn, 'Is Mr Death In?,' 12. See also Sotheby's, ed., *Damien Hirst's Pharmacy*, London: Sotheby's, 2004 (auction Sotheby's, London, 18 October 2004).
32. See Rosalind E. Krauss, *The Originality of the Avant-Garde and Other Modernist Myths*, Cambridge, MA: MIT Press, 1986.
33. Hirst, *I Want to Spend the Rest of My Life*, 250.
34. Tim Radford, 'Hirst Launches Spots into Space,' *Guardian* (29 November 2003); Paul Gauguin, *Where Do We Come From? What Are We? Where Are We Going?* (*D'où venonsnous? Que sommes-nous? Où allons-nous?*), 1897–98, oil on canvas, Boston Museum of Fine Arts, inv. 36.270.
35. See Philip Pocock's epilogue, Chapter 17 in this volume.
36. Radford, '*Hirst Launches Spots into Space*'; Hirst, *I Want to Spend the Rest of My Life*, 143–5.
37. Boris Groys, 'Geometrische Dekadenz,' *Parkett* 40.41 (1994), 71–3, here 72; Hirst, *Death of God*. See also Julian Stallabrass, *High Art Lite: British Art in the 1990s*, London: Verso, 1999, 160–3; and Kent A. Logan, ed., *Warhol/Koons/Hirst: Cult and Culture. Selections from the Vicki and Kent Logan Collection*, Aspen: Aspen Art Museum, 2001 (exhibition catalogue, Aspen Art Museum, 3 August–30 September 2001).
38. David Whitehouse, 'Beagle Mars Probe Awaits Sponsors,' *BBC News* (17 April 2003), http://news.bbc.co.uk/2/hi/science/nature/2950275.stm; Cicelyn, Codognato and D'Argenzio, *Hirst*, 98; Hirst and Burn, *On the Way to Work*, 220; Hirst, *I Want to Spend the Rest of My Life*, 246. See also Leonard Rau, 'Selling Space: Corporate Identity and Space Imagery,' in John Zukowsky, ed., *2001: Building for Space Travel*, New York: Harry N. Abrams, 2001, 124–9.
39. Damien Hirst, 'Why Cunts Sell Shit to Fools,' in Muir Gregor and Clarrie Wallis, eds, *In-a-gadda-da-vida: Angus Fairhurst, Damien Hirst, Sarah Lucas*, London: Tate, 2004, 82–5 (exhibition catalogue, Tate Britain, London, 3 March–31 May 2004); Hirst and Burn, *On the Way to Work*, 83.
40. http://www.beagle2.com.
41. John Gray, 'Leviathan: A Conversation with Damien Hirst and John Gray,' in Simon Baker, ed., *'Corpus': Damien Hirst: Drawings 1981–2006*, London: Other Criteria, 2006, 29–41 (exhibition catalogue, Gagosian Gallery, New York, 15 September–28 October 2006).
42. Clairmont, 'Hirst Sets the Tone on a Mars Mission': 'The *Spot Painting* is not just a target – it is a message from earthlings to Mars. [...] We don't believe we are communicating to Martians, but we are certainly communicating with a pattern.'

Epilogue

CHAPTER 17

Look Up! Art in the Age of Orbitization

Philip Pocock

> Did I create that sky? Yes, for, if it was anything other than a conception in my mind I wouldn't have said 'Sky.' – That is why I am the golden eternity. There are not two of us here, reader and writer, but one, one golden eternity, One-Which-It-Is, That-Which-Everything-Is.
>
> Jack Kerouac (1960)[1]

It was not the astrophysics and mathematics of seventeenth-century Pisan polymath Galileo Galilei (1564–1642) that saw him censured and imprisoned by Pope Urban VIII's Inquisition tribunal in 1634. The reigning theological milieu in Rome at the time considered technically enabled observation to be purely hypothetical, producing curious bagatelles of knowledge that posed no more threat to supreme religious doctrine than hyperbole. After all, 'mathematics was rated at the time as a thing for technicians and *virtuosi*, as they were called, with no claim to philosophical relevance,' Giorgio de Santillana posits in his 1955 book *The Crime of Galileo*.[2] Aristotelian colleagues of Galileo flatly refused even to peer through the 'Dutch trunk,' looker, spyglass, or telescope as it became commonly known in 1611. When Galileo asked the popular Paduan philosopher Cesare Cremonini (1550–1631) to observe what appeared to be lunar mountain ridges through a telescope, he refused, citing it would hurt his eye. 'Oh, my dear Kepler,' penned Galileo, 'how I wish that we could have one hearty laugh together. Here, at Padua, is the principal professor of philosophy, whom I have repeatedly and urgently requested to look at the moon and planets through my glass, which he pertinaciously refuses to do. Why are you not here? What shouts of laughter we should have

Philip Pocock (✉)
Berlin, Germany
e-mail: philip.pocock@gmail.com

Alexander C.T. Geppert (ed.), *Imagining Outer Space*
European Astroculture, vol. 1
https://doi.org/10.1057/978-1-349-95339-4_17

at this glorious folly! And to hear the professor of philosophy at Pisa laboring before the grand duke with logical arguments, as if with magical incantations, to charm the new planets out of the sky.'[3]

Unofficial Vatican voices with a seemingly 'humanistic' tone reinforced in their social milieu an already intuitive distrust in the veracity of technological measurement, assuming an anti-technological prosthetic, pro-human faculty argument shared centuries later by anti-automation Luddites and cyber-Luddite skeptics and detractors. The Inquisition's most appealing popular argument to disgrace the telescope's claim to be an indubitable extension of the naked eye went something like this: The virtual images conjured through this scopic contraption distort natural reality as sensed and then perceived by the human eye and brain. The telescope as impostor of innate vision may not interrogate God's master plan for His universe. This precept, while banishing technology in one fell swoop from the religious dominance of true wonder, subtly appealed to the grand public's pagan sense of wonder for its prosecution. 'Humanist' detractors of the telescope conveniently awakened popular proclivity toward occult lore, invoking 'crystal ball' sympathy, with its own communiqué – telescopic images may seem incredible, yet they certainly appeal to the credulous. Who is not dazzled by the effulgent character of curved glass lenses accepting God's honest light at one end, imagining it being wickedly bounced around inside a black box tube that manufactures misbegotten messages for the eye out the other end? How could His light ever be more highly resolved by technology for a God-given eye? Disregarding that Aristotle himself had relied on a pinhole *camera obscura* for his solar observations ca. 350 BCE, the telescope was characterized to be little more than a kaleidoscope, and Galileo to have been punishably hoisted by his own scopic petard.

Yet it was neither technical curio nor hypothetical observations that landed Galileo in non-secular hot water. It was the 'word,' a seethingly wry play, its actors masquerading as himself, a noble friend and the Pope. Its consequence – landing Galileo under life-long house arrest in his villa outside Florence in June 1634 – obviously attests to the play's rhetorically visceral blend of science treatise and autobiographical fiction (Figure 17.1). The play, a trialogue titled *Dialogue Concerning the Two Chief World Systems* (*Dialogo sopra i due massimi sistemi del mondo*) had passed the Roman censors in 1632 only to outrage his former admirer Cardinal Maffeo Barberini (1568–1644), elected Pope Urban VIII in 1623, enough to call Galileo before an Inquisition tribunal, threaten him with death by burning at the stake and compel him to falsely recant his Copernican science and, along with that, any vestige of scientific ethics. In his defense, legend has it Galileo muttered under his breath *Eppur si muove!* (And yet it moves!) as he was led out of court and 'looked up at the sky.'[4] Curiously, this was first vetted in a painting, attributed to Murillo or his school, dated 1643 or 1645, when the Belgian owners had it unframed in 1911 revealing a gloomy Galileo in a dungeon, his finger pointing across the canvas to an inscription – 'Rppur [*sic*] si muove.'[5]

DIALOGO
DI
GALILEO GALILEI LINCEO
MATEMATICO SOPRAORDINARIO
DELLO STVDIO DI PISA.
E Filoſofo, è Matematico primario del
SERENISSIMO
GR.DVCA DI TOSCANA.
Doue ne i congreſſi di quattro giornate ſi diſcorre
ſopra i due
MASSIMI SISTEMI DEL MONDO
TOLEMAICO, E COPERNICANO;
*Proponendo indeterminatamente le ragioni Filoſofiche, e Naturali
tanto per l'vna, quanto per l'altra parte.*

CON PRI  VILEGI.

IN FIORENZA, Per Gio:Batiſta Landini MDCXXXII.
CON LICENZA DE' SVPERIORI.

Figure 17.1 Galileo Galilei's dialogue between three characters: the Copernican Salviati representing Galileo; the wise nobleman Sagredo, named after a real life friend of Galileo, who is swayed by Salviati; and the Ptolemaic buffoon Simplicio arguing against the heliocentric model. Pope Urban VIII saw too much of himself in Simplicio, called Galileo to face the Inquisition tribunal, recant under threat of torture and death, and subsequently placed him under life-long house arrest.
Source: Galileo Galilei, *Dialogue Concerning the Two Chief World Systems, Ptolemaic and Copernican* (*Dialogo sopra i due massimi sistemi del mondo, Tolemaico e Copernicano*), published in Italian in Florence 1632, and in Latin as *Systema cosmicum* in 1635, translated by Matthias Bernegger, frontispiece and title page.

Galileo's provocative fusion of science and fiction was not without precedent. Through a mutual friend, Galileo and the mathematician and Copernican astronomer Johannes Kepler (1571–1630) began corresponding in 1610 upon the publication of Galileo's *Sidereal Messenger: Unfolding Great and Marvelous Sights, and Proposing Them to the Attention of Every One, but Especially Philosophers and Astronomers*, to which Kepler eagerly responded with his *Conversation with the Sidereal Messenger*. It may therefore be adduced that Galileo was also cognizant of Kepler's *Somnium* (The Dream), which he began at the University of Tübingen in 1593 as a shunned dissertation titled *What Would it Be Like to Stand on the Moon and Look at the Heavens?* It underwent a startling transformation between 1610 and 1630, becoming what Lewis Mumford, Carl Sagan and others consider the *fons et origo* of the Western science-fiction genre.[6]

In 1611, Kepler began circulating his *Somnium* account of the Danish astronomer Tycho Brahe's (1546–1601) Icelandic assistant *Duracotus*'s lunar voyage with his herbalist mother *Fiolxhilda*, propelled by occult knowledge she had gained from her instructor, the *Daemon of Lavania*. Although first published posthumously by his son Ludwig in 1634, Kepler's novel concoction of fantasy, legend, autobiography and scientific observation is an exemplar for Galileo's droll commentary and scientific theatrics in his *Dialogue*. Even Galileo's subsequent fate was foreshadowed by Katharina Kepler (1546–1622), Johannes's cantankerous mother, who was imprisoned in 1615 and putatively tortured on accusation of practicing witchcraft; she died shortly after her vindication in 1622. Such was the proscriptive *Zeitgeist* encircling Galileo Galilei's *Dialogue Concerning the Two Chief World Systems.*

I Space art

Such literary topoi not only set the stage for modern science fiction but for the genre of what is loosely categorized as 'space art.' In fact, the case may be made that the practice was initiated by Galileo in June and July 1612, with this hand-drawn prototype for what would be called cinema centuries later. And what better subject for a first movie than the sun itself! That summer Galileo produced meticulous and aesthetic *Sunspot* drawings from moments projected through a telescopic lens, each one a cell progressively tracking the shadowy migration of dark spots across the bright star's surface, recorded at regular diurnal intervals and published in 1613 as *Istoria e dimostrazioni intorno alle macchie solari e loro accidenti* (History and Demonstrations Concerning Sunspots and their Properties). Viewed now digitally in sequential flip-book fashion, film's trademarks – illusion of movement and dream-like duration – become apparent from Galileo's hand and lens predating the Lumière brothers by almost three centuries. The link between modern astronomical and imaginary pursuits and perceptions is eloquently explored by Steven Dick, both from a personal account and a professional socio-scientific perspective in his contribution to this volume.[7]

Noted space art advocate and artist Ron Miller considers its modern origin to be anchored in the mid-nineteenth century, specifically upon the publication of woodcuts by Emile Bayard (1837–91) illustrating Jules Verne's farcical *De la Terre à la Lune* (From the Earth to the Moon), a science-fiction novel released by noted Balzac, Hugo and Zola publisher Pierre-Jules Hetzel in 1865.[8] Verne's moon book inspired the Russian astronautics visionary Konstantin Eduardovich Tsiolkovsky (1857–1935), who included in his 1883 *Svobodnoe Prostranstvo* (Free Space) manuscript a sketch of what is considered the first depiction of human weightlessness. Another protagonist of modern space art is commonly considered to be Hollywood film matte painter Chesley Bonestell (1888–1986), when in 1944 he illustrated *Life* magazine with awesomely imaginative photo-real paintings of Saturn seen from the points of view of three of its orbiting moons. Space art marched on when Bertolt

Brecht's (1898–1956) critical play *Leben des Galilei* (Life of Galileo), written between 1937 and 1939, opened in Zurich, and was adapted with Charles Laughton as Galileo for an American audience just as the Second World War ended, with Brecht's pen slanting further left after the atomic bombing of Hiroshima ushering in the Atomic Age. Brecht stated at the time: 'The atomic age made its debut at Hiroshima in the middle of our work. Overnight the biography of the founder of the new system of physics read differently.'[9]

'Telecommunication Art' pioneer, the late Korean ex-pat Nam June Paik (1932–2006) titled a work from 1965, *Moon is the Oldest TV*. In former times, before backlit screens, people went outdoors on evenings and looked up at the moon for entertainment. In 1984 Paik suggested with some clairvoyance that electronic media was on the cusp of radically changing space art. With humor he described the coming of space-based 'social networks':

> Satellite art [...] must consider how to achieve a two-way connection between opposite sides of the earth; how to give a conversational structure to the art; how to master differences in time; how to play with improvisation, indeterminism, echoes, feedbacks [...] and how to instantaneously manage the differences in culture, preconceptions, and common sense that exist between various nations. Satellite art must make the most of these elements (for they can become strengths or weaknesses). [...] The satellite will accidentally and inevitably produce unexpected meetings of person and person. [...] Thanks to the satellite, the mysteries of encounters with others (chance meetings) will accumulate in geometric progression and should become the main nonmaterial product of post-industrial society.[10]

II *SpacePlace: Art in the Age of Orbitization*

At the ZKM Center for Art and Media Technology Karlsruhe in 2005, my lab proposed a sort of 'future satellite museum' prototype. Having produced *Beyond the Earth: The Orbital Age* event at Ars Electronica in 1986, Peter Weibel embraced my question: Had globalization distilled out from our cultural environment and, like the countless biannuals sprouting up, become commodity, creating in McLuhanist fashion a new climate for art, one I termed 'orbitization'? Together with a team of former students – Axel Heide, Onesandzeros, Heiko Hoos, who had experience on pioneering social network projects in my lab, as well as ZKM net-editor Heike Borowski – we developed a user-participatory mobile and hypercinematic 'app' exploring an inclusive space art database. Opening to a consuming or participating public at the ZKMax media space in an understreet passage in chic urban Munich on 7 June 2006, the day the forty-ninth session of the United Nations Committee on the Peaceful Uses of Outer Space convened in Vienna,[11] our experiment in conflating virtual curatorial and space art practices titled *SpacePlace: Art in the Age of Orbitization* became more art *and* exhibition than art *in* exhibition.

SpacePlace's social network system design was inspired by Paik's take on 'Satellite Art.' Its cyber-cinematic sensibility and user-generated mobile app, along with its title, was encouraged by cosmic jazz musician Sun Ra's 1974 cult science-fiction film classic *Space Is the Place*, which highlights space travelers' dual character as both colonists and exiles while it ponders the societal impact of outer space on popular culture. Himself victimized and vexed during the civil rights struggle, Sun Ra innovated an afrofuturist style personifying him as a citizen of the sky's brightest star, Sirius, the Dog Star. With free speech and racial tolerance seemingly absent from his experience on earth, he wondered if Afro-Americans would not be better off just rocketing off to another planet and take it from there. Pausing in an idyllic extraterrestrial jungle set in his movie, Sun Ra murmurs: '(*Humming softly* …) The music is different here. The vibrations are different. Not like planet Earth. Planet Earth, sound of guns, anger, frustration. There was no one to talk to on planet Earth who'd understand. We'll set up a colony for black people here. See what they can do on a planet all of their own without any white people there.'[12]

Now that every citizen with a switched-on mobile phone is an orbiting cell on a worldwide telecommunication network, new brands of low-altitude human 'satellites' appear on the glocal, that is, global *and* local, horizon. Coded at their fingertips in real time, interacting identities become virtually encapsulated into an exchange of streams of tiny text and image messages, circulating and encircling earth. *SpacePlace* channeled these modern impulses onto a participatory cultural field suggesting the future trajectory of its rich contents, space art. Presenting an interactive and participatory rather than an *interreactive* multiple choice scenario for its urban users or 'future museum' guests' participation, the anthology of space art expanded to include more radically contemporary and other forms of cultural production, aggregated by *SpacePlace*'s virtual exhibition system, calls on its mobile public to co-curate, contribute cell phone images, and actively participate in the cyber-social construction of culture (Figure 17.2).

It is the *SpacePlace* cell phone users' fingers that do the walking through the exhibition, unraveling, re-traveling and orbiting threads through contemporary astro-humanity aesthetic data: the medium itself communicating the importance of shared, consensual and intra-national space exploration and related endeavors. Every cell phone connecting guest leaves a trace, and should a passerby without a phone encounter *SpacePlace* and previous user threads through its database, as well as recent cell phone image uploads will parade across its screens.

More than 200 virtual entries in *SpacePlace* offer a host of artists' clues to the paradigmatic shift now underway from globalization to orbitization. Globalization may be checked and balanced and is in any event undergoing modification from a greater reach, the process of orbitization. In a post-Cold War timeframe, raising public awareness about how orbitization acclimatizes globalization may re-awaken to Cold War levels the public's awareness for

Figure 17.2 *SpacePlace: Art in the Age of Orbitization* was a virtual curatorial urban site installed at ZKMax in Munich in 2006. Its pioneering Bluetooth® mobile phone app encouraged passersby to participate, co-curating and uploading their images to a vast tagged Web 2.0 database of 'space art' projected on large screens downtown day and night. *SpacePlace* opened 7 June 2006, coinciding with the forty-ninth session of the United Nations Committee on the Peaceful Uses of Outer Space, convened in Vienna, Austria.

Source: Philip Pocock, director, with Peter Weibel, Axel Heide, Onesandzeros, Heiko Hoos and Heike Borowski. Courtesy of spaceplace.zkm.de (in need of restoration, 1 October 2017).

peaceful space programs and policies, identifying an alternative translocal subject within a global milieu under the influence of orbitization, which itself has been taking 'giant leaps' as consumers become *prosumers* (producer-consumers). Apps and publics operate provisionally in locally grounded yet within ubiquitous social networks, swarming with satellitic signals and voices.

SpacePlace installed in giant, underground passage vitrines in downtown Munich represents one 'future museum' experience governed on site by roaming guests. The installation, emitting ethereal sounds transduced by creative radio telescope engineers online, invites any passerby to stop and attend two large projections, one a grid of artists' headshots silhouetted in the hacker style known as the 'Gotchi' and another playing video and image sequences that wobble in concert with the surrounding sonorous impulses channeled from outer space. That projection's flow of images appear to hover and land on its screen with the fragility that a transmitted view from a descending lunar landing module might evoke (Figure 17.3). Paired to this cyber-cinematic display, the 'Gotchi' grid of artists' faces represents semantic categories as a visual *tag cloud* available for toggling and selection by *SpacePlace* guests who have agreed to the *SpacePlace* app broadcast message: 'Join SpacePlace. Take control of the screen. Upload if you like.'

Figure 17.3 *SpacePlace: Art in the Age of Orbitization* content was served from ZKM to both web and mobile web users. This screenshot displays Chinese artists Peng Yu and Sun Yuan's DIY spacecraft apparently having just landed at the art venue. Titled *Farmer Du Wenda's Flying Saucer*, installed at the China Pavilion, Venice Biennale, 2005, it presents a rural legend of space utopia. Head-shaped photos, left, are visual comments uploaded by mobile phone by guests on site at ZKMax, Munich.
Source: Courtesy of spaceplace.zkm.de (in need of restoration, 1 October 2017).

Clicking *OK*, a small app is downloaded to a guest's phone, allowing it to navigate various artists and works by keyword, gain deeper information, and select an individual work's media to stream on the paired cyber-cinematic screen. As such, the public virtually co-curates the display of *SpacePlace* in real time. At any point, the phone app toggler may take a photo and upload it as a visual comment directed at any preferred art entry. From there it is transmitted to the Web archive of *SpacePlace* at ZKM and integrated alongside its targeted entry. As well, space-related news aggregated from the Web in real time is forwarded from ZKM to *SpacePlace* in Munich and scrolls intermittently, sometimes machine-spoken, subtitling the bottom of the cyber-cinematic screen. The entire virtual installation is based on what Web 2.0 subculture calls the *Mashup*, a cyber extension of William S. Burroughs (1914–97) and Brion Gysin's (1916–86) 1950s and 1960s *Cut-Up* media technique, itself contingent upon collective Surrealist collage aesthetics. The Fluxus artist

ZKM

www.orbit.zkm.de

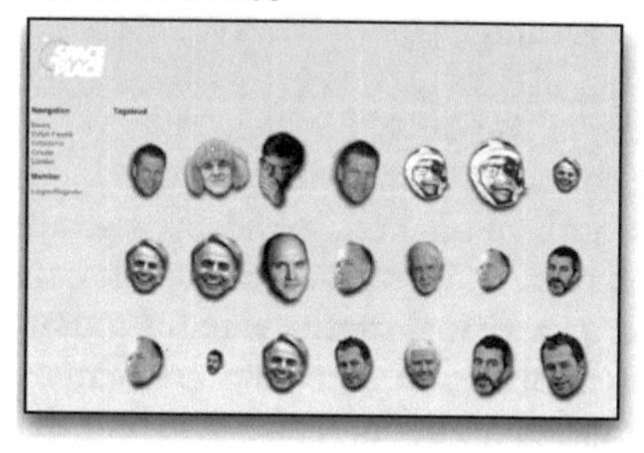

mobile.orbit.zkm.de

online access with standard browser on desktop and mobile

ZKMax

Projector

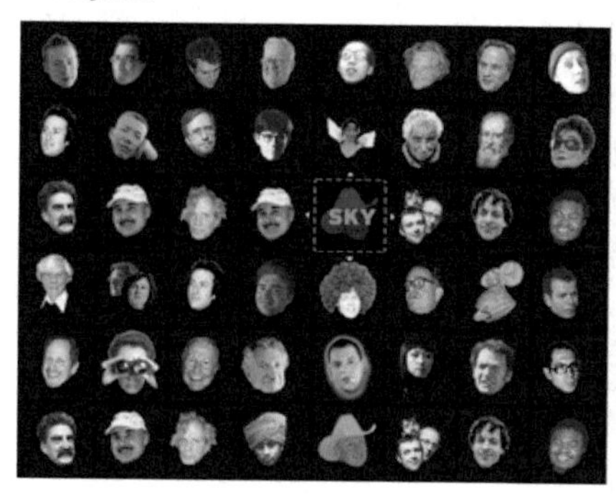

access via Java Client that has to be downloaded from mobile.orbit.zkm.de
or is delivered via Bluetooth by the machines in munich
(using Bluetooth media server http://www.consola.org/)

Java Client on users mobile

Figure 17.4 *SpacePlace: Art in the Age of Orbitization.* Datatecture (interface of data and dwelling) flowchart for mobile access to *SpacePlace* database, with app designer Axel Heide, 2005.

Source: Courtesy of spaceplace.zkm.de (in need of restoration 1 October 2017).

Joseph Beuys (1921–86) first declared in 1967 '*Jeder Mensch ist ein Künstler*' (everyone is an artist). That same year, futurist architect R. Buckminster Fuller (1895–1983) proclaimed, in *Operating Manual for Spaceship Earth*: 'We are all astronauts!' Perhaps now space art might quip: 'Everyone, a satellite?' It would seem so (Figure 17.4).

Complimenting the humanities program at the 2008 Bielefeld conference, *Imagining Outer Space*, *SpacePlace*'s art contents, some user-generated, add artists' perspectives to the space-related anthropological, philosophical and sociological texts presented by the conferees.[13] *SpacePlace* exemplifies the necessity facing cultural institutions to act as a virtual mothership, encouraging public engagement, recognizing that information is a valid art material able to 'behave' virtually with anyone, anywhere, anytime, and realizing that 'code is the current language of culture.' *SpacePlace* is scripted from open-source code libraries, creating a convocation of space art, computer algorithm and human interaction. As such, it updates the age-old search for a 'space for art' – in caves, in the desert, in the wilderness – which cities relocated over the course of centuries, making 'art about space.' Now, in the age of computer literacy, digital technology and satellite communication – orbitization – the time has come for 'art in space.'

Both sides of the digital divide cannot escape the accelerated distillation of globalization and the virtualization of economies, of identities, of aesthetic objects and cultural production.[14] Currently accepted notions of globalization seem to falter and are giving way. Globalization, as the seamless interdependence of regional economies, homogenizing political and cultural systems depicted as a spheroid bubble with a single static central point, overpowering all within a radius of influence, is being usurped. At the dawn of an age of orbitization with business and social data orbiting, aggregating, circulating, interlinking and speeding through communication satellite networks, information behaves differently, at times chaotic or spurious, at other times exhibiting self-organization and deliberation, either unpredictable or initially invisible to centers of power.[15] More and more breaches in global control erupt which, for whatever reason, may require unwarranted, unwanted or unavailable force to seal. Gracing the cover of the special 'Globalization' issue of Germany's *Der Spiegel* magazine in 2007 was an image of a shipping container on which a map of the earth had been painted. Exhibiting the commodity nature of globalization, the container was hovering on a hoist in orbit above a distant barren landscape under a pitch black sky. It says globalization is a done deal. It can be traded, bought and sold.[16]

By engaging its public interactively with space art, *SpacePlace* endeavors to support outer space policy adapted for the social network age of orbitization. As a weather map of cultural orbitization, *SpacePlace* is less global and more translocal as an experience. Whereas the global is pictured as a mighty orb, the spherical equivalent to a pyramid, the translocal looks cellular, mapped as a tangle of data webs, an organically forming mesh of human and computational events. Whereas global vectors for cultural power have radiated from mega-cities with financial clout, *translocal* vectors can pop up anywhere culture is undergoing review, exhibiting a more viral than radiant character. Globalization is now persistently visible, like a product on a shelf, as orbitization takes command of the market, both economic and cultural. *SpacePlace*, appealing as it does to the thinking eye and the curiosity that space art has always sparked, raises awareness of broader space concerns by focusing on cultural uses, and therefore peaceful uses of the near-earth orbit and the great beyond, addressing the needs of twenty-first century citizens on a spontaneously and rapidly changing world stage shaped by the process I call 'orbitization.'

SpacePlace: Art in the Age of Orbitization imagines 'orbit' as the raw unit of measurement for art in this age, as well as providing fine arts with memory, energy storage and a transmission system. As a model for the future of public museums and audience engagement, *SpacePlace* collapses the classically cultivated space between art spectators or subjects and art objects. Downsizing the art subject-object discrepancy by cyber-social means, *SpacePlace* and a growing number of kindred and relaunched space art projects create cultural membranes for art and audience based on a principle borrowed from astrophysics, the *trajective*.[17]

III Looking up and going up

One of the nimblest onset indicators of orbitization on the horizon during the height of the Cold War was a philosophically witty sculpture. Expanding the repertoire for space art and a touchstone for *SpacePlace*, a weighty metal object was produced in Denmark by an Italian conceptual artist not long after NASA's sputtering Mercury program response to the Soviets' 'co-traveler' Sputnik launch that shook the world. This work, Piero Manzoni's ironic *Socle du monde: Hommage à Galileo* (Base of the World: Homage to Galileo) was set outside the Herning Museum of Contemporary Art in 1961, during the artist's brief residency (Figure 17.5). Cleverly Manzoni (1933–63) underscored the provocative Space Race remark made in 1958 before a US Senate subcommittee by Senate Majority Leader and later President Lyndon B. Johnson (1908–73): 'Control of space means control of the world. From space, the masters of infinity would have the power to control the earth's weather, to cause drought and flood, to change the tides and raise the levels of the sea to divert the Gulf Stream and change temperate climates to frigid.'[18]

Yet Manzoni turns Johnson's remark somehow on its head. Like magic, Manzoni's *Base* employs very simple means. He inscribed one face of an iron cube with words written upside-down. They read 'Base of the World,' thereby identifying his object approximately one cubic meter in size as a pedestal. And on that pedestal rests the entire planet Earth, and all of us, all our actions, as his *Gesamtkunstwerk*. The artist only asks that we accept his cube as having no intrinsic imbalance and therefore no dominant top

Figure 17.5 Piero Manzoni, *Socle du monde: Hommage à Galileo* (Base of the World: Hommage to Galileo), 1961, iron, bronze, 82 × 100 × 100 cm, Herning Museum of Contemporary Art, Denmark. This Italian precursor to both Concept Art and Arte Povera called his corporate-sponsored residency at a shirt factory in Herning in 1960 and 1961 his 'paradise.' After creating textile-based Achrome temperature-sensitive paintings, Manzoni produced a work *Linea Lungi* (Line 7,200 m) at the local newspaper press as part of a global project to draw a line equal in length to the earth's circumference from segments rolled up and stored in cities around the world. Following this conceptual space art work, his tour de force *Base of the World*, an upside-down pedestal on which the entire planet rests, turning it and all earthlings into a work of art, was created a few hundred kilometers west of the sixteenth-century Danish geo-heliocentric astronomer Tycho Brahe's Stjerneborg observatory on the island of Hven, Sweden.
Source: Courtesy of Herning Museum of Contemporary Art, Herning, Denmark; photograph by Ole Bagger.

or bottom, and instead to accept the orientation of his upside-down words as orienting the pedestal's, and therefore our own place in the universe. But then on what ground does the pedestal rest? The firmament. The rail of earth orbit! Brilliantly situating the earth, all earthlings upon his pedestal in orbit, Manzoni lightheartedly reverberates with Galileo's iconoclastic sarcasm in his *Dialogue Concerning the Two Chief World Systems.* By elevating the entire planet as part of a living sculpture, Manzoni exchanges the classical Ptolemaic model of geocentricity not for Copernicus's heliocentricity or Brahe's geo-heliocentricity of planetary motion, but with a very modern *mediacentric* model, as if to say that it did not take the powers-that-be on earth very long after Galileo had proven Copernicus to be correct to put our planet back at the center of another universe, a media universe with satellites carrying FOX News standing in for Mars as they circle the earth and HBO, an orbiting understudy for Venus, and so on. Although Manzoni's *Base* is a mind-expanding work of art in its own right, it does act as a critique of *mediacentricity* as merely a modern update for geocentricity that replaces the earth with its idea, the Space Race distancing our connection with the natural world, mediated more and more exclusively by cultural remoteness.

Parallels between geocentric world views and mainstream media culture abound in contemporary fine arts. Museums and galleries, curators and collectors act as centers with cultural pull. One recent consequence of ubiquitous social networking and the influence of orbitization seems to be that consumerism has peaked to the point where, with almost every resource on earth having been depleted or overconsumed, now consumers consume consumers, for instance on YouTube, Facebook and Twitter. Manzoni apparently grasped, at the beginning of the 1960s, that the spectacle of the earth was almost complete in its medialization, his space art *Base* setting a chain reaction in motion across the art world with the French 'Situationist' Guy Debord's (1931–94) concerns with commodity fetishism in his 1967 publication *The Society of the Spectacle*, the same year that the Canadian media guru Marshall McLuhan (1911–80) exposed technological imperatives in his picture book *The Medium is the Massage* [*sic*], and a year later when another 'Pop' guru, Andy Warhol (1928–87), was quoted in his Moderna Museet Stockholm catalogue as supposing: 'In the future, everyone will be world-famous for 15 minutes.'[19] *Base of the World (Homage to Galileo)* is playfully making its beholders acutely aware of such an aesthetic future, when the whole world is a museum coordinated from space.[20]

Brilliantly offering up earth as a work of art, Manzoni's *Base* apparently groundless compels the observer to look up, suggesting that new grounds for art and life are to be found in our relationship with near-earth orbit and outer space. But looking up from Manzoni's site in Western Europe during the Cold War must have been a multi-pronged affair: a volatile mix of hope and optimism with military and political concerns. Manzoni's *Base* addresses the cultural component of a warning sent out in 1958 by the United Nations General Assembly's mandate creating an Ad Hoc Committee on the Peaceful Uses of Outer Space; that is, to achieve: 'the common aim that outer space should be used for peaceful purposes only.'[21] However, Vice-President Johnson's contradictory 'control of space' remarks from that same year continued to reverberate almost 40 years later when US Space Commander-in-Chief, General Joseph W. Ashy (1940–), expanded on Johnsonian space defense policy, telling *Aviation Week & Space Technology* flat out: 'It's politically sensitive, but it's going to happen. Some people don't want to hear this, and it sure isn't in vogue, but – absolutely – we're going to fight in space. We're going to fight from space and we're going to fight into space. That's why the US has development programs in directed energy and hit-to-kill mechanisms. We will engage terrestrial targets someday – ships, airplanes, land targets – from space.'[22]

What space art was happening behind the Iron Curtain at the time, or shortly after Piero Manzoni planted his paradigmatic *Base* in Western Europe? How did Manzoni's contemporaries in Russia respond? The Moscow-based dissident artist Ilya Kabakov (1933–), although active in the 1960s, was obviously only safely able to comment on that period directly through his work once he himself had been launched as an emigré out of the Soviet Union and into the West in the mid-1980s. He produced an installation with a major space art component that reflects on the 1960s Soviet space utopia in a New York gallery in 1988. It parallels well the irony and insight that Manzoni's European concept art cosmology offered. A tragicomic room in Kabakov's sprawling installation, a boarded-up, run-down bachelor flat, titled *The Man Who Flew Into Space From His Apartment* (Figure 17.6), took a leaf, or a leap, from the legendary manuscript of the Russian astronautics visionary Konstantin Tsiolkovsky's 1883 *Svobodnoe Prostranstvo* (Free Space), containing a drawing of what is considered to be the first depiction of human weightlessness in orbit. Distinctly cynical and critical of the restricted freedom of movement in the same Soviet Union that expelled him, Kabakov presents a situation that reveals the dark side of Soviet space utopianism fueled by the regime's anxious lead in the Space Race with NASA.

Barred from entering the room, Kabakov's public had to suffice themselves with peaking into what resembled a Georges Méliès-style set in an absurdist jail-house movie. There is no trace of the man who flew into space. Only signs of his dystopia remain, a shoddy pair of shoes on a dingy floor directly below a jerry-rigged escape vehicle, a farcical ejection seat of sorts, a crude contraption of cables, springs and a plank. The walls of the room are adorned with numerous Tsiolkovskian sketches for the crazy catapult the man had used to spring himself from the Khrushchev-Brezhnev era, under which that favorite Soviet propaganda medium, poster art, haphazardly hang (Figure 17.6). The Soviet art critic Boris Groys suggests that it is these powerful images of a Soviet space utopia that drove the man to impetuously transform himself into a home-made cosmonaut just to experience firsthand the utopic space freedom that the Soviet space program had come to emblemize during the Cold War. 'After Yury Gagarin became the first man in space in 1961, the country looked up into the sky with a different sense of pride,' writes Boris Groys in his essay on the Soviet context to this Kabakov work of space art.[23]

'Looking up' for Manzoni from Western Europe in 1963 appears diametrically opposed to Kabakov's 1988 perception of looking up from a perspective tainted by a promulgation of modernist utopianism and communist idealism.

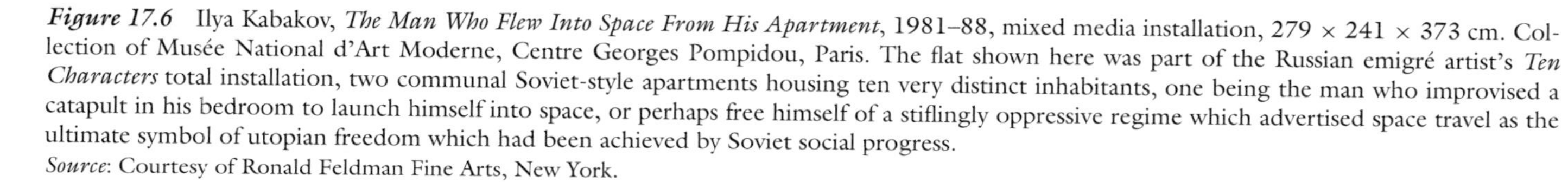

Figure 17.6 Ilya Kabakov, *The Man Who Flew Into Space From His Apartment*, 1981–88, mixed media installation, 279 × 241 × 373 cm. Collection of Musée National d'Art Moderne, Centre Georges Pompidou, Paris. The flat shown here was part of the Russian emigré artist's *Ten Characters* total installation, two communal Soviet-style apartments housing ten very distinct inhabitants, one being the man who improvised a catapult in his bedroom to launch himself into space, or perhaps free himself of a stiflingly oppressive regime which advertised space travel as the ultimate symbol of utopian freedom which had been achieved by Soviet social progress.

Source: Courtesy of Ronald Feldman Fine Arts, New York.

The gaping hole in the man's apartment ceiling testifies to the man's neurotic impatience with the infinite freedom the Soviet space program had come to symbolize. Not having first cut an exit hole, but instead violently blasting his body through his apartment's ceiling communicates Kabakov's own frustrations with Cold War Soviet attempts to employ utopian space art graphics as a stopgap surrogate for the lack of any individual freedom of movement on the ground, beyond their nation and satellite nations' borders. And in doing so, Kabakov's Rube Goldberg-style contraption communicates his distinct distaste for the inhumane impoverishment that technical utopian machinations offer deeply dystopic societies.

However, after a decade in the West and the fall of the Berlin Wall and the Iron Curtain, Ilya Kabakov's perspective on society's relationship to space transcended his earlier enraged irony. Rather than looking up at futile rips and tears in a squalid apartment ceiling left by a DIY cosmonaut, in 1997 Kabakov planted in Münster, Germany, an outdoor sculpture *Looking up, Reading the Words ... (Antenna)* with words woven between its ribs which 'looking up' spell out as if to whisper: 'My love! You lie in the grass, your head thrown back; around you: not a soul. You hear only the wind, and look up into the open sky – into the blue above, where the clouds float by – perhaps that is the loveliest thing that you have ever in your life done, and seen.'[24] Kabakov's 'Lettrist' sculpture, its European location out of doors and its visual integration of the heavens as part of its space art character offers all sorts of comparisons with Piero Manzoni's *Base of the World* planted three decades earlier. The main difference seems to be that while Manzoni appropriates the earth and us as his space art, Kabakov imagines air waves and an expanse of sky as key elements of his. They both ask their viewers to 'look up.' Kabakov speaks about the perceptual phenomenon of looking up, in an interview in Europe in 1997: 'Always, whenever we look up at the sky, we involuntarily have a presentiment, unconsciously we anticipate some sort of communication from there, it seems to us that something is occurring there, in that cosmos, in its infinite heights.'[25]

The preceding remarks on the space art dimensions in works by Galileo, Kepler, Verne, Tsiolkovsky and Brecht provide but a glimpse into the aesthetic topoi available to contemporary European cultural producers such as Manzoni and Kabakov. But how do approaches by their artistic counterparts overseas in US-America compare? Is there a disparity between contemporary European and US-American space art that sheds light on national attitudes concerning human spaceflight and space exploration in general?

Neither Manzoni nor Kabakov seems to exhibit the slightest desire to actually dispatch themselves or material art in actual earth orbit or permanently in outer space. They preserve an abstract remoteness to outer space. They appeal to a Kantian transcendent ideal associated with the act of looking up. It may be argued that Manzoni's *Base* and Kabakov's *Antenna* occupy outer space from a geo-eccentric point of view. Alexander von Humboldt's nineteenth-century classic, *Cosmos: A Sketch of the Physical Description of the Universe*, provides clues to Manzoni's metaphysical and Kabakov's pataphysical ideals motivating an aversion to launching physical art objects into orbit and outer space. In *Cosmos*, Humboldt outlines a geocentered method for aesthetically describing the universe as:

> Beginning with the depths of space and the regions of the remotest nebulæ, we will gradually descend through the starry zone to which our solar system belongs, to our own terrestrial spheroid, circled by air and ocean, there to direct our attention to its form, temperature, and magnetic tension, and to consider the fullness of organic life unfolding itself upon its surface beneath the vivifying influence of light. In this manner a picture of the world may, with a few strokes, be made to include the realms of infinity.[26]

Further insight into the 'Look Up!' phenomenology of Cold War European space art, in particular Manzoni and Kabakov, may be found in Immanuel Kant's idea of the 'Copernican Revolution' as expressed in the preface to his *Critique of Pure Reason*, in which he interrogates the assumption that objects precede their idea and proposes that it is the other way round, and the external world conforms to an *a priori* perception of it. Kant writes:

> It has hitherto been assumed that our cognition must conform to the objects [...] Let us then make the experiment whether we may not be more successful in metaphysics if we assume that the objects must conform to our cognition. [...] We here propose to do just what Copernicus did in attempting to explain the celestial movements. When he found that he could make no progress by assuming that all the heavenly bodies revolved round the spectator, he reversed the process, and tried the experiment of assuming that the spectator revolved, while the stars remained at rest. We may make the same experiment with regard to the intuition of objects. If the intuition must conform to the nature of the objects, I do not see how we can know anything of them *a priori*. If, on the other hand, the object conforms to the nature of our faculty of intuition, I can then easily conceive the possibility of such an *a priori* knowledge.[27]

Herein perhaps lies the rub among Cold War European, Soviet and US-American space art practice, which may be reflected in some differences between national space program policy. Manzoni's *Base* may be interpreted from a purely European transcendental idealist angle, that the art object is created from its idea, and remote interaction with outer space is sufficient. Kabakov, the emigré, straddles Soviet, US-American and European space art positions, advertising a desire to launch a work of art into space, in his case a living art object; yet despite an impulse of empirical realism, his protagonist does not actually follow through on his desire.[28]

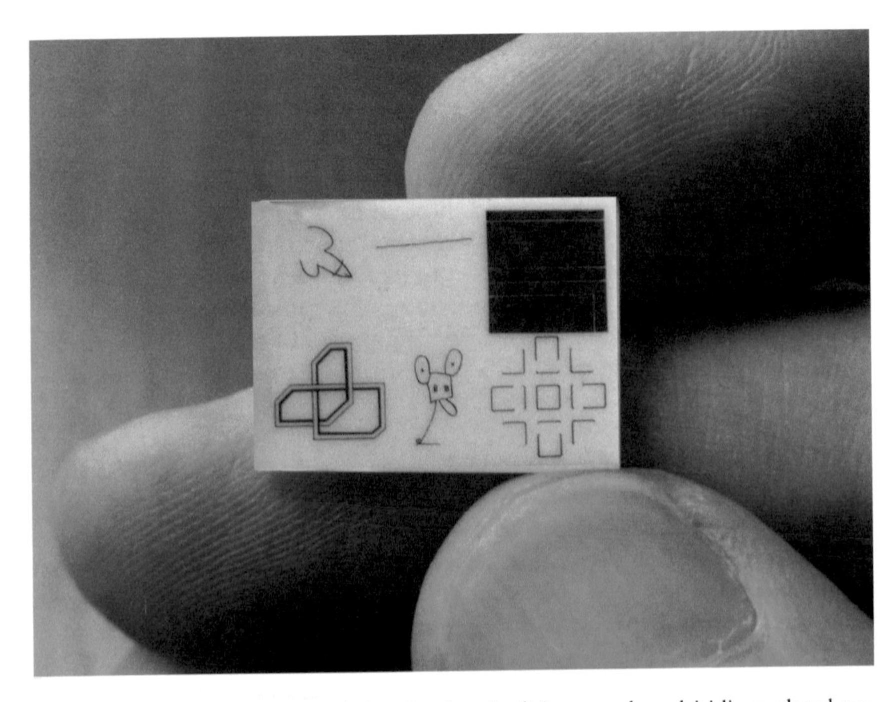

Figure 17.7 *Moon Museum*, miniaturized and edition-numbered iridium-plated tantalum, gold, anodized alumina ceramic wafer bonded to silicon chip, 1.9 × 1.25 × 0.06 cm, 1969. *Moon Museum* address: Apollo 12 lunar descent module lander leg, Ocean of Storms, moon, opened ca. 22 November 1969. This collaboration between Bell Labs Experiments in Art and Technology engineering scientists Fred Waldhauer and Billy Kluver, and noted American artists Andy Warhol, Robert Rauschenberg, Claes Oldenburg, John Chamberlain, David Novros and Forrest Myers, was anonymously attached at the Cape to the LEM lunar lander which remained on the lunar surface after the module lifted off and shuttled the astronauts to their orbiting Apollo spacecraft.
Source: Courtesy of Jade Dellinger.

While major European space art under the influence of the Cold War exhibited Dionysian imagination, US-American space art associated with NASA's Apollo missions crossed its namesake Apollonian frontier in 1969. On the second Apollo human moon-landing mission, New York artist Forrest Myers and NASA contract engineers smuggled a small art-inscribed ceramic wafer on board the craft. The US-American *Moon Museum* opened on the Ocean of Storms, moon, ca. 22 November 1969 (Figure 17.7).[29] Due to the surreptitious nature of this space art project, the record of *Moon Museum* is somewhat speculative. Apparently, it was installed on the moon without lunar module pilot Alan Bean's knowledge.[30]

The US-American space art represented on the *Moon Museum*'s ceramic wafer are from Andy Warhol, Robert Rauschenberg, Claes Oldenburg, David Novros, Forrest Myers and John Chamberlain. Their collaborative hieroglyphic etching was attached to one of the legs of Apollo 12's lunar descent module. The drawings sketch out the art currents in New York at the time: Warhol's micro-play between his monogram AW, a rocket and a penis; Chamberlain's trademark grid template; Oldenburg's geometered pop Mickey Mouse; Novros's chiaroscuro minimalism; Rauschenberg's baroque straight line; Myers's computer-generated graphic. Seemingly insignificant as art works, the thought behind the *Moon Museum* is nevertheless important in relation to corresponding Cold War European space art. Curated by the New York City's Soho pioneer and minor Minimalist sculptor Forrest 'Frosty' Myers (1941–), the selected all-star US-American Pop and Minimalist artists defined whatever fracture existed between European, Soviet and US-American space art at the time. The US-American unofficial Apollo art crew packed up a material art object and hand-delivered it to the moon, yet another gesture signing the universality of individual free expression, extending the Cold War art game that had pitted 1950s US 'Abstract Expressionism' against Soviet 'Social Realism' into space. Whereas the space art of Manzoni and Kabakov transcended space, US-American space art traveled in space. The *Moon Museum* was the first colonial claim to cultural territory in outer space. While Europe was saying 'Look up!,' America was saying 'Go up!'

Two years after Apollo 12's space art launch, the Apollo 15 mission added a second, even more primal cultural institution on the moon, a graveyard, again secreted on board and delivered by hand to the moon. A mini memorial site titled *Fallen Astronaut*, dedicated to 14 deceased astronauts and cosmonauts, was erected on 2 August 1971 at the base of the Apennine Mountains near Hadley Rille on the near north side of the moon. On the private request of Apollo 15 Commander David Scott (1932–), a Belgian artist, Paul van Hoeydonck (1925–), molded a small tin statuette to adorn an aluminum plaque carrying the names of the international space explorers who had fallen in the line of discovery (Figure 17.8).

Fallen Astronaut was laid and dedicated in a small ceremony conducted by the Apollo 15 crew, when Commander Scott took the lunar roving vehicle for its last spin to park it at a vantage point where its camera could record the lift-off of the landing module approximately 90 meters away. Leaving a Bible on the rover's dashboard, he reverently laid the plaque and the figurine in a small hollow six meters from the rover. By saluting the service to science for which both US-American astronauts as well as Soviet cosmonauts had given their lives, this poignant cultural ritual, only disclosed upon Commander Scott's return from the moon, helped set the world stage in the post-Cold War years for international space program cooperation. Aside from being on the sight-seeing checklist for any future lunar tourist, *Fallen Astronaut* clearly appeals to the power of peaceful and cultural uses of space.

After the fall of the Berlin Wall, some shades of European space art took a much more hands-on approach, sending art into orbit. After rigid ESA inspections in 2001, Pierre Comte managed to send a small kinetic sculpture, *Prisma*, up to the International Space Station in the kit bag of compatriot

Figure 17.8 Paul van Hoeydonck, *Fallen Astronaut*, 1971. Tin figure, aluminum plaque, names of 14 deceased NASA astronauts and USSR cosmonauts, Hadley-Apennine landing site, Ocean of Storms, moon. Three Apollo missions after the ***Moon Museum*** another art work was secreted to the moon. Houston control was kept in the dark while a small plaque and figurine laying ceremony in remembrance of the trailblazing American astronauts and Soviet cosmonauts who had given their lives was conducted by the Apollo 15 astronauts at the end of their moon mission.
Source: Courtesy of NASA.

spacionaut Claudie Haigneré.[31] Much to the amusement of its international crew, *Prisma*'s rainbow-colored array of seven armatures that had seemed somewhat lame under gravity came alive in earth orbit, sending an art message to European cultural producers that now is the time for subtly transcendent and emphatically empiricist European space art which complimented US pragmatic idealist and empirical realist space art, launched for the long run into space three decades earlier.

However, it must be noted that some years earlier, in 1993, the Swiss-based US-American Arthur Woods (1948–) had opened the hatch for a more physical empirical space art interventionist approach for Pierre Compte with a more figurative kinetic work, *Cosmic Dancer*. This whimsical doll-like dance partner for Russian cosmonauts had moving parts and also proved useful in kineasthetic and vascular exercising in the small confines of the space station. With corporate European cooperation, this US-American European space art ballerina had been transported via Soyuz rocket to the orbiting Mir space station on 22 May 1993, officially an international effort to promote the cultural use of near-earth orbit. Speaking to the International Astronautical Congress the year *Cosmic Dancer* took flight, Arthur Woods said: 'Visual artists and writers have created fictional images and scenarios on the development of space. Such visions are the primary way that the general public is introduced to ideas about space exploration. Artists and writers, in fact, lay the foundation which makes future space activities understandable by the general public and thus secures the necessary political support.'[32]

IV A space for art, spaces in art, and art in space

What goes up, or who goes up, must come down. The US-American 'Postmodernist' Ronald Jones (1952–) reminds his audience that human spaceflight is perhaps only – to borrow pioneering video artist Les Levine's catalogue title from 1979 – *a biotech rehearsal for leaving the body*. Jones titled his readymade space art exactly what it is, *The Bed Neil Armstrong Slept in his First Night Back from the Moon* (Figure 17.9). Dated 1969–98, the bed's woven-aluminum coverlet glistened with an eerie lunar radiance in the New York gallery where it was first exhibited, although its metallic structure was intended to shield others from any radiation Armstrong may emit while comforting the first man on the moon on his first night back home. Seen in relation to *The Man Who Flew Into Space From His Apartment*, Jones's space traveling protagonist is also missing; unlike Kabakov's impromptu cosmonaut as Odyssean exile, Jones's professional astronaut is a superpower hero back from a colonial sojourn in space. Neil Armstrong's bed on the night of 24 to 25 July 1969 attested to the fact that the moon is not an extraterrestrial diaspora but a potential colony.

President Obama raised the NASA budget slightly in 2011. In general, however, this turn-around has received rather low-level media spin compared to the peak levels of media attention at the height of the Cold War years. Mainstream media reports on space art have also been few and far between since the golden years of space travel in the 1950s and 1960s, with one high-profile exception, one that received major news coverage but conveyed a message not unlike that of the Space Shuttle tragedy in the 1980s. As part of the European Space Commission's Mars Express mission in 2003, the British built Beagle 2 Mars lander, set to bring science and 'Britart's' Damien Hirst's *Spot Painting* to the Red Planet, apparently crash-landed on the craterous

Figure 17.9 Ronald Jones, *The Bed Neil Armstrong Slept in his First Night Back From the Moon*, 1969–98. Wooden headboard and bed frame, mattress, box springs, pillow, pillowcase, sheets, woven aluminum bed spread, carpet (bed) 78.75 × 77.5 × 160 cm; (carpet) 105 × 192.5 × 0.6 cm. Taken out of its context and placed in a New York gallery, the shimmering hi-tech bedspread atop our typical place of dreams, the bed, encourages the memory of the first Apollo moon expedition to mix with personal, even fantastical, memories of the human relationship to outer space.
Source: Courtesy of the artist and Metro Pictures, New York.

plain Isidis Planitia on Mars on 25 December 2003 (Figure 16.4).[33] All communication with the landing probe was lost, and with it the fate of what would have been a European work of space art to follow US-America's *Moon Museum* and *Fallen Astronaut* as the third major cultural site transported from earth for permanent installation on the surface of a neighboring celestial body, the difference being that the European contribution was to be delivered by robotic means, unlike the preceding US-American works that were delivered by human beings.

ZKM Karlsruhe's *SpacePlace: Art in the Age of Orbitization*, along with the international *Leonardo Space Art Project* and *The Arts Catalyst* in the United Kingdom, as well as space humanities symposia and events worldwide, have taken on the challenging task of raising public awareness. They have introduced social networking and new media as outreach to younger tech-savvy audiences in order to bolster awareness of space art activities, furthering the cultural and peaceful use of outer space, contingent human spaceflight activities as well as space humanities and scientific research and exploration programs. The term 'space art' is invested with binaries, transcendent and empirical, ecologic and economic, aesthetic and technological, human and remote, Dionysian and Apollonian – pairs of meaning as fluid as the constantly shifting relationships between earth and outer space, now and certainly in the future.

A space for art has been the preoccupation of cultural producers since time immemorial. Nazca Lines (ca. 800–200 BCE) in the Peruvian desert, giant animistic geoglyphs wholly visible only from a bird's eye view and above, that is, from near-earth orbit, are an early example of art in search of a space on earth to install and show itself. Chauvet and Lascaux, late Pleistocine, Great Ice Age cave paintings, dating back 30,000–17,000 years respectively, act as impressive analogue instruction manuals for survival and societal practice, made unforgettable to pilgriming Cro-Magnon peoples by rituals accompanying their exposition, and are another instance of art in search of a terrestrial home, the cave. In the beginning, art went in search of space.

With the advent of civilizations and cities, temples, cathedrals, exposed rock faces, royal abodes, outlying territories, and, after the French Revolution, the exploding number of secular public museums, private merchant collections, the quandary of locating spaces for art had been more or less solved. The issue of art and space then shifted from exo- to endo-, and finding spaces in art became the *modus operandi* of the artist. History painting, landscape and abstraction, all the twentieth-century -isms, were concerned with identifying and traversing new frontiers. Cartographies, topographies, composition, kinesis and telepresence: each locates a new space in art. With the rapid recent proliferation of computational interaction in our age of orbitization and its dependents – real time cultural cyber-social networks beyond the reach of globalization controls – such spaces in art remain focused on what Debbora Battaglia poetically and aptly terms 'the lure of the aesthetic sublime.'[34] With millennia of art in search of space, and centuries of art about space, now is the time for art in space.

Notes

1. Jack Kerouac, 'Sutra 1,' in Jack Kerouac and Anthony Sampatakos, eds, *The Scripture of the Golden Eternity*, San Francisco: City Lights Publishers, 1960, 23.
2. Giorgio de Santillana, *The Crime of Galileo*, Chicago: University of Chicago Press, 1955, 4.
3. John Elliot Drinkwater Bethune, *Life of Galileo Galilei: With Illustrations of the Advancement of Experimental Philosophy*, Boston: William Hype, 1832, 92.
4. The story first appeared in print in English, in *The Italian Library* published by an Italian man of letters, Giuseppe Baretti (1719–89), in London in 1757. Baretti wrote: 'The moment he was set at liberty, he looked up at the sky and down to the

ground, and, stamping with his foot, in a contemplative mood, said Eppur si m[u] ove; that is, "still it moves," meaning the earth.' Stillman Drake, *Galileo at Work: His Scientific Biography*, Chicago: University of Chicago Press, 1978, 357.

5. Ibid.
6. The second-century Syrian satirist Lucian of Samosata (ca. 125–80) penned a work of literary criticism, *A True Story*, an elliptical parody of Homer's *Odyssey*, recounting a tale of alien and interplanetary warfare, which may in fact be the very first work of science fiction anywhere recorded. On Kepler's *Somnium*, see most recently Johannes Kepler, *Der Traum, oder: Mond-Astronomie. Mit einem Leitfaden für Mondreisende*, ed. Beatrix Langner, Berlin: Matthes & Seitz, 2011.
7. See Chapter 2.
8. Ron Miller, 'The Archaeology of Space Art,' *Leonardo* 29.2 (1996), 139–43, here 140.
9. Bertolt Brecht, *Galileo* [1939/1942], English version by Charles Laughton, introduction by Eric Bentley, New York: Grove, 1966, 16.
10. Nam June Paik, quoted in Kristine Stiles and Peter Selz, *Theories and Documents of Contemporary Art: A Sourcebook of Artist's Writings*, Berkeley: University of California Press, 1996, 435.
11. See United Nations Office of Outer Space Affairs (UNOOSA) web site, http://www.unoosa.org/oosa/en/COPUOS/copuos.html (accessed 1 August 2011).
12. From the film *Space Is the Place*, directed by John Corey, script and soundtrack by Sun Ra, starring Sun Ra, USA 1972 (released 1974, North American Star System).
13. As anecdote, this author might add here that the *Imagining Outer Space* conference was so positively intriguing and eye-opening that, although planning to set up *SpacePlace* in the foyer outside the main conference hall, present it on opening day and depart, I stayed on until I absolutely had to leave three days into the conference. It dawned on me that European public attitudes toward recognizing and supporting future space exploration missions and human spaceflights are and will be partially formed and informed by relevant findings contributed by humanities researchers at large.
14. As Paul Virilio argues in 'Speed and Information: Cyberspace Alarm!': 'The very word globalization is a fake. There is no such thing as globalization, there is only virtualization. What is being effectively globalized by instantaneity is time. Everything now happens within the perspective of real time: henceforth we are deemed to live in a one-time-system.' This article originally appeared as 'Vitesse et information: Alerte dans le cyberespace!,' *Le Monde Diplomatique* (August 1995), 28; translated by Patrice Riemens, University of Amsterdam, and available at http://www.ctheory.net/articles.aspx?id=72 (accessed 1 October 2017).
15. See Jean Baudrillard, *Screened Out*, London: Verso, 2002, 135: 'Outside of this gravitational pull which keeps bodies in orbit, all the atoms of meaning lose themselves or self-absorb in space. Our true artificial satellites are the global debt, the flows of capital and the nuclear loads that circle around the earth in an orbital dance. Debt circulates on its own orbit, with its own trajectory made up of capital which from now on is free of any economic contingency and moves about in a parallel universe. It is not even an orbital universe. It is rather ex-orbital, ex-centered. The only way to avoid this is to place ourselves straightaway on an alternative temporal orbit, to take an elliptic shortcut and go beyond the end by not allowing it time to take place.'
16. See 'Globalisierung: Die Neue Welt,' *Der Spiegel* 7 (2005), cover image; available at http://shop.spiegel.de/shop/action/productDetails/5156976/7_2005_globalisierung.

html (accessed 1 October 2017). To interpret the visualization of planet Earth as a shipping container implies that globalization has become a commodity pictured within a greater, relatively unrecognized, almost alien environment of orbitization.

17. See Paul Virilio, *Open Sky*, London: Verso, 1997, 24: 'Between the subjective and the objective it seems we have no room for the "trajective," that being of movement from here to there, from one to the other, without which we will never achieve a profound understanding of the various regimes of perception of the world that have succeeded each other throughout the ages – regimes of visibility of appearances related to the history of techniques and modalities of displacement long-distance communications, the nature of the speed of movement in transport and transmission entailing a transmutation in the "depth of field" and so in the optical density of the human environment, and not just an evolution in systems of migration or of populating this or that region of the globe.'
18. Lyndon B. Johnson quoted in James N. Giglio and Stephen G. Rabe, *Debating the Kennedy Presidency*, Lanham: Rowman and Littlefield, 2003, 132.
19. Guy Debord, *The Society of the Spectacle* [1967], New York: Zone Books, 1994. Marshall McLuhan foreshadowed mediacentric orbitization: 'Since Sputnik and the satellites, the planet is enclosed in a manmade environment that ends *Nature* and turns the globe into a repertory theater to be programmed. Shakespeare at the Globe mentioning "All the world's a stage, and all the men and women merely players" (*As You Like It*, Act II, Scene 7) has been justified by recent events in ways that would have struck him as entirely paradoxical. The results of living inside a proscenium arch of satellites is that the young now accept the public spaces of the earth as role-playing areas. Sensing this, they adopt costumes and roles and are ready to *do their thing* everywhere.' See Marshall McLuhan, *From Cliché to Archetype*, New York: Viking, 1970, 9–10.
20. The Tate museum network in the United Kingdom imagines a Tate in Space dependence envisioned by ETALAB (Extra-Terrestrial Architecture Laboratory): 'ETALAB envisages Tate in Space as both a real and virtual experience. Representing the ultimate synthesis of artistic and scientific endeavor, it would dock at the International Space Station (ISS) as its cultural component. It will also be able to detach itself from the ISS and travel to the moon or follow its own path through the solar system. Tate in Space is designed to respond to the environment of outer space and to the unpredictable needs of artists, curators and visitors engaged with post-millennial extra-terrestrial art.' See www.tate.org.uk/space/etalab.htm (accessed 1 August 2011).
21. United Nations Office for Outer Space Affairs, *Question of the Peaceful Use of Outer Space: Resolution Adopted by the General Assembly, 792nd Plenary Meeting*, New York: United Nations, 1958; available at http://www.oosa.unvienna.org/oosa/SpaceLaw/gares/html/gares_13_1348.html (accessed 1 August 2011).
22. Joseph W. Ashy quoted in William B. Scott, 'USSC Prepares for Future Combat Missions in Space,' *Aviation Week & Space Technology* (5 August 1996), 51.
23. Boris Groys, *Ilya Kabakov: The Man Who Flew Into Space From His Apartment*, London: Afterall Books, 2006, 4.
24. 'Mein Lieber! Du liegst im Gras, den Kopf im Nacken, um dich herum keine Menschenseele, du hörst nur den Wind und schaust hinauf in den offenen Himmel – in das Blau dort oben, wo die Wolken ziehen – das ist vielleicht das

Schönste, was du im Leben getan und gesehen hast'; see https://www.skulptur-projekte-archiv.de/de-de/1997/projects/18/ (accessed 1 October 2017).

25. Ilya Kabakov in an interview concerning *Blickst Du hinauf und liest die Worte...* (Looking Up, Reading the Words... or Antenna), Skulpturprojekte Münster 1997; see http://ilya-emilia-kabakov.com (accessed 1 October 2017).
26. Alexander von Humboldt, *Cosmos: A Sketch of the Physical Description of the Universe*, vol. 1 [1845], Baltimore: Johns Hopkins University Press, 1997, 79.
27. Immanuel Kant, *Critique of Pure Reason* [1781], ed. Brandt V.B. Dixon, Charleston: Forgotten Books, 2008, 12.
28. It might be noted that parallel to Manzoni and Kabakov's remote aesthetic sensing of outer space, concurrently European space programs, such as ELDO in the 1960s and ESA subsequently, also exhibit in both their mission statements a certain reticence for developing a human spaceflight program, opting for outer space missions designed for telematic remote sensing rather than eyewitness accounting.
29. See Bell Labs veteran engineer Burt Unger in electronic correspondence with Melissa Terras, University College London, Department for Digital Humanities, available at http://melissaterras.blogspot.com/2010/01/moon-museum-redux.html (accessed 1 October 2017): 'I've done some research on the *Moon Museum*, especially before I got involved in the project, so I think I can cover most of the details. Forrest dubbed the collective art the *Moon Museum*. Forrest Meyers knew two engineers/scientists at Bell Labs, Fred Waldhauer and Bill Kluver. They had worked together in a group named Experiments in Art and Technology. The six sketches were given to Fred Waldhauer who worked at Bell Labs in Holmdel, NJ. He in turn gave them to Bob Merkle, an engineer at Holmdel who worked in a thin film processing laboratory. I was the supervisor of the laboratory.

    In 1969 the Thin Film Lab was built in Holmdel to support circuit designers with microcircuits. The Lab was a large clean room with laminar flow hoods that had equipment for metal deposition, photolithography, and etching, plating and bonding. The circuits were made on alumina (aluminum oxide) ceramic with thin film resistors and capacitors made from tantalum and conductors from gold. The resistors were adjusted to exacting tolerance by anodization. We bonded silicon chips to complete the circuits.

    Bob had the six sketches photo-reduced and arranged in a three by two pattern on a glass mask that we used in our lithography process. The patterns were replicated in photo resist in tantalum that covered the ceramic surface and then etched to provide the sketches. Multiple patterns were made on three ceramics. They were then sawed apart and oxidized in a 500 degree centigrade oven for one hour. The patterns came out a vibrant purple color that is very hard and durable. I then broke the glass mask to prevent the wholesale processing of the *Moon Museum*.

    Fred Waldhauer took most of the *Museums* and distributed them to the artists and I think he knew someone at the Cape that attached one to the lunar lander. I took some of the *Museums* and gave them to my engineers as mementos. I don't know who the contacts at the Cape were and who attached it to the LEM. There will be a television program on the *Moon Museum*, called *Histories Mysteries*, sometime next summer. I'm told they are video taping it now.'

30. Astronaut Captain Alan Bean (1932–), the fourth man on the moon, now offers textured acrylic with 'Moondust' on aircraft plywood paintings for sale or on commission; see his *Painting Apollo: First Artist on Another World*, Washington, DC: Smithsonian Books, 2009.
31. Pierre Compte, *Prisma*, 1995–2001, 14 painted and varnished spheres 2.5 cm in diameter, seven articulated axial joints; size folded: 24.5 × 12 × 2.5 cm; International Space Station, earth orbit.
32. Arthur Woods, *Cosmic Dancer*, 1993, painted, welded aluminum tubing measuring approximately 35 × 35 × 40 cm; see http://www.ekac.org/levitation.html (accessed 1 October 2017).
33. See Tristan Weddigen's contribution, Chapter 16 in this volume, for a discussion of Hirst's *Spot Paintings* and the calibration target attached to Beagle 2.
34. See Debbora Battaglia's contribution, Chapter 11 in this volume.

BIBLIOGRAPHY

Journals

Acta Astronautica: Journal of the International Academy of Astronautics, 1974–.
Astropolitics: The International Journal of Space Politics and Policy, 2003–.
Flying Saucer Review, 1955–.
Journal of the British Interplanetary Society, 1934–.
Quest: The History of Spaceflight Quarterly, 1995–.
Die Rakete: Offizielles Organ des Vereins für Raumschiffahrt e.V. in Deutschland, 1927–29.
Science Fiction Studies, 1973–.
Space Policy: An International Journal, 1985–.
Spaceflight, 1956–.
The Space Review: Essays and Commentary About the Final Frontier, 2003–.
Weltraumfahrt: Beiträge zur Weltraumforschung und Astronautik, 1950–66.

Bibliographies

Andersen, Per, *Bibliografi over dansk ufo-litteratur 1950–1985*, Vanløse: Andersen Bogservice, 1986.

Beard, Robert Brookes, *Flying Saucers, U.F.O.'s and Extraterrestrial Life: A Bibliography of British Books, 1950–1970*, Swindon: R. Beard, 1971.

Bibliothèque nationale de France, direction des collections, département sciences et techniques, *Une Histoire des représentations de l'espace: Bibliographie*, Paris: BN, 2007.

Bloch, Robert N., *Bibliographie der Utopie und Phantastik 1650–1950 im deutschen Sprachraum*, Hamburg: Achilla, 2002.

Buike, Bruno, *UFOs: Geschichte und Naturwissenschaft. Bibliographie mit deutsch-englischen Anmerkungen und Adressen*, Marburg: Tectum, 1996.

Alexander C.T. Geppert (ed.), *Imagining Outer Space*
European Astroculture, vol. 1
https://doi.org/10.1057/978-1-349-95339-4

Caillens, Pierre and Alain Mauret, *L'Argus de la science-fiction*, 6th edn, 2 vols, Libourne: L'Annonce-Bouquins, 2004.

Catoe, Lynn E. and Kay Rodgers, eds, *UFOs and Related Subjects: An Annotated Bibliography*, Detroit: Gale/Book Tower, 1978.

Ciancone, Michael L., ed., *The Literary Legacy of the Space Age: An Annotated Bibliography of Pre-1958 Books on Rocketry and Space Travel*, Houston: Amorea, 1998.

Dag Hammarskjold Library/Bibliothèque Dag Hammarskjold, ed., *Outer Space: A Selective Bibliography/L'espace extra-atmosphérique: Bibliographie sélective*, New York: United Nations, 1982.

Delmas, Henri and Alain Julian, *Le Rayon SF: Catalogue bibliographique de science-fiction. Utopies, voyages extraordinaires*, Toulouse: Milan, 1985.

Dotzler, Bernhard J., Peter Gendolla and Jörgen Schäfer, eds, *MaschinenMenschen: Eine Bibliographie*, Frankfurt am Main: Peter Lang, 1992.

Eberhart, George M., ed., *UFOs and the Extraterrestrial Contact Movement: A Bibliography*, 2 vols, Metuchen: Scarecrow, 1986.

Emme, Eugene M., 'An Eclectic Bibliography on the History of Space Futures,' in idem, ed., *Science Fiction and Space Futures: Past and Present*, San Diego: American Astronautical Society, 1982, 213–45.

Hall, Halbert Weldon, ed., *Science Fiction and Fantasy Reference Index, 1992–1995: An International Subject and Author Index to History and Criticism*, Englewood: Libraries Unlimited, 1997.

Hübner, Thomas, ed., *Raumfahrt-Bibliographie: Ein Verzeichnis nichttechnischer deutschsprachiger Literatur von 1923 bis 1997*, Hörstel: Raumfahrt-Info-Dienst, 1998.

Illmer, Horst, ed., *Bibliographie Science Fiction & Fantasy: Buch-Erstausgaben 1945–1995. 50 Jahre alternative Weltentwürfe in Deutschland*, Wiesbaden: Harrassowitz, 1998.

Koelle, H.H. and H.J. Kaeppeler, eds, *Literaturverzeichnis der Astronautik*, Tittmoning: Walter Pustet, 1954.

Kok, G.J., *Ufo's: Een geannoteerde bibliografie met registers*, Uithuizermeeden: Nobovo, 1980.

Landrum, Larry N., 'A Checklist of Materials about Science Fiction Films of the 1950's,' *Journal of Popular Film* 1.1 (Winter 1972), 61–3.

———, 'Science Fiction Film Criticism in the Seventies: A Selected Bibliography,' *Journal of Popular Film and Television* 6.3 (1978), 287–9.

Locke, George, ed., *Voyages in Space: A Bibliography of Interplanetary Fiction 1801–1914*, London: Ferret Fantasy, 1975.

Looney, John J., *Bibliography of Space Books and Articles from Non-Aerospace Journals, 1957–1977*, Washington, DC: NASA, 1980.

Lundwall, Sam J., *Illustrerad bibliografi över Science fiction & fantasy 1741–1973*, Stockholm: Lindqvist, 1974.

———, *Bibliografi över Science fiction & fantasy 1974–1983*, Bromma: Delta, 1985.

Mallove, Eugene F., Robert L. Forward, Zbigniew Paprotny and Jürgen Lehmann, 'Interstellar Travel and Communication: A Bibliography,' *Journal of the British Interplanetary Society* 33.6 (June 1980), 201–48.

Marotta, Michael E., *Space Colonization: An Annotated Bibliography*, Mason: Loompanics, 1979.

Neumann, Hans-Peter, Ivo Gloss and Erik Simon, *Die große illustrierte Bibliographie der Science Fiction in der DDR*, Berlin: Shayol, 2002.
Phillips, Mark and Frank Garcia, *Science Fiction Television Series: Episode Guides, Histories, and Casts and Credits for 62 Prime Time Shows, 1959 through 1989*, Jefferson: McFarland, 1996.
Pisano, Dominick A. and Cathleen S. Lewis, eds, *Air and Space History: An Annotated Bibliography*, London: Garland, 1988.
Rasmussen, Richard Michael, ed., *The UFO Literature: A Comprehensive Annotated Bibliography of Works in English*, Jefferson: McFarland, 1985.
Reeken, Dieter von, ed., *Bibliographie der selbständigen deutschsprachigen Literatur über Außerirdisches Leben, UFOs und Prä-Astronautik, Zeitraum 1703–1995*, 4th edn, Lüdenscheid: Gesellschaft zur Erforschung des UFO-Phänomens, 1996.
Smith, Marcia S., ed., *Extraterrestrial Intelligence and Unidentified Flying Objects: A Selected, Annotated Bibliography*, Washington, DC: Library of Congress, 1976.
Spehner, Norbert, *Ecrits sur la science-fiction: Bibliographie analytique des études et essais sur la science-fiction publiés entre 1900 et 1987*, Longueuil: Préambule, 1988.
Tobias, Russell R., ed., *America in Space: An Annotated Bibliography*, Pasadena: Salem, 1991.
Wegner, Willy, *Dansk Ufo-litteratur: En bibliografi*, vol. 1: *1946–1970*, vol. 2: *1971–1979*, Copenhagen: Københavns Universite, 1972–1981.
———, *Skandinavisk UFO-litteratur, 1950–1982: En bibliografi*, 2nd edn, Hjallerup: Skeptica, 1982.
Wimmer, Heinrich, ed., *Bibliographisches Lexikon der utopisch-phantastischen Literatur: Verlags- und Reihenbibliographien*, Meitingen: Corian, 1987.

Dictionaries and Encyclopedias

Bali, Mrinal, *Space Exploration: A Reference Handbook*, Santa Barbara: ABC-CLIO, 1990.
Clark, Jerome, ed., *The UFO Encyclopedia: The Phenomenon from the Beginning*, Detroit: Omnigraphics, 1998.
———, ed., *Extraordinary Encounters: An Encyclopedia of Extraterrestrials and Otherworldly Beings*, Santa Barbara: ABC-CLIO, 2000.
Clute, John and Peter Nichols, eds, *The Encyclopedia of Science Fiction*, London: Orbit, 1999.
Dasch, E. Julius, ed., *A Dictionary of Space Exploration*, Oxford: Oxford University Press, 2005.
Dickson, Paul, *A Dictionary of the Space Age*, Baltimore: Johns Hopkins University Press, 2009.
Engelhardt, Wolfgang, *Enzyklopädie Raumfahrt*, Frankfurt am Main: Harri Deutsch, 2001.
Gatland, Kenneth, *The Illustrated Encyclopedia of Space Technology: A Comprehensive History of Space Exploration*, New York: Harmony, 1981.
Gerhards, Winfried, *Handbuch der phantastischen Fernsehserien*, Hamburg: Gryphon, 2001.
Henderson, C.J., *The Encyclopedia of Science Fiction Movies*, New York: Facts on File, 2001.

Herbert, A.P., ed., *Watch this Space (Six Years of It): An Anthology of Space (Fact), 4 October 1957–4 October 1963*, London: Methuen, 1964.
Johnson, Stephen B., ed., *Space Exploration and Humanity: A Historical Encyclopedia*, Santa Barbara: ABC-CLIO, 2010.
Koebner, Thomas, ed., *Science Fiction*, Stuttgart: Reclam, 2007.
Lewis, James R., ed., *UFOs and Popular Culture: An Encyclopedia of Contemporary Myth*, Santa Barbara: ABC-CLIO, 2000.
McLaughlin, Charles, ed., *Space Age Dictionary*, Princeton: Van Nostrand, 1959.
Plant, Malcolm, *Dictionary of Space*, Harlow: Longman, 1986.
Rabkin, Eric S., ed., *Science Fiction: A Historical Anthology*, Oxford: Oxford University Press, 1983.
Story, Ronald D. and J. Richard Greenwell, eds, *The Encyclopedia of UFOs*, London: New English Library, 1980.
Verger, Fernand, Isabelle Sourbès-Verger and Raymond Ghirardi, eds, *The Cambridge Encyclopedia of Space: Missions, Applications and Exploration*, Cambridge: Cambridge University Press, 2003.
Versins, Pierre, *Encyclopédie de l'utopie, des voyages extraordinaires et de la science fiction*, 2nd edn, Lausanne: L'Age d'Homme, 1984.
Walsh, Patrick J., *Spaceflight: A Historical Encyclopedia*, 3 vols, Santa Barbara: Greenwood Press, 2010.
Warren, Bill, *Keep Watching the Skies! American Science Fiction Movies of the Fifties*, 2nd edn, Jefferson: McFarland, 2009.
Willis, Donald C., ed., *Variety's Complete Science Fiction Reviews*, New York: Garland, 1985.
Wright, Gene, *The Science Fiction Image: The Illustrated Encyclopedia of Science Fiction in Film, Television, Radio and the Theatre*, London: Columbus Books, 1983.
Zimmerman, Robert, ed., *The Chronological Encyclopedia of Discoveries in Space*, Phoenix: Oryx, 2000.

Literature

Abret, Helga, 'Literatur und Technik: Von Brauns *Marsprojekt* und Nehers *Menschen zwischen den Planeten*,' in Hans Esselborn, ed., *Utopie, Antiutopie und Science Fiction im deutschsprachigen Roman des 20. Jahrhunderts*, Würzburg: Königshausen & Neumann, 2003, 118–32.
Abret, Helga and Lucian Boia, *Das Jahrhundert der Marsianer: Der Planet Mars in der Science Fiction bis zur Landung der Viking-Sonden 1976*, Munich: Heyne, 1984.
Adams, John, 'Outer Space and the New World in the Imagination of Eighteenth-Century Europeans,' *Eighteenth-Century Life* 19.1 (February 1995), 70–83.
Adamski, George, *Inside the Space Ships*, New York: Abelard-Schuman, 1955.
Agar, Jon, *Science and Spectacle: The Work of Jodrell Bank in Post-War British Culture*, Amsterdam: Harwood, 1998.
Agel, Jerome, ed., *The Making of Kubrick's* 2001, New York: Signet, 1970.
Ailleris, Philippe, 'The Lure of Local SETI: Fifty Years of Field Experiments,' *Acta Astronautica* 68.1–2 (January–February 2011), 2–15.
Allen, Michael, *Live from the Moon: Film, Television and the Space Race*, London: I.B. Tauris, 2009.

Allingham, Cedric [pseud. Patrick Moore?], *Flying Saucer from Mars*, London: Frederick Muller, 1955.

Almond, Gabriel A., 'Public Opinion and the Development of Space Technology,' *Public Opinion Quarterly* 24.4 (Winter 1960), 553–72.

Almond, Philip, 'Adam, Pre-Adamites, and Extra-Terrestrial Beings in Early Modern Europe,' *Journal of Religious History* 30.2 (June 2006), 163–74.

Ananoff, Alexandre, *Les Mémoires d'un astronaute ou l'Astronautique Française*, Paris: Albert Blanchard, 1978.

Anders, Günther, *Der Blick vom Mond: Reflexionen über Weltraumflüge* [1970], 2nd edn, Munich: C.H. Beck, 1994.

Andrews, James T., *Science for the Masses: The Bolshevik State, Public Science, and the Popular Imagination in Soviet Russia, 1917–1934*, College Station: A&M University Press, 2003.

———, 'In Search of a Red Cosmos: Space Exploration, Public Culture, and Soviet Society,' in Steven J. Dick and Roger D. Launius, eds, *Societal Impact of Spaceflight*, Washington, DC: NASA, 2007, 41–52.

Andrews, James T. and Asif A. Siddiqi, eds, *Into the Cosmos: Space Exploration and Soviet Culture*, Pittsburgh: University of Pittsburgh Press, 2011.

Anker, Peder, 'The Ecological Colonization of Space,' *Environmental History* 10.2 (April 2005), 239–68.

Arendt, Hannah, 'Man's Conquest of Space,' *American Scholar* 32.4 (Fall 1963), 527–40.

Armstrong, Rachel, ed., *Space Architecture*, New York: John Wiley, 2000 (= *Architectural Design* 70.2).

Armytage, W.H.G., *Yesterday's Tomorrows: A Historical Survey of Future Societies*, London: Routledge & Kegan Paul, 1968.

Arnheim, Rudolf, 'Outer Space and Inner Space,' *Leonardo* 24.1 (1991), 73–4.

Arnould, Jacques, 'Does Extraterrestrial Intelligent Life Threaten Religion and Philosophy?,' *Theology & Science* 6.4 (2008), 439–50.

Asimov, Isaac, 'A Science in Search of a Subject,' *New York Times Magazine* (23 May 1965), 52–8.

———, 'Plädoyer für Science-fiction,' *Der Spiegel* 11 (6 March 1972), 138–9.

———, 'The Next Frontier?,' *National Geographic* 150.1 (July 1976), 76–98.

Atwill, William D., *Fire and Power: The American Space Program as Postmodern Narrative*, Athens: University of Georgia Press, 1994.

Bailey, James Osler, *Pilgrims Through Space and Time: Trends and Patterns in Scientific and Utopian Fiction*, New York: Argus, 1947.

Bainbridge, William Sims, *The Spaceflight Revolution: A Sociological Study*, New York: John Wiley, 1976.

———, *Dimensions of Science Fiction*, Cambridge, MA: Harvard University Press, 1986.

———, 'Extraterrestrial Tales,' *Science* 279 (1998), 671.

———, 'The Spaceflight Revolution Revisited,' in Stephen J. Garber, ed., *Looking Backward, Looking Forward: Forty Years of U.S. Human Spaceflight Symposium*, Washington, DC: NASA, 2002, 39–64.

Balch, Robert W. and David Taylor, 'Seekers and Saucers: The Role of the Cultic Milieu in Joining a UFO Cult,' *American Behavioral Scientist* 20.6 (July–August 1977), 839–60.

Barth, Hans, *Hermann Oberth: Vater der Raumfahrt*, Esslingen: Bechtle, 1991.

———, ed., *Hermann Oberth: Briefwechsel* [1979], 2 vols, Bucharest: Kriterion, 1984.

Bartholomew, Robert E., 'The Quest for Transcendence: An Ethnography of UFOs in America,' *Anthropology of Consciousness* 2.1–2 (March–June 1991), 1–12.

Bartos, Adam and Svetlana Boym, *Kosmos: A Portrait of the Russian Space Age*, New York: Princeton Architectural Press, 2001.

Basalla, George, *Civilized Life in the Universe: Scientists on Intelligent Extraterrestrials*, Oxford: Oxford University Press, 2006.

Battaglia, Debbora, ed., *E.T. Culture: Anthropology in Outerspaces*, Durham: Duke University Press, 2005.

Baumunk, Bodo-Michael and Ralf Bülow, eds, *Weltraum: Sonnen, Monde, Galaxien: Aufbruch ins Unbekannte*, Berlin: Henschel, 2000.

Baxter, Stephen, 'A Human Galaxy: A Prehistory of the Future,' *Journal of the British Interplanetary Society* 58.2 (March–April 2005), 138–42.

———, 'Imagining the Alien: The Portrayal of Extraterrestrial Intelligence and SETI in Science Fiction,' *Journal of the British Interplanetary Society* 62.4 (April 2009), 131–8.

Beattie, Donald A., *Taking Science to the Moon: Lunar Experiments and the Apollo Program*, Baltimore: Johns Hopkins University Press, 2001.

Bell, Daniel, *The Winding Passage: Essays and Sociological Journeys, 1960–1980*, Cambridge, MA: Abt Books, 1980.

Bell, David and Martin Parker, eds, *Space Travel and Culture: From Apollo to Space Tourism*, Oxford: Wiley-Blackwell, 2009 (= *Sociological Review* 57.s1).

Bender, Hans, 'Zur Psychologie der UFO-Phänomene,' *Zeitschrift für Parapsychologie und Grenzgebiete der Psychologie* 3.1 (1959), 32–58.

Benjamin, Marina, *Rocket Dreams: How the Space Age Shaped Our Vision of a World Beyond*, New York: Free Press, 2003.

Bennett, Jeffrey, *Beyond UFOs: The Search for Extraterrestrial Life and Its Astonishing Implications for Our Future*, Princeton: Princeton University Press, 2008.

Benz, Ernst, *Außerirdische Welten: Von Kopernikus zu den Ufos* [1979], Freiburg im Breisgau: Aurum, 1990.

Bergaust, Erik, *Colonizing Space*, New York: Putnam, 1978.

Bergaust, Erik and William Beller, *Satellite!*, London: Lutterworth, 1957.

Berger, Albert I., 'Science-Fiction Critiques of the American Space Program, 1945–1958,' *Science Fiction Studies* 5.2 (July 1978), 99–109.

Bergermann, Ulrike, Isabell Otto and Gabriele Schabacher, eds, *Das Planetarische: Kultur – Technik – Medien im postglobalen Zeitalter*, Munich: Fink, 2010.

Beyer, Friedemann, 'Die Erfindung des Countdowns,' *Frankfurter Allgemeine Zeitung* (14 October 2004), 37.

Bignier, Michel, 'Les programmes d'application spatiale et la coopération internationale,' *Défense Nationale* 29.10 (October 1973), 69–91.

Billig, Otto, *Flying Saucers: Magic in the Skies. A Psychohistory*, Cambridge, MA: Schenkman, 1982.

Billingham, John, 'Cultural Aspects of the Search for Extraterrestrial Intelligence,' *Acta Astronautica* 42.10–12 (1998), 711–19.

Binkley, Sam, 'The Seers of Menlo Park: The Discourse of Heroic Consumption in the "Whole Earth Catalog,"' *Journal of Consumer Culture* 3.3 (November 2003), 283–313.

Bizony, Piers, *2001: Filming the Future*, London: Aurum Press, 1994.
Blühm, Andreas, ed., *Der Mond*, Ostfildern: Hatje Cantz, 2009 (exhibition catalogue, Wallraf-Richartz-Museum, Cologne, 26 March–16 August 2009).
Blumenberg, Hans, *Die Legitimität der Neuzeit*, Frankfurt am Main: Suhrkamp, 1966.
———, *Die Genesis der kopernikanischen Welt*, Frankfurt am Main: Suhrkamp, 1981.
———, *Die Vollzähligkeit der Sterne*, Frankfurt am Main: Suhrkamp, 1997.
———, *Geistesgeschichte der Technik*, Frankfurt am Main: Suhrkamp, 2009.
Bobrowsky, Peter T. and Hans Rickman, eds, *Comet/Asteroid Impacts and Human Society: An Interdisciplinary Approach*, Heidelberg: Springer, 2007.
Boia, Lucian, 'L'Image des autres mondes (deuxième moitié du XIXe siècle – début du XXe siècle),' *Analele Universitatii Bucuresti: Istorie* 33 (1984), 45–58.
———, *L'Exploration imaginaire de l'espace*, Paris: La Découverte, 1987.
———, *Pour une histoire de l'imaginare*, Paris: Les Belles Lettres, 1998.
Bond, Alan, ed., *Project Daedalus: The Final Report on the BIS Starship Study*, London: British Interplanetary Society, 1978 (= *Journal of the British Interplanetary Society Supplement*).
Bonestell, Chesley and Willy Ley, *The Conquest of Space*, New York: Viking, 1949.
Bonnet, Roger M. and Vittorio Manno, *International Cooperation in Space: The Example of the European Space Agency*, Cambridge, MA: Harvard University Press, 1994.
Bonting, Sjoerd L., 'Teilhard de Chardin en buitenaards leven,' *Gamma* 5.2 (April 1998), 41–3.
———, 'Theological Implications of Possible Extraterrestrial Life,' *Zygon* 38.3 (September 2003), 587–602.
Bormann, Natalie and Michael Sheehan, eds, *Securing Outer Space*, London: Routledge, 2009.
Boss, Alan, *The Crowded Universe: The Search for Living Planets*, New York: Basic Books, 2009.
Boyer, Paul S., *By the Bomb's Early Light: American Thought and Culture at the Dawn of the Atomic Age*, New York: Pantheon, 1985.
———, *Fallout*, Columbus: Ohio State University Press, 1998.
Boyle, Charles P., *Space Among Us: Some Effects of Space Research on Society*, Washington, DC: Aerospace Industries Association of America, 1974.
Brand, Illo, 'Das UFO-Sichtungsspektrum,' *Zeitschrift für Parapsychologie und Grenzgebiete der Psychologie* 17.2–3 (1975), 89–124.
Brand, Stewart, ed., *Space Colonies*, Harmondsworth: Penguin, 1977.
Braun, Wernher von, 'Lunetta,' *Leben und Arbeit: Zeitschrift der Bürger und Freunde der Deutschen Landerziehungsheime* 23.2–3 (1930–31), 88–92.
———, 'Das Problem einer interplanetarischen Expedition,' *Universitas* 7.7 (July 1952), 733–9.
———, 'Why I Chose America,' *The American Magazine* 154.1 (July 1952), 15, 111–15.
———, *The Mars Project*, Urbana: University of Illinois Press, 1953.
———, 'Space Superiority as a Means for Achieving World Peace,' *Ordnance* 37 (March–April 1953), 770–5.
———, 'Reminiscences of German Rocketry,' *Journal of the British Interplanetary Society* 15.3 (May–June 1956), 125–45.

———, 'Raketen verlängern die dritte Dimension,' *Frankfurter Allgemeine Zeitung* (9 September 1959), 9.
———, *First Men to the Moon*, New York: Holt, Rinehart & Winston, 1961.
Braun, Wernher von, Joseph Kaplan, Heinz Haber, Oscar Schachter, Fred L. Whipple and Cornelius Ryan, *Across the Space Frontier*, New York: Viking, 1952.
Bredekamp, Horst, Matthias Bruhn and Gabriele Werner, eds, *Imagination des Himmels*, Berlin: Akademie, 2007.
Brewer, John, *The Pleasures of the Imagination: English Culture in the Eighteenth Century*, London: HarperCollins, 1997.
Brügel, Werner, ed., *Männer der Rakete: In Selbstdarstellungen*, Leipzig: Hachmeister & Thal, 1933.
Brunner, Bernd, *Moon: A Brief History*, New Haven: Yale University Press, 2010.
Bulkeley, Rip, *The Sputniks Crisis and Early United States Space Policy: A Critique of the Historiography of Space*, Bloomington: Indiana University Press, 1991.
———, 'Harbingers of Sputnik: The Amateur Radio Preparations in the Soviet Union,' *History and Technology* 16.1 (1999), 67–102.
Bullard, Thomas E., 'UFO Abduction Reports: The Supernatural Kidnap Narrative Returns in Technological Guise,' *Journal of American Folklore* 102 (April–June 1989), 147–70.
———, *The Myth and Mystery of UFOs*, Lawrence: University Press of Kansas, 2010.
Bürgisser, Thomas, '"Im Banne des Satelliten": Zur medialen Rezeption des Sputnik-Schocks in der Schweiz,' *Schweizerische Zeitschrift für Geschichte* 57.4 (2007), 387–416.
Burrows, William E., *This New Ocean: The Story of the First Space Age*, New York: Random House, 1998.
Burton, Dan and David Grandy, *Magic, Mystery, and Science: The Occult in Western Civilization*, Bloomington: Indiana University Press, 2004.
Buthmann, Reinhard, 'Die DDR im Weltraum: Kosmosforschung im Licht der MfS-Akten,' *Deutschland-Archiv* 32.2 (1999), 223–32.
Cade, C. Maxwell, *Other Worlds than Ours*, London: Museum Press, 1966.
Calder, Nigel, *Spaceships of the Mind*, New York: Viking, 1978.
Callon, Michel, Pierre Lascoumes and Yannick Barthes, *Agir dans un monde incertain: Essai sur la démocratie technique*, Paris: Editions du Seuil, 2001.
Cantril, Hadley, Hazel Gaudet and Herta Herzog, *The Invasion from Mars: A Study in the Psychology of Panic*, Princeton: Princeton University Press, 1940.
Caroti, Simone, 'Defining Astrosociology from a Science Fiction Perspective,' *Astropolitics* 9.1 (April 2011), 39–49.
Carter, Dale, *The Final Frontier: The Rise and Fall of the American Rocket State*, New York: Verso, 1988.
Carter, John, *Sex and Rockets: The Occult World of Jack Parsons*, Venice: Feral House, 1999.
Carter, Leonard James, ed., *Realities of Space Travel: Selected Papers of the British Interplanetary Society*, London: Putnam, 1957.
Carter, Paul A., 'Rockets to the Moon, 1919–1944: A Dialogue Between Fiction and Reality,' *American Studies* 15 (Spring 1974), 31–46.
Castoriadis, Cornelius, *L'Institution imaginaire de la société*, Paris: Editions du Seuil, 1975.
Centre National d'Etudes Spatiales, *Les trente premières années du CNES: L'Agence française de l'espace, 1962–1992*, Paris: CNES, 1994.

Chadeau, Emmanuel, 'Etat, industrie, nation: La formation des technologies aéronautiques en France (1900–1950),' *Histoire, Economie et Société* 4 (1985), 275–99.

Chaikin, Andrew, *A Man on the Moon: The Voyages of the Apollo Astronauts*, New York: Penguin, 1998.

Chaisson, Eric, *Cosmic Dawn: The Origins of Matter and Life*, Boston: Little, Brown, 1981.

———, *Cosmic Evolution: The Rise of Complexity in Nature*, Cambridge: Harvard University Press, 2001.

Chakrabarty, Dipesh, *Provincializing Europe: Postcolonial Thought and Historical Difference*, Princeton: Princeton University Press, 2000.

Chun, Clayton K.S., 'Expanding the High Frontier: Space Weapons in History,' *Astropolitics* 2.1 (Spring 2004), 63–78.

Claeys, Gregory, *Searching for Utopia: The History of an Idea*, London: Thames & Hudson, 2011.

Clarke, Arthur C., 'Extra-Terrestrial Relays: Can Rocket Stations Give World-Wide Radio Coverage?,' *Wireless World* 51.10 (October 1945), 305–8.

———, 'The Challenge of the Spaceship (Astronautics and Its Impact upon Human Society),' *Journal of the British Interplanetary Society* 6.3 (December 1946), 66–81.

———, *Interplanetary Flight: An Introduction to Astronautics*, London: Temple, 1950.

———, 'The Conquest of Space,' *Fortnightly Review* (March 1950), 161–7.

———, 'Space-Travel in Fact and Fiction,' *Journal of the British Interplanetary Society* 9.5 (September 1950), 213–30.

———, *The Exploration of Space*, New York: Harper, 1951.

———, '"Flying Saucers,"' *Journal of the British Interplanetary Society* 12.3 (May 1953), 97–100.

———, *The Making of a Moon: The Story of the Earth Satellite Program*, London: Frederick Muller, 1957.

———, *Interplanetary Flight: An Introduction to Astronautics*, 2nd edn, New York: Harper, 1960.

———, *The Challenge of the Spaceship*, New York: Ballantine Books, 1961.

———, *Voices from the Sky: Previews of the Coming Space Age*, London: Gollancz, 1965.

———, ed., *The Coming of the Space Age: Famous Accounts of Man's Probing of the Universe*, New York: Meredith, 1967.

———, *The Promise of Space*, New York: Harper & Row, 1968.

———, *Prelude to Space: Science Fiction*, London: Sidgwick & Jackson, 1970.

———, *Astounding Days: A Science Fictional Autobiography*, New York: Bantam, 1990.

———, *Greetings, Carbon-Based Bipeds! Collected Essays, 1934–1998*, New York: St. Martin's Press, 1999.

Clarke, Arthur C. and R.A. Smith, *The Exploration of the Moon*, London: Frederick Muller, 1954.

Clarke, David, *The UFO Files: The Inside Story of Real-Life Sightings*, Kew: The National Archives, 2009.

Clarke, David and Andy Roberts, *Flying Saucerers: A Social History of UFOlogy*, Loughborough: Heart of Albion Press, 2007.

Clausberg, Karl, ed., *Zwischen den Sternen: Lichtbildarchive*, Berlin: Akademie, 2006.
Cleator, Philip Ellaby, *Rockets Through Space: The Dawn of Interplanetary Travel*, New York: Simon & Schuster, 1936.
———, *An Introduction to Space Travel*, London: Museum Press, 1961.
Cleaver, Arthur Valentine, 'The Interplanetary Project,' *Journal of the British Interplanetary Society* 7.1 (January 1948), 21–39.
———, 'Rocketry and Space Flight,' *Journal of the Royal Aeronautical Society* 70 (January 1966), 288–93.
———, 'The Case for Space,' *Aeronautical Journal* 73 (December 1969), 1007–18.
———, 'European Space Activities Since the War: A Personal View,' *Spaceflight* 16.6 (June 1974), 220–38.
———, 'On the Realisation of Projects: With Special Reference to O'Neill Space Colonies and the Like,' *Journal of the British Interplanetary Society* 30.8 (August 1977), 283–8.
Cocconi, Giuseppe and Philip Morrison, 'Searching for Interstellar Communications,' *Nature* 184 (19 September 1959), 844–6.
Cockroft, John, Russell Bertrand and Henry A. Wallace, *Atomic Challenge: A Symposium*, London: Winchester, 1947.
Codignola, Luca and Kai-Uwe Schrogl, eds, *Humans in Outer Space: Interdisciplinary Odysseys*, Vienna: Springer, 2008.
Cole, Dandridge M., *Beyond Tomorrow: The Next 50 Years in Space*, Amherst: Amherst Press, 1965.
Collins, Guy, *Europe in Space*, Basingstoke: Macmillan, 1990.
Collins, Martin, 'One World... One Telephone: Iridium, One Look at the Making of a Global Age,' *History and Technology* 21.3 (September 2005), 301–24.
Collins, Martin and Douglas Millard, eds, *Showcasing Space*, London: Science Museum, 2005.
Condon, Edward Uhler, *Final Report of the Scientific Study of Unidentified Flying Objects*, New York: Dutton, 1969.
Cooper, Richard N. and Richard Layard, eds, *What the Future Holds: Insights from Social Science*, Cambridge, MA: MIT Press, 2002.
Corn, Joseph J., ed., *Imagining Tomorrow: History, Technology, and the American Future*, Cambridge, MA: MIT Press, 1986.
Corn, Joseph J. and Brian Horrigan, eds, *Yesterday's Tomorrows: Past Visions of the American Future*, New York: Summit Books, 1984.
Cornea, Christine, *Science Fiction Cinema: Between Fantasy and Reality*, Edinburgh: Edinburgh University Press, 2007.
Cosgrove, Denis, 'Contested Global Visions: *One-World*, *Whole-Earth*, and the Apollo Space Photographs,' *Annals of the Association of American Geographers* 84.2 (June 1994), 270–94.
———, *Apollo's Eye: A Cartographic Genealogy of the Earth in the Western Imagination*, Baltimore: Johns Hopkins University Press, 2001.
Crary, Jonathan, *Techniques of the Observer: On Vision and Modernity in the Nineteenth Century*, Cambridge, MA: MIT Press, 1990.
Creedon, Jeremiah, 'ISY, SETI and the New Search for El Dorado,' *Space Policy* 8.3 (August 1992), 191–4.
Cressy, David, 'Early Modern Space Travel and the English Man in the Moon,' *American Historical Review* 111.4 (October 2006), 961–82.

Crouch, Tom D., *Aiming for the Stars: The Dreamers and Doers of the Space Age*, Washington, DC: Smithsonian Institution Press, 1999.

Crowe, Michael J., *The Extraterrestrial Life Debate, 1750–1900: The Idea of a Plurality of Worlds from Kant to Lowell*, Cambridge: Cambridge University Press, 1986.

———, 'A History of the Extraterrestrial Life Debate,' *Zygon* 32.2 (June 1997), 147–62.

———, ed., *The Extraterrestrial Life Debate, Antiquity to 1915: A Source Book*, Notre Dame: University of Notre Dame, 2008.

Crowley, David and Jane Pavitt, eds, *Cold War Modern: Design 1945–1970*, London: V&A, 2008 (exhibition catalogue, Victoria and Albert Museum, London, 25 September 2008–11 January 2009).

Csicsery-Ronay, Istvan, Jr., 'Science Fiction and Empire,' *Science Fiction Studies* 30.2 (July 2003), 231–45.

Däniken, Erich von, *Erinnerungen an die Zukunft: Ungelöste Rätsel der Vergangenheit*, Düsseldorf: Econ, 1968 (Eng. *Chariots of the Gods? Unsolved Mysteries of the Past*, London: Souvenir, 1969).

———, 'Interview: Erich von Däniken. A Candid Conversation with that Publicist for Ancient Astronauts, the Best-Selling Author of the Cult Classic "Chariots of the Gods?",' *Playboy* 21.8 (August 1974), 51–64, 151.

Darlington, David, *Area 51: The Dreamland Chronicles*, New York: Henry Holt, 1997.

Darrach, H. Bradford and Robert Ginna, 'Have We Visitors From Space?,' *Life* 32.14 (7 April 1952), 80–4, 90–6.

Daston, Lorraine and Peter Galison, *Objectivity*, New York: Zone, 2007.

Daum, Andreas W., *Wissenschaftspopularisierung im 19. Jahrhundert: Bürgerliche Kultur, naturwissenschaftliche Bildung und die deutsche Öffentlichkeit 1848–1914*, Munich: Oldenbourg, 1998.

Davies, Paul, *Are We Alone? Philosophical Implications of the Discovery of Extraterrestrial Life*, New York: Basic Books, 1995.

———, *The Eerie Silence: Are We Alone in the Universe?*, London: Allen Lane, 2010.

Day, Dwayne A., 'The Von Braun Paradigm,' *Space Times* (November–December 1994), 12–15.

———, 'Boldly Going: Star Trek and Spaceflight,' *The Space Review* (28 November 2005), http://www.thespacereview.com/article/506/1.

———, 'Exploring the Social Frontiers of Spaceflight,' *The Space Review* (25 September 2006), http://www.thespacereview.com/article/713/1.

Day, Peter, ed., *The Search for Extraterrestrial Life: Essays on Science and Technology*, Oxford: Oxford University Press, 1998.

Dean, Jodi, 'The Truth is Out There: Aliens and the Fugitivity of Postmodern Truth,' *Camera Obscura* 14.40–41 (May 1997), 43–74.

———, *Aliens in America: Conspiracy Cultures from Outerspace to Cyberspace*, Ithaca: Cornell University Press, 1998.

Deese, R.S., 'The Artifact of Nature: "Spaceship Earth" and the Dawn of Global Environmentalism,' *Endeavour* 33.2 (June 2009), 70–5.

Degroot, Gerard J., *Dark Side of the Moon: The Magnificent Madness of the American Lunar Quest*, New York: New York University Press, 2006.

Denzler, Brenda, *The Lure of the Edge: Scientific Passions, Religious Beliefs, and the Pursuit of UFOs*, Berkeley: University of California Press, 2001.

Dethloff, Henry C. and Ronald A. Schorn, *Voyager's Grand Tour: To the Outer Planets and Beyond*, Washington, DC: Smithsonian Books, 2003.
Deubel, Walter, 'Die Religion der Rakete,' *Deutsche Rundschau* 55 (October 1928), 63–70.
Devine, Robert A., *The Sputnik Challenge*, New York: Oxford University Press, 1993.
Diamond, Edwin, *The Rise and Fall of the Space Age*, Garden City: Doubleday, 1964.
Dick, Steven J., 'The Origins of the Extraterrestrial Life Debate and Its Relation to the Scientific Revolution,' *Journal of the History of Ideas* 41.1 (January–March 1980), 3–27.
———, *The Biological Universe: The Twentieth-Century Extraterrestrial Life Debate and the Limits of Science*, Cambridge: Cambridge University Press, 1996.
———, *Life on Other Worlds: The Twentieth-Century Extraterrestrial Life Debate*, Cambridge: Cambridge University Press, 1998.
———, ed., *Many Worlds: The New Universe, Extraterrestrial Life, and the Theological Implications*, Philadelphia: Templeton Foundation Press, 2000.
———, 'Cultural Evolution, the Postbiological Universe and SETI,' *International Journal of Astrobiology* 2.1 (January 2003), 65–74.
———, 'Anthropology and the Search for Extraterrestrial Life: An Historical View,' *Anthropology Today* 22.2 (April 2006), 3–7.
———, 'NASA and the Search for Life in the Universe,' *Endeavour* 30.2 (June 2006), 71–5.
———, ed., *Remembering the Space Age*, Washington, DC: NASA, 2008.
Dick, Steven J. and Roger D. Launius, eds, *Critical Issues in the History of Spaceflight*, Washington, DC: NASA, 2006.
———, eds, *Societal Impact of Spaceflight*, Washington, DC: NASA, 2007.
Dick, Steven J. and Mark L. Lupisella, eds, *Cosmos and Culture: Cultural Evolution in a Cosmic Context*, Washington, DC: NASA, 2010.
Dick, Steven J. and James Strick, *The Living Universe: NASA and the Development of Astrobiology*, New Brunswick: Rutgers University Press, 2004.
Dick, Steven J., Stephen J. Garber and Jane H. Odom, eds, *Research in NASA History: A Guide to the NASA History Program*, 3rd edn, Washington, DC: NASA, 2009.
Dickson, Paul, *Sputnik: The Shock of the Century*, New York: Walker, 2001.
Dolman, Everett C., Astropolitik: *Classical Geopolitics in the Space Age*, London: Frank Cass, 2002.
Dolman, Everett C. and John B. Sheldon, 'Editorial,' *Astropolitics* 1.1 (2003), 1–3.
Dornberger, Walter R., 'European Rocketry after World War I,' *Journal of the British Interplanetary Society* 13.5 (September 1954), 245–62.
Drees, Willem B., 'Theologie over buitenaardse personen: Verkenning en voorlopige afweging van standpunten,' *Tijdschrift voor Theologie* 27.3 (July–September 1987), 259–76.
Dreikausen, Margret, *Aerial Perception: The Earth as Seen from Aircraft and Spacecraft and Its Influence on Contemporary Art*, Philadelphia: Art Alliance, 1985.
Dreyfus, Bernard, 'L'Organisation européenne de recherches spatiales,' *Annuaire Européen/European Yearbook* 10.1 (1963), 151–75.
Dryden, Hugh L., 'Future Exploration and Utilization of Outer Space,' *Technology and Culture* 2.2 (Spring 1961), 112–26.
Ducrocq, Albert, *L'Humanité devant la navigation interplanétaire*, Paris: Calmann-Lévy, 1947.

———, *Victoire sur l'espace: La leçon des satellites et de la conquête lunaire*, Paris: Julliard, 1959.

———, *L'Homme dans l'espace: Les engins spatiaux de seconde génération*, Paris: Julliard, 1961.

———, *Demain l'espace*, Paris: Julliard, 1967.

Dunnett, Oliver, 'Identity and Geopolitics in Hergé's *Adventures of Tintin*,' *Social & Cultural Geography* 10.5 (August 2009), 583–98.

Dyson, Freeman J., 'The Search for Extraterrestrial Technology,' in Robert Eugene Marshak, ed., *Perspectives in Modern Physics: Essays in Honor of Hans A. Bethe*, New York: John Wiley, 1966, 641–55.

Dyson, George, *Project Orion: The Atomic Spaceship 1957–1965*, London: Allen Lane, 2002.

Edgerton, David, *England and the Aeroplane: An Essay on a Militant and Technological Nation*, Basingstoke: Macmillan, 1991.

———, *The Shock of the Old: Technology and Global History Since 1900*, London: Profile Books, 2006.

———, 'Innovation, Technology, or History: What is the Historiography of Technology About?,' *Technology and Culture* 51.3 (July 2010), 680–97.

Ehricke, Krafft A., 'The Anthropology of Astronautics,' *Astronautics* 2.4 (November 1957), 26–8, 65–8.

———, 'Astropolis: The First Space Resort,' *Playboy* 15.11 (November 1968), 96–8, 222.

———, 'Extraterrestrial Imperative,' *Bulletin of the Atomic Scientists* 27.9 (November 1971), 18–26.

———, 'Industrializing the Moon: The First Step into a New Open World,' *Fusion* 5.2 (December 1981), 21–31.

Eire, Carlos M.N., *A Very Brief History of Eternity*, Princeton: Princeton University Press, 2010.

Eisfeld, Rainer, *Mondsüchtig: Wernher von Braun und die Geburt der Raumfahrt aus dem Geist der Barbarei*, Reinbek: Rowohlt, 1996.

———, *Die Zukunft in der Tasche: Science Fiction und SF-Fandom in der Bundesrepublik*, Lüneburg: Dieter von Reeken, 2007.

Eisfeld, Rainer and Wolfgang Jeschke, *Marsfieber: Aufbruch zum Roten Planeten. Phantasie und Wirklichkeit*, Munich: Droemer, 2003.

Emme, Eugene M., ed., *The Impact of Air Power: National Security and World Politics*, Princeton: Van Nostrand, 1959.

———, ed., *Science Fiction and Space Futures: Past and Present*, San Diego: American Astronautical Society, 1982.

Erichsen, Johannes and Bernhard M. Hoppe, eds, *Peenemünde: Mythos und Geschichte der Rakete 1923–1989*, Berlin: Nicolai, 2004.

Essers, Ilse, *Max Valier: Ein Vorkämpfer der Weltraumfahrt 1895–1930*, Düsseldorf: VDI-Verlag, 1968.

———, *Hermann Ganswindt: Vorkämpfer der Raumfahrt mit seinem Weltenfahrzeug seit 1881*, Düsseldorf: VDI-Verlag, 1977.

———, *Max Valier: Ein Pionier der Raumfahrt*, Bozen: Athesia, 1980.

European Space Agency, *History Studies Reports*, 40 vols, Noordwijk: ESA, 1992–.

———, *The Impact of Space Activities upon Society*, Noordwijk: ESA, 2005.

Fallaci, Oriana, *Se il sole muore*, Milan: Rizzoli, 1965 (Eng. *If the Sun Dies*, New York: Atheneum, 1966).

Ferro, David L. and Eric G. Swedin, eds, *Science Fiction and Computing: Essays on Interlinked Domains*, Jefferson: McFarland, 2011.

Festinger, Leon, Henry W. Riecken and Stanley Schachter, *When Prophecy Fails: A Social and Psychological Study of a Modern Group that Predicted the Destruction of the World*, Minneapolis: University of Minnesota Press, 1956.

Fetscher, Justus and Robert Stockhammer, eds, *Marsmenschen: Wie die Außerirdischen gesucht und erfunden wurden*, Leipzig: Reclam, 1997.

Fewer, Greg, 'Conserving Space Heritage: The Case of Tranquility Base,' *Journal of the British Interplanetary Society* 60.1 (January 2007), 3–8.

Finney, Ben R. and Eric M. Jones, eds, *Interstellar Migration and the Human Experience*, Berkeley: University of California Press, 1985.

Finzsch, Norbert and Hermann Wellenreuther, eds, *Visions of the Future in Germany and America*, Oxford: Berg, 2001.

Fischer, William B., 'German Theories of Science Fiction: Jean Paul, Kurd Lasswitz, and After,' *Science Fiction Studies* 3.3 (November 1976), 254–65.

———, *The Empire Strikes Out: Kurd Lasswitz, Hans Dominik, and the Development of German Science Fiction*, Bowling Green: Bowling Green University Popular Press, 1984.

Fisher, Peter S., *Fantasy and Politics: Visions of the Future in the Weimar Republic*, Madison: University of Wisconsin Press, 1991.

Flammonde, Paris, *The Age of Flying Saucers: Notes on a Projected History of Unidentified Flying Objects*, New York: Hawthorn, 1971.

Flechtheim, Ossip K., *History and Futurology*, Meisenheim am Glan: Anton Hain, 1966.

Flechtner, Hans-Joachim, 'Die phantastische Literatur: Eine literarästhetische Untersuchung,' *Zeitschrift für Ästhetik und allgemeine Kunstwissenschaft* 24.1 (January 1930), 37–46.

Foerstner, Abigail, *James Van Allen: The First Eight Billion Miles*, Iowa City: University of Iowa Press, 2007.

Forschungsinstitut der Deutschen Gesellschaft für Auswärtige Politik (Bonn), Institut Français des Relations Internationales (Paris), Istituto Affari Internazionali (Rome), Nederlands Instituut voor Internationale Betrekkingen 'Clingendael' (The Hague) and Royal Institute of International Affairs (London), *Europe's Future in Space: A Joint Policy Report*, London: Routledge & Kegan Paul, 1988.

Freedman, Carl, 'Kubrick's *2001* and the Possibility of a Science-Fiction Cinema,' *Science Fiction Studies* 25.2 (July 1998), 300–18.

Freeman, Marsha, *Krafft Ehricke's Extraterrestrial Imperative*, Burlington: Apogee, 2009.

Freitas, Robert A., Jr., *Xenology: An Introduction to the Scientific Study of Extraterrestrial Life, Intelligence, and Civilization*, Sacramento: Xenology Research Institute, 1979.

Freud, Sigmund, 'Das Unbehagen in der Kultur' [1930], *Gesammelte Werke*, vol. 14, Frankfurt am Main: Fischer, 1999, 419–506.

Frewin, Anthony, ed., *Are We Alone? The Stanley Kubrick Extraterrestrial-Intelligence Interviews*, London: Elliott & Thompson, 2005.

Fries, Sylvia D., '*2001* to 1994: Political Environment and the Design of NASA's Space Station System,' *Technology and Culture* 29.3 (July 1988), 568–93.

Fritzsche, Peter, *A Nation of Fliers: German Aviation and the Popular Imagination*, Cambridge, MA: Harvard University Press, 1992.

Fritzsche, Sonja, 'A Natural and Artificial Homeland: East German Science-Fiction Film Responds to Kubrick and Tarkovsky,' *Film & History* 40.2 (Fall 2010), 80–101.

Fruth, Bryan, Alicia Germer, Keiko Kikuchi, Anamaria Mihalega, Melanie Olmstead, Hikaru Sasaki and Jack Nachbar, 'The Atomic Age: Facts and Films from 1945–1965,' *Journal of Popular Film and Television* 23.4 (Winter 1996), 154–60.

Fuller, R. Buckminster, *Operating Manual for Spaceship Earth*, Carbondale: Southern Illinois University Press, 1969.

Gaddis, John Lewis, 'The Cold War, the Long Peace, and the Future,' *Diplomatic History* 16.2 (1992), 234–46.

———, *The Cold War: A New History*, New York: Penguin, 2005.

Garber, Stephen J., ed., *Looking Backward, Looking Forward: Forty Years of U.S. Human Spaceflight Symposium*, Washington, DC: NASA, 2002.

Gardner, Martin, *Fads and Fallacies in the Name of Science*, New York: Putnam, 1952.

Gartmann, Heinz, *Träumer, Forscher, Konstrukteure: Das Abenteuer der Weltraumfahrt*, Düsseldorf: Econ, 1955.

———, *Künstliche Satelliten*, Stuttgart: Kosmos, 1958.

Gartmann, Heinz, ed., *Raumfahrtforschung*, Munich: Oldenbourg, 1952.

———, ed., *Das Bildbuch der Weltraumfahrt: Von Raketen und künftigen Dingen*, Frankfurt am Main: Umschau, 1960.

Gatland, Kenneth William, ed., *Project Satellite*, London: Allan Wingate, 1958.

Gehlhar, Fritz, 'Kometendeutung und Auseinandersetzung um das Weltbild,' *Deutsche Zeitschrift für Philosophie* 34.3 (1986), 218–27.

Geppert, Alexander C.T., 'Flights of Fancy: Outer Space and the European Imagination, 1923–1969,' in Steven J. Dick and Roger D. Launius, eds, *Societal Impact of Spaceflight*, Washington, DC: NASA, 2007, 585–99.

———, 'Space in Europe, Europe in Space: Symposium on 20th-Century Astroculture,' *NASA History News & Notes* 25.2 (May 2008), 1–12.

———, 'Space *Personae*: Cosmopolitan Networks of Peripheral Knowledge, 1927–1957,' *Journal of Modern European History* 6.2 (2008), 262–86.

———, 'Anfang – oder Ende des planetarischen Zeitalters? Der Sputnikschock als Realitätseffekt, 1945–1957,' in Igor J. Polianski and Matthias Schwartz, eds, *Die Spur des Sputnik: Kulturhistorische Expeditionen ins kosmische Zeitalter*, Frankfurt am Main: Campus, 2009, 74–94.

———, 'Wo geht's denn hier zum Weltraum? Der Historiker Alexander Geppert über Mondflüge, das Space Age und die Zukunft von gestern,' *Frankfurter Allgemeine Sonntagszeitung* (3 January 2010), 21 (interview with Harald Staun).

———, 'Storming the Heavens: Soviet Astroculture, 1957–1969,' *H-Soz-u-Kult* (17 July 2013), http://www.hsozkult.de/publicationreview/id/rezbuecher-17657.

———, ed., *Astroculture and Technoscience*, London: Routledge, 2012 (= *History and Technology* 28.3).

———, ed., *Limiting Outer Space: Astroculture After Apollo*, London: Palgrave Macmillan, 2018 (= *European Astroculture*, vol. 2).

Geppert, Alexander C.T. and Andrea B. Braidt, eds, *Orte des Okkulten*, Vienna: Turia + Kant, 2003 (= *Österreichische Zeitschrift für Geschichtswissenschaften* 14.4).

Geppert, Alexander C.T. and Till Kössler, eds, *Wunder: Poetik und Politik des Staunens im 20. Jahrhundert*, Berlin: Suhrkamp, 2011.

Geppert, Alexander C.T. and Tilmann Siebeneichner, eds, *Berliner Welträume im frühen 20. Jahrhundert*, Baden-Baden: Nomos, 2017 (= *Technikgeschichte* 84.4).
Geppert, Alexander C.T., Daniel Brandau and Tilmann Siebeneichner, eds, *Militarizing Outer Space: Astroculture, Dystopia and the Cold War*, London: Palgrave Macmillan, forthcoming (= *European Astroculture*, vol. 3).
Geppert, Alexander C.T., Uffa Jensen and Jörn Weinhold, eds, *Ortsgespräche: Raum und Kommunikation im 19. und 20. Jahrhundert*, Bielefeld: transcript, 2005.
Gerovitch, Slava, *From Newspeak to Cyberspeak: A History of Soviet Cybernetics*, Cambridge, MA: MIT Press, 2002.
Geyer, Martin H. and Johannes Paulmann, eds, *The Mechanics of Internationalism: Culture, Society, and Politics from the 1840s to the First World War*, Oxford: Oxford University Press, 2001.
Giarini, Orio, *L'Europe et l'espace*, Lausanne: Centre Recherches Européennes, 1968.
Gimbel, John, 'German Scientists, United States Denazification Policy, and the "*Paperclip* Conspiracy,"' *International History Review* 12.3 (August 1990), 441–65.
———, 'Project Paperclip: German Scientists, American Policy, and the Cold War,' *Diplomatic History* 14.3 (July 1990), 343–65.
Goldberg, Carl, 'The General's Abduction by Aliens from a UFO: Levels of Meaning of Alien Abduction Reports,' *Journal of Contemporary Psychotherapy* 30.3 (2000), 307–20.
Goldsen, Joseph M., ed., *Outer Space in World Politics*, New York: Praeger, 1963.
———, ed., *Research on Social Consequences of Space Activities*, Santa Monica: Rand Corporation, 1965.
Goldsen, Joseph M. and Leon Lipson, *Some Political Implications of the Space Age*, Santa Monica: Rand Corporation, 1958.
Golowin, Sergius, *Götter der Atom-Zeit: Moderne Sagenbildung um Raumschiffe und Sternenmenschen*, Bern: Francke, 1967.
Gooden, Brett, *Spaceport Australia*, Kenthurst: Kangaroo, 1990.
Goodman, Allen E., 'Diplomacy and the Search for Extraterrestrial Intelligence (SETI),' *Acta Astronautica* 21.2 (February 1990), 137–41.
Goodwin, Harold L., *Space: Frontier Unlimited*, Princeton: Van Nostrand, 1962.
Gordon, Joan, 'Ad Astra Per Aspera,' *Science Fiction Studies* 32.3 (November 2005), 495–502.
Gorman, Alice, 'The Cultural Landscape of Interplanetary Space,' *Journal of Social Archaeology* 5.1 (February 2005), 85–107.
Graf, Otto, ed., *Die Epoche des überfließenden Sehvermögens: Der Mensch im Weltraum*, Vienna: Österreichischer Bundesverlag, 1970 (exhibition catalogue, Museum des 20. Jahrhunderts, Vienna, 25 August–31 October 1970).
Grant, Barry K., '*Invaders from Mars* and the Science Fiction Film in the Age of Reagan,' *CineAction!* 8 (1987), 77–83.
Grazia, Alfred de, ed., *The Velikovsky Affair*, London: Sidgwick & Jackson, 1966.
Greco, Pietro, *L'astro narrante: La Luna nella scienza e nella letteratura italiana*, Milan: Springer Italia, 2009.
Green, Roger Lancelyn, *Into Other Worlds: Space-Flight in Fiction, from Lucian to Lewis*, London: Abelard-Schuman, 1958.
Gregory, Derek, *Geographical Imaginations*, Oxford: Blackwell, 1994.
Grünschloß, Andreas, *Wenn die Götter landen... Religiöse Dimensionen des UFO-Glaubens*, Berlin: Evangelische Zentralstelle für Weltanschauungsfragen, 2000.

Guthke, Karl S., 'The Idea of Extraterrestrial Intelligence,' *Harvard Library Bulletin* 33.2 (Spring 1985), 196–210.

———, *Der Mythos der Neuzeit: Das Thema der Mehrheit der Welten in der Literatur- und Geistesgeschichte von der kopernikanischen Wende bis zur Science Fiction*, Bern: Francke, 1983; (Eng. *The Last Frontier: Imagining Other Worlds from the Copernican Revolution to Modern Science Fiction*, Ithaca: Cornell University Press, 1990).

———, 'Nightmare and Utopia: Extraterrestrial Worlds from Galileo to Goethe,' *Early Science and Medicine* 8.3 (August 2003), 173–95.

———, 'Kolonialphantasien in der populären Naturwissenschaft der frühen Neuzeit,' *Early Science and Medicine* 9.1 (February 2004), 20–36.

Haber, Heinz, *Man in Space*, London: Sidgwick & Jackson, 1953.

———, 'Space Satellites: Tools of Earth Research,' *National Geographic* 109.4 (April 1956), 486–509.

———, *Brüder im All: Die Möglichkeit des Lebens auf fremden Welten*, Frankfurt am Main: Büchergilde Gutenberg, 1971.

Hacker, Barton C., 'The Idea of Rendezvous: From Space Station to Orbital Operations in Space-Travel Thought, 1895–1951,' *Technology and Culture* 15.3 (July 1974), 373–88.

———, 'The Gemini Paraglider: A Failure of Scheduled Innovation, 1961–1964,' *Social Studies of Science* 22.2 (May 1992), 387–406.

Hackett, Herbert, 'The Flying Saucer: A Manufactured Concept,' *Sociology and Social Research* 32.5 (May–June 1948), 869–73.

Hagner, Michael and Erich Hörl, eds, *Die Transformation des Humanen: Beiträge zur Kulturgeschichte der Kybernetik*, Frankfurt am Main: Suhrkamp, 2008.

Hallet, Marc, *Le Cas Adamski*, Paris: L'Œil du Sphinx, 2010.

Hanrahan, James Stephen and David Bushnell, *Space Biology: The Human Factors in Space Flight*, New York: Basic Books, 1960.

Hansen, James R., 'Aviation History in the Wider View,' *Technology and Culture* 30.3 (July 1989), 643–56.

Hantke, Steffen, '*Raumpatrouille*: The Cold War, the "Citizen in Uniform," and West German Television,' *Science Fiction Studies* 31.1 (March 2004), 63–80.

Harder, James A., J. Allen Hynek, Philip J. Klass, Frank B. Salisbury, R. Leo Sprinkle, Ernest H. Taves and Jacques Vallée, 'Playboy Panel: UFOs,' *Playboy* 25.1 (January 1978), 67–98, 128.

Harper, Harry, *Dawn of the Space Age*, London: Sampson Low, 1946.

Harries, Karsten, *Infinity and Perspective*, Cambridge, MA: MIT Press, 2001.

Harrison, Albert A., *After Contact: The Human Response to Extraterrestrial Life*, New York: Plenum, 1997.

———, *Spacefaring: The Human Dimension*, Berkeley: University of California Press, 2001.

———, *Starstruck: Cosmic Visions in Science, Religion, and Folklore*, New York: Berghahn, 2007.

———, 'The Search for Extraterrestrial Intelligence: Astrosociology and Cultural Aspects,' *Astropolitics* 9.1 (April 2011), 63–83.

Harrison, Albert A., ed., *From Antarctica to Outer Space: Life in Isolation and Confinement*, New York: Springer, 1991.

Hart, Bushnell Albert, 'Imagination in History,' *American Historical Review* 15.2 (January 1910), 227–51.

Hart, Michael H., 'An Explanation for the Absence of Extraterrestrials on Earth,' *Quarterly Journal of the Royal Astronomical Society* 16.2 (June 1975), 128–35.

Hartlaub, G.F., *Bewußtsein auf anderen Sternen? Ein kleiner Leitfaden durch die Menschheitsträume von den Planetenbewohnern*, Munich: Ernst Reinhardt, 1951.

Harvey, Brian, *Europe's Space Programme: To* Ariane *and Beyond*, London: Springer Praxis, 2003.

Hauser, Linus, 'Außerirdisches Leben: Herausforderung für die Theologie?,' in Heinz-Hermann Peitz, ed., *Der vervielfachte Christus: Außerirdisches Leben und christliche Heilsgeschichte*, Stuttgart: Akademie der Diözese Rottenburg-Stuttgart, 2004, 73–103.

Heacock, Raymond L., 'The Voyager Spacecraft,' *Proceedings of the Institution of Mechanical Engineers* 194 (September 1980), 211–24.

Head, Tom, ed., *Conversations with Carl Sagan*, Jackson: University Press of Mississippi, 2006.

Hecht, Gabriele, *The Radiance of France: Nuclear Power and National Identity after World War II*, Cambridge, MA: MIT Press, 1998.

Heffernan, William C., 'The Singularity of Our Inhabited World: William Whewell and A.R. Wallace in Dissent,' *Journal of the History of Ideas* 39.1 (January–March 1978), 81–100.

———, 'Percival Lowell and the Debate Over Extraterrestrial Life,' *Journal of the History of Ideas* 42.3 (July–September 1981), 527–30.

Heimann, Thomas and Burghard Ciesla, '*Die gefrorenen Blitze*: Wahrheit und Dichtung. FilmGeschichte einer "Wunderwaffe,"' *Apropos: Film. Das Jahrbuch der DEFA-Stiftung* (2002), 158–80.

Heimpel, Hermann, 'Der Versuch mit der Vergangenheit zu leben: Über Geschichte und Geschichtswissenschaft,' *Frankfurter Allgemeine Zeitung* (25 March 1959), 11.

Hein-Weingarten, Katharina, *Das Institut für Kosmosforschung der Akademie der Wissenschaften der DDR*, Berlin: Duncker & Humblot, 2001.

Heinrich, Christoph and Markus Heinzelmann, eds, *Rückkehr ins All*, Ostfildern-Ruit: Hatje Cantz, 2005 (exhibition catalogue, Kunsthalle, Hamburg, 23 September 2005–12 December 2006).

Held, George, 'Men on the Moon: American Novelists Explore Lunar Space,' *Michigan Quarterly Review* 18.2 (Spring 1979), 318–42.

Helmreich, Stefan, 'The Signature of Life: Designing the Astrobiological Imagination,' *Grey Room* 23 (Spring 2006), 66–95.

Hendershot, Cyndy, *Paranoia, the Bomb, and 1950s Science Fiction Films*, Bowling Green: Bowling Green State University Popular Press, 1999.

Hennessey, Roger, *Worlds Without End: The Historic Search for Extraterrestrial Life*, Stroud: Tempus, 1999.

Henry, Richard C., 'UFOs and NASA,' *Journal of Scientific Exploration* 2.2 (1988), 93–142.

Herrmann, Joachim, *Das falsche Weltbild: Astronomie und Aberglaube*, Munich: Deutscher Taschenbuch Verlag, 1973.

Hersch, Matthew H., 'Apollo's Stepchildren: New Works on the American Lunar Program,' *Technology and Culture* 49.2 (April 2008), 449–55.

———, 'Return of the Lost Spaceman: America's Astronauts in Popular Culture, 1959–2006,' *Journal of Popular Culture* 44.1 (February 2011), 73–92.

Hetherington, Norriss, 'The Extraterrestrial Life Debate: A Productive Perspective,' *Historical Studies in the Physical and Biological Sciences* 20.1 (1989), 179–82.

Hilger, Josef, *Raumpatrouille: Die phantastischen Abenteuer des Raumschiffes Orion*, 2nd edn, Berlin: Schwarzkopf & Schwarzkopf, 2005.

Hill, C.N., *A Vertical Empire: The History of the UK Rocket and Space Programme, 1950–1971*, London: Imperial College Press, 2001.

Hillegas, Mark R., 'Martians and Mythmakers, 1877–1938,' in Ray B. Browne, Larry N. Landrum and William K. Bottorf, eds, *Challenges in American Culture*, Bowling Green: Bowling Green University Popular Press, 1970, 150–77.

———, 'Victorian "Extraterrestrials,"' in Jerome H. Buckley, ed., *The Worlds of Victorian Fiction*, Cambridge, MA: Harvard University Press, 1975, 391–414.

Hoch, David G., 'Mythic Patterns in *2001: A Space Odyssey*,' *Journal of Popular Culture* 4.4 (Spring 1970–71), 961–5.

Hodgens, Richard, 'A Brief, Tragical History of the Science Fiction Film,' *Film Quarterly* 13.2 (Winter 1959), 30–9.

Hoffmann, Horst, *Die Deutschen im Weltraum: Zur Geschichte der Kosmosforschung in der DDR*, Berlin: Edition Ost, 1998.

Hohmann, Walter, *Die Erreichbarkeit der Himmelskörper: Untersuchungen über das Raumfahrtproblem*, Munich: Oldenbourg, 1925.

Hölscher, Lucian, 'Utopie,' in Otto Brunner, Werner Conze and Reinhart Koselleck, eds, *Geschichtliche Grundbegriffe: Historisches Lexikon der politisch-sozialen Sprache in Deutschland*, vol. 6, Stuttgart: Klett-Cotta, 1990, 733–88.

———, *Die Entdeckung der Zukunft*, Frankfurt am Main: Fischer, 1999.

———, ed., *Das Jenseits: Facetten eines religiösen Begriffs in der Neuzeit*, Göttingen: Wallstein, 2007.

Holton, Gerald, *The Scientific Imagination: Case Studies*, Cambridge: Cambridge University Press, 1978.

———, 'Imagination in Science,' in idem, *Einstein, History and Other Passions: The Rebellion Against Science at the End of the Twentieth Century*, New York: Addison-Wesley, 1996, 78–102.

Horrigan, Brian, 'Popular Culture and Visions of the Future in Space, 1901–2001,' in Bruce Sinclair, ed., *New Perspectives on Technology and American Culture*, Philadelphia: American Philosophical Society, 1986, 49–67.

Huebner, Andrew J., 'Lost in Space: Technology and Turbulence in Futuristic Cinema of the 1950s,' *Film & History* 40.2 (Fall 2010), 6–26.

Hunley, J.D., 'The Legacies of Robert H. Goddard and Hermann J. Oberth,' *Journal of the British Interplanetary Society* 49.1 (January 1996), 43–8.

Hynek, J. Allen, 'The UFO Gap,' *Playboy* 14.12 (December 1967), 144–6, 267–71.

———, *The UFO Experience: A Scientific Inquiry*, Chicago: Regnery, 1972.

Hynek, J. Allen and Jacques Vallée, *The Edge of Reality: A Progress Report on Unidentified Flying Objects*, Chicago: Regnery, 1975.

Innerhofer, Roland, *Deutschsprachige Science Fiction 1870–1914: Rekonstruktion und Analyse der Anfänge einer Gattung*, Vienna: Böhlau, 1996.

Jacobs, David Michael, *The UFO Controversy in America*, Bloomington: Indiana University Press, 1975.

James, Edward, '*Per ardua ad astra:* Authorial Choice and the Narrative of Interstellar Travel,' in Jaś Elsner and Joan-Pau Rubiés, eds, *Voyages and Visions: Towards a Cultural History of Travel*, London: Reaktion, 1999, 252–71.

Jameson, Fredric, *Archaeologies of the Future: The Desire Called Utopia and Other Science Fictions*, London: Verso, 2005.

Jaumotte, A.L., 'A Decade of Consolidation, 1971–1981,' *Acta Astronautica* 32.7–8 (July–August 1994), 509–13.

Jessup, Philip C. and Howard J. Taubenfeld, 'The United Nations Ad Hoc Committee on the Peaceful Uses of Outer Space,' *American Journal of International Law* 53.4 (October 1959), 877–81.

Jones, Harold Spencer, *Life on Other Worlds*, London: English Universities Press, 1940.

Jones, Robert A., 'They Came in Peace for all Mankind: Popular Culture as a Reflection of Public Attitudes to Space,' *Space Policy* 20.1 (February 2004), 45–8.

Josephson, Paul R., 'Rockets, Reactors, and Soviet Culture,' in Loren R. Graham, ed., *Science and the Soviet Social Order*, Cambridge, MA: Harvard University Press, 1990, 168–91.

Jung, Carl Gustav, 'Ein moderner Mythus: Von Dingen, die am Himmel gesehen werden' [1958], in *Zivilisation im Übergang*, Düsseldorf: Walter, 1995, 337–474 (Eng. *Flying Saucers: A Modern Myth of Things Seen in the Skies*, New York: Harcourt, Brace, 1959).

———, 'A Visionary Rumour,' *Journal of Analytical Psychology* 4.1 (January 1959), 5–19.

Jungk, Robert, *Die Zukunft hat schon begonnen: Amerikas Allmacht und Ohnmacht*, Stuttgart: Scherz & Goverts, 1952.

Junker, Christof, *Das Weltraumbild in der deutschen Lyrik von Opitz bis Klopstock*, Berlin: Matthiesen, 1932.

Kaiser, Karl and Stephan Freiherr von Welck, eds, *Weltraum und internationale Politik*, Munich: Oldenbourg, 1987.

Kant, Immanuel, *Allgemeine Naturgeschichte und Theorie des Himmels oder Versuch von der Verfassung und dem mechanischen Ursprunge des ganzen Weltgebäudes nach Newtonischen Grundsätzen abgehandelt* [1755], 4th edn, Frankfurt am Main: Harri Deutsch, 2009.

Kardashev, Nikolai S., 'Transmission of Information by Extraterrestrial Civilizations,' *Soviet Astronomy* 8.2 (September–October 1964), 217–21.

Kauffman, James L., *Selling Outer Space: Kennedy, the Media, and Funding for Project Apollo, 1961–1963*, Tuscaloosa: University of Alabama Press, 1994.

Kautzleben, Heinz, 'Das Koordinierungskomitee Interkosmos (1967–1990): Beteiligung der DDR an der Erforschung und Nutzung des Weltraumes für friedliche Zwecke,' *Sitzungsberichte der Leibniz-Sozietät der Wissenschaften zu Berlin* 96 (2008), 149–64.

Kay, W.D., 'Democracy and Super Technologies: The Politics of the Space Shuttle and Space Station *Freedom*,' *Science, Technology, & Human Values* 19.2 (Spring 1994), 131–51.

———, 'NASA and Space History,' *Technology and Culture* 40.1 (January 1999), 120–7.

Keel, John A., 'The Flying Saucer Subculture,' *Journal of Popular Culture* 8.4 (Spring 1975), 871–96.

Kern, Stephen, *The Culture of Time and Space, 1880–1918*, Cambridge, MA: Harvard University Press, 1983.

Keyhoe, Donald E., 'The Flying Saucers Are Real,' *True* (January 1950), 11–13, 83–7.

———, *The Flying Saucers Are Real*, New York: Fawcett, 1950.
———, *Flying Saucers from Outer Space*, New York: Henry Holt, 1953.
Khuon, Ernst von, ed., *Waren die Götter Astronauten? Wissenschaftler diskutieren die Thesen Erich von Dänikens*, Munich: Knaur, 1972.
Kilgore, De Witt Douglas, 'Engineers' Dreams: Wernher von Braun, Willy Ley, and Astrofuturism in the 1950s,' *Canadian Review of American Studies* 27.2 (March 1997), 103–31.
———, *Astrofuturism: Science, Race, and Visions of Utopia in Space*, Philadelphia: University of Pennsylvania Press, 2003.
Kirby, David A., 'Science Consultants, Fictional Films, and Scientific Practice,' *Social Studies of Science* 33.2 (April 2003), 231–68.
———, 'The Future is Now: Diegetic Prototypes and the Role of Popular Films in Generating Real-World Technological Development,' *Social Studies of Science* 40.1 (February 2010), 41–70.
———, *Lab Coats in Hollywood: Science, Scientists, and Cinema*, Cambridge, MA: MIT Press, 2011.
Kirk, Andrew G., *Counterculture Green: The Whole Earth Catalog and American Environmentalism*, Lawrence: University Press of Kansas, 2007.
Knobloch, Eberhard, 'Vielheit der Welten – extraterrestrische Existenz,' in Wilhelm Voßkamp, ed., *Ideale Akademie: Vergangene Zukunft oder konkrete Utopie?*, Berlin: Akademie, 2002, 165–86.
Koelle, H.H., 'Astronautical Activities in Germany,' *Journal of the British Interplanetary Society* 14.3 (May–June 1955), 121–31.
Kolker, Robert Phillip, ed., *Stanley Kubrick's* 2001: A Space Odyssey. *New Essays*, Oxford: Oxford University Press, 2006.
Koselleck, Reinhart, *Vergangene Zukunft: Zur Semantik geschichtlicher Zeiten*, Frankfurt am Main: Suhrkamp, 1979 (Eng. *Futures Past: On the Semantics of Historical Time*, Cambridge, MA: MIT Press, 1985).
Koyré, Alexandre, *From the Closed World to the Infinite Universe*, Baltimore: Johns Hopkins University Press, 1957.
Kraemer, Robert S., *Beyond the Moon: A Golden Age of Planetary Exploration, 1971–1978*, Washington, DC: Smithsonian Institution Press, 2000.
Kraemer, Sylvia K., 'Opinion Polls and the US Civil Space Program,' *Journal of the British Interplanetary Society* 46.11 (November 1993), 444–6.
Krah, Hans, 'Atomforschung und atomare Bedrohung: Literarische und (populär)wissenschaftliche Vermittlung eines elementaren Themas 1946–1959,' *Kodikas/Code* 24.1–2 (January–June 2001), 83–114.
———, '"Der Weg zu den Planetenräumen": Die Vorstellung der Raumfahrt in Theorie und Literatur der Frühen Moderne,' in Christine Maillard and Michael Titzmann, eds, *Literatur und Wissen(schaften) 1890–1935*, Stuttgart: J.B. Metzler, 2002, 111–64.
Krämer, Peter, '"Dear Mr. Kubrick": Audience Responses to *2001: A Space Odyssey* in the Late 1960s,' *Participations* 6.2 (November 2009), 240–59.
———, *2001: A Space Odyssey*, Basingstoke: Palgrave Macmillan, 2010.
Kretzmann, Edwin M.J., 'German Technological Utopias of the Pre-War Period,' *Annals of Science* 3.4 (October 1938), 417–30.
Krige, John, ed., *Choosing Big Technologies*, Amsterdam: Harwood, 1992 (= *History and Technology* 9.1–4).

———, *American Hegemony and the Postwar Reconstruction of Science in Europe*, Cambridge, MA: MIT Press, 2006.

Krige, John and Dominique Pestre, eds, *Science in the Twentieth Century*, Amsterdam: Harwood, 1997.

Krige, John and Lorenza Sebesta, 'US-European Co-Operation in Space in the Decade after Sputnik,' in Giuliana Gemelli, ed., *Big Culture: Intellectual Cooperation in Large-Scale Cultural and Technical Systems*, Bologna: Clueb, 1994, 263–85.

Krige, John, Arturo Russo and Lorenza Sebesta, *A History of the European Space Agency*, 2 vols, Noordwijk: ESA, 2000.

Kristof, Ladis K.D., 'Teilhard de Chardin and the Communist Quest for a Space Age World View,' *Russian Review* 28.3 (July 1969), 277–88.

Kropf, Carl R., 'Douglas Adams's *Hitchhiker* Novels as Mock Science Fiction,' *Science Fiction Studies* 15.1 (March 1988), 61–70.

Krugman, Herbert E., 'Public Attitudes Toward the Apollo Space Program, 1965–1975,' *Journal of Communication* 27.4 (Fall 1977), 87–93.

Kuberski, Philip, 'Kubrick's *Odyssey*: Myth, Technology, Gnosis,' *Arizona Quarterly* 64.3 (Fall 2008), 51–73.

Kubrick, Stanley, 'Interview: A Candid Conversation with the Pioneering Creator of "2001: A Space Odyssey," "Dr. Strangelove" and "Lolita,"' *Playboy* 15.9 (September 1968), 85–96, 180–95.

Kuhn, Annette, ed., *Alien Zone: Cultural Theory and Contemporary Science Fiction Cinema*, London: Verso, 1990.

———, ed., *Alien Zone II: The Spaces of Science Fiction*, London: Verso, 1999.

Kuhn, Thomas S., *The Copernican Revolution: Planetary Astronomy in the Development of Western Thought*, Cambridge, MA: Harvard University Press, 1957.

Kühn, Rudolf, Arnold W. Frutkin, Jean Coulomb and Max Mayer, 'Herausforderung "Weltraum" – Europas Antwort,' *Dokumente: Zeitschrift für übernationale Zusammenarbeit* 20.3 (1964), 201–22.

Laak, Dirk van, *Weiße Elefanten: Anspruch und Scheitern technischer Großprojekte im 20. Jahrhundert*, Stuttgart: Deutsche Verlags-Anstalt, 1999.

Lagrange, Pierre, *La Rumeur de Roswell*, Paris: La Découverte, 1996.

———, *La Guerre des mondes a-t-elle eu lieu?*, Paris: Laffont, 2005.

Lagrange, Pierre and Hélène Huguet, *Sur Mars: Le Guide du touriste spatial*, Paris: EDP Sciences, 2003.

Lamb, David, *The Search for Extraterrestrial Intelligence: A Philosophical Inquiry*, London: Routledge, 2001.

Lambright, W. Henry, 'The Political Construction of Space Satellite Technology,' *Science, Technology, & Human Values* 19.1 (January 1994), 47–69.

Lane, K. Maria D., *Geographies of Mars: Seeing and Knowing the Red Planet*, Chicago: University of Chicago Press, 2011.

Lasby, Clarence G., *Project Paperclip: German Scientists and the Cold War*, New York: Atheneum, 1971.

Latour, Bruno, *Science in Action: How to Follow Scientists and Engineers Through Society*, Cambridge, MA: Harvard University Press, 1987.

Latour, Bruno and Peter Weibel, eds, *Making Things Public: Atmospheres of Democracy*, Cambridge, MA: MIT Press, 2005 (exhibition catalogue, ZKM Zentrum für Kunst und Medientechnologie, Karlsruhe, 20 March–3 October 2005).

Launius, Roger D., 'NASA History and the Challenge of Keeping the Contemporary Past,' *The Public Historian* 21.3 (Summer 1999), 63–81.

———, 'The Historical Dimension of Space Exploration: Reflections and Possibilities,' *Space Policy* 16.1 (February 2000), 23–38.
———, 'Perfect Worlds, Perfect Societies: The Persistent Goal of Utopia in Human Spaceflight,' *Journal of the British Interplanetary Society* 56.5 (September–October 2003), 338–49.
———, 'Public Opinion Polls and Perceptions of US Human Spaceflight,' *Space Policy* 19.3 (2003), 163–75.
———, *Space Stations: Base Camps to the Stars*, Washington, DC: Smithsonian Books, 2003.
———, 'Perceptions of Apollo: Myth, Nostalgia, Memory or all of the Above?,' *Space Policy* 21 (May 2005), 129–39.
———, 'Interpreting the Moon Landings: Project Apollo and the Historians,' *History and Technology* 22.3 (September 2006), 225–55.
———, 'Underlying Assumptions of Human Spaceflight in the United States,' *Acta Astronautica* 62 (March–April 2008), 341–56.
———, 'United States Space Cooperation and Competition: Historical Reflections,' *Astropolitics* 7.2 (2009), 89–100.
Launius, Roger D. and Howard E. McCurdy, *Imagining Space: Achievements, Predictions, Possibilities, 1950–2050*, San Francisco: Chronicle, 2001.
———, *Robots in Space: Technology, Evolution, and Interplanetary Travel*, Baltimore: Johns Hopkins University Press, 2008.
Launius, Roger D., John M. Logsdon and Robert W. Smith, eds, *Reconsidering Sputnik: Forty Years Since the Soviet Satellite*, Amsterdam: Harwood, 2000.
Lazier, Benjamin, 'Earthrise; or, The Globalization of the World Picture,' *American Historical Review* 116.3 (June 2011), 602–30.
Lear, John, 'Dr. Mead and the Red Moons,' *New Scientist* 2.52 (14 November 1957), 20.
Lefebvre, Henri, *La Production de l'espace*, Paris: Anthropos, 1974 (Eng. *The Production of Space*, Oxford: Blackwell, 1991).
Leffler, Melvyn P., 'The Cold War: What Do "We Know Now"?,' *American Historical Review* 104.2 (April 1999), 501–24.
Lenoir, Timothy, 'Practice, Reason, Context,' *Science in Context* 2.1 (March 1988), 3–22.
———, ed., *Instituting Science: The Cultural Production of Scientific Disciplines*, Stanford: Stanford University Press, 1997.
———, *Inscribing Science: Scientific Texts and the Materiality of Communication*, Stanford: Stanford University Press, 1998.
Lepselter, Susan, 'From the Earth Native's Point of View: The Earth, the Extraterrestrial, and the Natural Ground of Home,' *Public Culture* 9.2 (Winter 1997), 197–208.
Leslie, Desmond and George Adamski, *Flying Saucers Have Landed*, London: Werner Laurie, 1953.
Leslie, Desmond and Patrick Moore, *How Britain Won the Space Race*, London: Mitchell Beazley, 1972.
Levy, Lillian, ed., *Space: Its Impact on Man and Society*, New York: Norton, 1965.
Lewis, James R., ed., *The Gods Have Landed: New Religions from Other Worlds*, Albany: State University of New York Press, 1995.
Ley, Willy, *Rockets, Missiles, and Space Travel* [1944], 3rd edn, New York: Viking, 1951.

———, 'Die Geschichte des Raumfahrtgedankens,' in Karl Schütte and Hans K. Kaiser, eds, *Handbuch der Astronautik*, Constance: Athenaion, 1958, 1–28.
———, *Space Stations*, Poughkeepsie: Guild Press, 1958.
———, *Engineers' Dreams*, New York: Viking, 1959.
———, ed., *Die Möglichkeit der Weltraumfahrt: Allgemeinverständliche Beiträge zum Raumschiffahrtsproblem*, Leipzig: Hachmeister & Thal, 1928.
Ley, Willy and Wernher von Braun, *The Exploration of Mars*, New York: Viking, 1956.
Liessmann, Konrad Paul, *Zukunft kommt! Über säkularisierte Heilserwartungen und ihre Enttäuschung*, Vienna: Styria, 2007.
Limerick, Patricia, 'Imagined Frontiers: Westward Expansion and the Future of the Space Program,' in Radford Byerly Jr., ed., *Space Policy Alternatives*, Boulder: Westview Press, 1992, 249–61.
Linke, Felix, *Die Verwandtschaft der Welten und die Bewohnbarkeit der Himmelskörper*, Leipzig: Quelle & Meyer, 1925.
———, *Das Raketen-Weltraumschiff: Wanderung zum Monde und zu anderen Planeten*, Hamburg: Auer, 1928.
Locke, Richard Adams, *The Moon Hoax: Or, a Discovery that the Moon has a Vast Population of Human Beings* [1859], Boston: Gregg Press, 1975.
Logsdon, John M., *The Decision to Go to the Moon: Project Apollo and the National Interest*, Chicago: University of Chicago Press, 1970.
———, *John F. Kennedy and the Race to the Moon*, Basingstoke: Palgrave Macmillan, 2011.
Logsdon, John M. and Alain Dupas, 'Was the Race to the Moon Real?,' *Scientific American* 270 (June 1994), 36–43.
Lord, M.G., *Astroturf: The Private Life of Rocket Science*, New York: Walker, 2005.
Lovell, Bernard, *The Story of Jodrell Bank*, Oxford: Oxford University Press, 1968.
Lucanio, Patrick, *Them or Us! Archetypal Interpretations of Fifties Alien Invasion Films*, Bloomington: Indiana University Press, 1987.
Lucanio, Patrick and Gary Coville, *Smokin' Rockets: The Romance of Technology in American Film, Radio, and Television, 1945–1962*, Jefferson: McFarland, 2002.
Luckhurst, Roger, 'The Science-Fictionalization of Trauma: Remarks on Narratives of Alien Abduction,' *Science Fiction Studies* 25.1 (March 1998), 29–52.
———, 'Bruno Latour's Scientifiction: Networks, Assemblages, and Tangled Objects,' *Science Fiction Studies* 33.1 (March 2006), 4–17.
Ludwiger, Illobrand von, *Der Stand der UFO-Forschung*, Frankfurt am Main: Zweitausendeins, 1992.
Luhmann, Niklas, 'The Future Cannot Begin: Temporal Structures in Modern Society,' *Social Research* 43.1 (Spring 1976), 130–52.
Lüst, Reimar and Paul Nolte, *Der Wissenschaftsmacher: Reimar Lüst im Gespräch mit Paul Nolte*, Munich: C.H. Beck, 2008.
Macauley, William R., *Picturing Knowledge: NASA's Pioneer Plaque, Voyager Record and the History of Interstellar Communication, 1957–1977*, PhD thesis, University of Manchester, 2010.
MacDonald, Fraser, 'Anti-*Astropolitik*: Outer Space and the Orbit of Geography,' *Progress in Human Geography* 31.5 (2007), 592–615.
———, 'Space and the Atom: On the Popular Geopolitics of Cold War Rocketry,' *Geopolitics* 13.4 (Winter 2008), 611–34.
Mack, John E., *Abduction: Human Encounters with Aliens*, New York: Charles Scribner's Sons, 1994.

Mack, Pamela E., 'Space History,' *Technology and Culture* 30.3 (July 1989), 657–65.

MacLeish, Archibald, 'A Reflection: Riders on Earth Together, Brothers in Eternal Cold,' *New York Times* (25 December 1968), 1.

Madders, Kevin, *A New Force at a New Frontier: Europe's Development in the Space Field in the Light of Its Main Actors, Policies, Law and Activities from Its Beginnings up to the Present*, Cambridge: Cambridge University Press, 1997.

Magin, Ulrich, *Von Ufos entführt: Unheimliche Begegnungen der vierten Art*, Munich: C.H. Beck, 1991.

Maier, Charles S., 'Consigning the Twentieth Century to History: Alternative Narratives for the Modern Era,' *American Historical Review* 105.3 (June 2000), 807–31.

Mailer, Norman, *Of a Fire on the Moon*, Boston: Little, Brown, 1970.

Mandel, Siegfried, 'The Occult: The Great Saucer Hunt,' *Saturday Review* 38 (6 August 1955), 28–9.

Markley, Robert, *Dying Planet: Mars in Science and the Imagination*, Durham: Duke University Press, 2005.

Markowitz, William, 'The Physics and Metaphysics of Unidentified Flying Objects,' *Science* 157 (15 September 1967), 1274–80.

Marsiske, Hans-Arthur, *Heimat Weltall: Wohin soll die Raumfahrt führen?*, Frankfurt am Main: Suhrkamp, 2005.

Maruyama, Magoroh and Arthur M. Harkins, eds, *Cultures Beyond the Earth: The Role of Anthropology in Outer Space*, New York: Vintage Books, 1975.

Massey, Harrie and Malcolm O. Robins, *History of British Space Science*, Cambridge: Cambridge University Press, 1986.

Matt, Gerald A. and Cathérine Hug, eds, *Weltraum: Die Kunst und ein Traum/Space: About a Dream*, Nürnberg: Verlag für moderne Kunst, 2011 (exhibition catalogue, Kunsthalle, Vienna, 1 April–1 August 2011).

Maurer, Eva, Julia Richers, Monica Rüthers and Carmen Scheide, eds, *Soviet Space Culture: Cosmic Enthusiasm in Socialist Societies*, Basingstoke: Palgrave Macmillan, 2011.

Mazlish, Bruce, ed., *The Railroad and the Space Program: An Exploration in Historical Analogy*, Cambridge, MA: MIT Press, 1965.

———, 'Comparing Global History to World History,' *Journal of Interdisciplinary History* 28.3 (Winter 1998), 385–95.

———, *The Idea of Humanity in a Global Era*, Basingstoke: Palgrave Macmillan, 2009.

McAleer, Neil, *Odyssey: The Authorised Biography of Arthur C. Clarke*, London: Gollancz, 1992.

McCarthy, Patrick A., 'The Genesis of "Star Maker,"' *Science Fiction Studies* 31.1 (March 2004), 25–42.

McCray, W. Patrick, 'Amateur Scientists, the International Geophysical Year, and the Ambitions of Fred Whipple,' *Isis* 97.4 (December 2006), 634–58.

———, *Keep Watching the Skies! The Story of Operation Moonwatch and the Dawn of the Space Age*, Princeton: Princeton University Press, 2008.

McCurdy, Howard E., *Space and the American Imagination*, Washington, DC: Smithsonian Institution Press, 1997.

McDannell, Colleen and Bernhard Lang, *Heaven: A History*, New Haven: Yale University Press, 1988.

McDermott, Mary and W.R. Robinson, '*2001* and the Literary Sensibility,' *Georgia Review* 26 (1972), 21–37.
McDougall, Walter A., 'Technocracy and Statecraft in the Space Age: Toward the History of a Saltation,' *American Historical Review* 87.4 (October 1982), 1010–40.
———, 'Space-Age Europe: Gaullism, Euro-Gaullism, and the American Dilemma,' *Technology and Culture* 26.1 (January 1985), 179–203.
———, *...The Heavens and the Earth: A Political History of the Space Age*, New York: Basic Books, 1985.
McIver, Shirley, *The UFO Movement: A Sociological Study of UFO Groups*, PhD thesis, University of York, 1983.
McLaughlin, William I., 'The Potential of Space Exploration for the Fine Arts,' *Journal of the British Interplanetary Society* 46.11 (November 1993), 421–30.
McLeod, Ken, 'Space Oddities: Aliens, Futurism and Meaning in Popular Music,' *Popular Music* 22.3 (October 2003), 337–55.
McQuaid, Kim, 'Selling the Space Age: NASA and Earth's Environment, 1958–1990,' *Environment and History* 12 (May 2006), 127–63.
———, 'Sputnik Reconsidered: Image and Reality in the Early Space Age,' *Canadian Review of American Studies* 37.3 (2007), 371–401.
———, 'Earthly Environmentalism and the Space Exploration Movement, 1960–1990: A Study in Irresolution,' *Space Policy* 26.3 (August 2010), 163–73.
Meadows, Donella H., Dennis L. Meadows, Jørgen Randers, William W. Behrens and the Club of Rome, *The Limits to Growth: A Report for the Club of Rome's Project on the Predicament of Mankind*, New York: Universe Books, 1972.
Meehan, Paul, *Saucer Movies: A UFOlogical History of the Cinema*, Lanham: Scarecrow, 1998.
Meerloo, Joost A.M., 'The Flying Saucer Syndrome and the Need for Miracles,' *Journal of the American Medical Association* 203.12 (18 March 1968), 170.
Méheust, Bertrand, *Science-fiction et soucoupes volantes*, Paris: Mercure de France, 1978.
Mellor, Felicity, 'Colliding Worlds: Asteroid Research and the Legitimization of War in Space,' *Social Studies of Science* 37.4 (August 2007), 499–531.
Mesarovic, Mihajlo D. and Eduard Pestel, *Mankind at the Turning Point: The Second Report to the Club of Rome*, New York: Dutton, 1974.
Meyer, Werner, ed., *Klaus Bürgle: Zurück in die Zukunft. Technische Fantasien und Visionen*, Göppingen: Kunsthalle Göppingen, 2010 (exhibition catalogue, Kunsthalle Göppingen, 19 February–25 April 2010).
Michael, Donald N., 'Man-Into-Space: A Tool and Program for Research in the Social Sciences,' *American Psychologist* 12.6 (June 1957), 324–8.
———, 'The Beginning of the Space Age and American Public Opinion,' *Public Opinion Quarterly* 24.4 (Winter 1960), 573–82.
Michaud, Michael A.G., 'Spaceflight, Colonization and Independence: A Synthesis,' *Journal of the British Interplanetary Society* 30.3/6/9 (March/June/September 1977), 83–95, 203–12, 323–31.
———, *Contact with Alien Civilizations: Our Hopes and Fears About Encountering Extraterrestrials*, New York: Copernicus, 2007.
Michel, Aimé, *Lueurs sur les soucoupes volantes*, Paris: Maison Mame, 1954 (Eng. *The Truth About Flying Saucers*, New York: Criterion, 1956).
———, *Mystérieux objets célestes*, Paris: Arthaud, 1958.

Miller, Ron, *The Dream Machines: An Illustrated Chronology of the Spaceship in Art, Science and Literature*, Malabar: Krieger, 1993.

———, 'The Archaeology of Space Art,' *Leonardo* 29.2 (1996), 139–43.

Miller, Ryder W., ed., *From Narnia to a Space Odyssey: The War of Ideas Between Arthur C. Clarke and C.S. Lewis*, New York: Ibooks, 2003.

Minois, Georges, *Histoire de l'avenir: des Prophètes à la prospective*, Paris: Fayard, 1996.

Mitter, Rana and Patrick Major, eds, *Across the Blocs: Cold War Cultural and Social History*, London: Frank Cass, 2004.

Moffitt, John F., *Picturing Extraterrestrials: Alien Images in Modern Culture*, Amherst: Prometheus, 2003.

Monchaux, Nicholas de, *Spacesuit: Fashioning Apollo*, Cambridge, MA: MIT Press, 2011.

Montgomery, Scott L., *The Moon and the Western Imagination*, Tucson: University of Arizona Press, 2001.

Morrison, Philip, John Billingham and John Wolfe, eds, *The Search for Extraterrestrial Intelligence*, Washington, DC: NASA, 1977.

Morton, Oliver, *Mapping Mars: Science, Imagination, and the Birth of a World*, New York: Picador, 2002.

Morton, Peter, *Fire Across the Desert: Woomera and the Anglo-Australian Joint Project 1946–1980*, Canberra: Australian Government Publishing Service, 1989.

Moulin, Hervé, 'The International Geophysical Year: Its Influence on the Beginning of the French Space Program,' *Acta Astronautica* 66.5–6 (March 2010), 688–92.

Mudgway, Doug, *William H. Pickering: America's Deep Space Pioneer*, Washington, DC: NASA, 2008.

Müller, Wolfgang D., *Du wirst die Erde sehn als Stern: Probleme der Weltraumfahrt*, Stuttgart: Deutsche Verlags-Anstalt, 1955.

Murray, Charles and Catherine Bly Cox, *Apollo: The Race to the Moon*, New York: Simon & Schuster, 1989.

Myrach, Thomas, Tristan Weddigen, Jasmine Wohlwend and Sara Margarita Zwahlen, eds, *Science & Fiction: Imagination und Realität des Weltraums*, Bern: Haupt, 2010.

Nadel, Alan, *Containment Culture: American Narratives, Postmodernism, and the Atomic Age*, Durham: Duke University Press, 1995.

Nagl, Manfred, *Science Fiction: Ein Segment populärer Kultur im Medien- und Produktverbund*, Tübingen: Narr, 1981.

———, 'The Science-Fiction Film in Historical Perspective,' *Science Fiction Studies* 10.3 (November 1983), 262–77.

Narjes, Karl-Heinz, 'Space and the European Community,' *Space Policy* 5.1 (February 1989), 59–64.

Neal, Valerie, 'Bumped from the Shuttle Fleet: Why Didn't *Enterprise* Fly in Space?,' *History and Technology* 18.3 (September 2002), 181–202.

Nehring, Holger, 'National Internationalists: British and West German Protests against Nuclear Weapons, the Politics of Transnational Communications and the Social History of the Cold War, 1957–1964,' *Contemporary European History* 14.4 (November 2005), 559–82.

Nelson, Stephanie and Larry Polansky, 'The Music of the Voyager Interstellar Record,' *Journal of Applied Communication Research* 21.4 (November 1993), 358–76.

Neufeld, Michael J., 'Weimar Culture and Futuristic Technology: The Rocketry and Spaceflight Fad in Germany, 1923–1933,' *Technology and Culture* 31.4 (October 1990), 725–52.

———, *The Rocket and the Reich: Peenemünde and the Coming of the Ballistic Missile Era*, New York: Free Press, 1995.

———, 'The Excluded: Hermann Oberth and Rudolf Nebel in the Third Reich,' *Quest* 5.4 (1996), 22–7.

———, 'The Reichswehr, the Rocket, and the Versailles Treaty: A Popular Myth Reexamined,' *Journal of the British Interplanetary Society* 53.3 (May–June 2000), 163–72.

———, 'Wernher von Braun, the SS, and Concentration Camp Labor: Questions of Moral, Political, and Criminal Responsibility,' *German Studies Review* 25.1 (February 2002), 57–78.

———, '"Space Superiority": Wernher von Braun's Campaign for a Nuclear-Armed Space Station, 1946–1956,' *Space Policy* 22.1 (February 2006), 52–62.

———, *Von Braun: Dreamer of Space, Engineer of War*, New York: Alfred A. Knopf, 2007.

———, 'Von Braun and the Lunar-Orbit Rendezvous Decision: Finding a Way to Go to the Moon,' *Acta Astronautica* 63.1–4 (July–August 2008), 540–50.

Neufeld, Michael J. and Alex M. Spencer, eds, *Smithsonian National Air and Space Museum: An Autobiography*, Washington, DC: National Geographic, 2010.

Nicolson, Marjorie Hope, 'The Telescope and Imagination,' *Modern Philology* 32.3 (February 1935), 233–60.

———, *A World in the Moon: A Study of the Changing Attitude Toward the Moon in the Seventeenth and Eighteenth Centuries*, Northampton: Smith College, 1935–36.

———, *Voyages to the Moon*, New York: Macmillan, 1948.

———, *Science and Imagination*, Ithaca: Cornell University Press, 1956.

Nieman, H.W. and C.W. Nieman, 'What Shall We Say to Mars?,' *Scientific American* 122 (March 1920), 298, 312.

Noble, David F., *The Religion of Technology: The Divinity of Man and the Spirit of Invention*, New York: Alfred A. Knopf, 1997.

Noordung, Hermann, *Das Problem der Befahrung des Weltraums: Der Raketen-Motor*, Berlin: Richard Carl Schmidt, 1929.

Nye, David E., *American Technological Sublime*, Cambridge, MA: MIT Press, 1994.

———, 'Don't Fly Us to the Moon: The American Public and the Apollo Space Program,' *Foundation: The Review of Science Fiction* 66 (Spring 1996), 69–81.

Oberg, James E., *Red Star in Orbit*, New York: Random House, 1981.

Oberth, Hermann, *Die Rakete zu den Planetenräumen*, Munich: Oldenbourg, 1923.

———, *Wege zur Raumschiffahrt*, Munich: Oldenbourg, 1929.

———, 'Flying Saucers Come from a Distant World,' *American Weekly* (24 October 1954), 4–5.

———, *Menschen im Weltraum: Neue Projekte für Raketen- und Raumfahrt*, Düsseldorf: Econ, 1954.

———, *Katechismus der Uraniden: Haben unsere Religionen eine Zukunft?*, Wiesbaden: Ventla, 1966.

———, *Parapsychologie: Schlüssel zur Welt von morgen*, Kleinjörl: Schroeder, 1976.

———, *Der Weltraumspiegel*, Bucharest: Kriterion, 1978.

O'Hagan, Andrew, 'Goodbye Moon,' *London Review of Books* 32.4 (25 February 2010), 13–14.

O'Neill, Gerard K., 'The Colonization of Space,' *Physics Today* 27.9 (September 1974), 32–40.

———, *The High Frontier: Human Colonies in Space*, New York: William Morrow, 1977.

Opt, Susan K., 'American Frontier Myth and the Flight of Apollo 13: From News Event to Feature Film,' *Film & History* 26.1–4 (February 1996), 40–51.

Ordway, Frederick I., 'Collecting Literature in the Space and Rockets Fields,' *Space Education* 1.4–6 (September–October 1982–83), 176–82, 279–87.

———, *Visions of Spaceflight: Images from the Ordway Collection*, New York: Four Walls Eight Windows, 2001.

Ordway, Frederick I. and Mitchell R. Sharpe, *The Rocket Team*, New York: Crowell, 1979.

Ordway, Frederick I. and Randy Liebermann, eds, *Blueprint for Space: Science Fiction to Science Fact*, Washington, DC: Smithsonian Institution Press, 1992.

Ordway, Frederick I., Mitchell R. Sharpe and Ronald C. Wakeford, 'Project Horizon: An Early Study of a Lunar Outpost,' *Acta Astronautica* 17.10 (October 1988), 1105–21.

Pagden, Anthony, ed., *The Idea of Europe: From Antiquity to the European Union*, Cambridge: Cambridge University Press, 2002.

Palmer, Susan, *Aliens Adored: Raël's UFO Religion*, New Brunswick: Rutgers University Press, 2004.

Papp, Desiderius, *Was lebt auf den Sternen? Ein Buch über die Bewohner anderer Welten*, Zurich: Amalthea, 1929.

Parkinson, Bob, ed., *Interplanetary: A History of the British Interplanetary Society*, London: British Interplanetary Society, 2008.

Parks, Lisa, *Cultures in Orbit: Satellites and the Televisual*, Durham: Duke University Press, 2005.

Parrinder, Patrick, *Science Fiction: Its Criticism and Teaching*, London: Methuen, 1980.

Parsons, William B., 'The Oceanic Feeling Revisited,' *Journal of Religion* 78.4 (October 1998), 501–23.

Partridge, Christopher, ed., *UFO Religions*, London: Routledge, 2003.

Pass, Jim, 'Astrosociology as the Missing Perspective,' *Astropolitics* 4.1 (2006), 85–99.

———, 'Examining the Definition of Astrosociology,' *Astropolitics* 9.1 (2011), 6–27.

Patterson, David W., 'Music, Structure and Metaphor in Stanley Kubricks's *2001: A Space Odyssey*,' *American Music* 22.3 (Fall 2004), 444–74.

Peebles, Curtis, *Watch the Skies! A Chronicle of the Flying Saucer Myth*, Washington, DC: Smithsonian Institution Press, 1994.

Pendle, George, *Strange Angel: The Otherworldly Life of Rocket Scientist John Whiteside Parsons*, Orlando: Harcourt, 2004.

Pendray, G. Edward, 'Pioneer Rocket Development in the United States,' *Technology and Culture* 4.4 (Fall 1963), 384–92.

Penley, Constance, *NASA/TREK: Popular Science and Sex in America*, London: Verso, 1997.

Penley, Constance and Andrew Ross, eds, *Technoculture*, Minneapolis: University of Minnesota Press, 1991.

Perkowitz, Sidney, *Hollywood Science: Movies, Science and the End of the World*, New York: Columbia University Press, 2010.

Petri, Winfried, 'Raumfahrt in der Sowjetunion: Historisch-ideologischer Rahmen und praktische Erwartungen,' *Studies in Soviet Thought* 18.1 (February 1978), 45–55.

Pickering, W.H., 'Blossoming of the Space Age, 1961–1971,' *Acta Astronautica* 32.7–8 (July–August 1994), 501–8.

Piel, Gerard, Richard F. Babcock, Isaac Asimov, Buckminster Fuller and Edmund N. Bacon, 'Five Noted Thinkers Explore the Future,' *National Geographic* 150 (July 1976), 68–74.

Pincio, Tommaso, *Gli alieni: Dove si racconta come e perché sono giunti tra noi*, Rome: Fazi, 2006.

Pinvidic, Thierry, ed., *OVNI: Vers une anthropologie d'un mythe contemporain*, Bayeux: Heimdal, 1993.

Polianski, Igor J. and Matthias Schwartz, eds, *Die Spur des Sputnik: Kulturhistorische Expeditionen ins kosmische Zeitalter*, Frankfurt am Main: Campus, 2009.

Poole, Robert, *Earthrise: How Man First Saw the Earth*, New Haven: Yale University Press, 2008.

Pössel, Markus, *Phantastische Wissenschaft: Über Erich von Däniken und Johannes von Buttlar*, Reinbek: Rowohlt, 2000.

Prelinger, Megan, *Another Science Fiction: Advertising the Space Race 1957–1962*, New York: Blast, 2010.

Price, David, 'Political and Economic Factors Relating to European Space Co-operation,' *Spaceflight* 4.1 (January 1962), 6–15.

Pugsley, Derek, 'The Recognition of Extraterrestrial Intelligence: Are Humans Up to It?,' *Journal of the British Interplanetary Society* 61.1 (January 2008), 20–3.

Pynchon, Thomas, *Gravity's Rainbow*, New York: Viking, 1973.

Pyne, Stephen J., 'The Extraterrestrial Earth: Antarctica as Analogue for Space Exploration,' *Space Policy* 23.3 (August 2007), 147–9.

Rabinowitch, Eugene I. and Richard S. Lewis, eds, *Men in Space: The Impact on Science, Technology, and International Cooperation*, Aylesbury: Medical and Technical Publishing, 1970.

Rabkin, Eric S., 'The Composite Fiction of Olaf Stapledon,' *Science Fiction Studies* 9.3 (November 1982), 238–48.

Rabkin, Eric S., ed., *Science Fiction: A Historical Anthology*, Oxford: Oxford University Press, 1983.

Ramet, Sabrina P., 'UFOs over Russia and Eastern Europe,' *Journal of Popular Culture* 32.3 (Winter 1998), 81–99.

Rauchhaupt, Ulf von, 'Colorful Clouds and Unruly Rockets: Early Research Programs at the Max Planck Institute for Extraterrestrial Physics,' *Historical Studies in the Physical and Biological Sciences* 32.1 (2001), 115–24.

———, *Der neunte Kontinent: Die wissenschaftliche Eroberung des Mars*, Frankfurt am Main: Fischer, 2009.

Ratcliff, J.D., 'Italy's Amazing Amateur Space Watchers,' *Reader's Digest* (April 1965), 110–14.

Rauschenbach, Boris, *Hermann Oberth 1894–1989: Über die Erde hinaus. Eine Biographie*, Wiesbaden: Böttiger, 1995.

Rawer, Karl, *Möglichkeiten und Grenzen europäischer Weltraumforschung*, Munich: Fraunhofer-Gesellschaft, 1965.

Redfield, Peter, 'Beneath a Modern Sky: Space Technology and Its Place on the Ground,' *Science, Technology, & Human Values* 21.3 (Summer 1996), 251–74.

———, *Space in the Tropics: From Convicts to Rockets in French Guiana*, Berkeley: University of California Press, 2000.

———, 'The Half-Life of Empire in Outer Space,' *Social Studies of Science* 32.5–6 (October–December 2002), 791–825.

Reece, Gregory L., *UFO Religion: Inside Flying Saucer Cults and Culture*, London: I.B. Tauris, 2007.

Reeken, Dieter von, *Hermann Oberth und die UFO-Forschung*, 2nd edn, Lüdenscheid: Gesellschaft zur Erforschung des UFO-Phänomens, 1994.

Regis, Edward, Jr., ed., *Extraterrestrials: Science and Alien Intelligence*, Cambridge: Cambridge University Press, 1985.

Rehm, Georg W., 'Die Behandlung der Weltraumfrage in den Vereinten Nationen 1957–1964,' *Weltraumfahrt* 15.6 (1964), 172–7.

Reinke, Niklas, *Geschichte der deutschen Raumfahrtpolitik: Konzepte, Einflußfaktoren und Interdependenzen 1923–2002*, Munich: Oldenbourg, 2004.

Reuter, Karl-Egon and Johann Oberlechner, 'The ESA History Project,' *ESA Bulletin* 119 (August 2004), 48–54.

Reynolds, Glenn H. and Robert P. Merges, *Outer Space: Problems of Law and Policy*, 2nd edn, Boulder: Westview Press, 1998.

Rhodes, Gary D., ed., *Stanley Kubrick: Essays on His Films and Legacy*, Jefferson: McFarland, 2008.

Rickman, Gregg, ed., *The Science Fiction Film Reader*, New York: Limelight, 2003.

Ricœur, Paul, 'Ideology and Utopia as Cultural Imagination,' in Donald M. Borchert and David Stewart, eds, *Being Human in a Technological Age*, Athens: Ohio University Press, 1979, 107–25.

Rietz, Frank E., *Die Magdeburger Pilotrakete 1933: Auf dem Weg zur bemannten Raumfahrt?*, Halle: Mitteldeutscher Verlag, 1998.

Riordan, Maurice and Jocelyn Bell Burnell, eds, *Dark Matter: Poems of Space*, London: Calouste Gulbenkian Foundation, 2008.

Ritner, Peter, *The Society of Space*, New York: Macmillan, 1961.

Ritter, Gerhard A., *Großforschung und Staat in Deutschland: Ein historischer Überblick*, Munich: C.H. Beck, 1992.

Roberts, Adam, *The History of Science Fiction*, Basingstoke: Palgrave Macmillan, 2006.

Robertson, Frances, 'Science and Fiction: James Nasmyth's Photographic Images of the Moon,' *Victorian Studies* 48.4 (Summer 2006), 595–623.

Robinson, George S., 'Space Law, Space War and Space Exploitation,' *Journal of Social and Political Studies* 5.3 (1980), 163–78.

Roland, Alex, 'Celebration or Education? The Goals of the U.S. National Air and Space Museum,' *History and Technology* 10.1 (1993), 77–89.

Rose, Mark, *Alien Encounters: Anatomy of Science Fiction*, Cambridge, MA: Harvard University Press, 1981.

Rosenfelder, Andreas, 'Medien auf dem Mond: Zur Reichweite des Weltraumfernsehens,' in Irmela Schneider, Torsten Hahn and Christina Bartz, eds, *Medienkultur der 60er Jahre: Diskursgeschichte der Medien nach 1945*, Wiesbaden: Westdeutscher Verlag, 2003, 17–33.

Rosenfield, Stanley B., 'Where Air Space Ends and Outer Space Begins,' *Journal of Space Law* 7.2 (Fall 1979), 137–48.

Ross, Andrew, ed., *Science Wars*, Durham: Duke University Press, 1996.

Ross, H.E., 'Gone with the Efflux,' *Journal of the British Interplanetary Society* 9.3 (May 1950), 93–101.

Rottensteiner, Franz, 'SF in Germany: A Short Survey,' *Science Fiction Studies* 27.1 (March 2000), 118–23.

Rovin, Jeff, *Aliens, Robots, and Spaceships*, New York: Facts on File, 1995.

Rudoff, Alvin, *Societies in Space*, New York: Peter Lang, 1996.

Rushing, Janice Hocker, 'Mythic Evolution of "The New Frontier" in Mass Mediated Rhetoric,' *Critical Studies in Mass Communication* 3.3 (September 1986), 265–96.

Russell, Miles, ed., *Digging Holes in Popular Culture: Archaeology and Science Fiction*, Oxford: Oxbow, 2002.

Rüthers, Monica, 'Kindheit, Kosmos und Konsum in sowjetischen Bildwelten der 1960er Jahre: Zur Herstellung von Zukunftsoptimismus,' *Historische Anthropologie* 17.1 (2009), 56–74.

Ryan, Cornelius, Wernher von Braun, Fred L. Whipple and Willy Ley, *Conquest of the Moon*, New York: Viking, 1953.

Sacha-Eisleb, Katharina, *Engel, Menschen, Monstren: Außerirdische in der viktorianischen Science Fiction*, Berlin: Wissenschaftlicher Verlag, 2000.

Sachs, Wolfgang, 'Satellitenblick: Die Ikone vom blauen Planeten und ihre Folgen für die Wissenschaft,' in Ingo Braun and Bernward Joerges, eds, *Technik ohne Grenzen*, Frankfurt am Main: Suhrkamp, 1994, 305–46.

Sadoul, Jacques, *Histoire de la science-fiction moderne, 1911–1971*, Paris: Albin Michel, 1973.

Saethre, Eirik, 'Close Encounters: UFO Beliefs in a Remote Australian Aboriginal Community,' *Journal of the Royal Anthropological Institute* 13.4 (December 2007), 901–15.

Sagan, Carl, ed., *Communication with Extraterrestrial Intelligence*, Cambridge, MA: MIT Press, 1973.

———, *The Cosmic Connection: An Extraterrestrial Perspective*, London: Hodder & Stoughton, 1974.

———, *Broca's Brain: Reflections on the Romance of Science*, New York: Random House, 1979.

———, *Cosmos: The Story of Cosmic Evolution, Science and Civilisation*, London: Abacus, 1980.

———, *Pale Blue Dot: A Vision of the Human Future in Space*, New York: Random House, 1994.

Sagan, Carl and Thornton Page, eds, *UFO's: A Scientific Debate*, Ithaca: Cornell University Press, 1972.

Sagan, Carl, Linda Salzman Sagan and Frank D. Drake, 'A Message from Earth,' *Science* 175 (25 February 1972), 881–4.

Sagan, Carl, Frank D. Drake, Ann Druyan, Timothy Ferris, Jon Lomberg and Linda Salzman Sagan, *Murmurs of Earth: The Voyager Interstellar Record*, New York: Random House, 1978.

Saler, Benson, Charles A. Ziegler and Charles B. Moore, *UFO Crash at Roswell: The Genesis of a Modern Myth*, Washington, DC: Smithsonian Institution Press, 1997.

Salewski, Michael, *Zeitgeist und Zeitmaschine: Science Fiction und Geschichte*, Munich: Deutscher Taschenbuch Verlag, 1986.

Salisbury, Frank B., 'Martian Biology,' *Science* 136 (6 April 1962), 17–26.

———, 'The Scientist and the UFO,' *Bio-Science* 17.1 (January 1967), 15–24.

Salkeld, Robert, 'Space Colonization Now?,' *Astronautics and Aeronautics* 13.9 (September 1975), 30–4.

Sänger, Eugen, *Raketen-Flugtechnik*, Munich: Oldenbourg, 1933.

———, *Zur Mechanik der Photonen-Strahlantriebe*, Munich: Oldenbourg, 1956.

———, 'Raumfahrt: Einige politische Aspekte,' *Weltraumfahrt* 9.1 (March 1958), 12–26.

———, ed., *Raumfahrt wohin? Was bringt uns der Vorstoß ins All? Weltraumforscher aus acht Ländern antworten*, Munich: Bechtle, 1962.

Sarantakes, Nicholas Evan, 'Cold War Pop Culture and the Image of U.S. Foreign Policy: The Perspective of the Original *Star Trek* Series,' *Journal of Cold War Studies* 7.4 (Fall 2005), 74–103.

Sardar, Ziauddin and Sean Cubitt, eds, *Aliens R Us: The Other in Science Fiction Cinema*, London: Pluto, 2002.

Schäfer, Herbert, 'Die "Flying Saucer Story": Eine neue Form der Okkultkriminalität,' *Sterne und Weltraum* 1.7 (October 1962), 140–3.

Schauer, William H., *The Politics of Space: A Comparison of the Soviet and American Space Programs*, New York: Holmes & Meier, 1976.

Schetsche, Michael, 'Reale und virtuelle Probleme: "UFO abduction experiences" als Testfall für die (Problem-)Soziologie,' *Berliner Journal für Soziologie* 8.2 (1998), 223–44.

———, ed., *Der maximal Fremde: Begegnungen mit dem Nichtmenschlichen und die Grenzen des Verstehens*, Würzburg: Ergon, 2004.

———, 'Rücksturz zur Erde? Zur Legitimierung und Legitimität der bemannten Raumfahrt,' in Christoph Heinrich and Markus Heinzelmann, eds, *Rückkehr ins All*, Ostfildern-Ruit: Hatje Cantz, 2005, 24–7.

Schetsche, Michael and Martin Engelbrecht, eds, *Von Menschen und Außerirdischen: Transterrestrische Begegnungen im Spiegel der Kulturwissenschaft*, Bielefeld: transcript, 2008.

Schmidt-Gernig, Alexander, 'Die gesellschaftliche Konstruktion der Zukunft: Westeuropäische Zukunftsforschung und Gesellschaftsplanung zwischen 1950 und 1980,' *WeltTrends* 18 (Spring 1998), 63–84.

———, 'The Cybernetic Society: Western Future Studies of the 1960s and 1970s and their Predictions for the Year 2000,' in Richard N. Cooper and Richard Layard, eds, *What Next? The Future of Human Life in the Light of the Social Science*, Cambridge, MA: MIT Press, 2002, 233–59.

Schmidt, Stanley and Robert Zubrin, eds, *Islands in the Sky: Bold New Ideas for Colonizing Space*, New York: John Wiley, 1996.

Schmölders, Claudia, 'Fenster ins All: Über Sprache und Weltraum,' *Lettre International* 45 (Summer 1999), 53–4.

Scholes, Robert and Eric S. Rabkin, *Science Fiction: History, Science, Vision*, Oxford: Oxford University Press, 1977.

Schröder, Iris and Sabine Höhler, eds, *Welt-Räume: Geschichte, Geographie und Globalisierung seit 1900*, Frankfurt am Main: Campus, 2005.

Schwartz, Matthias, *Die Erfindung des Kosmos: Zur sowjetischen Science Fiction und populärwissenschaftlichen Publizistik vom Sputnikflug bis zum Ende der Tauwetterzeit*, Frankfurt am Main: Peter Lang, 2003.

Schwarz, Michiel, 'European Policies on Space Science and Technology 1960–1978,' *Research Policy* 8.3 (July 1979), 204–43.

Schwoch, James, 'Satellites, Rocketry, Security and Space Policy: A Comparative History?,' *Historical Journal of Film, Radio and Television* 18.2 (1998), 295–9.

———, *Global TV: New Media and the Cold War, 1946–69*, Urbana: University of Illinois Press, 2009.

Schwonke, Martin, *Vom Staatsroman zur Science Fiction: Eine Untersuchung über Geschichte und Funktion der naturwissenschaftlich-technischen Utopie*, Stuttgart: Ferdinand Enke, 1957.

Sebesta, Lorenza, 'La Science, instrument politique de securité nationale? L'espace, la France et l'Europe, 1957–1962,' *Revue d'histoire diplomatique* 106.4 (1992), 313–41.

———, 'The Politics of Technological Cooperation in Space: US-European Negotiations on the Post-Apollo Programme,' *History and Technology* 11.2 (1994), 317–41.

———, *Alleati competitivi: Origini e sviluppo della cooperazione spaziale fra Europa e Stati Uniti, 1957–1973*, Rome: Laterza, 2003.

Seeßlen, Georg and Fernand Jung, *Science Fiction: Geschichte und Mythologie des Science-Fiction-Films*, 2 vols, Marburg: Schüren, 2003.

Sheldon, Charles S., 'L'Ere de l'espace,' *Analyse et Prévision* 2.4 (October 1966), 733–43.

Shepherd, Leslie R., 'Prelude and First Decade, 1951–1961,' *Acta Astronautica* 32.7–8 (July–August 1994), 475–99.

Siddiqi, Asif A., 'The Rockets' Red Glare: Technology, Conflict, and Terror in the Soviet Union,' *Technology and Culture* 44.3 (July 2003), 470–501.

———, 'Deep Impact: Robert Goddard and the Soviet "Space Fad" of the 1920s,' *History and Technology* 20.2 (June 2004), 97–113.

———, 'Imagining the Cosmos: Utopians, Mystics, and the Popular Culture of Spaceflight in Revolutionary Russia,' *Osiris* 23.1 (2008), 260–88.

———, *The Red Rockets' Glare: Spaceflight and the Soviet Imagination, 1857–1957*, Cambridge: Cambridge University Press, 2010.

———, 'Competing Technologies, National(ist) Narratives, and Universal Claims: Toward a Global History of Space Exploration,' *Technology and Culture* 51.2 (April 2010), 425–43.

Siefarth, Günter, *Geschichte der Raumfahrt*, Munich: C.H. Beck, 2001.

Sivier, David J., 'SETI and the Historian: Methodological Problems in an Interdisciplinary Approach,' *Journal of the British Interplanetary Society* 53.1 (January–February 2000), 23–5.

———, 'The Develoment of Politics in Extraterrestrial Colonies,' *Journal of the British Interplanetary Society* 53.5 (September–October 2000), 290–6.

Slade, Joseph W., 'Rockets and Rationalization: Science and Technology in Nazi Germany and Thomas Pynchon,' *Dimensions* 10.2 (1996), 39–44.

Slotten, Hugh R., 'Satellite Communications, Globalization, and the Cold War,' *Technology and Culture* 43.2 (April 2002), 315–50.

Smith, Michael L., 'Selling the Moon: The U.S. Manned Space Program and the Triumph of Commodity Scientism,' in Richard Wightman Fox and T.J. Jackson Lears, eds, *The Culture of Consumption: Critical Essays in American History, 1880–1980*, New York: Pantheon, 1983, 175–209.

Smith, Ralph Andrew and Bob Parkinson, *High Road to the Moon: From Imagination to Reality*, London: British Interplanetary Society, 1979.

Smith, Robert, *The Expanding Universe: Astronomy's Great Debate, 1900–1931*, Cambridge: Cambridge University Press, 1982.

Sobchack, Vivian, 'The Alien Landscapes of the Planet Earth,' *Film Journal* 2.3 (1974), 16–21.

———, 'Surge and Splendor: A Phenomenology of the Hollywood Historical Epic,' *Representations* 29 (Winter 1990), 24–49.

———, *Screening Space: The American Science Fiction Film*, 2nd edn, New Brunswick: Rutgers University Press, 1997.

Soldovieri, Stefan, 'Socialists in Outer Space: East German Film's Venusian Adventure,' *Film History* 10.3 (1998), 382–98.

Sombart, Werner, 'Technik und Kultur,' *Archiv für Sozialwissenschaft und Sozialpolitik* 33 (1911), 305–47.

Sontag, Susan, 'The Imagination of Disaster,' *Commentary* 40.4 (October 1965), 42–8.

———, *Regarding the Pain of Others*, New York: Farrar, Straus & Giroux, 2003.

Soojung-Kim Pang, Alex, '"Stars Should Henceforth Register Themselves": Astrophotography at the Early Lick Observatory,' *British Journal for the History of Science* 30.2 (June 1997), 177–202.

Sparks, Glenn G. and Marianne Pellechia, 'The Effect of News Stories About UFOs on Readers' UFO Beliefs: The Role of Confirming or Disconfirming Testimony from a Scientist,' *Communication Reports* 10.2 (Summer 1997), 165–72.

Sparks, Glenn G., Marianne Pellechia and Chris Irvine, 'Does Television News About UFOs Affect Viewers' UFO Beliefs? An Experimental Investigation,' *Communication Quarterly* 46.3 (Summer 1998), 284–94.

Spengler, Oswald, *Der Untergang des Abendlandes: Umrisse einer Morphologie der Weltgeschichte*, Munich: C.H. Beck, 1919/1922 (Eng. *The Decline of the West*, New York: Alfred A. Knopf, 1939).

Stapledon, Olaf, 'Interplanetary Man?,' *Journal of the British Interplanetary Society* 7.6 (November 1948), 213–33.

Staudenmaier, John M., 'Recent Trends in the History of Technology,' *American Historical Review* 95.3 (June 1990), 715–25.

Strehl, Rolf, *Fliegende Untertassen: Ein Geheimnis geistert um die Welt*, Lüneburg: Oldenkott-Rees, 1979.

Strick, James E., 'Creating a Cosmic Discipline: The Crystallization and Consolidation of Exobiology, 1957–1973,' *Journal of the History of Biology* 37.1 (Spring 2004), 131–80.

Strick, Philip, *Science Fiction Movies*, London: Octopus, 1976.

Stucke, Andreas, 'Die Raumfahrtpolitik des Forschungsministeriums,' *Leviathan* 20.4 (March 1992), 544–62.

Sullivan, Walter S., *We Are Not Alone: The Search for Intelligent Life on Other Worlds*, New York: McGraw-Hill, 1964.

Suvin, Darko, *Positions and Presuppositions in Science Fiction*, Basingstoke: Macmillan, 1988.

Suzuki, Kazuto, 'Space and Modernity: 50 Years On,' *Space Policy* 23.3 (August 2007), 144–6.

Swiarski, Peter, *A Stanislaw Lem Reader*, Evanston: Northwestern University Press, 1997.

Swift, David W., *SETI Pioneers: Scientists Talk About Their Search for Extraterrestrial Intelligence*, Tucson: University of Arizona Press, 1990.

Swords, Michael D., 'A Guide to UFO Research,' *Journal of Scientific Exploration* 7.1 (1993), 65–87.

Sykora, Fritz, 'Pioniere der Raketentechnik aus Österreich,' *Blätter für Technikgeschichte* 22 (1960), 189–204.

Syon, Guillaume de, *Zeppelin! Germany and the Airship, 1900–1939*, Baltimore: Johns Hopkins University Press, 2002.

Szöllösi-Janze, Margit, 'Wissensgesellschaft in Deutschland: Überlegungen zur Neubestimmung der deutschen Zeitgeschichte über Verwissenschaftlichungsprozesse,' *Geschichte und Gesellschaft* 30.2 (2004), 277–313.

Tarter, Donald E., 'Peenemünde and Los Alamos: Two Studies,' *History of Technology* 14 (1992), 150–70.

Teilhard de Chardin, Pierre, *Le Phénomène humain*, Paris: Editions du Seuil, 1955 (Eng. *The Phenomenon of Man*, New York: Harper Perennial, 2008).

———, *L'Avenir de l'homme*, Paris: Editions du Seuil, 1959.

Telotte, J.P., *Distant Technology: Science Fiction Film and the Machine Age*, Middletown: Wesleyan University Press, 1999.

———, *Science Fiction Film*, Cambridge: Cambridge University Press, 2001.

———, 'Disney in Science Fiction Land,' *Journal of Popular Film and Television* 33.1 (April 2005), 12–21.

———, 'Animating Space: Disney, Science, and Empowerment,' *Science Fiction Studies* 35.1 (February 2008), 48–59.

Thomas, Shirley, *Men of Space: Profiles of the Leaders in Space Research, Development, and Exploration*, 8 vols, Philadelphia: Chilton, 1960–68.

Tietenberg, Annette and Tristan Weddigen, eds, *Planetarische Perspektiven: Bilder der Raumfahrt*, Marburg: Jonas, 2009 (= *Kritische Berichte* 37.3).

Topham, Sean, *Where's My Space Age? The Rise and Fall of Futuristic Design*, Munich: Prestel, 2003.

Tough, Allen, ed., *When SETI Succeeds: The Impact of High-Information Contact*, Bellevue: Foundation for the Future, 2000.

Trimborn, Hermann, 'Außerirdische Raumfahrer in Amerika,' *Saeculum* 30.2–3 (1979), 226–39.

Trischler, Helmuth, *Luft- und Raumfahrtforschung in Deutschland 1900–1970: Politische Geschichte einer Wissenschaft*, Frankfurt am Main: Campus, 1992.

———, *The 'Triple Helix' of Space: German Space Activities in a European Perspective*, Noordwijk: ESA, 2002.

Trischler, Helmuth and Kai-Uwe Schrogl, eds, *Ein Jahrhundert im Flug: Luft- und Raumfahrtforschung in Deutschland 1907–2007*, Frankfurt am Main: Campus, 2007.

Tumminia, Diana G., *When Prophecy Never Fails: Myth and Reality in a Flying-Saucer Group*, Oxford: Oxford University Press, 2005.

———, ed., *Alien Worlds: Social and Religious Dimensions of Extraterrestrial Contact*, Syracuse: Syracuse University Press, 2007.

Turner, Fred, *From Counterculture to Cyberculture: Stewart Brand, the Whole Earth Network, and the Rise of Digital Utopianism*, Chicago: University of Chicago Press, 2006.

Turner, Frederick Jackson, *The Frontier in American History*, New York: Henry Holt, 1920.

Tuzet, Hélène, *Le Cosmos et l'imagination*, Paris: José Corti, 1965.

Uhl, Matthias, 'Stalins V-2: Zum Transfer der deutschen Raketentechnologie in die UdSSR 1945–1958,' *Osteuropa* 51.7 (July 2001), 847–66.

———, *Stalins V-2: Der Technologietransfer der deutschen Fernlenkwaffentechnik in die UdSSR und der Aufbau der sowjetischen Raketenindustrie 1945 bis 1959*, Bonn: Bernard & Graefe, 2001.

Ulivi, Paolo, 'ESRO and the Deep Space: European Planetary Exploration Planning Before ESA,' *Journal of the British Interplanetary Society* 59.5 (June 2006), 204–23.

United Nations, *United Nations Treaties and Principles on Outer Space: Text of Treaties and Principles Governing the Activities of States in the Exploration and Use of Outer Space, Adopted by the United Nations General Assembly*, New York: United Nations, 2002.

Unwin, Timothy, 'Jules Verne: Negotiating Change in the Nineteenth Century,' *Science Fiction Studies* 32.1 (March 2005), 5–17.

Vakoch, Douglas A., 'Constructing Messages to Extraterrestrials: An Exosemiotic Perspective,' *Acta Astronautica* 42.10–12 (May 1998), 697–704.

Valentine, Burl, 'Obstacles to Space Cooperation: Europe and the Post-Apollo Experience,' *Research Policy* 1.2 (April 1972), 104–21.

Vallat, Francis, 'The Outer Space Treaties,' *Aeronautical Journal* 73 (September 1969), 751–8.

Vallée, Jacques and Janine Vallée, *Challenge to Science: The UFO Enigma*, London: Neville Spearman, 1966.

Van Riper, A. Bowdoin, 'From Gagarin to Armageddon: Soviet-American Relations in the Cold War Space Epic,' *Film & History* 31.2 (September 2001), 45–51.

Varvarov, N.A. and Y.T. Fadeyev, 'Philosophical Problems of Astronautics,' *The Soviet Review* 3.6 (June 1962), 21–38.

Veit, Karl L., *Planetenmenschen besuchen unsere Erde: Grundsätzliches zum Verständnis weltbewegender Luftphänomene*, Wiesbaden: Ventla, 1961.

———, *Erforschung außerirdischer Weltraumschiffe: Ein wissenschaftliches Anliegen des 20. Jahrhunderts*, Wiesbaden: Ventla, 1963.

Velikovsky, Immanuel, *Earth in Upheaval*, Garden City: Doubleday, 1955.

———, *Worlds in Collision*, London: Gollancz, 1959.

———, *Stargazers and Gravediggers: Memoirs to Worlds in Collision*, New York: Morrow, 1983.

Vieth, Errol, *Screening Science: Contexts, Texts, and Science in Fifties Science Fiction Film*, Lanham: Scarecrow Press, 2001.

Virilio, Paul, *La Vitesse de libération: essai*, Paris: Galilée, 1995.

Vizzini, Bryan E., 'Cold War Fears, Cold War Passions: Conservatives and Liberals Square Off in 1950s Science Fiction,' *Quarterly Review of Film and Video* 26.1 (January 2009), 28–39.

Wabbel, Tobias Daniel, ed., *Leben im All: Positionen aus Naturwissenschaft, Philosophie und Theologie*, Düsseldorf: Patmos, 2005.

Wachhorst, Wyn, 'Seeking the Center at the Edge: Perspectives on the Meaning of Man in Space,' *Virginia Quarterly Review* 69.1 (Winter 1993), 1–23.

Wallace, Helen, 'Building a European Space Policy,' *Space Policy* 4.2 (May 1988), 115–20.

Wallace, Jennifer, *Digging the Dirt: The Archaeological Imagination*, London: Gerald Duckworth, 2004.

Walters, Helen B., *Hermann Oberth: Father of Space Travel*, New York: Macmillan, 1962.

Warren, Donald I., 'Status Inconsistency Theory and Flying Saucer Sightings,' *Science* 170 (6 November 1970), 599–603.

Weart, Spencer R., *Nuclear Fear: A History of Images*, Cambridge, MA: Harvard University Press, 1988.

Weber, Ronald, *Seeing Earth: Literary Responses to Space Exploration*, Athens: Ohio University Press, 1985.

Weber, Thomas P., ed., *Science & Fiction II: Leben auf anderen Sternen*, Frankfurt am Main: Fischer, 2004.

Weibel, Peter, ed., *Jenseits der Erde: Kunst, Kommunikation, Gesellschaft im orbitalen Zeitalter*, Vienna: Hora, 1987.

Weitekamp, Margaret A., *Right Stuff, Wrong Sex: America's First Women in Space Program*, Baltimore: Johns Hopkins University Press, 2004.

Welck, Stephan Freiherr von, 'Weltraumnutzung als politisches Konfliktpotential,' *Europa-Archiv* 39.24 (December 1984), 729–40.

———, 'Outer Space and Cosmopolitics,' *Space Policy* 2.3 (August 1986), 200–5.

Weldes, Jutta, ed., *To Seek Out New Worlds: Science Fiction and World Politics*, Basingstoke: Palgrave Macmillan, 2003.

Wendt, Alexander and Raymond Duvall, 'Sovereignty and the UFO,' *Political Theory* 36.4 (August 2008), 607–33.

Wenger, Christian, *Jenseits der Sterne: Gemeinschaft und Identität in Fankulturen. Zur Konstitution des Star Trek-Fandoms*, Bielefeld: transcript, 2006.

Werth, Karsten, 'A Surrogate for War: The U.S. Space Program in the 1960s,' *Amerikastudien/American Studies* 49.4 (Winter 2004), 563–87.

———, *Ersatzkrieg im Weltraum: Das US-Raumfahrtprogramm in der Öffentlichkeit der 1960er Jahre*, Frankfurt am Main: Campus, 2006.

Wertheim, Margaret, *The Pearly Gates of Cyberspace: A History of Space from Dante to the Internet*, New York: Norton, 1999.

Westad, Odd Arne, ed., *Reviewing the Cold War: Approaches, Interpretations, Theory*, London: Frank Cass, 2000.

Westfahl, Gary, 'The Case Against Space,' *Science Fiction Studies* 24.2 (July 1997), 193–206.

———, *Space and Beyond: The Frontier Theme in Science Fiction*, Westport: Greenwood, 2000.

Westfahl, Gary and George Slusser, eds, *Science Fiction and the Two Cultures: Essays on Bridging the Gap Between the Sciences and the Humanities*, Jefferson: McFarland, 2009.

Westphal, Peter G., *UFO UFO: Das Buch von den Fliegenden Untertassen*, Stuttgart: Deutsche Verlags-Anstalt, 1968.

Westrum, Ron, 'Social Intelligence about Anomalies: The Case of UFOs,' *Social Studies of Science* 7.3 (August 1977), 271–302.

———, 'Science and Social Intelligence About Anomalies: The Case of Meteorites,' *Social Studies of Science* 8.4 (November 1978), 461–93.

Weyer, Johannes, 'European Star Wars: The Emergence of Space Technology through the Interaction of Military and Civilian Interest Groups,' in Emmeret Mendelsohn, Merrit R. Smith and Peter Weingart, eds, *Science, Technology and the Military*, Dordrecht: Kluwer, 1988, 243–88.

———, *Akteurstrategien und strukturelle Eigendynamiken: Raumfahrt in Westdeutschland 1945–1965*, Göttingen: Otto Schwartz, 1993.

———, ed., *Technische Visionen – politische Kompromisse: Geschichte und Perspektiven der deutschen Raumfahrt*, Berlin: Sigma, 1993.

———, *Wernher von Braun*, Reinbek: Rowohlt, 1999.

White, Frank, *The Overview Effect: Space Exploration and Human Evolution*, Boston: Houghton Mifflin, 1987.

White, Luise, 'Alien Nation: The Hidden Obsession of UFO Literature. Race in Space,' *Transition* 63 (1994), 24–33.

Whitfield, Stephen J., *The Culture of the Cold War*, Baltimore: Johns Hopkins University Press, 1991.

Whyte, Neil and Philip Gummett, 'The Military and Early United Kingdom Space Policy,' *Contemporary Record* 8.2 (Fall 1994), 343–69.

———, 'Far Beyond the Bounds of Science: The Making of the United Kingdom's First Space Policy,' *Minerva* 35.2 (Summer 1997), 139–69.

Wiemer, Annegret, 'Utopia and Science Fiction: A Contribution to their Generic Description,' *Canadian Review of Comparative Literature* 19.1–2 (March–June 1992), 171–200.

Wiesenfeldt, Gerhard, 'Dystopian Genesis: The Scientist's Role in Society, According to Jack Arnold,' *Film & History* 40.1 (Spring 2010), 58–74.

Wilford, John Noble, *Mars Beckons: The Mysteries, the Challenges, the Expectations of Our Next Great Adventure in Space*, New York: Alfred A. Knopf, 1990.

Wille, Holger, *Kant über Außerirdische: Zur Figur des Alien im vorkritischen und kritischen Werk*, Münster: Monsenstein und Vannerdat, 2005.

Willhite, Irene Powell, 'The British Interplanetary Society: Val Cleaver and Wernher von Braun,' *Journal of the British Interplanetary Society* 54.5 (September–October 2001), 291–9.

Williams, Beryl and Samuel Epstein, *The Rocket Pioneers: On the Road to Space*, New York: Messner, 1958.

Winter, Frank H., *Prelude to the Space Age: The Rocket Societies, 1924–1940*, Washington, DC: Smithsonian Institution Press, 1983.

———, *Rockets into Space*, Cambridge, MA: Harvard University Press, 1990.

Witkin, Richard, ed., *The Challenge of the Sputniks*, New York: Doubleday, 1958.

Wolfe, Tom, *The Right Stuff*, New York: Farrar, Straus & Giroux, 1979.

Wolverton, Mark, *The Depths of Space: The Story of the Pioneer Probes*, Washington, DC: Joseph Henry, 2004.

Wormbs, Nina, 'A Nordic Satellite Project Understood as a Trans-National Effort,' *History and Technology* 22.3 (September 2006), 257–75.

Wunder, Edgar, 'UFO-Sichtungserfahrungen aus der Perspektive der Sozialwissenschaften: Literatur-Überblick, aktueller Forschungsstand, offene Fragen,' *Zeitschrift für Anomalistik* 6.1 (2006), 163–211.

Wunder, Edgar and Rudolf Henke, 'Menschen mit UFO-Sichtungserfahrungen: Eine Umfrage unter 447 Besuchern von Volkshochschul-Vorträgen zum UFO-Phänomen,' *Zeitschrift für Anomalistik* 3.1–2 (2003), 112–33.

Zabusky, Stacia E., *Launching Europe: An Ethnography of European Cooperation in Space Science*, Princeton: Princeton University Press, 1995.

———, 'Food, National Identity, and Emergent Europeanness at the European Space Agency,' *European Studies* 22.1 (October 2006), 203–36.

Zeiler, Thomas W., 'The Diplomatic History Bandwagon: A State of the Field,' *Journal of American History* 95.4 (March 2009), 1053–73.

Zenner, Elmar, 'Per Tele live auf den Mond,' in Annette Deeken, ed., *Fernsehklassiker*, Alfeld: Coppi, 1998, 121–39.

Zhu, Yilin, 'China's Early Space Activities,' *Journal of the British Interplanetary Society* 47.5 (May 1994), 195–8.

Zill, Rüdiger, 'Im Wendekreis des Sputnik: Technikdiskurse in der Bundesrepublik Deutschland der 50er Jahre,' in Irmela Schneider and Peter M. Spangenberg, eds, *Medienkultur der 50er Jahre: Diskursgeschichte der Medien nach 1945*, Wiesbaden: Westdeutscher Verlag, 2002, 25–49.

———, 'Zu den Sternen und zurück: Die Entstehung des Weltalls als Erfahrungsraum und die Inversion des menschlichen Erwartungshorizonts,' in Michael Moxter, ed., *Erinnerung an das Humane: Beiträge zur phänomenologischen Anthropologie Hans Blumenbergs*, Tübingen: Mohr Siebeck, 2011, 301–29.

Zimmerman, Robert, *Leaving Earth: Space Stations, Rival Superpowers, and the Quest for Interplanetary Travel*, Washington, DC: Joseph Henry, 2003.

Zinsmeister, Annett, ed., *Welt[stadt]raum: Mediale Inszenierungen*, Bielefeld: transcript, 2008.

Zukowsky, John, ed., *2001: Building for Space Travel*, New York: Harry N. Abrams, 2001 (exhibition catalogue, Art Institute of Chicago, 24 March–21 October 2001).

Index

Note: Page numbers appearing in *italics* refer to illustrations. A page reference in the form 'n' indicates a note number; for example, 204n59 is note 59 on page 204.

Alexander C.T. Geppert (ed.), *Imagining Outer Space*
European Astroculture, vol. 1
https://doi.org/10.1057/978-1-349-95339-4

F

G

M

Q

R

S

T

X

Y

Z

Printed in the United States
By Bookmasters